Intruder

A-6
INTRUDER

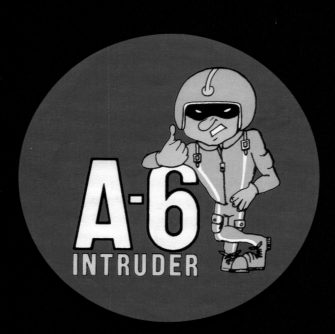

Intruder

The Operational History of Grumman's A-6

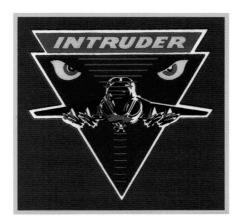

Mark Morgan & Rick Morgan

Schiffer Military History
Atglen, PA

Dedicated to the "Men Who Knew the Power," those who flew, fought, and fixed the mighty A-6 *Intruder*, and to the master aircraft builders of Grumman who built the machine at Calverton, New York.

Book design by Robert Biondi.

Library of Congress Catalog Number: 2004106328.

Printed in China.
ISBN: 0-7643-2100-5

Published by Schiffer Publishing Ltd.
4880 Lower Valley Road
Atglen, PA 19310
Phone: (610) 593-1777
FAX: (610) 593-2002
E-mail: Info@schifferbooks.com.
Visit our web site at: www.schifferbooks.com
Please write for a free catalog.
This book may be purchased from the publisher.
Please include $3.95 postage.
Try your bookstore first.

In Europe, Schiffer books are distributed by:
Bushwood Books
6 Marksbury Ave.
Kew Gardens
Surrey TW9 4JF
England
Phone: 44 (0)20 8392-8585
FAX: 44 (0)20 8392-9876
E-mail: Bushwd@aol.com.
Free postage in the UK. Europe: air mail at cost.
Try your bookstore first.

Contents

Foreword

Mark and Rick have compiled an A-6 *Intruder* encyclopedia, and one of the best histories of an aircraft that is very spe-cial to thousands of Naval Aviators and many more aviation enthusiasts. They take you from the beginning to the end of the *Intruder*. They have included the growing pains; growing pains that most – if not all – new aircraft go through before they are fixed through the "trial and error" method and become acceptable to the fleet operators. Not only is this book a history of the Vietnam era aircraft, but it provides another view into "Why the Vietnam War," and a view from the front that could answer questions about the dedication of the thousands of men that were shot at day, after day, after day.

Even though the Vietnam War will probably go down as one of our biggest blunders, history appears to be very clear: American involvement was caused by an administration that approached the conflict in a way that seemed to trivialize war and killing. Our faith teaches us forgiveness, but it also teaches us that members of that administration will have to answer for the 58,000 American lives lost in that terrible war. Secretary McNamara even confessed he was wrong about Vietnam and wrote a book about his mistake. Imagine the magnitude of his blunder, and then he has the guts to try and capitalize on it.

You will read words in this book that show how Vietnam war fighters felt about the directions coming out of Washington (including not only those from McNamara, but also President Johnson); words such as "stupid," "insane," "unreasonable," "uneducated," and of course the most used acronym, "AFU." Every military man I know believes strongly in the policy of civilians deciding when we should go to war, but they also believe the same civilians should allow the people they pay to fight a war decide how to fight and fight to win.

As I read through the book, I was never more aware that the key to the A-6's survival and successes was the people – all the people in the Navy/Grumman Team. I am very humbled by having my name listed with so many wonderful Americans, and to hell with McNamara.

You will read about the many bombing halts, but more importantly, you will understand how the North Vietnamese used those halts to reload their missile sites and guns, which caused more needless loss of life when we started bombing again. Targets in North Vietnam that were referred to by the aircrews as "milk runs" before the bombing halts had to be bombed back into "milk runs" after each bombing. That process always resulted in significant losses of aircrews. This process was another example of politicians playing unsuccessful games with young men's lives.

The authors give us a bonus by also looking at all the aircraft in a typical air wing on a carrier. You will come away with some knowledge of how the A-7, A-4, RA-5C, EA-6B, E-2C, F-4, and helos all play their part in making up this incomparable strike force.

The Marines received the A-6 also, but they were always at the "back of the bus" when it came to supporting the aircraft. However they, in typical fashion, found a way to support their A-6s, not always by the book, but the results were successful and to the enemy's disadvantage.

The Marines taught the Navy many lessons during my 37 years in the Navy. One was during the Zumwalt era, when Admiral Zumwalt made changes in the Navy to better match the direction society was heading. He relaxed many rules, such as sailors being allowed to have beards and beer in the barracks, among many other changes. He demolished "chain of command" by setting up a phone line where any sailor could dial the CNO direct. He directed the changes, usually in the form of "Z Grams".

The Marine Commandant, in contrast, said when asked by the media if he was going to follow Admiral Zumwalt's lead "absolutely not." In fact, he said he was going to tighten up the existing rules. He proved the military did not have to succumb to the whims of society. He proved the military should not be used as an experiment for or reflect the direction of society. The Commandant knew his job was to prepare his troops to defend the country, and discipline was one of the primary prerequisites for success. It boiled down to teaching young people the difference from right and wrong, and surprisingly, to the Zumwalts of the world, right and wrong did not always follow the whims of society.

The Marines also believed in the importance of combat experience because combat was their business. It was told to me that they emphasized that importance by the Commandant directing selection boards to give priority to combat experience. In fact, it was rumored that if you were in the Marines during the seven plus Vietnam War period and did not warrant wearing the Vietnam Service Medal, you did not get promoted.

The Navy did not follow the wisdom of the Commandant with respect to the importance of combat experience and continued to promote tailhook aviators to Admiral, even though those tailhook aviators could not seem to find the opportunity to go to Vietnam. However, these carrier aviators caused great Americans like Admiral Stan Arthur to end up with over 500 missions in Vietnam (everyone knew if you asked to go, you got your wish). There are two possible deductions: Admiral Arthur had more than his fair share of missions because these other tailhook aviators did not do their duty, or the Navy may have promoted cowards that could send our young men into

Rear Admiral Lyle Bull is seen here in 1987 flanked by his sons, Midshipman 1st Class Dell Bull (Left) and Lt. Bruce Bull (Right), of VA-196. Admiral Bull was ComCarGru SEVEN at the time and preparing to fly with his sons off *Constellation* (CV-64). (U.S. Navy)

harm's way. The Marines' policy eliminated the possibility of these two problems.

During the first part of the Vietnam War there were combat limits. It started out at 100 missions, then 125, then 150, and finally no limit, because the politicians could not see an end to the war. It was evident the policies coming from the White House were going to result in the war going on for a long time. If they had limits, they would soon run out of combat aircrews.

Mark and Rick have provided the reader with insight into the "war fighter" with their compilation of some great sea stories, and it allows the reader to better understand what makes a man a good warrior and what ended up spoiling him for the peacetime Navy. Combat was the only Navy experience some of the young aviators had. Combat bonds mean more than most experiences. It generates maximum adrenaline flow, and when the threat subsides, it releases a generous amount of euphoria – long periods of "highs and lows." Combat reduces the emphasis on paperwork and cuts through red tape.

When the war is over, a problem precipitates with these young men that know nothing but flying and fighting. After a few months of peacetime the "war avoiders" jump out from behind the bushes and start taking charge of these men. These combat aviators could not handle the lack of logic that starts being the peace time way of running the Navy, and they opt out of the service because they can't adapt. Those that had the peacetime Navy experience before the war usually are better at the transition back to the paper world and reinstitution of doing bureaucratic type work.

This book is a wonderful addition to any library because it was well researched and gives the reader one of the best pictures of the Vietnam era and the *Intruder* because it is from the eyes of many warriors who were there. I started flying the A-6 in 1065 and thought I knew the history of the A-6. I was wrong! Thank you Mark and Rick for taking the time to put this book together – properly.

Rear Adm. Lyle F. Bull
USN, (Ret).

Rear Admiral Lyle F. Bull is a native of Port Byron, Illinois. He enlisted in the Navy in April 1956, graduated from Iowa State University in February 1960, and was commissioned an Ensign in September 1960. Following flight training at NAS Pensacola, FL, he went to VAH-123 at NAS Whidbey Island, WA, and in May 1961 received his designation as a bombardier/navigator in the Douglas A-3 *Skywarrior*. In January 1965, following fleet and instructor tours in the heavy attack community, Rear Admiral Bull was selected as one of the first west coast B/Ns to undergo A-6 training at NAS Oceana, VA. In May 1965 he and the other members of the training detachment returned to Whidbey and established the A-6 training program at VAH-123, and ultimately VA-128.

Assigned to VA-196 from 1967 to 1970, Rear Admiral Bull flew 237 combat missions over three cruises, two in *Constellation* (CVA-64) and one in *Ranger* (CVA-61). In April 1970 he reported to the

Office of the Chief of Naval Operations, where he helped initiate the A-6E and A-6E TRAM programs; he subsequently reported back to VA-196 as executive officer, and in June 1975 became the Main Battery's 26th commanding officer. Later tours included assignment in *Enterprise* (CVAN-65) as Air Operations Officer, command of VA-128, and an assignment as Assistant Chief of Staff for Operations, Carrier Group Seven. In February 1981 he assumed command of the combat stores ship *San Jose* (AFS-7), followed by command of *Constellation* (CV-64) in September 1982.

In July 1984 Rear Admiral Bull assumed duties as Executive Assistant and Senior Naval Aide to the Assistant Secretary of the Navy (Research, Engineering, and Systems). He assumed command of Carrier Group Seven in June 1986, then became Commander, Battle Force U.S. Seventh Fleet/Commander, Carrier Strike Force Seventh Fleet/ Commander Carrier Group Five in July 1988. In November 1990 he became Deputy Chief of Staff, U.S. Pacific Fleet; in October 1992 he fleeted up to Assistant Deputy Chief of Naval Operations (Naval Warfare), followed by a final tour as Deputy Director of Naval Training. Rear Admiral Bull retired on 1 April 1993.

Among his many awards are the Navy Cross, Distinguished Service Medal (with gold star in lieu of second award), Legion of Merit (three gold stars in lieu of fourth award), Distinguished Flying Cross, Meritorious Service Medal, 19 Strike/Flight medals, seven Navy Commendation Medals (with Combat V), the Imperial Order of the Rising Sun (Third Class) from the Japanese Maritime Self Defense Force, and various unit awards. He is married to the former Diana Kay Stone; they have four children, 13 grandchildren, and currently reside on Whidbey Island, Washington.

Acknowledgements

Naval Aviator and NFO Wings

To put it simply, this book is about the operational history of the A-6 Intruder. If we appear to have glossed over the development and design history of the type we apologize only so far. Our goal has been to deal with the type's long and remarkable flight history and combat record. Along with this we hope we've put an appropriate spotlight on the men who flew and fought the type in three decades of service. If we have shorted anyone, it's the maintenance crews, for without them none of the flying heroics would have been possible. For this frequently unheralded group of men we have nothing but undying admiration.

A quick look at our work will show that it is heavy in the language of Naval Aviation. We do this without apology, as it was most fitting to leave our interviews, for the most part, in their own words. It goes without saying that it will help to have a working knowledge of the jargon of our business, but we've also provided what we hope is a helpful glossary as an appendix.

On the subject of names, the authors have tried to use multiple sources to confirm the names used in this book. For aircrew lost in combat or mishap we have tried to find at least two sources. For personal interviews we have tried to use other sources to help confirm; unfortunately, memories are not always correct, and we sincerely regret any errors made in recording people's names. Likewise, all names used in combat losses are as they appear in official records. It needs to be pointed out that we had no access to official loss records – specifically the privileged Mishap Investigation Reports (MIRs) – and information provided on peacetime losses is based solely on interviews, media, author's notes, or other *unofficial* information.

Along with the story of the *Intruder* we cover the previous history of the many distinguished squadrons that flew the type. It has been said by more than one historian that the Navy does not always pay attention to its own heritage. In spite of having a superb history office, the service frequently doesn't seem to have an active interest in its own combat heritage or tradition. The termination of many valorous Vietnam-era A-6, A-4, and F-4 squadrons in lieu of brand new "rookie" FA-18 units would seem to confirm this.

We have included many squadrons' historical backgrounds for this reason, for they forged the legacy of attack. Indeed, the last attack squadrons in the Navy – not strike-fighter, but by God *attack* – were two proud Intruder outfits, VAs-75 and -196. Their final passage in early 1997 marked the end of an incredible 51 years of attack aviation and ensured the name *Intruder* will live long in Naval Aviation's hall of honor. For the designers, the maintainers, and the A-6s' crews, this is their legacy.

Finally, this book would not have been possible without the support and assistance of a large number of dedicated individuals, and to them we owe undying thanks.

At the top of the list are the Tailhook Association and the editors and staff of the association's magazine, *The Hook*. This effort began as a three-part history of the A-6 for that publication that was published in Spring, Fall, and Winter 1997 issues under the title "Pride of the Ironworks." PHCS(AC) Bob Lawson, USN(Ret); Capt. Steve Milliken, USN(Ret); Cmdr. Jan Jacobs, USNR(Ret); and Cmdr. Doug Siegfried, USN(Ret) provided historical, technical, and photographic support at every turn of the project, as well as regular boosts to the author's morale when it seemed this project would never end. For that and their ongoing friendship we thank them.

We are proud to have Rear Admiral Lyle Bull – *Intruder* hero, warrior, and legend, as well as Navy Cross winner – write the forward for our book. We are honored by his interest in our work.

Other major support came from several members of the A-6 community who continue their work to ensure the *Intruder's* legacy is passed on to subsequent naval aviation generations. They include Lt. Cmdr Ted Been, USN(Ret), one of the first A-6 B/Ns and the founder of the A-6 Intruder Association; former Marine Steve Dumovitch, creater of the All Weather Attack web page; Rear Adm. Bob Mandeville, USN(Ret); and Rear Adm. "Rupe" Owens, USN(Ret).

A few of our contributors and friends were "cleared to higher holding" during the production of this history. We remember them with great fondness and thanks, and hope this book will help stand as a record of their contributions to our nation: Lt.Col. Larry Beasley, USAF(Ret); Lt.Cmdr. John F. Diselrod, USN(Ret); Fred "Ferd" House; Cmdr. Roger Lerseth, Capt. Leo T. Profilet, Cmdr. Phil Schuyler, USN(Ret); Cmdr. Howard G. "Bud" White, USN(Ret); and Lt.Cmdr. Floyd "Pink" Cordell, USN(Ret).

Finally, we thank the following who played a part in producing this history of the *Intruder*: Capt Ron Alexander, USN(Ret); Mrs. Betty Anderson; Cmdr Terry Anderson, USN (Ret); Bill Angus, Capt. Art Barie, USN(Ret), Cmdr. Steve "Boots" Barnes, USN(Ret); Lt.Cmdr. Steve Beales, USNR; Lt.Col. Paul Bless, USMC; David Brown, Mrs. Diana K. Bull; LtCmdr. "Steve Jet" Bulwicz, USN(Ret); Lt.Cmdr. Rick Burgess, USN(Ret); Col. William D. "Charlie" Carr, USMC(Ret); Tony Cassanova, Charles Clark, USMC; Cmdr. Pete Clayton, USN(Ret); Cmdr. Michael "Casey" Collins, USN; Capt. J.B. Dadson, USN(Ret); Lee DeHaven; Col. Don Diederich, USMC(Ret); Capt. Lou Dittmar, USN(Ret); Lt.Cmdr. William DuBois, USN; AE1(AW) Damon Duncan, USNR; Maj. N.Craig "Nellie" Dye, USMC(Ret);

No matter what the aircrew accomplished, it could not have been done without the tireless work of thousands of troopers on the flight deck, in the shops, and in the squadron offices. (U.S. Navy)

Lt. Denny Franklin, USN(Ret); Cmdr. Pete Frano, USN; Cmdr. Dave Frederick, USN; Phil Friddell; R.James "Diamond Jim" Garing, Robert Guerra, USMC; Lt.Col. Joe Hahn, USMC(Ret); Capt. J.R. Haley, USN; Col. Jim Henshaw, USMC(Ret); LtCmdr. Tom Hickey, USN(Ret); Lt. Col. Bob Holloway; USAF(Ret); Rear Adm. Charlie Hunter, USN(Ret); Lt.Col. Earl Jacobson, USMC(Ret); Cmdr. John Juan, USN(Ret); Tom Kaminski; Cmdr. Patrick Keller, USN; Capt. Daryl F. Kerr, USN(Ret); John Kerr, USAF(Ret); Bill Kretzschmar, USMC; John MacPherson; Lt.Cmdr Walt Martin, USN; Frank McBaine; Capt James A. McKenzie, USN(Ret), Cmdr. John S. McMahon Jr, USN(Ret); Cmdr Dee Mewborne, USN; Lt.Cmdr. Jerry Menard, USN(Ret); Sam Merkel; Lt. Guy Miller, USN; Lt.Col. Richard H. Morgan, USAF(Ret); Capt. Larry Munns, USN(Ret); Capt. Gordon R. Nakagawa, USN(Ret); Lt.Cmdr. Steve Nakagawa, USN; The Naval Historical Center; MSG Al Parks, USMC(Ret); Cmdr. Jerry Patterson, USN(Ret); Jim Perso, USMC; Capt. John K. Pieguss, USN(Ret); Kay Pieguss; Capt. Roger Pierce, USN(Ret); Lt.Cmdr. Gary Poe, USN(Ret); SM1(SW) Bryan Potter, USN; Cmdr. Ken Pyle, USN, Capt. Don Quinn, USN; Stan Richardson; Lt Col Brian Rogers, USAF(Ret); Lt.Cmdr. Robert Rubery, USNR(Ret): Capt. Roger Sheets, USN(Ret); Cmdr. Richard "Simo" Simon, USN; Maj. Clyde Smith, USMC(Ret); Lt.Cmdr Tim Sparks, USN(Ret); Lt.Col. Ray Springfield, USMC(Ret); AVCM(AW) Dave Staley, USN; Rear Adm. George H. Strohsahl, USN(Ret); John Stubbs; Lt.Cmdr. R.G. "Tugg" Thompson III, USN(Ret); Col. John Thornell, USMC(Ret); Dick Tinsley, USMC; Capt. Richard J. Toft, USN(Ret); Capt. Terry Toms, USN(Ret); Tom Urgyu; Cmdr. Jim Vannice, USN(Ret); Mark Wagner; Maj. Charlie Walsh, USMC(Ret); Mike Weeks; Capt. Dan Wright, USN(Ret); and Mrs. Nancy Zick and Capt. Richard A. Zick, USN(Ret).

Mark Morgan
Issaquah, WA

Rick Morgan
Woodbridge, VA
April 2004

CHAPTER ONE

Requirements, Proposal, and Development

In November 1950 young *Skyraider* pilot Ens. Leo Profilet was making both his first fleet deployment *and* his first combat cruise. A recently commissioned Aviation Cadet, Profilet flew AD-4s with VA-115 in *Philippine Sea* (CV-47). North Korea had invaded South Korea with seven divisions earlier that year, on 25 June. The first carriers to respond were *Valley Forge* (CV-45) – with Carrier Air Group 5 (CVG-5) embarked – and HMS *Triumph*; they went into combat on 3 July. By the end of the month *Phil Sea* with CVG-11 joined them. Through the desperate early months of the conflict other units piled into Japan and South Korea, and by the end of the year they had fought their way out of the Pusan Perimeter, landed at Inchon, and pushed their way to the Yalu River under a blanket of land and carrier-based air power. Now they were falling back again, in the face of the Chinese People's Liberation Army and the fierce Korean winter.

For the carrier aviators, the missions were rough and options few in case they ran into trouble. Adding to the difficulties was the Chinese proclivity for human wave attacks at night, when the carriers were least able to respond. "We canceled a lot of missions in those days," Profilet recalls. "We didn't fly at night because it was a straight-deck carrier. The VCs (Composite Squadrons, including VC-35 and VC-3, operating specially equipped *Skyraiders* and *Corsairs*) flew at night, but it wasn't that much easier for them."

Design and technology limitations, both with the aircraft and the carriers, precluded extensive night combat operations. Flying in the typically rotten mid-winter weather during daylight hours was also questionable, although the aviators gave it their best shot. Profilet vividly remembers one mission in late November 1950:

> "The Marines were trying to get down the hill from Chosin Reservoir. We were trying to do Close Air Support (CAS); we did quite a bit, but when the weather came in it filled the passes up. On one mission I remember myself and another starry-eyed ensign said, 'Hey, no problem. ' We went out to sea, dropped under the clouds, and flew in underneath into this valley. We managed to get a FAC (Forward Air Controller) and managed to hit the targets – the *Spad* carried a lot, four to five tons of bombs – but it was strictly an eyeball operation.
>
> "I remember getting coached in. The Marines were on a two-lane road coming down from the reservoir, and they were coming under fire. The FAC put us across the valley. Now, the textbook for CAS has all these fancy grid numbers, but that's not how it was really done. 'See that tall tree?' 'Yeah.' 'There's a big rock near it; come in between the rock and the tree.' We rolled on in, and I remember the FAC almost had an orgasm. That was the way he sounded, because apparently we hit *right* between the rock and tree. It was one of the biggest adrenaline pumps in my life, because we hit the targets and we got to see the column get started down the road."

Now a retired captain, Profilet remembers most other missions didn't work out as well during that hard winter of 1950-1951. Along with the other carrier forces assigned to the theater, the *Air Group 11* team went after other targets, occasionally with unplanned or unexpected results:

> "We were the second carrier out there," he continues. "So we were at Pusan and the Inchon landings, dropped the bridge at Seoul over the Han River, then went up to Pyongyang, where we lost CAG (Cmdr. R.W.) 'Sully' Vogel. The bridges over the Yalu were a bitch. There was one place that was really, really hairy. Our ground rules said the target was officially the south end, but we had a couple of guys who couldn't bomb worth a damn, and their bombs always ended up on the north end of the bridges.
>
> "Those were tough targets, and at that time the MiGs came around; thank God for the F-86s. In many ways, it was a much tougher war than I found later in Vietnam. There were many times when you just couldn't go in. That was frustrating for the squadron. We helped a lot, actually, but one or two times the weather moved in and we were helpless. I thought a night-flying ability was needed. It would have been great if we could have gone in at night, low level, after those bridges, one plane, maybe two. "

Lessons such as these from the Korean War proved the need for a new carrier attack aircraft, one that could operate effectively at night and in all known weather and visibility conditions. Getting to the target was one requirement; any new design would also have to accurately deliver – without external targeting assistance – a large payload at long range.

Starting with its wartime experiences and continuing past the Korean Armistice of 27 July 1953, the Navy took a long, hard look at developing such an aircraft. In 1955, the Navy's Long Range Objective Study Group looked at what was available, balanced requirements with what was technologically achievable, and determined the service should have its new attack aircraft.

Existing Options

At the time no appropriate aircraft existed. The attack mission of the 1950s Navy rested on the AD *Skyraider*, which had entered service

Above: Douglas' A-1 *Skyraider* set the standard for the Medium Attack community and was a tough act to follow. Rugged, long-legged, and able to carry prodigious amounts of ordnance, the "Able Dog" became a legendary tailhook aircraft. The first ten Navy *Intruder* squadrons would transition from the A-1. (Douglas Aircraft)

Opposite: The world's first supercarrier, *Forrestal* (CVA-59), steams in August 1959 with a good part of CVG-8 on deck. "Medium Attack" during this period was defined as the AD-6 *Skyraiders* present from VA-85, as well as specialist detachment AD-5W and AD-5Ns from VAW-12 and VA(AW)-33. Among the over 50 aircraft present are A4D-2s from VA-81 and -83 and F8Us from VF-103 and VFP-62, as well as ten – count 'em, ten A3D-2 *Skywarriors* from the *Mushmouths* of VAH-5. The ship's single TF-1 COD is also present. For whatever reason, VF-102 and its F4D-1 *Skyrays* are not onboard. (U.S. Navy)

with VA-19A in December 1946. While this product of Ed Heinemann's engineering group at Douglas had not yet achieved status as a classic of naval aviation by the end of the Korean War, it was well on its way. Unfortunately the AD was of limited or no use in bad weather conditions and darkness, although the AD-3N, -4N, and -5N night attack *Skyraider* variants helped pioneer night and all-weather operations. Still, the *Spad* would continue to serve as the mainstay of carrier-born attack aviation in both the conventional and nuclear roles through 1968.

Another future attack classic, the A4D *Skyhawk*, was coming on line during this period and would itself see a long and fruitful career. Designed as a light, high-speed, clear-weather attack jet, the first examples entered the fleet in late 1956 with the *Bluehawks* of VA-72 at NAS Oceana, VA. The *Scooter* – also known as "Heinemann's Hotrod," among other names – also proved capable of carrying a staggering amount of ordnance, but it was strictly a visual bomber with no avionics to speak of. Later models possessed greatly increased capability, culminating in the final U.S. version, the A-4M – still largely a VFR aircraft, but at the time of the Navy's first stirrings towards all-weather attack such capabilities were a long way off.

The Navy operated a mixed bag of other jet-powered attack aircraft into the late 1950s, several of which converted from other roles. These included the Grumman F9F *Cougar/Panther* series and F11F *Tiger*, McDonnell F2H *Banshee*, the Vought F7U-3 *Cutlass*, and the North American FJ *Fury* series. Some, such as the F9F and FJ, achieved some popularity with both the Navy and Marines; indeed, on 2 April 1951 two VF-191 F9F-2Bs operating from *Princeton* (CV-37/CVG-19) made history by bombing the railroad bridge near Songjin, Korea. The event marked the Navy's first use of jets in the attack role. On the other hand, other aircraft, such as the F11F-1 never really achieved success as either a fighter or an attack plane, and the F7U "Gutless" was an abject disaster. Limited payloads and range handicapped all of the planes, and each had one – perhaps ultimate – design drawback: they were single-seat aircraft, flown by one pilot doing the "aviating, navigating, and communicating." Flying on a dark night and/or in bad weather, dodging antiaircraft and terrain, all while attempting to locate and hit the target; the outcome oftentimes proved fatal.

At the large end of the jet aircraft scale Douglas designed and produced the A3D *Skywarrior* as a carrier-borne, multi-place, semi-all-weather heavy attack aircraft. Developed to replace the largely forgettable North American AJ *Savage* series in the nuclear delivery role, the first *Skywarriors* entered service with VAH-1 at NAS Jacksonville, FL, in 1956. The aircraft's exploits and longevity in the fleet also became legendary, but so did its shortcomings. The *Whale* – as it became nicknamed – was large, took up a lot of deck space, and was hard to handle around the boat. Its introduction to carrier operations was marked by accidents, low morale among the crews and maintainers, and a poor reputation with carrier and air wing staffs.

Its biggest drawbacks were probably its size and limited avionics. Douglas designed the *Skywarrior* as a high-speed, high-altitude *strategic* bomber for employment in probable one-way missions into the Soviet heartland. As with the U.S. Air Force's later experiences with the B-47, B-58, and B-52, the A3D proved difficult to convert to the low-level mission and saw only limited success in Vietnam as a tactical bomber. By the mid-1960s the Navy shifted the type to a primarily aerial refueling and electronic warfare role, where it achieved much success; EW variants provided the foundation for the Navy's tactical electronic warfare community and eventual development of the EA-6B *Prowler*.

The *Whale's* titular replacement in the nuclear strike mission, North American's A3J *Vigilante*, effectively was a non-starter. Designed for

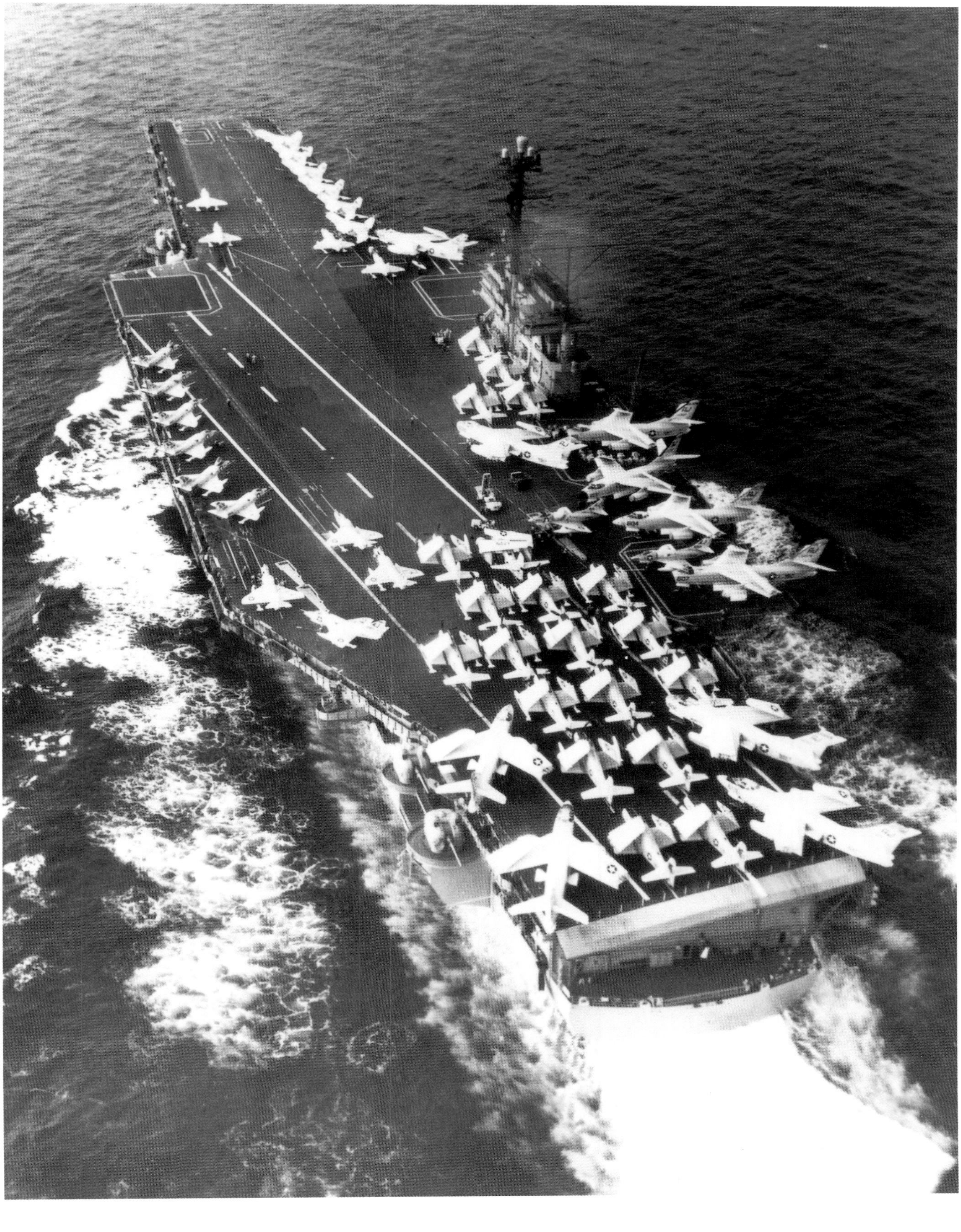

high-altitude missions and equipped with enhanced navigation and delivery systems, it did help prove the concept of the high-speed, two man crew (pilot and systems operator) with specific, delineated roles. However, by the time the *Vigi* finally made it to the fleet – with VAH-1 and -7 at NAS Sanford, FL, in 1961 – the A3J was found wanting in its primary assignment. The Navy moved away from having a dedicated long range nuclear strike aircraft, and instead placed the strategic mission with Fleet Ballistic Missile submarines carrying the *Polaris*. By 1963, after only a few cruises in the nuclear strike role, North American converted the *Vigis* into high-speed reconnaissance platforms. In that role – and redesignated the RA-5C – the aircraft continued to serve with gusto through 1979, scaring Landing Signal Officers (LSOs) everywhere.

Several of these aircraft saw long careers and achieved legendary status. Planes like the *Spad* and *Scooter* were award winners and are still talked about by warriors with pride and affection. Others, including the *Whale* and the *Vigi*, found their niche in other missions and achieved grudging admiration. However, to meet its requirements the Navy had to develop an entirely new design, one that would push the limits of the technology available at the time. The resulting design eventually supplanted several types and complimented others, but in its primary role of all-weather, long-range attack the new aircraft achieved primacy and its own measure of fame.

Requirements & Competition

On 2 October 1956 the Chief of Naval Operations' staff, having balanced naval aviation's needs with existing options and technology, issued an operational requirement for a new attack aircraft. The resulting specification (CA-10504) specified a two-man crew, all-weather avionics, and the ability to deliver a large payload at long range and low altitude.

The aircraft was to be able to make 500-knots at full load, yet be capable of low approach speeds; notably, the specification also required short takeoff and landing (STOL) capabilities for use at unimproved airfields. The requirement also specified a combat radius of 300-nm in the Close Air Support (CAS) mission and 1000-nm on long range interdiction missions. Finally, the new attack plane had to be capable of operating from all the carriers in the Navy's inventory, from the oldest *Essex* class ships to the newest, the USS *Forrestal* (CVA-59). This meant a length of no more than 56-feet to fit carrier elevators and a folded-wing width not to exceed 24-feet, set by a requirement for passing two planes through the hangar bay fire doors on the smaller aircraft carriers.

In early March 1957 the Navy delivered Type Specification 149 to Bell, Boeing, Douglas, Grumman, Lockheed, Martin, North American,

The frantic 1950s saw a bewildering number of aircraft introduced into the fleet as the Navy tried to keep up with the rapid advances in aviation. Two of the biggest disappointments are depicted here. The North American AJ *Savage* had three engines (two props and a jet buried in the fuselage) and a miserable reputation as a Heavy Attack aircraft. The radical Vought F7U *Cutlass* deployed with both fighter and attack units and failed at both missions. The "Ensign Eliminator" was out of the fleet by 1958. Here a VC-7 *Savage* refuels a "Gutless" from VX-3. (U.S. Navy)

Below: North American's beautiful A3J *Vigilante* was the great hope of Naval Aviation during the late fifties. It was the future of the heavy attack community, yet it never really clicked in that mission. Converted to the RA-5C, the type was arguably the finest conventional recon platform in the U.S. military during the Vietnam War. (North American Aviation)

and Vought. A BuAer letter followed spelling out what the Navy was looking for. Two months later the manufacturers received the contract design requirement; subsequent amendments provided detail specifications such as communications and avionics requirements, employment of drop tanks and other stores, and landing and arresting gear requirements.

The eight manufacturers responded with eleven proposals. Boeing, Douglas, Vought, and Martin each submitted two designs featuring either turboprop or turbojet propulsion, while Grumman, Bell, Lockheed, and North American submitted single designs featuring jet propulsion only. Bell's proposal was for a VSTOL (Vertical/Short Takeoff and Landing) aircraft and quickly fell by the wayside, as did the single-engine submissions. By the end of the year the Navy reduced the competition to the twin jet engine designs offered by Grumman, Douglas, and Vought.

On 30 December 1957 the Navy announced Grumman Aircraft Corporation's proposal, developed by the team of Lawrence J. Mead, Bob Nafis, and Bill Tuttle, as the winner of the design contract. Grumman's internal designation for their successful design was G-128; the Navy assigned the model designation A2F-1, for "Attack-type, 2nd produced by Grumman, 1st model."

Grumman's submission had gone through several design permutations – including M-Wing (forward-swept inboard, rear-swept outboard), wing-imbeded engines (similar to British aerospace practice) – and others before the company settled on a final configuration. The winning design was still radical: large, two-seat, twin-engine, with big nose radome and side-mounted intakes; swept, mid-fuselage-mounted wings; and tapering empennage. The powerplants were two 8500 pound thrust Pratt & Whitney J-52-P6 turbojets. To help accomplish the Navy's low-speed landing requirements, the plane featured fuselage-mounted air brakes and spoilers for roll control, the latter in lieu of conventional aileron control surfaces.

What undoubtedly struck observers most was the cockpit design, which featured a large, bulbous sliding canopy and side-by-side seating for the crew of two. Grumman's design formally divided the tasks between the pilot and a second crewman in the right seat called a bombardier/navigator (B/N). The B/N, placed slightly to the rear and below his partner for pilot visibility reasons, would operate the radars and weapons systems. Chief among the systems was *DIANE* (Digital Integrated Attack and Navigation Equipment), which employed a ground mapping radar, a track radar, an analog computer, and a doppler navigation set. With these systems the A2F's crew – sitting side-by-side and working together – could, in theory, find and destroy any radar significant target in any weather, without external visual cues.

On 3 February 1958 the Chief, Bureau of Aeronautics (BuAer) appointed a Weapons System Team (WST) to work with Grumman in getting the A2F to the fleet. The service designated a program manager and assigned a staff composed of personnel from subordinate BuAer groups, including Contracts, Maintenance, and Production, as well as a Research and Development (R&D) Project Manager. The latter also directed the activities of the R&D Project Team, manned by personnel from the bureau's Airborne Equipment, Airframe, and Power Plant divisions. The establishment of the Weapon System Team marked something new in aircraft development and procurement: the A2F was not just an airplane but a *weapon system*. All aspects of design, equipment, avionics, weapons, initial flight test and evaluation, and production were done on an integrated basis. Grumman, working with the WST, would not turn over just an airframe ready for government or other contractor provided engines, avionics, and weapons; it was responsible for the entire package. As a result the Navy would receive a complete (hopefully) seamless weapon system for fleet test, evaluation, and introduction.

The WST program and designation of program managers for each individual weapons development program was a direct result of the findings of the Robertson Committee. Chaired by Deputy Secretary of Defense Reuben W. Robertson, the committee was tasked with determining methods of streamlining the defense procurement process. Its findings, released in a report on 9 February 1957, pushed for an integrated design and development process and Navy/industry partnerships in procurement. The services could thereby reduce excess activities, waste, and duplication, while saving money and delivering a better product faster and more efficiently. Grumman's A2F became one of the first products of this new method of procurement.

On 14 February 1958 the Navy issued cost-plus-fixed fee contract No. 58-524C for $3,410,148 to fund the first mockup A2F; the Navy and Grumman signed the actual contract letter on 29 April. In March 1958 Grumman named Bill Cochran as program manager for the A2F, but when he died in the crash of an Air Force B-52 the following May near Westover AFB, MA, Grumman named Bruce Tuttle as his successor. Overcoming the tragedy, Grumman workers at Calverton and Bethpage continued their work. On 26 March 1959 Grumman received a $101,701,000 development and production contract covering eight initial aircraft in two batches of four each. The Navy assigned Bureau Numbers 147864 to 147867 and 148615 to 148618 to these first eight aircraft.

The production contract was the first ever award for a fully integrated weapon system on a cost-plus basis. In keeping with the tenets of the Robertson Committee, Grumman was on the hook for *everything* – from radome to rudder – and the company responded.

First Flight & Testing

The first of the many, YA2F-1 BuNo 147864, rolled out of Grumman Plant No. 5 at Bethpage, Long Island on 1 April 1960. Following initial taxi and other tests Grumman partially broke down the new bomber – now christened *Intruder* – and trucked it to Calverton for flight-testing. The first flight came on 19 April 1960 with Grumman test pilot Bob Smyth at the controls.

Writing in the July 1997 newsletter of the *Intruder* Association, he commented on the YA2F-1's initial hop:

> "After several days of engine runs and system checks we began taxi tests consisting of low speed testing of brakes and nose wheel steering, progressing to higher speeds and energy levels. When we were confident of the brakes we performed several lift-offs utilizing our 10,000-foot runway. A lift off is a lightweight take-off with an immediate cut in power to land straight ahead on the remaining runway. It does not count as a flight, since in the trade you have to make a turn to call it a flight. It does give the pilot a feel for longitudinal stability in ground effect. It also gives some assurance that nothing critical to flight is wired through the weight-on-wheels switch.
>
> "Satisfied that we had checked everything that we could on the ground, we scheduled the airplane to fly. This becomes something of a production. Flight test instrumentation must be checked and ready. The No. 1 airplane had no weapons system installed; instead it carried instrumentation in the form of a photopanel and oscillograph, which would record a significant number of airframe and engine parameters for post-flight perusal by the engineering staff, or by the accident board if such were required. Supporting personnel consisting of chase crews, engineers in a ground communication facility, crash crew, photographers, etc., had to be briefed and coordinated. When all are up and running and the weather is CAVU (perfectly clear: "Ceiling and Visibility Unlimited") the flight may proceed.
>
> "Now, the real purpose of a first flight is to make a successful landing. There is a tremendous level of interest at this point. Hundreds of people have worked long hours for months to reach this point; a large part of the company's future is tied to the airplane's success; and the customer is anxious to see what he's buying. All

A-6A number three, 147866, flies with 30 inert Mk.82 500-pound bombs in April 1964. The test airframe still has a nose probe, wing-mounted pitot, fuselage speed brakes, and rotating exhaust nozzles. The "max load" of 30 bombs would soon be reduced to 28 when it became apparent that the forward gear doors interfered with weapons loaded on the inner bomb racks. (U.S. Navy, PH1 Hughes)

this creates a great deal of pressure on all concerned. One person has it within his power to bring instant relief to all hands: the lucky guy who gets to make the first flight.

"This is not the time for an ambitious flight profile that touches all corners of the envelope. A simple plan that exercises the operable systems, looks at basic handling qualities, poses for some photos, makes a fly-by for the assembled crowd, and lands safely is appropriate. This accomplished, everyone knows great joy! The airplane is put in the hangar, everybody attends the party, the customer returns to Washington, and the next day you prepare for the real start of the Flight Test program on the second flight when peace and quiet prevail.

"The first flight is flown solo for two reasons, which seemed quite appropriate at the time. The tactical side of Grumman was strictly single-pilot aircraft with no professional non-pilot aircrew. Secondly, we thought there was no sense exposing other personnel to potential danger if it was not necessary. So, on 19 April 1960 we rolled down Runway 32 at Calverton on the maiden voyage of the A2F-1. Ernie Von der Hayden, the project pilot, was chasing in an F9F-8T with a photographer in back. Our rescue helo and a company G-21 (Grumman *Goose*) for water rescue were airborne. We climbed out to 10,000-feet before raising the landing gear as a precaution against any unusual trim change or some hydraulic mischief. Unfortunately, we got a barber pole (unsafe down landing gear indication) on one of the gear indicators and returned the gear to the down position. They all locked down, and prudence dictated leaving them that way.

"This interfered with the modest profile we had planned, so we just flew around in the 10,000 to 15,000-foot altitude range, getting a good feel for the handling qualities, which were quite good. We also exercised the engines, since this was the first time the P&W J-52 was in a manned airplane. Its prior experience was in the *Hound Dog* missile on the B-52. I can't remember a problem on this flight, but we soon found compressor stalls on rapid engine decelerations, which required a bleed valve modification. I believe we were the first to coin the word 'chugs' to describe the stalls.

"We also checked the tilting tail pipes, which worked beautifully. The A2F tail pipes normally bend down seven degrees with respect to the fuselage reference line. To satisfy a Marine Corps requirement for a short takeoff roll the tailpipes incorporated a hydraulic actuator that deflected them down to thirty degrees when activated by a knurled knob on the outboard throttle. There was no trim change as I recall, and they reduced the power-on stall speed by five knots, or for a given approach speed they reduced the angle of attack by three degrees.

"After some in-flight photos (unfortunately with the gear down) we came in for our fly-by and landed, one hour after takeoff. A photo taken on landing shows the airplane with tail pipes and fuselage speed brakes extended; an interesting configuration. Since the fuselage brakes were effective thrust spoilers we later found that Vmax with full throttle with gear down and brakes out was 109 knots! As you all know, the fuselage brakes were eventually bolted in and wing tip brakes were installed."

With this first flight of the *Intruder* under the belt, testing continued, as Smyth further relates:

> "On the second flight Ernie Von der Hayden took the left seat and I moved over to the right, and we would alternate left and right on subsequent flights. On this flight we got the gear up with no barber poles and began our speed build up. At 283 knots we felt a bang, and the chase pilot said our boarding ladder carried away and landed in Long Island Sound. The latching mechanism was redesigned before the next flight two days later. In all, we flew eight flights for 8.6 hours before the official Navy Acceptance day on 29 April. On this occasion the airplane had a new gray paint job (not the bare metal of the first flights) and the name *Intruder* painted on its nose."

Some test hops were decidedly more eventful than others. On 28 July 1960 Von der Hayden made a dead stick landing in airframe No. 2 (BuNo 147865) to end its first flight. Investigation later determined a fuel shutoff valve – which enabled fuel transfers for testing of the aircraft's center of gravity – was incorrectly installed. The rear or engine feed tank went dry and both engines flamed out as the *Intruder* approached Calverton. Fortunately, both pilot and plane survived the incident none the worst for wear.

Unplanned landings notwithstanding, the test program revealed several other deficiencies in the basic design, or at the least features of dubious utility. For example, only the first seven aircraft were completed with the tilting tailpipes to meet the STOL needs of Marine *Intruders*. The requirement was for a 1,500-foot takeoff over a 50-foot obstacle. To accomplish this the pipes vectored down 23 degrees from the cruise position, reducing liftoff speed for a lightly loaded plane from 86 to 78-knots. According to Bob Smyth, the feature worked, but:

"The first Navy Evaluation team, led by (Cmdr. Leonard A.) 'Swoose' Snead, was charged with seeing whether this feature was worthwhile. They found that they were only marginally effective for the Marine Corps requirement, and as for reducing approach speeds, the speeds with pipes up were lower than anything they had seen since propeller days. They recommended deleting the tilting tail pipes at a cost saving of $25,000 per airplane! Actually, the concept would have worked well on the heavyweight EA-6B much later on."

The Marines and Navy argued back and forth a bit, but the tailpipes and their complicated actuation mechanisms were deleted after aircraft No. 8, BuNo 148618.

Developing a combat aircraft – or any plane, for that matter – requires constant testing, evaluation, analysis, and engineering changes. The A2F was no exception, and periodically the airframes received modifications that were adopted as the standard. Before the first flight the design gained an additional two feet of wingspan, adding 1000-pounds of fuel capacity and improving the *Intruder's* cruise drag figures. The fourth plane, BuNo 147867, was the first to have the radars emplaced and a distinctive fixed refueling probe attached to the top of the radome. Other changes brought an enlarged rudder chord from the sixteenth aircraft on (BuNo 149482) for better spin recovery qualities.

A major glitch developed with the A2F's fuselage-mounted speed brakes. During a test flight in the fall of 1960, just prior to the type's Navy Preliminary Evaluation (NPE), pilot Ernie Van der Hayden noted sluggish pitch response at high G while the brakes were extended; this had not turned up in the wind tunnel or during prior tests. In response, in January 1961 Grumman started an intense redesign project. In an attempt to correct the deficiency the company moved the horizontal stabilators on airframe No. 3 rearward on the fuselage by 16 inches, but the problems continued. Further study indicated that with the brakes deployed the aircraft's center of pressure shifted forward and inboard, exceeding the capacity of the horizontal stab's hydraulic actuators; the actuators actually stalled in flight for several seconds. Fortunately nothing disastrous had happened, but a mod was needed.

Another nail in the coffin for the air brake design came in May 1961 following NPE-1A. The Navy determined the brakes were ineffective, both in dive-bombing and around the boat. Finally, in early 1963 the fuselage speed brake's disruption of engine exhaust airflow was cited as a causal factor in an accident at Patuxent River. As a result, Grumman disconnected the brakes in the early aircraft until the company could institute a permanent design modification. The company later added split wingtip speed brakes to the aircraft, and the fuselage brakes were permanently bolted to the fuselage.

Systems Development

Concurrent with the airframe flight test program, Grumman test crews and Navy personnel tested and developed DIANE and its associated systems. The entire package was considered a remarkable example of modern technology; it was one of the first – if not *the* first – integrated flight and weapons control systems.

The primary sensors were the AN/APQ-92 search and AN/APQ-88 track radars, stacked one over the other in the *Intruder's* bulbous radome. The system also employed an AN/APQ-61 ballistics computer, Litton AN/ASN-31 inertial platform, CP-729A air data computer, AN/APN-141 radar altimeter, AN/APN-153 Doppler navigation system, and AN/ASQ-57 integrated electronics control. The system provided the pilot with flight direction information, including steering/navigation, ordnance delivery, and terrain clearance displays. From his side of the cockpit the B/N worked the radars to navigate and locate the target, loaded the navigation and target data into the computer, and set up the weapons panel. The computer took the inputted and sensor information and calculated an automatic weapon release point while providing the pilot with his steering and G cues, resulting in the successful destruction of the target, first pass, every time. The system was designed to operate in any known weather condition and at any dive angle without the crew ever actually seeing the beneficiary of their "interest."

However, things rarely work as advertised, particularly when you're addressing major leaps in technology. The first all-up avionics plane, No. 4 (BuNo 147867), made its initial flight in December 1960, and it quickly became apparent some changes were needed. The initial avionics NPE held in November 1961 turned up deficiencies in the pilot's and B/N's displays in terms of brightness and resolution, as well as major problems with system reliability (despite its title, DIANE was not really digital; it was wholly analog). The first major Engineering Change Proposal (ECP), No. 100, was approved in 1962 to solve the display problems; it changed the B/N's display from a five-inch television display to a seven-inch cathode ray tube. The fix helped, but it delayed the avionics Board of Inspection and Survey (BIS) trials and pushed back the fleet delivery date by almost a year.

Grumman continued to work on these and other problems while adding other features, such as increased avionics air conditioning and doubling the computer memory. Also during this period the first Navy personnel were showing up at Bethpage for systems training and pilot transition.

Among the Naval Officers assigned during these early days was Limited Duty Officer (LDO) Ens. John Diselrod. A former enlisted bombardier with the *Savage Sons* (née *Mushmouths*) of VAH-5 at NAS Sanford, FL, Diselrod reported to Bethpage in the spring of 1962:

"I started in March 1962 as the Systems Acceptance Officer at BuWepsRep (Bureau of Weapons Representative) Bethpage, flying the A2F-1 Intruder, and retired (while attached to) *Medium Attack Wing One* on 1 September 1975. I had seven flight hours in the A2F-1 prior to the aircraft being redesignated as the A-6A. My introduction to the aircraft and system was through several of the outstanding Grumman test pilots: Ralph Donnell; Ernie Von der Hayden; John Norris; Vince Fascenella; Jim Filben; Jack Stephenson; Don King; and Bill Smyth. The Navy pilots assigned at the time were Cmdr. Jack Little, Lt. George Baxter, and Lt. j. g. LeRoy Bish.

"On one of my initial system flights, Ralph Donnell – who had been doing the development flying of the Search Radar Terrain Clearance (SRTC) – took me up a river valley in northern New York. As he demonstrated the Terrain Clearance mode of the Visual Display Indicator (VDI) he said, 'Look up at those high tension wires.' They were across the river from peak to peak where the river made a quick turn right, then left, and then again right. One could either turn hard to follow the river *or* follow the SRTC and pull up and over the little ridge, skim down the other side and be over the river again. He said he had flown that route for quite some time before he even knew there were power lines in the area. We went under them that day with plenty of room to spare, but it was flights like that in which I became a real believer in SRTC ... with a good pilot.

"It was during this tour of duty that – as the A2F-1 became the A-6A – NATOPS was introduced, the DIANE system underwent ECP 100, and things started in Southeast Asia. The ECP 100 change removed the map tracking capability from the system, and the artificial image transmitted from a four-inch tube to a display in the cockpit was replaced with a radarscope showing 'raw video.' This not only improved the radar target imaging capability, but allowed improved AMTI (Airborne Moving Target Indicator). One could now sit in the development lab at Norden in Connecticut and look at traffic moving on Long Island."

As time progressed and production continued, Diselrod and others occasionally got a "good deal," ie., delivering an aircraft to another location for evaluation, public affairs, or training purposes. One such trip found him flying with pilot Jack Little to Kirtland Air Force Base, NM, and the Naval Weapons Evaluation Facility (NWEF). Formerly the Naval Nuclear Ordnance Evaluation Unit, the command served as the Navy's primary air-delivered special weapons center in conjunction with the U.S. Air Force's activities at Kirtland, Sandia, and Manzano Bases:

> "Sounds like an easy trip, nice weather, new airplane. We landed at NAS Glenview, IL, for refueling. After refiling our flight plan the left engine would not start. The Constant Speed Starter/Generator (CSD) would not come out of Air Turbine Mode (ATM). Cmdr. Little called the plant and asked what to do. They got a troubleshooter on the line and he gave me instructions as follows: get a ten foot long piece of No. 9 electrical wire. Open the left engine bay door and remove the cover of the junction box on the bottom of the CSD. Go to the external electrical panel and unscrew the cap of the aux connector. Put one end of the wire in pin "E" and then hold the other end on the middle screw of the three screws of the CSD junction box. Have the pilot hold down the start button of the left engine, and then have the ground crewman apply air from the start unit. This electrical jolt will cause the internal breaker to the CSD to shift back to the start mode.
>
> "It worked like a charm, and we went merrily on to New Mexico. I tried this one other time, much to my undoing. When the brake clamped down on the shaft, the CSD was already turning in the ATM mode. As can be expected, the shaft broke. The squadron maintenance officer sent a new CSD, along with a mechanic to 'show me how to change the CSD' while laying on the hot ramp at NAS Cecil Field, FL."

Lt. Howard G. "Bud" White, a former *HATRON 6* bombardier, was one of the later BuWeps A-6 representatives. His experiences were a tad different, but still illustrated the enhanced prestige afforded the B/N in the new aircraft:

> "I requested A-6s and ended up going to Grumman at Bethpage, but BuWeps initially sent me as the assistant program manager for the E-2A. When I got there they set up a 'mini RAG,' just me, learning all there was to know about the *Hawkeye* from the engineers and designers. Well, the guy who had the A-6A program decided to quit the Navy and go to the airlines. I immediately volunteered to take his slot; in doing so, I became the first non-pilot to have the position, which was kind of neat.
>
> "After a period we got word one of the pilots in VA-42 would be up shortly to assume the job, which did not sit well with me. I went into the boss' office and told him, 'I'll be so valuable, you won't be able to afford losing me!' He took his glasses off, looked up at me, and said, 'Fair enough.' I worked my butt off, but I kept the job and got to know the aircraft very well. In fact, when I left, they rewrote the billet requirements so that only B/Ns could hold my job, *not* pilots."

Later White and Diselrod moved on to fleet tours, as did other Navy personnel who worked their way through Grumman or through flight test at Patuxent River. They provided the foundation for the A-6's fleet introduction and operational service.

Fleet Evaluation

The Navy let its first production order for *Intruders* under the Fiscal Year 1962 defense budget, dated 1 July 1961. Totaling $150.3 million, the contract authorized 24 A2F-1s beginning with BuNo 149475. The following year's FY63 budget called for an additional 43 planes at $159.3 million, starting with BuNo 149947. Thus, heading into the end of 1962 production aircraft were on the way, and it was time for the *Intruder's* final test prior to fleet introduction.

The process had started in September 1961 with the Naval Preliminary Evaluation 1. Involving several aircraft, Navy personnel, and Grumman specialists, the fleet evaluation program included a wide range of tests to determine the aircraft's appropriateness, capabilities, and performance levels in its designed missions. Along the way the *Intruder* had picked up a new designation: A-6A. On 18 September 1962, the Department of Defense instituted the tri-service designation system, by which all military aircraft types were identified using a single method.

Reportedly, the new system was the result of some well-publicized confusion on the part of Secretary of Defense Robert Strange McNamara. During a news conference, the secretary had introduced the Navy's new McDonnell F4H-1 *Phantom II* and its Air Force counterpart, the F-110A, as two totally different aircraft. Upon learning of his gaffe McNamara ordered a new means of designating aircraft based on the Air Force system. As a result, the A2F-1 became the A-6A, taking its place behind the A-1 (AD) *Skyraider*, A-2 (AJ) *Savage*, A-3 (A3D) *Skywarrior*, A-4 (A4D) *Skyhawk*, and A-5 (A3J) *Vigilante*.

In October 1962 the Airframe Board of Inspection and Survey convened at Naval Air Test Center, Patuxent River, MD. Two months later, two YA-6As were craned aboard *Enterprise* (CVAN-65) at Naval Station Norfolk for the *Intruder's* first carrier qualifications. The "Big E" was itself wrapping up initial qualifications before reporting to the Atlantic Fleet. Built by Newport News Shipbuilding and commissioned on 25 November 1961, the ship was the nation's first nuclear-powered aircraft carrier, its second nuclear-powered surface combatant – following *Long Beach* (CGN-9) – and was the world's largest and most expensive ship. Assigned *Carrier Air Group 6*, the new "Big E" completed its shakedown cruise to Guantanamo Bay, Cuba, the previous April, and was scheduled to make its first Med deployment in February 1963.

During operations on 19 December 1962 YA-6A No. 8 (BuNo 148618) made the Navy's second successful at sea test of the nose-tow catapult launch system; an E-2A, also undergoing carrier qualifications, had made the first launch a few minutes earlier. The nose-tow method – one of the prime features of the *Intruder's* landing and arresting gear design – used an integrated nose strut-mounted tow bar that fitted into a housing in the catapult shuttle. This method allowed the Navy to dispense with the catapult bridle method of launching aircraft – the use of a wire cable to fling the aircraft off the bow – and greatly reduced launch intervals.

The *Intruder's* carrier qualifications were a success, as were subsequent evaluations. In March 1963 the Navy wrapped up the aircraft's Avionics and Weapons System BIS following 18 months of intense testing and evaluation. On 10 October 1963 the service formally accepted aircraft No. 32, BuNo 149946, as its first *Intruder*. The A-6 was going to the fleet.

Going Operational

When the time came, the honor of becoming the first *Intruder* squadron fell to *Attack Squadron 42* at NAS Oceana. Established as VF-42 at NAS Oceana on 1 September 1950, the *Green Pawns* had served for eight years as a fleet *Corsair* and *Skyraider* squadron prior to becoming the AD replacement squadron – aka RAG, for Replacement Air Group – under *Readiness Carrier Air Group 4 (RCVG-4)*. With a new aircraft and new mission, its members now gathered additional personnel and prepared to assume the replacement training role with the A-6A *Intruder*.

When VF-42 stood up in late 1950 under Lt. Cmdr. Ross B. Spencer it equipped with F4U-4 *Corsairs* and adopted the *Green Pawns* name and emblem of the previous VF-42, which had disestablished on 8 June 1950. Eighteen days after it opened up shop in Virginia Beach the squadron moved to NAS Jacksonville, FL, where it joined CVG-6. Through the summer of 1951 the *Pawns* made a three-week "maximum cross-deck" deployment to the Caribbean – departing in USS

A VA-42 A-6A is shown in flight not long after the squadron started training crews in the new aircraft. The black nose would remain an *Intruder* trademark through the mid to late 1960s. (U.S. Navy, PH2 Holder)

Midway (CVB-41), transferring to *Wright* (CVL-49), and returning in *Cabot* (CVL-28) – while changing home stations twice. The first move across Jacksonville to NAAS Cecil Field came on 9 June 1951; the second transfer, on 27 August 1951, brought them back to Oceana. Two subsequent Mediterranean deployments in *Saipan* (CVL-48) and *Midway* took the squadron through May 1953.

AD-4 *Skyraiders* replaced the *Corsairs* in September 1953; AD-6s arrived in November, along with a redesignation as *Attack Squadron 42*. Following participation in the shakedown cruise of the Navy's newest carrier, *Forrestal* (CVA-59), in the spring of 1956, VA-42 joined *Air Task Group 181*. Three more deployments followed, including a single WestPac in *Bennington* (CVA-20). On 24 October 1958 VA-42 transferred to RCVG-4 and changed its mission to the training of fleet replacement AD ("Spad") aviators. Operating a mix of ADs and T-28Bs – the latter used for instrument training – the *Pawns* graduated its first *Skyraider* pilots in early 1959. The squadron's basic mission of providing qualified aviation personnel to the fleet would continue at Oceana for another 35 years.

As for NAS Oceana itself, it quickly adopted a new status as the home of medium attack aviation. The installation – located in farmland southwest of Virginia Beach proper – first opened in 1940 as an outlying field for NAS Norfolk. The Navy expanded Oceana early in World War II, and on 17 August 1943 redesignated it as an auxiliary air station. During the remainder of World War II a wide range of patrol and carrier-based units passed through for training, ranging from individual squadrons to entire air groups.

The station remained open postwar, albeit in a reduced status. However, in 1950 the Navy selected the field for development as a Master Jet Base; under this concept Oceana and several other air stations were upgraded to operate, maintain, and support large numbers of modern jet aircraft. In Oceana's case this meant extended runways and the construction of all new facilities on the south portion of the field, resulting in its redesignation as a naval air station on 1 April 1952. Within short order the base was bustling, hosting several fighter and attack squadrons, as well as support and training components. On 4 June 1957 the field was named for the late Vice Adm. Apollo Soucek, a noted Navy test pilot and holder of several altitude records, and Chief, Bureau of Aeronautics from 30 June 1953 to 4 March 1955. Oceana was truly booming by the time VA-42 was ready to receive its first A-6A. The SB2Cs, TBMs, and F4Us of the immediate post-World War II period had given way to A-1s, A-4s, F8 *Crusaders*, and the first few F-4A/B *Phantom IIs* of VF-101.

As VA-42 expanded into its new mission it gained quite a group of personnel with an extensive background in attack aviation. Then Lt. Robert S. "Rupe" Owens was one of the men assigned to the *Green Pawns* during its first days as the A-6 RAG and vividly recalls the officers he served with:

> "Cmdr. Jack Herman had previously commanded VA-66 and was a member of Flight Test, Pax River, where he completed the crosswind landing trials for the A-3. His executive officer was Don Ross, an A-4 pilot; Don was later rewarded for his performance with command of VA-35, VA-174 – the A-7 RAG – and fleet introduction of that aircraft. The Operations Officer was Lt. Cmdr. Bob Mandeville, an A-4 pilot. There were seven other A-4 pilots in the initial cadre: Lieutenants Bob Miles; myself; Jim Brooks; Jim Roth; Bill Westerman; Sonny Caldwell; and Jim Tabb. The original bombardier/navigators were Roger Smith, Jay Kohler, John Diselrod, Bob Morgan, Steve Stevenson, George Mallek, Ted Been, Irv Stahel, Don Hahn, and Jack Murphy. "

The big day for Jack Herman's *Green Pawns* finally came on 1 February 1963 when, under the watchful gaze of *Commander in Chief U.S. Atlantic Fleet* Adm. Frank O'Beirne, the squadron gathered on the Oceana ramp to welcome the first two fleet *Intruders*. However, Owens recalls the weather did not cooperate and the festivities quickly went downhill:

> "On the day the A-6 was scheduled to make its debut at Oceana the weather was poor, with rain, low ceilings, and fog. Jack Herman/Jay Kohler and Don Ross/Roger Smith were scheduled to ferry the A-6s from the Grumman Peconic River facility to arrive in time for the ceremony. The aircraft were not ''full-system' – not certified for instrument flight operations – and were unable to land with the field at minimums."

Big crowd, big anticipation, a lot of brass on the ramp in front of VA-42's hangar ... and no aircraft. However, the ceremony commenced on time; you could not hold up the admiral, even if the planes couldn't make it in. Owens continues:

> "The local Grumman Field Service Representative provided an A-6 model for *CINCLANTFLT*. The local Norfolk paper could not – and did not – let the opportunity pass to take a shot at the 'all new, all-weather' bomber at NAS Oceana. The two planes diverted to Patuxent River, and later that night Don Ross and Roger Smith finally made it into Oceana."

It was an inauspicious start for the A-6 and the new medium attack community. The first full systems aircraft, BuNo 149939, did not arrive at VA-42 until 12 June. However, its arrival allowed the squadron to start weapons system indoctrination training for the pilots and B/Ns – part of the Fleet Introduction Program (FIP) – and the *Pawns* would subsequently initiate the A-6A Fleet Replacement Aviation Maintenance program (FRAMP) for squadron maintenance personnel. In the meantime, VA-42 opened its doors and accepted its first fleet A-6 trainees.

The *Sunday Punchers* Check In

On 3 September 1963 FIP (Fleet Introduction Program) training convened with class 1-63 for the pilots and newly assigned bombardier/navigators of VA-75. Thus, the *Sunday Punchers*, led by Cmdr. William L. Harris, Jr., became the first of the fleet *Intruder* squadrons.

The squadron was decidedly "old Navy," dating to its 20 July 1943 establishment at NAS Alameda, CA, as VB-18. In August 1944, after training in several locations, the unit and its SB2C-3 *Helldivers* joined *Carrier Air Group 18* in *Intrepid* (CV-11). The *Punchers'* maiden deployment lasted four months and involved combat against Japanese forces in the Battles of Leyte Gulf and Cape Engano. At Leyte, VB-18 participated in the sinking of the battleship *Musashi*; at Cape Engano the squadron attacked a Japanese carrier task force, resulting in the awarding of the Navy Cross to CO Lt. Cmdr. George D. Ghesquiere and Lt. Benjamin G. Preston. On the day of the Japanese surrender the *Sunday Punchers* were stationed at NAS San Diego and getting ready for another go.

Postwar the squadron redesignated to VA-7A on 15 November 1946, to VA-74 on 27 July 1948, and to VA-75 on 15 February 1950. During this period it made several Med cruises from NAS Quonset Point as a component of CVG-7, in the process exchanging its SB2C *Beasts* for F4U-4 *Corsairs* and – in 1949 – *Skyraiders*. Mid-1952 found VA-75 embarked with CVG-7 in *Bon Homme Richard* (CV-31) and fighting in Korea. In 1953 the *Punchers* returned to the Mediterranean in *Bennington* (CVA-20); following transition to AD-6s in 1954 they made one more WestPac deployment in *Hornet* (CVA-12), followed by a short trip to the Northern Atlantic and four Med cruises. In late 1962 – still assigned to CVG-7 – VA-75 wrapped up its last *Spad* cruise in *Independence* (CVA-62) and turned in its aircraft. The squadron then moved over to VA-42's spaces and entered training. It received the first of its own *Intruders* on 14 November 1963.

During this period the *Green Pawns* were still settling into the A-6 RAG routine. On 8 September 1963 the squadron finally retired its last A-1H, BuNo 135324, thus formally ending its five years as an AIRLANT *Spad* replacement squadron. It kept one A-1E and two T-28Bs for use in instrument training. During this time VA-42 crews went to the boat for the *Intruder's* initial squadron carrier qualifications. The day quals were held over 8-11 July in *Forrestal*; unfortunately, they were not without incident. During one launch, B/N Lt. Jim Brooks' ejection seat malfunctioned on the cat stroke. His seat moved up the rails, causing the drogue gun to fire and deploying the drogue chute through the canopy. As the A-6A passed over the bow, Brooks' 26-foot diameter main parachute deployed.

According to his pilot, Rupe Owens, the carrier immediately transmitted "Your signal divert, Norfolk is 163 miles, steer 290." He rejected the order and requested an immediate recovery, as he felt Brooks would not survive the trip to the beach. Owens then safely landed the *Intruder* with the blossoming chute fully deployed behind the aircraft. He later commented it was the only occasion in his A-6 career that he had no trouble controlling the speed coming down the glide slope. Brooks survived his wild ride, and the squadron would return to *Forrestal* over 14-15 October 1963 for its initial night traps.

Training proceeded at Oceana as aircraft continued to arrive from Grumman, and in March 1964 several major events occurred in the small medium attack community. On the twelfth VA-42 transferred its last T-28B, formally ending its instrument training program for fleet A-1 pilots. Two days later the squadron's last A-1E *Skyraider* departed. On 13 March the *Sunday Punchers* officially completed A-6 transition training and reported back *to Carrier Air Wing 7* as a deployable squad-

Sunday Punchers cruise off Fort Story, VA, prior to their first deployment. Within months the squadron would take the type into combat. Both of these aircraft would survive Vietnam. AG510 would be lost with VA-35 in August 1975 when it had a mid-air collision with an A-7E in the tanker pattern. AG511 would be converted to B and E versions, and finally retired to the Arizona desert in 1995. (U.S. Navy)

ron. They immediately started workups for their next cruise; in their stead, the *Black Falcons* of VA-85 continued training with VA-42 as the second fleet *Intruder* squadron.

Black Falcons, Batmen, and *Tigers*

Another long time Oceana-based operator of the noble *Spad*, VA-85 found its origins as Naval Air Reserve VA-859. Called to active duty on 1 February 1951 for the Korean War, the NAS Niagara Falls-based squadron exchanged its TBMs for AD-2s. In November 1951 it departed in USS *Tarawa* (CV-40) for a Med cruise with *Carrier Air Group 8*. The squadron returned to the states at NAS Oceana in June 1952, transitioned to the AD-4, and on 4 February 1953 became the active duty Navy's VA-85. The subsequent eleven years were spent in a regular Med rotation with CVG-8, including one cruise each in USS *Coral Sea* (CVA-43), *Lake Champlain* (CVA-39), and *Intrepid*. January 1960 brought the first of three Med deployments in *Forrestal*.

Two notable events for the squadron came in May 1958. First, the Chief of Naval Operations approved VA-85's new patch design, its third. Featuring a stylized black falcon, it gave the squadron its famous name. The other event came about as part of exercise *LANTRAEX 1-58*, a regularly scheduled fleet training exercise. In a prestidigious feat – commonly referred to as a *Sandblower* mission – Lt. j.g.s Carl Strang and Carl Woods launched in their AD-6s from *Forrestal* off Jacksonville and flew all the way to NAS North Island, CA. They flew the entire 10 1/2-hour flight below 1,000-feet altitude. Two days later the *Spad* pilots returned to the carrier, again non-stop and this time carrying 50 copies of the *San Diego Union* for the ship's personnel.

The *Black Falcons* concluded their final A-1H cruise in *Forrestal* on 2 March 1963; led by Cmdr. John C. McKee; the squadron picked up its new B/Ns and quickly dove into training at VA-42. The first squadron A-6As arrived on 6 March 1964; one month later VA-85 formally transferred from CVW-8 to *Commander, Fleet Air Norfolk* for the duration of the training.

By now the *Green Pawns* were fully up to speed, as was the Grumman production line, and the squadrons were coming fast and furious, as were the Marines. The first two prospective *Intruder* pilots from the Corps, Capt. Bennie Sprier and 1st Lt. William McCutchen, flew their initial hops over 5-27 August 1964. The following month the *Batmen* (or *Bats*) of VMA-242 left their A-4Cs at MCAS Iwakuni, Japan, and returned to MCAS Cherry Point, NC. From there the squadron moved *en masse* to NAS Oceana to get in line with VA-42. Notably, the *Bats* were the first squadron selected to transition to *Intruders* from another jet aircraft.

The squadron possessed a proud history in attack aviation dating to its establishment on 1 July 1943 at MCAS El Centro, CA, as Marine Torpedo Bombing Squadron 242 (VMTB-242). Initially equipped with SNJs and a few SBD *Dauntlesses*, the *Bats* gained their first TBFs and TBMs in August 1943 and embarked on a six-month training program over the beautiful Salton Sea region of the Golden State. In January 1944 they shipped aboard USS *Kitkun Bay* (CVE-71) and departed CONUS for Espiritu Santo, arriving in the Southwest Pacific in February. First combat with their *Avengers* came on 11 April 1942 when the *Bats* raided Japanese facilities on Rabaul. Several moves around the Pacific followed, including operations from Guadalcanal, Eniwetok, Oahu, and Saipan-Tinian. The squadron completed its part in World War II by flying antisubmarine patrols from Iwo Jima and Tinian, returned to the United States and inactivated at San Diego on 23 November 1945.

The *Batmen* did not return to service for fifteen years, finally reactivating as VMA-242 at MCAS Cherry Point on 1 October 1960.

Equipped with A4D-2 *Skyhawks*, the squadron participated in various training and readiness activities for two years. They saw their first hint of combat during the Cuban Missile Crisis, departing Cherry Point on 23 October 1962 and taking up station at NAS Key West. After a month the crisis subsided and -242 returned to North Carolina.

Early September 1963 found the *Bats* on the road again, this time to Camp Pendleton, CA, in preparation for a move overseas. On 7 October VMA-242 formally transferred to MCAS Iwakuni. Their tour in the far east lasted approximately a year, during which time they made training and exercise deployments to Cubi Point, Taiwan, and Korea. On 1 October 1953 the squadron relocated to NAS Oceana and was redesignated as VMA(AW)-242, becoming the first of six fleet Marine A-6 units. Upon completion of training on 15 December they returned to North Carolina with a two-fold mission: train other Marine squadrons and prepare for deployment.

Finally, in late 1964 the *Tigers* of VA-65 returned home to Oceana from the USS *Enterprise's* famous around-the-world cruise, turned in their *Skyraiders*, and got in line for *Intruders*. Among the officers assigned to the squadron was Naval Aviation Observer John Diselrod, late of Grumman.

Bringing Up B/Ns

With his background as a bombardier in A-3 *Whales* and A-6 development B/N at Grumman, Diselrod quickly found himself in demand for training purposes:

> "Moving on to NAS Oceana in late 1964 I joined VA-65; they were in transition to the A-6A in VA-42, and I was utilized as an instructor there. Some of the pilots were from the training command, but in those days many were *Spad* or *Scooter* retreads. Bob Mandeville, Bob Miles, and Bill Westerman were A-4 pilots I knew from previous duty, along with Jack Fellows from ADs. All these people had been on their own while flying, and the first thing one had to teach was 'trust the B/N. '
>
> "Low-level flying was the main training priority. Both the pilots and the B/Ns had to learn to trust the aircraft and the system. In those days, low was as comfortable as the crew could handle, since the flight controllers had not gotten into the situation yet. To teach a pilot to stay below radar coverage when crossing a ridgeline got to be a chore. First one had to convince them that the SRTC was reliable. Then, if you ran at a ridgeline head on you tended to overshoot the top and far side. The best way was to approach head on with steering centered, turn slightly either way, and let the wing toward the ridge line lift up and over, turn back to steering going down the other side, and stay low. The skipper did not like any aircraft coming back from one of these flights with chlorophyll streaks on the underside of the plane from the Blue Ridge Mountains, but it happened."

It was a unique and occasionally an awkward situation in the Navy, placing two equal crew members side by side and having them work together with a definite delineation of duties; the growing F-4 *Phantom* community was making a similar adjustment, although in that case the pilot and Radar Intercept Officer sat in tandem. The *Whale* crews were familiar with the setup and adjusted well to the *Intruder*, but reportedly some of the single-seat pilots never did. John S. McMahon, Jr., a former *Skyhawk* pilot with VSF-1, remembers:

> "It was different flying with a B/N. We felt we could do anything and everything, but now there was a guy there in the cockpit with you who *knew* you couldn't do anything and everything, and you had to adjust.
>
> "I never flew combat in the A-4, but I learned that having a B/N with you in combat was good. Some pilots never adjusted; I have heard stories of the old 'line drawn down the center of the cockpit – you stay on your side I'll stay on mine' situations – but as a pilot I never ran into that myself."

For the bombardier/navigators the times were good, and their particular corner of Naval Aviation was growing by leaps and bounds. The

Not all early aircraft would end up with squadrons; 147867, the fourth *Intruder* airframe, was redesignated as an NA-6A and ended up at NAEC Lakehurst, where it was used to train Aviation Boatswain's Mates. (April 1979, Mark Morgan)

community actually dated to the 12 July 1921 authorization of the Naval Aviation Observer (NAO) designator by act of Congress. The same legislation created the Bureau of Aeronautics (BuAer) and specified its chief must qualify as either a pilot or observer; hence the first Chief of BuAer, Rear Adm. William A. Moffett, entered flight training, and on 17 June 1922 became Naval Aviation Observer No. 1.

Over the following decades, NAOs served in a wide range of aircraft that required qualified, non-pilot personnel. By the mid-1950s these included both officers and enlisted personnel serving as bombardiers, radar operators/intercept officers, airborne electronic countermeasures operators, navigators, radio operators, and flight engineers. As an example, during the Korean War Marine F3D *Skynight* radar intercept operators were almost uniformly enlisted men or warrant officers. Later in the decade the heavy attack bombardiers were all enlisted. Now with the arrival of aircraft such as the A-6A *Intruder*, F-4B *Phantom II*, and E-2A *Hawkeye* the NAOs were becoming equals in the cockpit, and this equality was reflected in the medium attack community, where a large proportion of the early B/Ns moved from enlisted to Limited Duty Officer status.

The Navy eventually recognized the changing circumstances, and on 8 February 1965 the Bureau of Personnel authorized the designator "Naval Flight Officer," replacing Naval Aviation Observer. After July 1968 NFOs received new wings, similar to those of Naval Aviators but with two crossed anchors. They still could not command squadrons (yet), but they could and did work the mission as full partners with the pilots.

More Squadrons

As 1965 opened VA-85 and VA-65 were well along in their training. The year would prove to be a hectic one at Oceana, with several squadrons rotating through and others preparing to deploy; adding to the pressure was a growing conflict in a little-known country on the opposite side of the world.

The *Tigers* of ATKRON 65 were another long-time *Skyraider* operator, dating to the 1 May 1945 establishment of VT-74 at NAAF Otis Field, MA. After equipping and training in SB2C-4Es and SBW-4Es (the Canadian Car and Foundry-built variant of the *Helldiver*) the squadron joined CVG-74 for the initial shakedown cruise of *Midway*. As with their counterparts at VA-85, the squadron subsequently went through several redesignations: VA-2B on 15 November 1946; VA-25 on 1 September 1948 (which brought a new patch and the name *Tigers*); and VA-65 on 1 July 1959. The squadron also went through a series of attack aircraft, culminating in the assignment of AD-6s in October 1953. By that time the *Tigers* were firmly ensconced at Oceana, under assignment to *Carrier Air Group 6*.

Heading into the 1960s VA-65 took part in several notable operations. In June 1961 the squadron deployed to the Caribbean onboard *Intrepid* in response to the assassination of Gen. Rafael Trujillo Molina and resultant domestic unrest in the Dominican Republic. Sixteen months later the outfit quickly pulled up stakes again and rushed back to the Caribbean – this time in *Enterprise* – for the Cuban Missile Crisis. Finally, from February through October 1964 the *Tigers* participated in the around-the-world cruise of *Enterprise*. The portion from Gibraltar through the Indian and Pacific Oceans to Norfolk – designated Operation *Sea Orbit* – lasted 65 days and marked the first deployment of a wholly nuclear-powered task force. The task force included cruisers *Long Beach* (CLGN-9) and *Bainbridge* (DLGN-25) and covered 30,216 miles. Afterwards VA-65 returned to Oceana, dumped off its A-1Hs, and reported to RCVW-4 on 1 January 1965. After completing training with VA-42 on 20 February 1966 the *Tigers* transferred to Carrier Air Wing 15 and prepared for deployment.

During this period the Navy tried to decide which unit would become its fourth fleet *Intruder* squadron. The first three squadrons – VAs -75, -85, and -65 – were all ex-*Spad* outfits, and the Navy now had to pick a fourth. As described by Lt. Cmdr. Jerry Menard – a former aviation machinist who survived a tour as an anti-submarine aircrewman with VS-21 and went on to a lengthy career as a professional maintainer – what came next was one of those weird situations that crops up every now and then.

According to Menard the "powers that be" decided to send the proud *Valions* of VA-15 to Oceana ... and it turned out to be a classic bureaucratic mess. The *Valions* (as in "VA-Lions") were A-1H operators who had completed their last *Spad* cruise in *Franklin D. Roosevelt* (CVA-42) in April 1963. In April 1964 the squadron detached from *Carrier Air Wing 1* and NAS Jacksonville and moved as a group to NAS Oceana for the next big thing. However, upon their arrival the Navy – in its infinite wisdom – decided it wanted somebody else to get in line. The obvious solution would have been to redesignate VA-15 in place – giving it the desired squadron's number – and transfer the VA-15 title elsewhere without personnel or equipment. Apparently, that made too much sense; instead, in August the squadron transferred once again to sunny Cecil Field where it transitioned to A-4Bs. In its place the Navy sent the *Black Panthers* of *Attack Squadron 35* to Oceana from Jacksonville.

"The whole evolution was incredibly expensive," recalls Menard. For his part, he was ready to go back to Florida with the squadron when he learned he'd screened for A-6 maintenance duty. "I protested to BuPers, but the detailer told me it was a high visibility, plum position. He told me, 'Don't worry, I'll take care of you.'" He went on to join VA-35, but remembers one of his fellow selectees later screwed up and quickly found himself "haze gray-underway" on a fleet oiler.

The Navy later repeated the process at least two more times, under different – and less expensive – circumstances. In August 1967 VA-115 retired its A-1Hs and A-1Js at NAS Lemoore and entered inactive status. Until January 1970, when it remanned under Cmdr. C. J. "Connie" Ward at NAS Whidbey Island, the *Arabs* were a squadron in name only, under the administrative control of VA-125, the west coast A-4 RAG. On 1 April 1969 the world famous *Saints* of VA-163 – an A-4E squadron recently of CVW-21 – also entered inactive status at NAS Lemoore. According to *The Dictionary of American Naval Aviation Squadrons, Volume 1* "the squadron probably was placed in an inactive status due to manpower and aircraft availability while awaiting transition to the A-6 *Intruder*." The call never came; the fighting *Saints* were removed from the Naval Aeronautical Organization list and disestablished on 1 July 1971.

The end result saw the *Black Panthers* formally report to RCVW-4 for duty with A-6s on 15 August 1965. At the time of its arrival ATKRON 35 was one of the oldest and most storied squadrons in naval aviation history, dating to its initial establishment at NAS Norfolk on 1 July 1934 as Bombing Squadron 3B (VB-3B). After initially equipping with the Martin BM-1 the squadron shifted to the Great Lakes BG-1 and made its initial deployment in USS *Ranger* (CV-4) over March-April 1935 through the Panama Canal to the west coast.

Over the following five years – during which time it adopted the name and historic Black Panther emblem – the squadron participated in several exercises and expositions, as well as a few movies while operating as a component of the *Ranger* and *Lexington* Air Groups. The *Panthers* underwent two redesignations – to VB-4 on 1 July 1937 and VB-3 on 1 July 1939 – and operated the BG-1 and SB2U-series of scout bombers. On 1 January 1939 VB-4 joined the *Saratoga* Air Group (later to become *Carrier Air Group 3* and *Carrier Air Wing 3*); the transfer marked the start of a remarkable 23-year affiliation with the group.

On the day of the Pearl Harbor attack VB-3 was operating SBD-3s in *Saratoga* in the vicinity of Hawaii. The following morning the carrier pulled out of the shattered Pacific Fleet homeport for the first of four years of operations against the Japanese. This subsequent period read like a "who's who" of combat in the Pacific, including operations in support of the Doolittle Raid against Tokyo while temporarily embarked in *Enterprise* (CV-6) and combat in the Battle of Midway from the deck of *Yorktown* (CV-5). In the latter action VB-3 crews participated in the sinking of the carriers IJN *Soryu* and *Hiryu*; 17 officers, including CO Lt. Cmdr. Max Leslie, received the Navy Cross for the action. Through the remainder of the war VB-3 operated from *Saratoga* and the second carrier *Yorktown*, participating in operations in the Solomons, Leyte, Formosa, the South China Sea, Iwo Jima, and against targets in the Japanese home islands. The squadron departed the Pacific Theater on 6 March 1945 in *Lexington* (CV-16) and finished the war with four Presidential Unit Citations.

VB-3 started the war in SBDs, transitioned to SB2Cs in 1943, and in November 1948 reequipped with the AD-2 *Skyraider*. By that time the squadron had again redesignated, to VA-3A on 15 November 1946, and to VA-34 on 7 August 1948. On 15 February 1950 – while still assigned to CVG-3 – the *Panthers* gained their final designation as VA-35. They remained with CVG-3 through the Korean War in *Leyte* (CV-32), the 1958 Lebanon crisis, and the Cuban Missile Crisis of October 1962. Their final departure from CVW-3 occurred on 15 August 1965 when the squadron retired its last A-1Hs. The transfer marked the end of 17 years of *Skyraider* operations and an incredible 34 years with *Carrier Air Wing 3* and its predecessors.

Line of Departure

In the spring of 1965, while the Navy was still flailing around over the transition schedule, one squadron departed NAS Oceana under different circumstances. On 10 May the *Sunday Punchers* of VA-75 deployed from the United States as a component of CVW-7 in *Independence*. The *Intruder* was going to war.

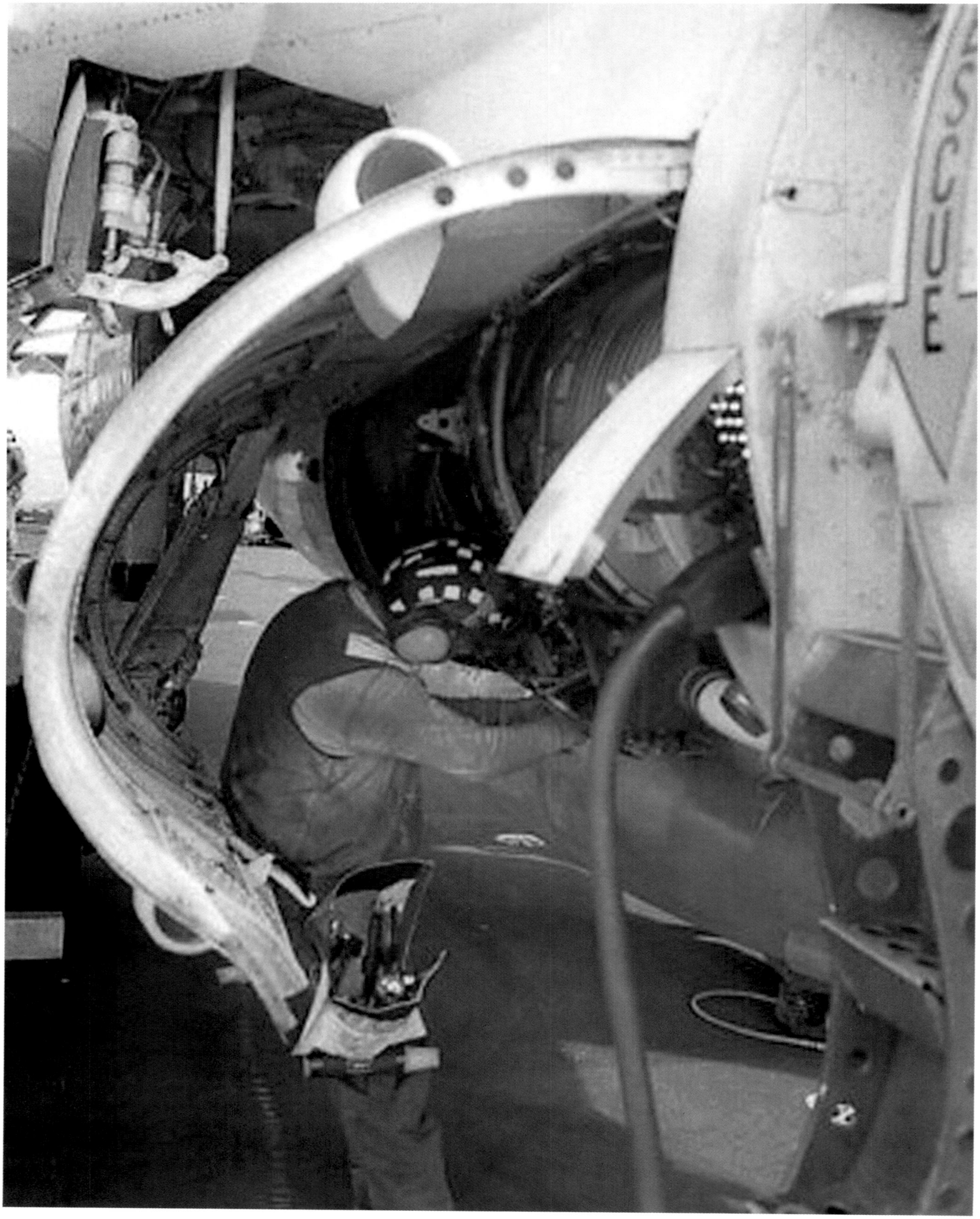
SCUE

The Heart of the *Intruder*

If the fuselage of an aircraft is its body, then the engines are certainly its heart(s). For the power plants not only provide the thrust necessary to defeat drag and induce lift from the wings, they also turn the hydraulic pumps and spin the generators that produce the blood and electrical synapses that control the beast. If an aircraft is only as good as its engines, then the *Intruder* was truly blessed to have two reliable Pratt & Whitney J52 engines buried in its tadpole fuselage.

First announced in late 1956, the J52 was designed under auspices of the Navy's BuWeps as a medium-sized turbojet for the next generation of attack aircraft. The new motor was an axial flow, non-afterburning design with 12 compressor and 2 turbine stages mounted on twin-spools. On the test bench the J52-P-8 produced 8,700lbs of thrust. In the *Intruder*, however, this was reduced to 7,700 pounds due to the S-shaped exhaust ducting.

As well as powering all fleet *Intruders* (only the F-model had other engines, F-400 turbofans), the J52-P-6 and P-8 were the power plants of choice for the A-4E and F model *Skyhawks*, replacing the English designed/Wright-built J-65 *Sapphire* engine used in earlier versions. Of some interest, the J52 had also been intended for two earlier *Skyhawks*, both stillborn – the A4D-3 and -4.

It was in the J52-P-408, developed in the late 1960s for the EA-6B *Prowler*, that the engine reached its zenith. The type featured an uninstalled thrust rating of over 11,000 pounds, largely through the addition of variable inlet guide vanes on the first stage stator. The "P408" would also be installed in the A-4M and retrofitted to at least 100 A-4Fs. Plans to install it in the A-6G ended with the cancellation of that program, as did attempts to place an even more capable "J52-P-409" in the stillborn ADVCAP EA-6B.

Although designed under Navy contract, the Air Force's primary use of the J52 engine would lead to one of the better "urban legends" in *Intruder* lore. As the J52-P-3 the motor was used in the AGM-28 *Hound Dog*, a late-50s era nuclear-tipped cruise missile carried in pairs under the wings of B-52s. The *Hound Dog* was a 42 foot long, five ton weapon intended to blast a path in air defenses in front of Strategic Air Command's penetrating aircraft, and was, as such, viewed as a "single use" weapon, hence the story that the A-6's engine was originally designed as a "use only once/throwaway" power plant. In reality, the *Hound Dog's* thrust could be used to supplement the parent *Stratofortress'* own eight engines for take off, with fuel being replenished in flight. The AGM-28 saw service in SAC from 1961 until retirement in 1976.

Above: The Air Force's primary use of the J-52 engine was in the AGM-28 *Hound Dog* air to surface missile. Most B-52s could carry two of these nuclear-tipped beauties, as shown on display here at the Castle AFB Museum. (May 1989, Mark Morgan)

Opposite: An *Intruder's* starboard engine gets the attention of a VA-155 Mech prior to launch on *Independence*. (U.S. Navy)

CHAPTER TWO

Vietnam, 1965-1967

On 10 May 1965 VA-75 left Tidewater in *Independence* for the carrier and *Intruder's* first combat deployment. Not since the introduction of the Curtis SB2C *Helldiver* in 1943 had a new Navy carrier aircraft made its first deployment directly into combat.

The leader of the *Sunday Punchers* was now Cmdr. Leonard "Swoose" Snead, who had relieved Cmdr. Bill Harris on 8 May 1964. A former VAH-6 *Whale* driver, Snead had directed the Navy's YA2F-1 evaluation team three years prior. His squadron's transit to the theater was a long one: Indy heading to the Philippines via an ORI in Puerto Rico and the Cape of Good Hope. Enroute, the crews of *Air Wing 7* flew, maintained, and studied the war to date.

By 1965, the use of air power in Southeast Asia was nothing new; neither was the use of naval aviation. During World War II, both Allied and Japanese carriers had roamed the region. During the French Indo-China War the French Navy acquired the former U.S. light carriers *Belleau Wood* (CVL-24) and *Langley* (CVL-29) on loan for use against the Viet Minh; they served their temporary owners as *Bois de Belleau* and *Lafayette*, respectively. The United States maintained an interest in the region following the French defeat at Dien Ben Phu in 1954 and subsequent signing of the Geneva Agreement. The pact granted independence to Laos and Cambodia, while creating two Vietnams: the Republic of Vietnam (RVN, south) and People's Democratic Republic of Vietnam (PDRV, north). It also called for elections to determine Vietnam's future, but they never took place, and by 1960 the region was rife with insurgency and guerrilla warfare.

In that year, the U.S. provided propeller-driven attack aircraft to the South Vietnamese Air Force, along with advisors. Continued violence in the RVN and neighboring Laos led to increased U.S. involvement, usually through the form of more advisors and American-operated military equipment. Navy and Air Force aircraft also participated in increased surveillance over flights of the region, resulting in several losses. Following Lyndon Baines Johnson's accession to the presidency in November 1963 the level of involvement went up several fold, including the stationing of Navy destroyers with modified electronics suites in the Tonkin Gulf. Termed *DeSoto* patrols, the operation saw the ships cruise up and down the coast performing signals and electronic intelligence (SIGINT/ELINT) activities.

At about 1600 on 2 August 1964, one of the *DeSoto* ships – destroyer *Maddox* (DD-731) – came under attack from North Vietnamese P-4 and *Swatow*-class torpedo boats. Returning fire, *Maddox* departed the area while calling in assistance. Aircraft from *Ticonderoga* (CVA-14/CVW—5) responded and sank one of the attackers. The following day *Constellation* prematurely ended its scheduled port visit in Hong Kong and headed for the Tonkin Gulf. On the evening of the 4th, *Ticonderoga* aircraft responded to reports of additional torpedo boat attacks on *Maddox* and *Turner Joy* (DD-951). The two destroyers expended a lot of ordnance and star shells, and claimed to have dodged several torpedoes, but one respondent, VF-51 skipper Cmdr. James Stockdale, found no evidence of communist surface craft. Still, the Johnson administration factored in the alleged second attack while considering possible retaliatory action against the government of North Vietnam.

On the afternoon of 5 August 1964 aircraft from CVW-14 in *Constellation* and CVW-5 in *Ticonderoga* struck naval targets in the vicinities of Hon Gai, Loc Chao, Quang Khe, and Vinh. The president ordered the strikes – designated Operation *Pierce Arrow* – in retaliation for the attacks on the U.S. destroyers. In a nationally televised address, broadcast two hours *before* the Navy aircraft arrived over their targets, the president stressed the nation's response was, "... limited and fitting. We still seek no wider war."

The reception afforded the Navy A-1, A-4, and F-8 crews over the targets was anything but limited, but they pressed the attack and wrecked several facilities. *Connie* lost two aircraft; at Loc Chao Lt.j.g. Richard Sather of VA-145 was gunned out of the sky, while at Hon Gai, Lt.j.g.Edward Alvarez was shot down on his second pass. Fortunately, Alvarez was able to eject from his burning A-4C *Skyhawk*, but became the first Naval Aviator to enter North Vietnamese captivity. Many others would follow Alvarez, as the Johnson administration's "fitting and limited" skirmishes grew into a full-fledged shooting war.

On 7 August 1965 the U.S. Senate passed the Tonkin Gulf Resolution. As requested by President Johnson, the resolution authorized the United States to take whatever means necessary to defend its interests in the region. Its passage – accomplished in a popular atmosphere of patriotism and support for the administration's efforts in the region – laid the foundation for rapid expansion of American combat and support activities in Vietnam.

The following December the United States kicked off *Barrel Roll* (armed reconnaissance missions over Laos). On Christmas Eve 1964 Viet Cong guerrillas bombed a hotel in Saigon serving as a Bachelor Officers' Quarters, killing or wounding 73 U.S. advisors. Following subsequent terrorist attacks on Pleiku Air Base and Camp Holloway – resulting in the death of 8 Americans and the wounding of 109 – the Johnson administration authorized retaliatory strikes. The first of these, *Flaming Dart*, was initiated on 7 February 1965. Roughly a month later (2 March 1965) Operation *Rolling Thunder* commenced with strikes on Xom Ban and Quang Khe. The first Navy involvement came on 15 March with an attack on an ammunition depot at Phu Qui.

VA-75 took the *Intruder* on its first deployment, straight into combat, in 1965. AG509 is shown on "Indy" in February 1965, three months before the ship and air wing started operations in Vietnam. (U.S. Navy, PH3 Fleischer)

The government's intent was as much diplomatic as military. By expanding the war northward in a logical, measured, and tightly controlled manner, the U.S. felt it could dissuade the North Vietnamese from further aggression in the south. In approving *Rolling Thunder*, the president stated the strikes were, "a program of limited and measured action against selected military targets." Four months later Secretary of Defense Robert Strange McNamara stressed he had taken a direct role in the strike planning, saying, "not since the Cuban Missile Crisis has such care been taken in making a decision." It was a clear indication of how Washington would run the air war. Later in the conflict pilots shared tales of being pulled from their cockpits at the last minute to deliver direct verbal confirmation to McNamara of routing, weapons loadouts, and rules of engagement parameters.

There were further indicators. On 5 April 1965 an RF-8A from VFP-63 Det D in *Coral Sea* took the first photographs of an SA-2 *Guideline* Surface-to-Air Missile (SAM) site under construction southeast of Hanoi. The government directed that no action be taken against the installation over concerns of possible fatalities to Soviet advisors. On 24 July 1965 an SA-2 bagged an Air Force F-4C from the *47th Tactical Fighter Squadron*; the pilot died and his backseater became a POW. As the air strikes escalated over the coming months, more aircraft would be fighting up North under tight constraints and in the face of an expanding threat. It was under these circumstances that the *Intruder* was put to the test.

VA-75 Enters Combat

The *Independence*/CVW-7 team finished its long transit from Norfolk on 5 June 1965 when it checked in with *Commander, Task Force 77 (CTF-77)*. The squadron had already lost the first *Intruder* in an operational mishap (21 May 1964), when one crashed while on an emergency approach to NAS Meridian, Mississippi. Both crew ejected safely from this aircraft, which was having electrical and hydraulic problems. Other incidents occurred in the normal scheme of work-ups, including one particularly intense night during pre-deployment carrier qualifications.

On the evening in question one VA-75 pilot left his fuel dumps on when departing marshal for recovery. By the time he called the ball he was well below the specified recovery fuel state and *well* below Bingo. Unfortunately, an RA-5C from RVAH-1 had somehow become lodged in the bow cat track and the ship was having trouble moving it clear of the landing area. The *Intruder* pilot was sent to the duty VAH-4 Det 62 A-3B tanker for a load of fuel, but the join-up was difficult and he failed to take on any gas.

By now he was *really* below Bingo; adding to the excitement, his right main landing gear was not indicating down and locked. Finally, the carrier decided to have him take the barricade. On final the pilot reported "fuel 0." His approach was fast, and he went high and the LSO waved him off. However, the barricade's top load strap snagged the A-6's nose gear and the plane descended to the deck with a resounding thunk. It quickly decelerated before colliding with the horizontal stab of the immovable *Vigilante*. Amazingly, the aircraft would be repaired in time.

By the time *Indy* arrived in the Tonkin Gulf *Rolling Thunder II* was underway, having started on 12 May 1965. On 27 June the carrier went on line for initial operations over South Vietnam; VA-75 entered combat in North Vietnam on 2 July with a mission against bridges at Bac Bang. On 14 July two *Intruders* flown by Lt.Cmdr. Bill Ruby and Lt. Don Boecker launched against a road target in the vicinity of Sam Nuc, northern Laos. When Boecker released his bombs – at about 5,500-feet – one of the Mk.82s exploded, shredding his aircraft. The right engine immediately caught fire, and the damaged wing started dumping fuel at a high rate. Boecker and B/N Lt. Don Eaton started climbing out on the remaining engine, but within short order its fire light also illuminated. With the stricken aircraft's controls locked the crewmen ejected. Boecker came down near a village, while Eaton disappeared over a hill. After some initial evasion it became dark and both men settled in for the night.

Early the next morning a variety of search aircraft started appearing over the crash site and the downed airmen established contact. At 0815 two H-34s moved in to attempt a rescue, but one helo was quickly shot up and both were forced to depart. By mid-morning squadron A-6s joined in on the recovery attempt, but problems with Boecker and Eaton's radios made communications difficult. Adding to the moment were injuries suffered by Eaton during ejection and the constant presence of bad guys searching for the *Intruder* crew. Finally – around noon and under the covering fire of A-1s and T-28s – the rescuers hoisted both men to safety. An Air America plane took them to Udorn Royal Thai Air Force Base, and from there they returned to *Indy*. It had been quite a night.

Three days later, VA-75's Lt.Cmdr. Cecil "Pete" Garber guided four VA-72 A-4Es to one of the early SA-2 sites near Kep Airfield and quickly bombed it out of existence. Garber later reported that as the strike departed the scene the radar vans, stored missiles, and nearby support structures were collapsed and burning. The raid marked one of the earliest coordinated attempts by CTF-77 aircraft against a SAM facility.

Tragedy struck again on 18 July when *Air Wing 7* Operations Officer Cmdr. Jerry Denton and B/N Lt.j.g. Bill Tschudy went down near the Thanh Hoa power plant. Tschudy was captured immediately upon landing; Denton came down in the Thanh Hoa River and attempted to swim to safety, but was intercepted by a boat and taken into captivity. Ironically Denton was to have assumed command of VA-75 the following day; doubly ironic, Secretary of Defense McNamara – onboard *Independence* during one of his early inspection and fact-finding tours of the war – had served as catapult launch officer for Denton and Tschudy's mission. Following his repatriation eight years later Denton would write:

> "Only later would I realize the fateful implications inherent in the U.S. Secretary of Defense seeing me off to work."

In Denton's stead, XO Cmdr. Leonard F. "Mike" Vogt, Jr., fleeted up as squadron commanding officer and VA-75 pressed on. However, on 24 July the *Punchers* lost their third *Intruder* in ten days. Pilot Lt.Cmdr. Richard P. "Deke" Bordone and B/N Lt.j.g. Pete Moffett had just released on the target in Laos when their plane exploded and immediately nosed over, out of control. Both men were able to eject and were retrieved, but were severely injured in the ejection process and required lengthy recuperation. Moffett never flew an ejection seat aircraft again; Bordone was able to fully recover and later commanded VA-75. Barely a month into its first line period VA-75 was down to twelve crews and nine aircraft. A subsequent investigation determined in-flight detonation of the electrically-fused Mk.82s caused the first three losses. The squadron corrected the problem by swapping their bomb racks out for Multiple Ejector Racks (MERs) and Triple Ejector Racks (TERs) and switching to mechanical fusing, and the problem did not return.

Still, there were growing rumblings from Navy leadership concerning the viability of the A-6 *Intruder*. On their part squadron personnel pointed out the losses had occurred on daylight missions over targets that were decidedly low value and high threat. They complained the A-6 was being used as a "Big A-4" and not in the night/all-weather role as designed. The aircraft regularly returned from their day missions with inoperative or marginal systems; when night came they were in the hangar deck under repair and could not fly. Conversely, the afloat leadership could not justify reserving for night and/or bad weather *only* an aircraft that could carry 28 500-pound bombs, and were not happy with the *Intruder's* high maintenance requirements and trouble-prone avionics. The controversy would continue for some time as squadrons and staffs debated the A-6's proper place in the air wings.

Unfortunately for VA-75 the situation got worse. On the night of 17-18 September Cmdr. Vogt and his B/N, Lt. Red Barber, were shot down and killed while attacking North Vietnamese patrol boats off Bach Long Vi Island. Squadron maintenance officer Cmdr. William B. Warwick moved up to the command slot, and the Navy sent two new *Intruders*; VA-42 provided a replacement crew, and the *Sunday Punchers* hunkered down and carried on.

The *Independence/Air Wing 7* team completed its third and final line period on 1 November 1965, and departed Seventh Fleet on the 21st, having spent 100 days in combat. VA-75's total for their seven-month cruise – which ended on 13 December with their return to Oceana – was four combat losses, four air crew recovered, two POWs – including the prospective commanding officer – and two killed in action, including the CO. These served as a precursor of the losses carrier air wings would sustain over the coming years.

The *Black Falcons* Go To War

While VA-75 was undergoing its baptism by fire VA-85 – under Cmdr. Billie J. Cartwright – had completed training and was preparing to make the A-6's second Vietnam combat cruise. Assigned to CVW-11 in *Kitty Hawk* (CVA-63), the squadron went on line on 26 November 1965, relieving the battered *Sunday Punchers*.

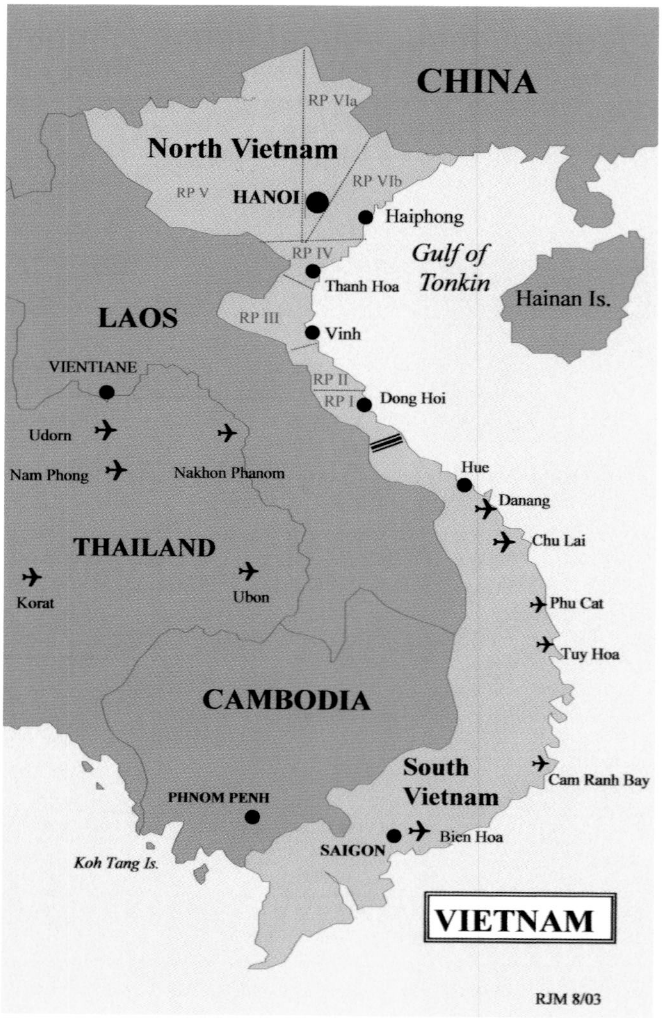

As with its predecessor, VA-85's learning curve with the *Intruder* was high. One incident occurred on 27 July during workups in *Kitty Hawk* off San Diego, when two squadron A-6As collided in midair. Pilots Lt.Cmdr. George B. Ball, Jr., and Lt.j.g. Frank G. Wagner and B/N Lt.Cmdr. John N. Spartz ejected and were pulled from the waters of the Pacific. The other bombardier, Alex Malinkoff, was lost, along with A-6As 151780 and 151786. The remainder of workups passed uneventfully, and *Kitty Hawk* departed NAS North Island on 19 October 1965.

The carrier arrived for duty with CTF-77 on 15 November; at the time *Rolling Thunder* was in its eighth month. During this period the U.S. systematically increased combat operations over the North, albeit under continued tight political control. In State Department terms the intent was to "get the signal through to Hanoi," but the stage was set for continued tragedy.

During initial strikes over South Vietnam Lt.j.g. Frank Wagner and B/N Lt.j.g. Jim Bobberton did the honors of dropping the first *Black Falcon* A-6 ordnance of the war. However, on 22 December 1965 *Buckeye One* went down on a night mission over North Vietnam. Cmdr. Cartwright was killed and B/N Lt. Ed Gold was listed as Missing in Action, leaving Cmdr. Jack E. Keller to assume command of the squadron. The New Year passed, and on 18 February 1966 another squadron

In May 1965 VA-42 painted one *Intruder* in an experimental camouflage paint scheme to compare to the factory-applied gloss gray and white scheme. 149948 received a dark green and light blue on the starboard, and green and white on the port side. CVW-11 and CVW-15 would soon test similar schemes in Vietnam. (U.S. Navy, PH1 Hendricks)

Intruder hit a mountain during a day mission under IFR conditions, killing Lt.j.g. Joseph V. Murray and Lt.j.g. Thomas A. Schroeffel.

April turned out to be a particularly hectic month. On the 17th, AAA tagged a squadron *Intruder* on a day mission over North Vietnam; fortunately Lt.Cmdrs Samuel L. Sayers and Charles D. Hawkins ejected and were rescued. The following day, two VA-85 A-6's bombed the Uong Bi thermal power plant, 15 miles north-northeast of Hanoi, in a fashion that spectacularly highlighted the A-6's capabilities. The U.S. government had authorized strikes on the plant in late 1965, making it the first targeted industrial complex in North Vietnam. CTF-77 responded on 22 December by sending in Alpha Strikes from *Enterprise*, *Kitty Hawk*, and *Ticonderoga*. The raid thoroughly hammered the facility, destroying the generator building, boiler house, and other supporting structures; however, the North Vietnamese quickly repaired the valuable facility and returned it to the national grid, hence the re-attack.

The two *Intruders* – *Black Falcon* XO Cmdr. Ron Hays with B/N Lt. Ted Been in one aircraft, with Lt. Eric M. "Bud" Roemish and Lt.Cmdr. Bill Yardbrough in the other – launched at midnight, separated at the coast, and coordinated their runs over the target. "The power plant had previously been attacked without success by multiple aircraft, including at least one Alpha strike in December 1965," writes Ted Been:

"We launched at 0100 under EMCON conditions. Only three aircraft were assigned to the mission, the two *Intruders* and one E-2 *Hawkeye* for airborne control. As was the common practice for night strikes we rendezvoused at 1,000-feet in EMCON and then turned off all our lights (this was before the addition of the low viz formation lights). The real heroes of this story were the two guys on our wing, who only had my red flashlight in the canopy rail to guide their way and keep us from bumping in mid-air.

"We flew from our carrier up to Haiphong Harbor on a NW heading at 500-feet. When we were 25 miles due south of the plant we swung due north for the final system run on the power plant. At ten miles we started a gradual ascent to 1,800-feet for safe separation from the bombs we were going to be dropping. The only radio communication we allowed ourselves between aircraft were 'click signals.' When we gave Bud's *Intruder* three clicks by keying the mike button Bud and Bill cracked their speed brakes and took separation from us for their own separate system run.

"The target gave us an outstanding radar picture on our scopes (DVRI and PHD). The plant stood out by itself with no other distracting targets around it. I was able to clearly identify the plant at about 12 miles and track it with the search radar. The track radar set was too unreliable, and no one in their right mind would use it on a 'real' mission. The system was operating flawlessly, for once. Our run on the target was textbook perfect. We hit our release point, and the system released all 13 1,000-pound bombs. The bombs went right into the plant and surrounding grid network. When I looked over my shoulder I could see the flashes of my bombs detonating and showers of white hot sparks engulfing the power plant. Bud and Bill's *Intruder* had system trouble on the inbound, so they made a fixed range line run and did a good job on the target."

The two *Buckeye* A-6s dumped a total of 26 1,000-lb bombs on the power plant. Photo-reconnaissance flights the following day revealed the extent of the damage:

"The RF-8 *Crusaders* and RA-5 *Vigilantes* BDA'd the plant the next day and clearly showed three holes in the roof of the plant, and total havoc in the switching grid. The lights in the area did go out and would remain that way for a while. In all there were 25 craters that could be seen in the post-strike photos. Apparently, we caught the North Vietnamese sleeping. We did not receive any flak until about 30 seconds after the bombs hit their target. By then we were well clear of the area."

Been adds there was a political response following the raid:

"In the days that followed the strike garnered a lot of publicity, including articles in *Time* magazine, numerous newspaper articles, and an announcement to the world by Ho Chi Minh himself that the United States had escalated the war in Viet Nam by introducing the B-52 into the conflict in the North.

"This was a delicate time for the new A-6 *Intruder*. It was fortunate that the strike was successful, as there were a lot of upper echelon Navy types who were trying very hard to cancel the A-6 program ... they questioned the high cost ($6.5 mil-plus) and mostly the reliability of the system. There was some validity in the second part of their complaint, as the state of the art electronic systems at that time were not very conducive to tactical naval aviation. The success of the strike shut up the detractors for a while, but not permanently. The A-6 *Intruder* was always a source of controversy throughout its service. The detractors were mainly from the single-seat, afterburning 'mafia' who resented being constantly shown up by the outstanding performance of the A-6 and her two-man crew."

Tough Missions and a Navy Cross

On 19 April 1966 *Air Wing 11* aircraft hit additional previously restricted targets in the vicinity of Cam Pha, including a water pumping station, railroad yards, and a coal treatment plant, dumping over 50 tons of ordnance on the selected targets. Afterwards CTF-77, Rear Adm. James R. "Sunshine Jim" Reedy, commented:

> "We had launched the strike against Cam Pha within 90 minutes after its appearance on the target list, for it was a key target we had repeatedly asked to hit, and always before the answer had been no."

Reportedly *this* time the howls of outrage emanated from Washington, DC. Several politicians, government analysts, and other experts expressed shock and outrage over the Navy's quick attack on a newly authorized target.

For VA-85, the situation took a downturn two days later (21 April) when Cmdr. Keller and Lt.Cmdr. Ellias E. "Ed" Austin went down, their aircraft exploding in flight for no apparent reason during an attack. The following day *Black Falcons* Lt.Cmdr. Robert F. Weimorts and Lt.j.g. William B. Nickerson hit the water and died after taking AAA. On the 27th VA-85 lost yet another aircraft in combat, the sixth for the cruise. However, the circumstances of the successful retrieval of its pilot and B/N became Navy legend; as related by Ted Been – who participated in the mission – and it also led to the first award of a Navy Cross to an A-6 crewman:

> "One afternoon Cmdr. Ron Hayes (CO VA-85) and myself in *Buckeye 1* and Lt. Bill Westerman and Lt.j.g. Brian Westin in *Buckeye 2* were scheduled for road recce in Route Package 2 in North Vietnam, armed with Mk.82s. Just prior to manning aircraft a report came in to the ship that a bunch of barges were massed up near the mouth of the Vinh River. The staff (CTF-77), being short of TacAir experience, went ballistic and changed our flight's target to these massed barges and our ordnance load to *napalm* of all things. In those days a napalm drop had to be executed at very low altitude and high airspeed in order to ensure detonation. We screamed bloody murder about our ordnance change (the target was great!), but to no avail, and so we launched on the one and only napalm mission (as far as I know) north of Route Package 1 in North Vietnam.
>
> "The launch and join up were normal with only one small problem with *Buckeye 2*. Brian's radio system evidently had a short somewhere in the circuit, and every time he pressed the ICS button he transmitted on the UHF. This was considered a minor annoyance, so we pressed on to the target. The briefed target tactic was that once the run was commenced Westy would assume a loose trail position and make his own independent run. We were then to rendezvous over the water and proceed back to marshal.
>
> "We got to the target area – which was only two or three miles inland – and the barges were there, as briefed, ready for the slaughter. We made our run (all ordnance on target, of course) and pulled off left to head out to the rendezvous area. At this point we heard Brian ask, "Are you hit?" Ron and I looked at each other, thinking he was asking us if we were hit and he was telling us something we did not know. Just as we started to reply Brian made his mayday call, stating that the pilot was hit and HE WAS FLYING THE AIRPLANE out to feet wet. Brian was flying from the right seat! Of course the flying was a little erratic, but effective, and we made a join up with them in a loose formation at about 7,000-feet over the water.
>
> "A single rifle bullet, about the size of a .30-30, had entered the aircraft through the lower left aft portion of the canopy and gone through Westy's shoulder. It probably would have been a clean wound, but the bullet shattered against Westy's koch parachute fastener and ripped his whole left chest open. Westy was able to fly the airplane to feet wet, but his left arm and hand were completely immobilized, so he told Brian to take over the stick while he used his right hand to manipulate the throttles. Brian was a very busy boy at this point, trying to fly, communicate, and take care of Westy.
>
> "At this point Westy was experiencing cycles of tunnel vision. His vision would blur and narrow down to a constricted tun-

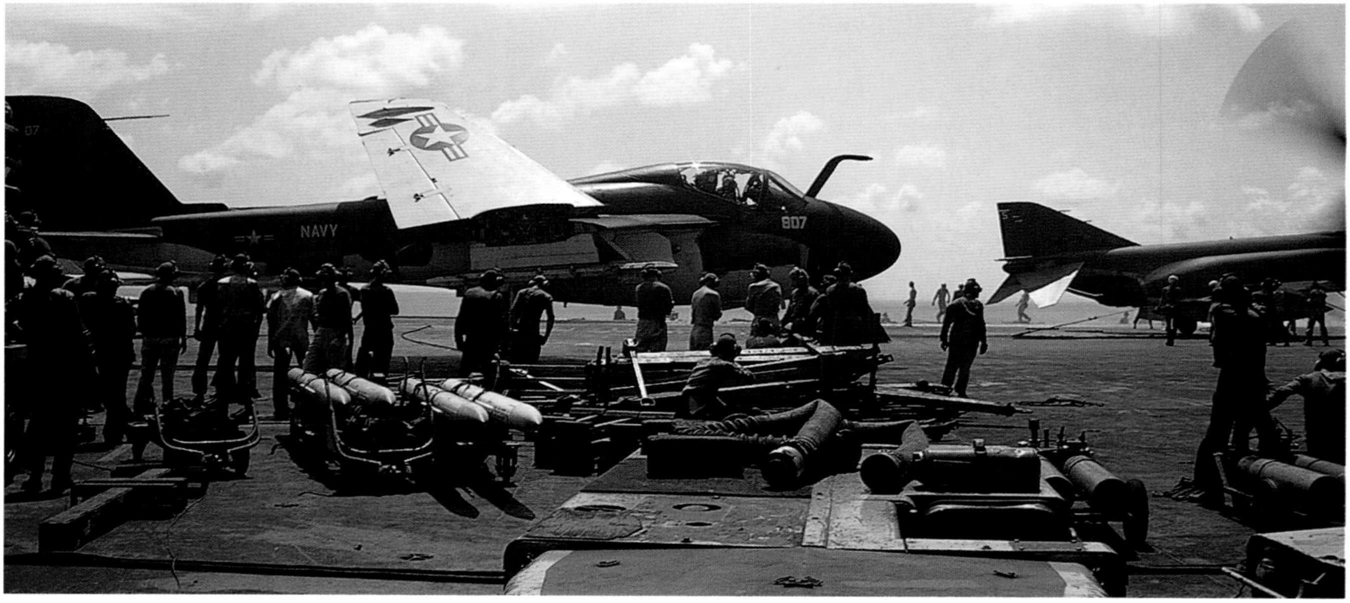

VA-85 was one of three squadrons that evaluated camouflage *Intruders* in combat during the 1965-66 time frame. NH807 is shown on *Kitty Hawk's* deck with a rare F-4G, also in green, in the background. (Guy Merkel)

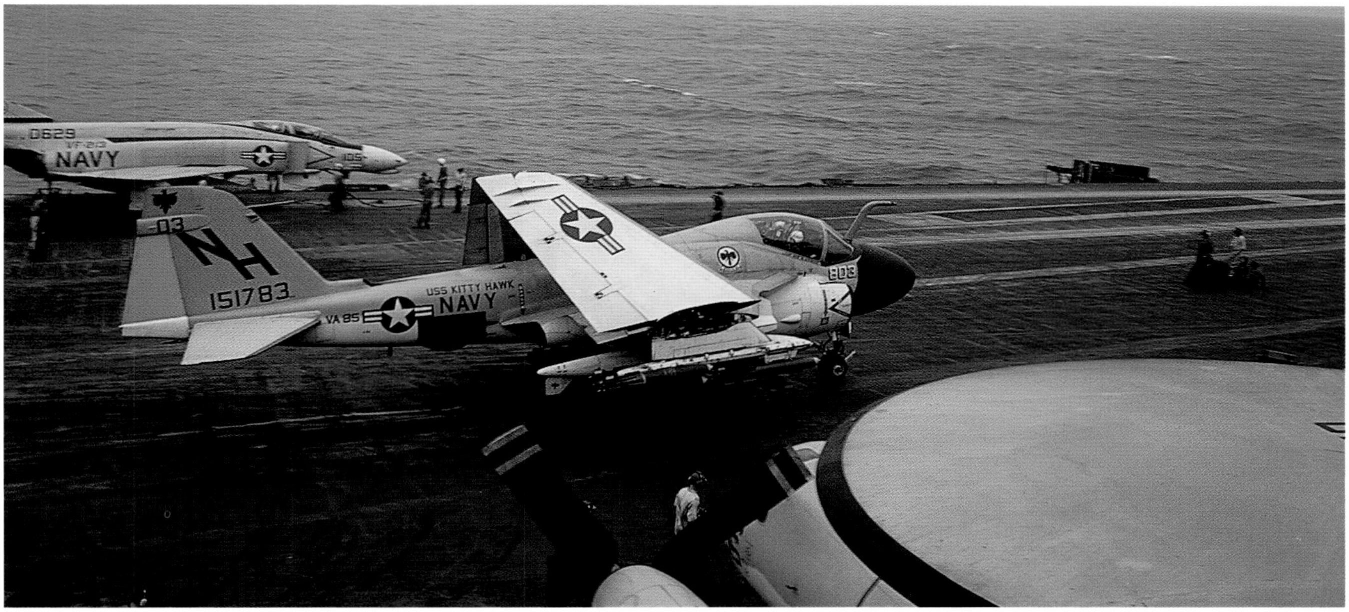

The *Black Falcons* were the second *Intruder* squadron to deploy to Vietnam, arriving with *Kitty Hawk* and CVW-11 on 26 November 1965. The squadron lost seven aircraft during the cruise, with eight aircrew killed or missing. NH805 taxis forward on a lousy day with a load of *Snakeyes*. The VF-213 *Phantom* to its left is one of the twelve unique F-4Gs assigned to the *Black Lions*. (Guy Merkel)

nel and then expand back to normal. Finally Brian remembered he had a miniature of the flight surgeon's brandy in his nav bag. He broke this open and fed it to Westy, who revived enough at this point to realize he had better do something quick. He told Brian to eject, but Brian chose to wait until Westy ejected. Westy then jettisoned the canopy and ejected. By the time Brian got himself ready for ejection, the airplane had gone on about five or six miles.

"Meanwhile, back in *Buckeye 1*, I switched to Guard and broadcast a mayday with our location, which was about 10 miles east of Tiger Island. *Red Crown* answered immediately and got the SAR helo on the way. We were still flying a loose formation and saw the canopy separate from *Buckeye 2*. We saw one ejection, lost the parachute, but finally picked up the raft and circled it. Since Ron and I had only seen one ejection we assumed that Westy had gone in with the airplane. What really happened was we saw Westy's ejection, lost him, and then picked up Brian in his raft, thinking the whole time it was the same person! At this point the SAR *Spad* flight arrived on the scene. The SARCAP was VA-115 (*Arabs*) from CVW-11 in *Kitty Hawk*. They were the last *Spad* squadron to deploy to North Vietnam prior to their transition to the A-6A. The flight leader was Lt.Cmdr. Cliff Johns, who relieved us as SAR Commander. He located Brian in his raft and vectored the helo in for pickup. We watched Brian's rescue. The helo pilot then came on the air to say he had successfully rescued Brian, and that Brian said the pilot had ejected before him. This was the first indication we had that Westy was still alive.

"Naturally, an intensive search by all the aircraft in the area was set up to find Westy. After about five minutes, Cliff Johns saw a tracer across his nose, dropped down, and there was Westy. What happened was that Westy, after he got in the water and out of his chute, was unable to locate his raft (It was lost in the ejection. Westy had tucked his dead arm under the lap belt to keep it from flailing and thinks this action may have inadvertently jettisoned his raft). In the meantime, he's watching all these planes and helos flying all over the Gulf of Tonkin, but none are anywhere near him. Getting desperate, Westy somehow managed to get his pencil flare out and get it armed with one round. He vowed to shoot down the first plane that came in range. When Cliff set out to find Westy, he happened to fly right over Westy, who aimed his pencil flare at Cliff, and that was the tracer Cliff saw.

"The helo was vectored in over Westy posthaste, and the sling was dropped for pickup. Westy had been bleeding profusely and was in the middle of a pool of blood. Due to his wound, Westy was unable to get into the sling. The helo did not have a swimmer on board, and Brian – who by this time had stripped off his torso harness and was wearing only his G-suit and flight suit – ordered the sling operator to lower him into the water. Brian then got Westy into the sling, and Westy was brought aboard the helo successfully. When they tried to lower the sling to pick up Brian the winch jammed and could not be operated. Westy was in desperate need of medical attention, so Brian waved the helo off and the helo called in the backup helo and went back to *Red Crown*, leaving Brian in the bloody water. He used his G-suit as a flotation device by unzipping it and manually blowing it up. Sharks had been sighted in the area, and Brian floated in that bloody pool of water for an eternity (in reality, about fifteen minutes) until the backup helo picked him up and returned him to *Kitty Hawk*."

Lt. Westerman – one of the original VA-42 *Intruder* pilot cadre – had a long and painful recuperation, but eventually logged over 1,000 carrier landings, commanded both an air wing and an LPH, and retired as a captain. For his actions that day in saving his pilot Brian Westin received the Navy Cross, the first of 14 in the medium attack community to receive the award during the Vietnam War. Somehow the familiar and clichéd words, "in keeping with the finest traditions of the Naval Service" did not seem to quite do enough justice in this case.

Now led by Cmdr. Ron Hays, VA-85 moved on. On 15 May 1966 the squadron lost one last *Intruder* under non-combat circumstances; the aircraft was unable to take on fuel from the tanker and ran out of gas. Fortunately, the crew of Lt.Cmdrs. "Buzz" Ellison and Robert G. Blackwood were able to step out. A week later, on 22 May, CVW-11 and *Kitty Hawk* finished their sixth and last line period in the Tonkin

Gulf. They outchopped on 6 June 1966 and returned home to San Diego on 13 June, ending the second *Intruder* combat cruise. On the plus side VA-85 dropped over six million pounds of ordnance during the deployment in all weather and visibility conditions. However, the squadron's successes were muted by total losses of seven aircraft, seven Killed in Action (two of which were commanding officers), one MIA, and four aircrew recovered.

While VA-75 and VA-85 had worked relentlessly to prove the all weather attack mission and aircraft, the A-6 was gaining a reputation that was anything but positive. Part of the problem was with the battle group staffs, as related previously, but another problem was with the *Intruder* itself. Constant exposure to 90 degree plus temperatures, 75 percent-plus humidity, and the open ocean environment under combat conditions rendered many of the A-6's systems ineffective. The aircraft's avionics proved to be fragile in this environment; and it was common to see reports of 26 percent full system effectiveness and radar reliability levels of only 40 percent. The staffs compounded the problem by keeping the A-6s in the air during the day, cutting into maintenance time; when night came many of the *Intruders* had down systems.

Then-Lt.Cmdr. James McKenzie made part of VA-85's first *Intruder* cruise as the squadron ops officer. A former AD pilot with VA-196, he was one of the early "converts" to the A-6 and easily recalls the early difficulties with the plane:

> "On this early cruise the A-6 took some particular care and feeding that the other aircraft in the wing did not require. What really caused people problems was the plane had to sit still until the B/N said the inertial platform was aligned and ready to go. That caused a lot of grumbling. We went on *Yankee Station* with twelve aircraft and returned with *two* of the original twelve. We lost two COs and their B/Ns; I lost my roommate, Joe Murray, and his B/N. It was a *very rough* cruise from that point of view. But, it was a good deployment because it helped bring the A-6 to maturity. VA-75 before us did a good job, the best they could with what they had. Big things were expected from both of us; it just took a little while."

McKenzie feels a large part of what the Falcons *did* accomplish was directly due to their Grumman technical representative, the legendary Bob McNeil:

> "One of the big differences was made by our squadron tech rep, Bob McNeil. He was a former A-3 enlisted B/N who later became one of the A-6 development B/Ns for Grumman. As a former 'E' he was warmly embraced in the shops, which helped them figure out the systems. The crews would spend two hours or more doing weapon systems debriefs, reviewing the readouts and results, and he would recommend the approaches and techniques that made the system work. This is when the A-6 started becoming a true all-weather aircraft. By the time the cruise ended, the *Kitty Hawk*'s CO (Red Carmody) stated he'd like an air wing made up of nothing but F-4s and A-6s. That felt good."

Still, the A-6 remained suspect with many staffs; McKenzie remembers recommendations the *Intruder* be pulled from carrier flight decks and replaced by "something worthwhile." It would take one more combat deployment to finally prove both the *Intruder* and the concept of all-weather attack.

The Next Squadrons Get In Line

Throughout 1965, while VA-75 and VA-85 prosecuted the war in Vietnam the *Intruder* ramp at NAS Oceana continued to bustle. As more squadrons wrapped up their training and made their own preparations for deployment others reported to VA-42 for their turn in transition. In addition, the *Green Pawns* were now turning out qualified Fleet Replacement Pilots and Bombardier/Navigators for the individual squadrons, which now numbered five: VA-75; VA-85; VA-65; VA-35; and VMA(AW)-242 at Cherry Point.

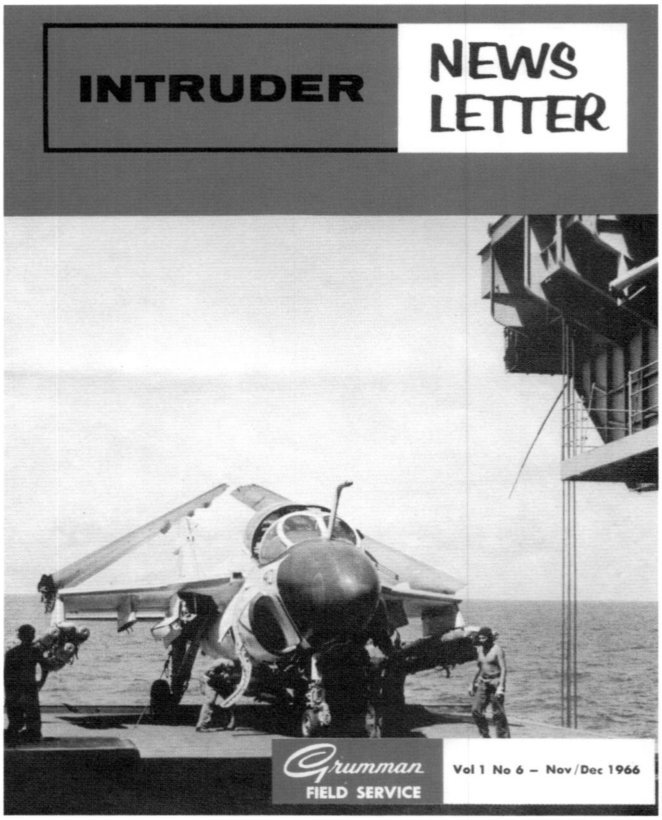

The Grumman Company provided extensive support for the *Intruder* with a dedicated group of Technical Representatives (TechReps) that deployed with squadrons. In addition, the company published a newsletter to help pass information and "lessons learned" about the new aircraft.

In November 1965, VA-42 turned its attention to the first 12 students from the west coast. The men – six each pilots and bombardiers from *The Professionals* of VAH-123 at NAS Whidbey Island – reported for six months duty with the A-6 RAG. Upon completion of training they would return to "The Rock" and establish AirPac's A-6A replacement training program within the existing A-3 RAG.

A few days after the group had departed for the east coast BuPers called Heavy 123's Commanding Officer concerning a personnel matter. Assistant Maintenance Officer and former *HATRON 11* pilot Lt.Cmdr. J.C. "Jerry" Patterson just happened to be in the skipper's office when the call came in:

> "'About that A-6 det at VA-42 ... one of the Lieutenants broke both of his wrists so you're going to have to send someone else.' I immediately grabbed some paper and scribbled 'SEND ME! SEND ME!' and shoved it in front of the skipper. After he hung up he pointed out that I had successfully screened for A-3 command and might not get another chance if I switched to A-6s. I told him I'd take my chances."

Patterson joined the group and, after promotion to commander, became Officer-in-Charge of the VAH-123 A-6 det. His group included Lt.Cmdrs. Buzz Eidsmoe, Charlie Hunter, and Don King, and Lts. Hugh Brainerd, Lyle Bull, Don Cotter, Bob Eychaner, Fred L. Holmes, John Smith, Jim Vester, and Carl Wiechart.

Meanwhile, down in North Carolina the *Bats* of VMA(AW)-242 were also busy, serving as the Marine Corps' A-6 RAG in preparation for the second Marine *Intruder* squadron. In early 1965, VMA-533 Sub-Unit 1 was established at MCAS Cherry Point for training under -242. The full squadron turned in its last A-4Cs and redesignated as an all-weather attack unit on 1 July.

The *Nighthawks* of -533 dated to their 1 October 1943 organization as Marine Night Fighter Squadron-533 at MCAS Cherry Point. Equipped with F6F-3N *Hellcats*, the squadron initially moved to Eniwetok in *Long Island* (CVE-1) in May 1944 before settling into Yontan Airfield, Okinawa, under assignment to *Marine Aircraft Group 31*. The squadron scored its first kill on 16 May 1945 when 1st Lt. Robert M. Wilhide shot down a Japanese *Betty* medium bomber. Three more *Betties* went down under the *Nighthawks'* guns on the 18th, followed by another three on 24 May. During June 1945 Cpt. Robert Baird knocked down five aircraft, making him the only Marine night fighter ace of the war. By VJ Day VMF(N)-533 rang up a total of 35 kills, making it the highest scoring Marine night fighter squadron.

Immediately postwar the squadron operated F7F-3N *Tigercats* from Peiping, China, followed by service during the Korean War as a night fighter training unit. In May 1953 it redesignated as VMF-533 and re-equipped with F2H-3 *Banshees*, subsequently making several short carrier deployments. The year 1956 found the *Hawks* transitioning to F2H-4s and adding "all-weather" to their designator; the following year 533 made two Med cruises in *Lake Champlain* (CVA-39) as a component of *Air Task Group 182* and MAG-26. Following their return to the states in October 1957 the squadron again redesignated, this time to VMA-533, and traded in its *Banshees* for F9F-8 *Cougars*. The Grumman products lasted until August 1959 when the squadron received its first A4Ds; -533's transition to the *Skyhawk* marked the last operational use of the F9F-8 by an active duty Marine squadron.

Scooter ops continued for six years, during which time VMA-533 set several records. These included the establishment of an 8-hour 25-minute endurance record for the A4D set by two *Nighthawk* pilots in January 1962, and the first aerial refueling of an entire deploying squadron, accomplished while enroute to Puerto Rico in June 1962. The squadron repeated the feat when it returned to ConUS the following August. Roughly three years later (28 August 1965), squadron CO Lt.Col. E.S. Maloney ferried the first *Nighthawk* A-6A to Cherry Point. The last *Skyhawk* went to VMA-225 on 12 October, and on 2 November -533 received its twelfth and final *Intruder*.

In and around training the second Marine *Intruder* squadron VMA(AW)-242 was also preparing for its own deployment to South Vietnam. The situation made for interesting times, according to B/N William D. "Charlie" Carr:

> "The crew makeup was a hodgepodge of A-4/*Spad*/F-9/F-8/ nugget pilots and RIOs, pilot washouts/ECMOs/nugget B/Ns. There were no USMC RAG/FRS until 1969 when newly formed VMAT(AW)-202 accepted its first students. Integration of USMC B/Ns into the attack mission was considered as initial training and preliminary situational awareness training. B/N training, as well as pilot training, occurred in the tactical squadrons. Of course, this type of training produced predictable results: accidents and combat training *in combat*."

Carr was perhaps not-quite typical of the first Marine B/Ns; as some would say, Charlie was definitely "one of a kind." He worked his way up the ranks from enlisted navigator to warrant officer ECMO with VMCJ-1, flying the EF-10B *Skynight*. Commissioned a second lieutenant in the spring of 1965, he would go on to serve in several Marine A-6 squadrons.

Marine training practice aside, VA-42 kept turning 'em out at Oceana. In the spring of 1966, the *Pawns* took in their first west coast A-1 squadron, VA-196. Known as the *Main Battery*, the outfit turned in its A-1Hs and Js at NAS Lemoore, CA, and moved to Virginia Beach effective 1 May 1966. First established at NAS Alameda on 15 July 1948 as the VF-153 *Thundercats*, the squadron had already seen combat in two wars. Its first deployment came with F8F *Bearcats* in *Boxer* in January 1950; one month into the cruise the squadron redesignated as VF-194, aligning it with *Carrier Air Group 19*. The following five deployments came as a component of Air Task Group 1, including Korean War cruises in both *Boxer* and *Valley Forge*.

On 4 May 1955 the squadron became VA-196 and, equipped with AD-6 *Skyraiders*, continued a regular WestPac deployment schedule with CVG-19. The *Main Battery's* tenth cruise, however – in *Bon Homme Richard* from 28 January to 21 November 1964 – took them smack into the Tonkin Gulf incident and their second shooting war. After a second *Bonnie Dick*/CVW-19 combat cruise, which concluded on 13 January 1966, the *Battery* got in line for the *Intruder*, in the process becoming the sixth A-6 squadron.

Thus, in the spring of 1966 the aircrews and men of -196 settled into the transition routine, while VA-85 was wrapping up in *Kitty Hawk* and preparing to come home. Elsewhere at Oceana the third and fourth fleet *Intruder* squadrons were making their final preparations for deployment.

The Third Combat Cruise

It fell to the *Tigers* of VA-65 to make the third A-6 combat deployment, and on 12 May 1966 they departed with *Carrier Air Wing 15* in *Constellation*. However, long before they had even started their final preparations the word had gone out: this was going to be the make or break cruise for the *Intruder*.

Connie's commanding officer at the time was Capt. William D. "Bill" Houser, who later served as Deputy Chief of Naval Operations (Air Warfare). In comments published in *The Hook* in 1985, retired Vice Adm. Houser discussed his marching instructions *vis a vis* the A-6:

> "During the workup period off Southern California in the spring of 1966, Vice Adm. Tom Connolly, ComNavAirPac, called me into his office and asked me to bring along my A-6 squadron CO, Cmdr. Bill Small. Tom told us he had just received word from Washington that the Secretary of Defense had reviewed the first two deployments of the new A-6 aircraft. In their opinion, it was being used only as a super A-4. Based on their calculations we could do better to buy more A-4s and no more A-6s, and thus, if the A-6 operations did not improve significantly, that was going to be the result.
>
> "Tom stated that he was the project director and I would be the manager, with Bill Small as my assistant, to make sure that the A-6 was tested in its proper role, as a sophisticated all-weather attack aircraft. If we failed, production would be terminated, and if we succeeded, it would continue. Bill and I returned to the ship with those orders and nothing more. We decided that we would treat the A-6 not as just another aircraft, but as something special, which indeed it was."

"Prior to the deployment the squadron was made aware of what was expected," says Rupe Owens, who by this time was assigned to VA-65:

> "The question: could the A-6, properly supported, operate effectively as an all-weather system for a full combat cruise? Capt. Bill Houser ... left no doubt in any of our minds that he expected the A-6 to carry the mail. And, he took the unheard of position that the flight deck – and the whole ship – would do whatever it took to establish an operational routine that would enable the *Intruder* to reach its potential.
>
> "We asked that we be given priority on the elevator so that upon return from a mission, if the aircraft was not fully opera-

tional it got an express trip to the hangar bay for maintenance. We planned one full weapons system strike for each aircraft every day. We flew additional sorties with aircraft without full operational readiness. Our definition of full up included search radar, track radar, computer, inertial, doppler, ECM and ESM equipment, and the complete armament system, along with airframe components. Cmdr. Jake Ward, *CAG 15*, also fully supported our full up system plans and provided the leadership for the air wing during this critical period. Cmdr. Nello Perrozi, as Ship's Operations, capably directed the actions of the ship's team so collectively the ship/air wing team concept worked beautifully to produce – for the first time – an operational environment in which the new all-weather attack bomber could operate effectively. And the *Intruder* delivered."

Tiger B/N John Diselrod concurs:

> "When we started flying combat in June the *Connie* gave us the royal treatment. Even if we returned with a full-up system aircraft, we were struck below for a complete post-flight and system checkout before the aircraft would be rescheduled."

Some of the work took place before the squadron deployed, including system upgrades to boost the aircraft's capabilities. These included replacement of the AN/APQ-88 track radar by the Norden AN/APQ-112, addition of a Radar Warning Receiver, and multiple systems reliability enhancements. According to one squadron member, each deploying aircraft went up to Calverton for a complete system evaluation, followed by two system hops. Once the plane passed the hops it headed west. These activities helped boost the *Intruder's* performance in the tough climate of Southeast Asia.

The intensity of the situation and the high standard of performance expected of the *Tigers* was apparent to everyone in the squadron. Lt.j.g. Robert "Rube" Rubery had joined VA-65 in June 1965 as the Air Intelligence Officer; he remembers the seniors strongly pushed the team concept:

> "When I reported to VA-65 we were a new squadron which had been built up by the Navy to make or break the A-6 program," he writes. "Most of the pilots were senior single-seat attack jocks and most of the B/Ns were – with a few exceptions – right out of training command. A great deal of effort was put into matching the pilots and B/Ns so that we had the right mix, personality, confidence, etc. There was a lot of moving around initially. Most of the time a senior pilot was put with a junior B/N or vice versa, but not in every case.
>
> "One crew I recall that was very good was Lt.j.g. Ron Zlatoper and Lt.j.g. Dick Schram. 'Zap' was the first pilot right out of training command to go A-6s and Bob Williams was second. Bob – or 'Willard' – teamed with John Diselrod. As a side note, Zap went on to become an Admiral and CINCPACFLT. Schram was the son of the 'Flying Professor' who used to perform at airshows as the 'drunk' in the Piper *Cub*."

In July 1965 the squadron went to SERE (POW) training at Brunswick, ME. In order to get everyone through, the Tigers "attended" in two groups, with half of the crews going one week and the others following about two weeks later. According to Rube, XO Bob Mandeville decided to liven things up for the skipper Bill Small's first group:

> "While the first group was in Maine, Mandy came to me and asked if we could pinpoint the exact location of the POW camp, which we did. He then said we needed a systems cross-country training flight. We made up hundreds of 5x5 'flyers' with the squadron Tiger on the front and the following on the back:
>
> 'TIGERS: AMERICAN TROOPS HAVE SURROUNDED THIS AREA AND ARE ADVANCING STEADILY. ADVISE YOUR CAPTORS THAT ESCAPE IS IMPOSSIBLE. THE CAPTIVE WILL BECOME CAPTOR. THE SENIOR OFFICER PRESENT IS DIRECTED TO ASSUME COMMAND AND DEMAND THE SURRENDER OF THE CAMP TO ALLIED CONTROL. THE AGRESSORS WILL BE HELD ACCOUNTABLE FOR THEIR ACTIONS.'
>
> "These flyers went in the speed brakes of a division of A-6s that flew to Maine and came in right over the camp on the deck and dropped them. That's all the guys talked about when they came back. I was in the second group, and I found some when I had to go out on a 'wood detail.' We talked to the instructors at the end of the exercise, told them who we were, and asked them about the fly over. They loved it, but said it had blown the whole exercise, and they were never able to gain complete control again. *Great* Team building!"

It was January 1966, and the team found itself in *Lexington* for Carrier Quals. John Diselrod recalls there was nothing exciting about the evolution until he and pilot Lt. Jim Joyner flew off for Oceana:

> "Something came loose in the birdcage during the cat stroke and hit the tailhook cable, which dropped to the deck and about ripped the ship's nose tow shuttle out of the track. Sure didn't slow us down.
>
> "Being an east coat squadron assigned to a west coast ship is always nice. We flew to NAS North Island in February 1966 for a June deployment and worked with squadrons that were home every night or weekend. Plus, shuttling the A-6As back and forth to Grumman on Long Island to 'tweak' the systems, we made it through ORI and on to Hawaii."

It was a lot of hard work, but the *Tigers* believed in themselves. The men of VA-65 *knew* the *Intruder* could do the job; they just had to convince everyone else, which led to a couple more humorous encounters, recalled by Diselrod:

> "One night off San Diego Bob Williams and I got caught by a fog bank. We had an excellent system, and even asked the *Connie* if we could shoot a landing mode to the ball since their radar was giving them a problem. Naturally, they called us idiots and told us to Bingo to North Island. I came back and asked them to authenticate 'Juliet-Delta.' Long pause, then the E-2 came up, authenticated, and told us to go to North Island. Wouldn't you know, when we got there the BOQ was full. The manager finally let us have an admiral's suite if we were out by 0600. We were ... after drinking all the miniature bottles of booze in the reefers.
>
> "Another night bingo was with Capt. Ray Edens, USMC. Everyone called him 'Pygmy' because he was short, but feisty. Upon arrival at North Island I called transportation for a car and was told, 'I'm sorry, lieutenant, nothing is available.' The Pygmy picked up the telephone, called the same place, and said, 'This is *Captain* Edens off *Constellation*, we need transportation.' A car and female driver showed up shortly thereafter."

In this atmosphere the *Intruder* would finally start to live up to its advance press, albeit with some continued resistance from staff types. According to Adm. Houser:

> "We ... had great cries of alarm from the Air Boss, flight deck officers, and other squadron COs, but their pleas went unheeded. We stuck to our program, and it soon became a way of life with no undue hardships on other aircraft or ship operations."

Adds another VA-65 officer:

> "Previous COs had been regularly dumped on and overruled by the CAG and other staff types; it was a new aircraft, the CO was only a commander, and he didn't have the horsepower. There was also a major problem with parts; and logistics just wasn't supporting the aircraft over in the Tonkin Gulf. This cruise, we worked closely with the task force commander and CAG. Every aircraft was scheduled for at least one system's hop each day; when the aircraft trapped they were immediately struck below for avionics work. If their system was up they flew night systems missions. If their systems were down, the wing was allowed to play with them during the day, alpha strikes and whatever. It worked out; and we had the best system and availability record yet."

Before the *Tigers* went on line the skippers of VA-85 and VA-65 (Ron Hays and Bill Small) met for a quick verbal passdown. Their B/Ns, Lt. Ted Been and Marine 1st Lt N. Craig "Nellie" Dye, also participated; Dye, now a retired major, writes:

> "Early in 1966 a face to face debrief was held. We discussed their missions and their successes and losses. It was obvious that a complete review of tactics and employment of the A-6 was imperative. It was becoming critical to the future of the A-6, which was already being questioned at higher levels.
>
> "Within days, VA-65 departed for the South China Sea aboard the USS *Constellation*. We were more concerned than ever about the future of the A-6 and, more importantly, the survival of the flight crews. Many hours were devoted to planning and reviewing the debriefings. Our main concerns were the enemy order of battle in the area that our flights would be taking us, our tactics, our weapons employment, and the new electronic counter-countermeasures equipment, that were being installed in the aircraft to increase the survivability of the crews.
>
> "We arrived in the South China Sea off the coast of Vietnam on the 13th of June 1966. Our first assignments would be to do some warm-up flights in support of ground troops in South Vietnam. This gave everyone a chance to become familiar with combat operations in a lower-risk environment. On June 22d, *Constellation* sailed north to take up position on Yankee Station for flights into the 'jaws of the tiger,' North Vietnam. Our first flight over North Vietnam was June 23rd, and the reception was anything but friendly."

Still, to a man VA-65 felt it was ready. Now it was time to prove the point.

VA-65 Takes the Final Exam

Two of VA-65's personnel had made previous combat deployments and found the now two-year-old war to be a curious mix of old and new. On one hand Washington still tightly controlled air operations against the North, which now continued as *Rolling Thunder IV*; if anything, the Rules of Engagement (ROE) was even more restrictive. Conversely, the powers back in D.C. *had* shown a tad more flexibility in target selection, releasing several important complexes and other facilities for strikes:

> "We found ourselves flying against a variety of targets during the cruise with emphasis on night, systems hops," remembers John Diselrod. "We had a lot of success in obtaining hard targets and a lot of night work. The CTF-77 staff had come to recognize the ability of the A-6, and the target selection just kept getting better (apparently the word had got out). There were a few daylight raids, usually due to 'target importance and surrounding enemy resistance,' and the occasional *Iron Hand* mission. The A-4s were generally the *Iron Hands*, but we got to be *Shrike* shooters, too. My idea of *Shrike/Iron Hand* is like trying to fight a nest of rattlesnakes with a barbecue fork: the SAM radar was good to 30 miles, the SAM was good between 17 and 21 miles, and the *Shrike* was optimum around nine miles. Not the best percentages in your favor. We did get to shoot some, though, but not in an intense SAM environment; I was *very* happy to see *Standard ARM* development."

One major change was in the delineation of operational responsibility over North Vietnam. The country's airspace was now divided into six route packages split between the Air Force and the Navy, with the latter service holding responsibility for Route Packages II, III, IV, and IVB, covering the coastal areas. However, while the Air Force and Navy now had specific areas of responsibility the White House still – through

Black-nosed *Tigers* head inland at low altitude during the squadron's first *Intruder* cruise. The load is 18 500-pound bombs, and the pilot is flying with his sleeves rolled up and sans mask. (Dye, via Steve Dumovich)

Commander-in-Chief, Pacific, Commander Seventh Fleet, Commander U.S. Military Assistance Command, Vietnam, and Commander, Seventh Air Force – kept a very close watch on what they were doing:

> "From the debriefs we were sometimes able to warn 'Higher Authority' of things that may come to pass," says Rubery. "One such incident occurred during an Alpha Strike.
>
> "A lot of planning and coordinating went into these day time strikes with the intent of getting 20 or 30 aircraft into and out of their run on the target in as little as time as possible – maybe 40 seconds. The lead aircraft were the A-6s, followed by the A-4, and on the outside going in with the A-6s were the F-4s for flak suppression. Each aircraft had an assigned target within the complex. When we were in the 'POL' phase the target that came down from Washington was the POL plant and field at Haiphong. On any of these Alpha strikes it was very precise how, when, and where everyone would approach the target and drop their ordnance. Most of the time they were more concerned about the other aircraft in the flight than the AAA. However, after the run was made it was everybody for themselves to get feet wet!
>
> "On this Alpha strike against Haiphong, during the briefing it was noted that there were 'friendly' merchant ships in the roadstead and at the docks in Haiphong, and they were to be avoided at all cost. The POL facility was right on the roadstead. The 'roadstead' was the waterway leading from the Tonkin Gulf to downtown Haiphong.
>
> "Anyway, the strike went as planned, everyone got in, and XO Cmdr. Frank Cramblett and Lt. Don Hahn chose to egress on the deck right along the roadstead. Obviously the AAA lowered their guns to shoot at them primarily from the left because the water was to the right. As he was coming out Frank said he looked to his left and all the AAA that was coming from his right was missing them and raking the merchant ships anchored in the roadstead. When this came out in the debrief it was passed up the line, and sure enough, in the press the next day the 'American Air Pirates' were accused of bombing and strafing the neutral merchant ships in Haiphong. Because we had already gotten the word out it didn't turn into a flap; all inquiries were taken care of promptly."

Several events came in quick succession at the end of June. On the 27th Lt. Dick Weber and B/N Lt.j.g. Charles W. Marik were tagged by anti-aircraft artillery during a day strike on the Hoi Thoung Barracks near North Vietnam's coast. During the bombing run their *Intruder* took accurate anti-aircraft fire in the tail, damaging its controls, and Weber called for the ejection. SAR forces rescued the pilot, but regrettably Marik disappeared; Weber did talk to him on the way down, but lost sight of the B/N once he hit the water. The Navy called off the search after four hours. According to Craig Dye:

> "This was a wake-up call for the squadron. The message was: do it right, adhere to mission briefing, and don't get complacent. This was no drill!"

The loss of Chuck Marik *was* a wake up for everyone, although it affected some of the squadron members in different ways. John Diselrod comments:

> "He (Marik) was my roommate, along with Wes Starr and another lieutenant junior grade B/N. That B/N decided it was something he didn't want to do, gave up his wings, and was off the ship within 24 hours."

On 29 June 1966 Air Force and Navy aircraft executed the first authorized major strikes against North Vietnamese POL (Petroleum/Oil/Lubricant) facilities. The Air Force went after installations near Hanoi, while aircraft from *Constellation* and CVW-15 in *Ranger* hit Do Son and Haiphong, respectively. The attacks – followed by additional strikes at Bac Gang on the 30th – had been recommended by CINCPAC, Adm. U.S. Grant Sharp, for months. Adm. Houser commented the missions were:

> "A bold decision. The raids were a huge success and caught the North Vietnamese off guard, as previously these areas had been considered sanctuaries."

On 1 July the North Vietnamese sent three PT boats out into the Tonkin Gulf. The vessels were making a bee line for *Seventh Fleet* units when spotted by the F-4 BARCAP and quickly came under fire. Lacking guns, the *Phantoms* were unable to do any damage to the small, high-speed targets so CTF-77 sent in the *Intruders*. The responding A-6s sent the three boats to the bottom of the gulf well before they could come within torpedo range of USS *Rogers* (DD-876), *Coontz* (DLG-9), and *King* (DLG-10), operating on the North SAR station:

> "During the 'PT Boat' period – our first line period – Lt.Cmdr. Jack Fellowes and Lt.j.g. George Coker were diverted to a 'high speed contact' off of Hon Gai," Rube Rubery comments. "Jack said the PT boat was really moving and had a rooster tail like the *Miss Budweiser* hydroplane. They made a couple of runs at it, couldn't hit it, but a near miss caved in its side and it went dead in the water. The captain and some crewmembers were captured and really spilled the beans with regard to the location of other PT Boats. I painted a PT Boat on the door of Jack's stateroom."

Adm. Houser recalls 19 North Vietnamese crewmen were taken prisoner, and they revealed the existence of additional boats at several small Vietnamese islands. Again the A-6s went in; Diselrod recalls the PT boats as:

> "Evasive little things, unless you can catch them tied up, then you have a field day. Lt. Dick Weber and Lt.Cmdr. Nels Gillette, along with their respective B/Ns, spotted a PT boat camouflaged against a rock formation. The A-6s made five runs at the boat, leaving it resting on the bottom when they left. In the same time frame, Lt.j.g. Ron 'Zap' Zlatoper and his wingman left eight barges burning during a coastal reconnaissance mission."

Diselrod also remembers occasional switchology problems in the cockpit, what with the sudden wealth of targets:

> "Going on this type of mission, they started loading the A-6s with eighteen Mk.82s on stations 1, 3, and 5, and *Shrike* on stations 2 and 4. When a PT boat or barge was spotted, the B/N would select a bomb station and then turn on the Master Arm. About the time the pilot got on the perch for roll in a *Fan Song* would come up. The pilot would say, 'Give me a missile,' so off with the Master Arm, off with the bomb station, on with the missile station, and on with the Master Arm. As the crew turned in the radar would shut down. A couple of times of this and the B/N would have the switchology such that going from missiles to bombs – forgetting to switch off the Master Arm – the aircraft would be about upside down at the roll in point when six Mk.82s would just go airborne. Frustrating."

Still, there were the occasional minor victories, as Rube Rubery recalls:

> "During the 'PT Boat period' all of the aircraft in the air wing were getting involved, of course. One night a section of

two A-4s were on a recce mission up off of Hon Gay looking for PT boats and their docking facilities. They made a couple of runs and all of a sudden the lead lost radio contact with his wingman and couldn't raise him. The procedure for lost radio was to proceed to Marshal and make a lost radio approach. The wingman never showed up at Marshall.

"During the debriefing they got everyone together – squadron CO, etc. – and Capt. Houser was there also, and it was decided to send two A-6s up there using the 'System' – as they had pretty good coordinates of where they were – to arrive at first light. Maj Edens was to lead (his B/N was Steve Bernstein), and I don't remember who was in the other aircraft. They arrived just as it was getting light, and over the air comes, 'Hey, you guys in the A-6s, where have you been? Come get me!' The A-6 those days smoked big time from the exhaust, and that was the first thing the downed pilot saw, the four plumes of smoke from the two A-6s. He had dislocated his hip, but had still managed to get into his raft and was floating just off an island. Pygmy Edens called the 'Big Mother' rescue helicopter to coordinate the pickup, while the other A-6 and the TARCAP F-4s dropped ordnance on the island, which by this time was starting to fire on the raft and the pilot. The communications between Pygmy and the helicopter became confusing, as the helicopter did not know which way to head. Pygmy got so frustrated he brought the A-6 down on the deck, put it in slow flight, dropped the gear, flaps, etc., and flew a race track pattern right alongside, leading the helicopter to the pilot. The pilot survived, and it was a new successful use of the A-6 system."[1]

Speaking of "target identification," John Diselrod and Ray Edens experienced another problem that apparently took place while *Connie* was enroute to Southeast Asia. It did not get resolved until the carrier and air wing were well into their Yankee Station operations:

"Ray Edens and I were together again in Honolulu. We left the ship and started at the Monkey Bar in Pearl City. From there we went to the beach, and on the way back through downtown we ran into some of the troops. We stopped at a downtown bar and bought them a round of drinks. The mistake was that they returned the favor. Being pretty well on the step walking down the street, the Pygmy wanted to stop, get a drink, and go to the head. We just walked into the bar on the corner, asked for a drink, and noticed the place was pretty empty. Behind the bar were three (???) cross-dressers or whatever. We got our drinks; Pygmy hit the head and came back. We were laughing and getting ready to leave, and I decided I'd better go to the head. As I exited the restroom, one of the male/females was standing at the door. It took me out the back door and down the alley, because the Shore Patrol and police had the Pygmy for being in an 'out of bounds' establishment. They were out front writing up the captain as I came around the corner. I asked what the problem was, stated I knew the man, was on my way back to the ship, and would escort him back if they wanted. He was turned back to my custody, and we went on to the ship.

"The report showed up months later. The skipper said since we were on Yankee Station, and the Pygmy was our LSO, that he was confined to ship for 20 days. Are we having fun yet?"

Morale was high among the *Tigers*; despite a few "situations" like the one mentioned above the squadron had successfully integrated into the air wing and were achieving a high sortie rate and high systems "up" rate. The squadron's progress was reflected through its daily newspaper, the *Tiger Rag*. Edited by Lt.j.g. "Zap" Zlatoper, the *Rag* provided information on ordnance expended, a summary of the primary missions, reasons for canceled sorties, and bombing results. For example, on 27 June 1966 the *Rag* selected the crews of Lt.Cmdr. Bernie Deibert/ Lt.Cmdr. Dale Purdy and Lt.Cmdr. Jack Fellowes/Lt.j.g. George Coker for the "Hit of the Day" after they plastered the Pho Can railroad yard. The newsletter indicated:

"Sixteen boxcars were destroyed, and billowing black smoke was observed when our planes retired from the target (indicating the possibility of a POL explosion)."

An additional note provided:

"A special 'thanks' from a VA-155 flight deck crewman to W.F. Chagnon, AE3 of VA-65. Jet blast blew the *Skyhawk* mechanic over the #1 elevator rail while the elevator was lowered, and he was left hanging to a line 70 feet above the water. In an instant Chagnon was on the scene holding on to the unfortunate sailor. A well done."

More Strikes

On 30 June 1966, VA-65 had a short break for a change of command as Cmdr. Bob Mandeville relieved Cmdr. Bill Small in Capt. Houser's import cabin onboard *Constellation* on Yankee Station. Mandeville was dressed in a flight suit and was headed for a flight briefing immediately after taking the squadron's helm. The previous day the outgoing skipper had led 14 CVW-15 aircraft on a successful strike against the Do Son POL facility, 10 miles southeast of Haiphong. It marked a fitting end to Bill Small's tour as *Tiger One*. Then it was back to the war, as John Diselrod continues:

"After Cmdr. Bill Small departed and Cmdr. Bob Mandeville became CO we continued to get selected hard targets. One such target was the railway bridge at Ninh Binh. Because it was located north of the city and just south of Nam Dinh it was decided to send two A-6s with five 2,000-pound bombs each and have about eight A-4s and eight F-4s for flak suppression. Mandy and Craig Dye were in the lead A-6, and Willard and I had the wing. We coasted in just southeast of the target with a pop-up to the perch, followed by a left-hand roll to take separation.

"Just as we separated from Mandy some 37mm decided to fire a quick burst ... an A-4 with a *Bullpup* saw him, and I remember seeing the *Bullpup* missile cross our windscreen between us and No. 1. I don't know what the flak suppressors dropped against it, but I do know that an F-4 *Phantom* loaded with empty MERs doesn't go too fast. The A-6 going downhill with five 2000-lb bombs is like a runaway freight train, and when we salvoed the bombs and were clean wing we went by the two F-4s like they were sitting still. *Constellation's* skipper, Capt. Houser, was pleased because the bridge was dropped, but boy, what an adrenaline rush!"

John's regular pilot for the cruise was Lt.j.g. Robert "Willard" Williams, the junior stick in the squadron. This method of pairing experience with inexperience – flying junior pilots with experienced B/Ns and vice versa – was already "Standard Operating Procedure" (SOP) in the medium attack community. It allowed the squadrons to put the best possible combinations of crews over the targets, and for once there were plenty of targets to choose from. Diselrod continues:

"Port Redone, a transhipment point north of Haiphong – one night with the EF-10 jammers of VMCJ-1 from Da Nang, started up Haiphong Harbor, and when I cycled from 75-mile PPI to 30-mile PPI my scope went blank. The *Cottonpickers* were jamming, and the flak from Haiphong was heavy, but we turned back to sea and I got the radar under control. We turned back north and started

[1] The rescued *Skyhawk* pilot was Lt. N.E. Holben, of VA-155.

"What Bomb Shortage?" Another Alpha Strike – more VA-65 *Intruders* bound for Vietnam. While not a great shot, this backlit photo shows a *Tiger* A-6A with five World War II era M-66 2000 pound box fin bombs that the type occasionally carried for a period in the mid-1960s. Other CVW-15 aircraft are visible in the background, including F-4Bs and A-3Bs. (Dye, via Steve Dumovich)

the gauntlet again. When I cycled steering from 'Nav Checkpoint' to 'Target,' the cursor went wild, and we turned back out again. We told *Cottonpicker* to stand by and we were going again. Finally getting things settled down, we again went up Haiphong Harbor – with the *Cottonpickers* jamming and laughing – and would you believe, there was no flak on this run.

"Another time, same target. I'll outsmart them this time and we'll go feet dry south of Cat Bi airfield, turn north, and go west of Haiphong, then run the target from the west. All was going fine until we were between Haiphong and Hanoi – crossing the rail line – when we got illuminated by search lights. Bob, with one quick snap forward then aft on the stick, moved us from 1500-feet to 500-feet, and we continued on our way. Sometimes he beat me up back on the ship at debrief, but he never said, 'No, we won't do that, or go that way.'"

One evening during the cruise John and Willard were outbound after evading SAMs when they picked up I-band emissions, indicating a pursuing fighter:

"Target, railway bridge between Hanoi and Nam Din. Five 2,000-lb free-fall set to salvo. Coast in smooth, twenty-eight miles to the IP, turn southwest. 'I have the target, track lock, tracking good, ten miles, in attack, everything looks good.' Willard hasn't said a word. He kept the steering centered, altitude 2200-feet, and was watching a SAM lift off from Nam Din coming our way. 'Thump,' the bombs come off, he breaks hard right and says, 'SAM your side.' I yelled, 'YES, BREAK LEFT!' and watched the radar altimeter stop at 300-feet as the SAM crossed behind us and impacted on the ground. I cycled steering to the coast out, thirty-eight miles east, and he picked up the heading. That's when the X-Tail (warning) light came on and we got a hissing sound from the ECM receivers.

"As Willard started jinking, we were both looking back into the darkness to see if we could get a glimpse of anything. We also called the BARCAP over the Northern SAR DD and said we had something on our tail; they said to bring him out. In our effort to evade we strayed over Nam Din; I guess we woke up the gunners, because they fired a few rounds. About one or two minutes later the flak started again, but behind us. That was when the X-Tail light went out. We proceeded on to coast out and on back to the ship; reports were that an unidentified came out behind us and turned north to Cat Bi. Phew."

The remainder of the crew's trip back to *Connie* was uneventful. There, Williams and Diselrod's wholly unusual method of shaking a North Vietnamese fighter became known throughout *Air Wing 15* as the "Willard Egress."

On 10 August 1966 VA-65 paid a return visit to the newly rebuilt – for the third time – Uong Bi Power Plant. In the words of pilot Rupe Owens, the attack was done:

"In a typical four-plane coordinated attack, a VA-65 trademark. A-6s coming from various directions released their bombs at the same time, confusing and distracting enemy gunners."

According to Marine B/N Craig Dye, who was now crewed with Cmdr. Mandeville, the planning for this mission started the previous month:

"The squadron flew through the month of July with one small side trip to Hong Kong for some needed R&R. The month of July introduced us to the mission called the Alpha Strike, a daylight

VFR gaggle of 40-plus aircraft attempting to put bombs on a target complex. This was obviously an Air Force-conceived plan with the sole purpose of reducing the size of the current fleet of aircraft. It was certainly HIGH-risk.

"It was about this time that Cmdr. Mandeville and I would talk for many hours about improved tactics that could accomplish the desired results of target destruction and reduce the exposure of the aircraft to the high SAM and AAA threat. The A-6 had a much greater capability of putting bombs on target in any weather than any other aircraft in the theater, so why expose it to such a high-risk area in daytime VFR environment? The challenge became to achieve the desired destruction level of any given target. On many targets, in order to accomplish this, it would require more bombs than one A-6 could carry, and the A-6 could carry a great deal of ordnance. A new tactic had to be devised.

"The previous A-6 squadrons had attempted the 'bomber stream' type tactics with very little success. Hitting the target was like smacking a hornet's nest. The attacking *Intruders* would follow each other into the target in trail, ie, one aircraft after another at a specific interval. Once the hornets' nest was hit, it caused a tremendous amount of hostile activity by the North Vietnamese, and the number three and four in a bomber steam normally took all the heat (and most of the hits). Several aircraft were lost or damaged using this tactic. This proved to be not the best attack plan in a small, high-risk area. The results were very similar to the Alpha Strike. There was not an abundance of excitement on the part of the aircrews to participate in these missions, especially if they were slotted for the number three and four spots.

"It was after our visit to Hong Kong that we started to practice a new tactic that we conceived in the Cubi Point O'Club. It had to be wild, innovative, and a maneuver the North Vietnamese had not seen before. Thus, the 'coordinated attack' was born. The plan was simple: if the *Blue Angels* could join up over an airfield from different directions at the same moment, why couldn't we? We certainly had the technology. We started flying two-plane strikes, then three and then four using this tactic. By the time August rolled around we had developed confidence that this tactic was working. With the help of the Carrier Air Group Commander and the ship's captain we were able to put a muzzle on the Air Force target planning folks long enough to schedule a coordinated attack on a major target. We planned the entire mission around the A-6. We were able to use our own talent, aircraft capabilities, and ingenuity to employ the A-6 the way it was designed to be used in that environment."

Rupe Owens concurs; he later wrote:

"After two other A-6A deployments to West Pac and the ongoing combat action in Vietnam, Navy officials were not completely impressed with the *Intruder's* performance. Due to the relatively new electronic systems inherent in the A-6A and the lack of an experienced base squadrons were struggling to meet assigned combat strike requirements. Staff officers responsible for making combat targeting decisions did not understand the aircraft's capabilities. Due to the A-6A's ability to carry 28 Mk82 general-purpose bombs on a single mission it was expedient for staff planners to assign the A-6As with a maximum bomb load targeted on low priority missions.

"During the early spring of 1966 a sister-squadron successfully convinced targeting officials of the aircraft's combat capabilities when operated at night or during instrument conditions. They enjoyed successful prosecution of selected night targets, achieving serious damage to alpha strike quality targets. Attack execution employed either single aircraft missions or 'bomber-stream' tactics. The bomber stream usually resulted in more serious AAA threat to the later aircraft over the target than the first ones. This tactic was archaic and unnecessarily dangerous, and was frowned upon by many aircrews. The squadrons continued to discuss the requirements and tactics that could overcome target defenses. These discussions pointed toward a simultaneous multiple-plane strike as the most effective tactic.

"Development of the VA-65 All Weather Coordinated Strike Tactic stemmed from these simple observations:

- The A-6 weapons system enabled more precise navigation than heretofore experienced in tactical naval aviation;
- An A-6A could find and accurately attack radar significant targets at night or in adverse weather;
- Bomber streams usually resulted in the latter planes receiving concentrated and accurately aimed fire;
- A flight of four A-6As could deliver 56,000 pounds of ordnance on a target.

"And, there was one other important element: VA-65 would deploy with more experienced aircrews and maintenance expertise, drawing heavily from the instructor pilots and bombardier/navigators who had established the A-6 RAG squadron. Significantly, an experienced combat pilot from VA-75 requested assignment adding to the squadron's overall experience base. The two Maintenance Officers, both former enlisted men, were particularly skilled in the field of avionics, where their talents provided important advantages. This experience base enabled development of procedures and tactics from a higher position on the learning curve than the first two squadrons possessed upon entering combat."

On the 10th, the *Tigers* received the frag for the Uong Bi plant. Using the tactics they had devised in house, they launched. Dye describes the remainder of the mission:

"On the night of 10 August 1966 we were assigned a target by the Air Force. They wanted a high probability of destruction and suggested an Alpha Strike to take out a major North Vietnamese power plant. This was our chance to prove the capability of the coordinated strike. We launched a four-plane strike, two A-6s with five 2000-pound bombs to fly broadside to the target and two A-6s with 28 500-pound bombs to string through the transformer yard and support facilities. All attack paths were to achieve maximum levels of destruction on the target complex, all bombs arriving on the target at the same precise time from four different directions. The timing of the strike went flawlessly. All the *Intruders* crossed exactly as planned. Our bombs released perfectly.

"Cmdr. Mandeville and I both made the comment at the moment the bombs detonated, 'I wonder if the other guys aborted the mission?' We had all made the appropriate call at our initial point, about three minutes before drop, but we only saw one explosion, albeit one big explosion. Moments after we released our bombs the other three *Intruders* call in that their bombs dropped as advertised. What we had seen was the cumulative effect of all the ordnance exploding in a near simultaneous detonation."

The next day's BDA gave the results, which were impressive according to Dye:

"The next morning a Navy RA-5C Vigilante photo-recon aircraft took a sweep through the area. The pictures he brought back showed in detail the success of our coordinated attack. The target was no longer standing. It was a pile of rubble. The target we destroyed was in one of the most highly defended areas in North

Vietnam; not a single SAM was fired, and anti-aircraft fire didn't commence until after we were on our way safely out of the target area."

Success built upon success. On another occasion during August the team of Williams and Diselrod got to work with Army OV-1 *Mohawks* engaged in truck-busting on route 1 near Vinh. According to the B/N, the A-6's capabilities once again proved a surprise to the North Vietnamese:

"The trucks would run down the road at night until launch time, and then they would accumulate in towns and under trees for about an hour. They knew the 1 1/2 hour cycle, so they would hide from the A-4's flares. The Army said they could see them on the Side-Looking Infrared and requested someone to be available between cycles. We launched with a load of Mk.82s and joined the OV-1 over the coast. They were right; about an hour after launch the A-4s had expended their flares and weapons on suspected transit sites, and as they departed for marshal, the roads came alive with moving targets. As we turned north from Cape Mui Ron I told the Army that I had about 18 miles of movers on the AMTI. Willard and I just moseyed along, pickling a bomb here and a bomb there, disrupting traffic. It was a shame we didn't have the CBUs available until later, in 1967."

Sometimes the strikes did not go as planned. One particularly confused evening led to Capt. Houser paying a "friendly" visit to Diselrod:

"My assignment in VA-65 was 'Targeting Officer.' I reviewed the target list daily and fed the selected targets to Strike Operations for loadout on the Air Plan, and also to the squadron flight officer for crew assignments. During the monsoon season the A-6s were the only aircraft flying over the beach; the normal schedule was two per cycle. One night the E-2 was launched, the KA-3B tanker, along with two F-4s for BARCAP, two F-4s for RESCAP, and coordination with the *Cottonpicker* jammers, when just as the first A-6 was launched the second went down due to a hydraulic leak. Later that evening, while I was in Air Ops doing my targeting, Capt. Houser came by and said, 'John D, if you are going to only send *one* A-6 over the beach, please cancel it so we don't have to launch so many support aircraft.' What can one say? 'Yes sir.'"

Despite the run-in the skipper was obviously well pleased with the Intruders' efforts. Capt. Houser later wrote:

"Probably the peak accomplishment came during a week of bad weather, in October 1966, in which the 12 A-6s of VA-65 flew 77 percent of all Navy combat sorties over North Vietnam and 49 percent of the total U.S. effort. Some Air Force aircraft had slipped into the lower part of Route Package I under the clouds, but essentially all other U.S. air activity had ceased."

A big part of the reason for VA-65's success was the oft-used but sometime forgotten word, teamwork. *Constellation*, CVW-15, and everyone from the bridge down to the shaft alleys worked hard to make this cruise work. It was obvious to everyone by now that the A-6 would work, and it was certainly apparent in the *Tiger's* small intelligence office; they were working just as hard as anyone else. Rube Rubery states:

"I need to mention also that there was another AIO. He actually was called an RTIO, or Radar Targeting Intelligence Officer; his name was Wes Starr. He was an ensign when the cruise started, and in fact was JOPA (Junior Officer Present Afloat) on the *Connie* for the first portion of the cruise. He was a 33-year-old ensign that had been an enlisted Chinese linguist in the Air Force, had gotten out, went to work for an insurance company, hated it, and came back in as an officer in the Navy.

"During the work up period we, as AIOs, built up a great rapport with the flight crews. Wes and I had to go to NAS Lemoore to put together the target folders for all the flight crews. The squadron was together all the time, particularly from January to May 1966 when we transferred to *Connie* at North Island. We only got back to Oceana for a one week leave during that time. We did everything together, work and play. In everything we did it was work hard and play hard."

During this period VA-65 also built a reputation as bridge busters, usually employing one aircraft at night to take out a particularly troublesome span. On the evening of 12 August 1966 a single *Tiger* A-6 crewed by Lt.Cmdrs. Bernie Deibert and Dale Purdy dropped the center span of the Hai Duong bridge between Hanoi and Haiphong. Two months earlier it had taken the combined assets of two *air wings* to collapse the heavily defended bridge:

"We did successfully complete some coordinated strikes on a single target with multiple aircraft," comments Diselrod. "Lt. Cmdr. Rupe Owens came up with the idea that if two or more aircraft ran the same target, at the same altitude, same airspeed, and from opposite directions, that upon bomb release they would not come within one-quarter mile of each other if all aircraft turned (in the same direction off target)."

Lt. Pete Garber was already something of a legend in the air wing for his single-minded determination to take out the Thanh Hoa, or "Dragon's Jaw" bridge, south of Hanoi. He had started his quest during a prior tour with VA-75 during its first combat deployment. However, according to Diselrod, during this trip Garber and his B/N Lt. Don Hahn:

"Made some coordinated strikes against the Thanh Hoa Bridge, to no avail. What bothered Don was that Pete said he was going to get that bridge if he had to kamikaze the thing."

Rubery says there were other occasions where Don Hahn returned to the boat after another exciting night with Pete Garber and was *very* quiet:

"I'll relate one humorous incident that occurred with the MIDS (Mission Debrief Sheets). We were required to complete MIDS after every mission. The MIDS included who the crew was, aircraft, weather at the target, where (coordinates), and how much ordnance was dropped, what kind of target it was, time on and off target, altitude they were at when they dropped, system run or not, radar detected, ECM used, type and amount of AAA or SAMs, BDA, etc. They were quite extensive, with very detailed information; somewhere in the bowels of Washington there must be millions of these reports.

"I was debriefing Pete Garber and his B/N Don Hahn after a particularly tough night mission up in the Nam Dinh area," he continues. "This was well into the cruise, probably the second or third line period. I knew it had been tough, because Don was very quiet, and they both were trying to 'unwind.' At a point in the MIDS they ask for type and amount of AAA encountered. When I asked Pete the question he mumbled, 'Yeah.' I asked him again – 'Yea.' Again – 'Yea.' I asked him was it Heavy, Medium, or Light? In a raised voice he said 'Goddam, it was f'ing heavy!' Then he said, 'Sorry.' I said 'OK Pete, that's what I'm putting down, 'F'ing heavy!' We finished the debrief, they left, and I finished everything up and put the MIDS in the basket.

"Two or three days later the Head of IOIC (Integrated Intelligence Operations Center) came to me and asked if I had debriefed Pete and Don on that particular mission – I said yes – and he said that Adm. Reedy, CTF-77, wanted to see me, as he periodically reviewed some of the MIDS and had a question for me. We went down to the CTF 77 spaces, and you know what he asked me, and all I could do was say 'Yes Sir, Yes Sir. I had forgotten to erase the F'ing heavy!' I tried to explain to him why it was done and that it was a mistake etc., etc. He said these MIDS are *very* important and they must be completed in a *professional manner*. I, of course, said 'Yes Sir!' and was dismissed. It was a stupid mistake, and we laughed about it for the rest of the cruise. I used it in training sessions as an example or what not to do."

Another strike took the *Tigers* to the Thai Binh railroad bridge. The crews were XO Frank Cramblet and Lt.Cmdr. Leon M. Stevenson, Jr., with Willard and John in the second aircraft. Diselrod recalls the XO planned this one, which took place at night:

"The two A-6s would attack from a 30-degree angle to the bridge. He would come in from the southeast and turn right back out to sea. We would attack from the northeast and turn inland. In all, it went pretty well. We each had five 2000-pound bombs, but they set theirs for 'train' and I set ours for 'salvo.' The bridge was dropped, but the nice thing about it was as we turned inland, no one on the ground thought anything about it. As the XO turned right for the coast-out he was flying right down a flak corridor. Boy, is it pretty when the 'puffballs' are going the other way!

"Rupe Owens, Roger Smith, Willard, and I also got to run Cat Bi airfield during this cruise. Rupe and Smitty had a load of Mk.83s for the runway and Willard, and I had a load of Mk.82s for the hangars. BDA showed the Mk.83 had less disbursement than the Mk.82s. Smitty's run down the runway left a lot of craters, but my Mk.82s, due to the direction of ejection off the MERs, left a snake pattern through the hangar area. Oh well, live and learn. The next cruise the weapons had *Snakeye* tails and could be dropped lower with better accuracy."

Occasionally, other missions did not turn out "as advertised" with more long-term results. On 27 August 1966, Jack Fellowes with George T. Coker and Robert Williams with John Diselrod headed out to destroy a "strategic bridge." According to John the "high value target" turned out to be an eight-foot culvert, and while making their run Fellowes and Coker were shot down. Recalls Diselrod:

"The target was north of Vin Son. We thought it was a SAM that shot them down, but it turned out to be AAA. It hit them right in their No. 3 weapons station, taking the right wing off."

Needless to say, the squadron heard about it almost before the two men's chutes hit the ground. Rubery remembers:

"We (the AIs) tried to get outside in the sun once a day. We had finished the briefing and the debriefing of the returning flights, and a few of us decided to go outside and watch the next launch. We had a place we would go up on top of the island, outside where we were out of people's way. We watched Jack Fellowes/George Coker and Bob Williams/John Diselrod man their aircraft. I remember Jack was kidding with his crew chief, wearing that towel around his neck that he always wore. We watched the launch and went back down to IOIC, as we had a debrief to do.

"As we got into IOIC someone came up to me and said 'one of your A-6s is down.' I said which one? He said he didn't know. I ran over to StrikeOps – which was just down the passageway – to get the current info. I asked which aircraft and where was it down, and was told it was Jack and George, they had two good chutes and they gave me the coordinates – we plotted it, and it was in the water. I initially thought, 'Good, they're okay, and we're never going to hear the end of this based on Jack's personality.' Then revised coordinates came in – they were on land, and then we couldn't get them back. Tough!"

Fortunately, both men did safely eject, but they would spend the next seven years as "guests" of the North Vietnamese. On other occasions the crews were able to return, thoroughly scared but having survived. Even in the intensity of combat, some could even find cause for a chuckle ... after the fact, as one B/N recalls:

"Bob Mandeville and Rupe Owens were assigned for a section night strike. Unusual for the A-6, but they launched for the target. As the story goes Mandy was letting down toward the coast-in and Rupe was flying a close wing by watching Craig Dye's red map light, turned to the right. As Mandy leveled around 300-feet on the radar altimeter, he decided (for some reason) to check the taxi light. This was before it was made dual purpose as a refueling light, so the beam reflected off the water. Well, as one can guess, Rupe went straight up as the whole world lit up."

It was also during this period that VA-65 and the other squadrons of CVW-15 participated in a short-lived experiment in aircraft camouflage. The tests were an outgrowth of several experiments done by the CO of VA-42, Cmdr. W.S. Nelson, during the spring of 1965. After much study and senior-level back and forth the Navy directed *Air Wing 11* in *Kitty Hawk* and CVW-15 in *Constellation* to paint several aircraft in combinations of green, brown, and sky blue, and evaluate the results.

According to VA-65 C.O. Bob Mandeville, the camo birds:

"Looked great from the air; it was the neatest thing. But on one mission I looked up and was immediately able to pick out the camouflaged A-4s. They stuck out like big, black thumbs in the sky, where I couldn't see the other A-4s. Obviously, with our biggest threat being on the ground, this didn't work. We had several air wing aircraft shot down and shot up; got back to the boat, and stripped the paint off the planes."

For their part, the wing personnel did not like the idea of evaluation of camouflage paint schemes in combat and listed other concerns, including maintenance and flight deck visibility problems. In a 5 July 1966 message to *ComSevenFlt Connie* advised the experiment was not working out. Four camouflaged CVW-15 aircraft had been downed by visually-aimed AAA to date; the ship intended to repaint the remaining test aircraft and resume combat as before. Thus, the experiment came to a close. While Naval Air Systems Command eventually performed other evaluation programs back in the states, no additional carrier-based tests would ever take place.

John and Willard's Bogus Journey

Towards the end of September 1966 *Constellation* was finishing up its third line period, and to this point operations with the *Intruders* had gone exceedingly well. However, there were a couple of busted aircraft that needed repair. The question was how to get one of the A-6s to Atsugi for the necessary work. John Diselrod and his pilot Willard were selected and became the beneficiaries of an unexpected "vacation":

"Seems we had an A-6 at Atsugi with a busted radome; we also had an A-6 taxi up behind the JBD with its wings folded and no jury struts. The blast shaking the wing caused the spar to crack around the locking fitting. The maintenance crew spread the wing, drove the locking pin home for security, and now needed someone to fly the aircraft from Yankee Station to Cubi Point, then on to Atsugi for a wing change. Guess who got the good deal. We launched off *Connie* and the wing held together. Landed at Cubi Point and took on five drop tanks and a full load of fuel to get us to Atsugi. Departed Cubi OK, and got on top of a heavy overcast and storm. We lost radio contact departing Luzon and approaching Okinawa, so we decided to fly two triangles to see if we could get assistance. Yea! Someone showed up, and we shot an approach to Naha.

"As we came screaming across the fence line in a driving rain and touched down the base was a mess. When we got to Ops we were informed that there were two (2) typhoons approaching that were expected to join and everyone was on alert to go to Taiwan. We asked for some hangar space to dry out. HA! They wanted us to go to Taiwan, but the ops officer offered just enough space for an A-6. The next morning we got up, filed our flight plan for Atsugi, and were on our way.

"Got to NAS Atsugi in fine shape, just ahead of the typhoons; we tied down the A-6 and beat feet for the BOQ. I couldn't believe the mess that weekend; most of the windows were blown out of the BOQ, and the press box at the stadium was torn loose, tossed across the field, and through the roof of the gym. Max destruction everywhere, and there sat our soaking wet A-6. What had been scheduled as a four to five day trip was now degrading quickly. We finally got out of Atsugi a week late. With a soaking wet A-6 during the five hours from Japan back to Cubi we progressively lost just about everything. The inertial quit shortly after takeoff, the search radar went about an hour later, slowly the pitot system froze up, and we lost flight instruments one at a time. We were flying on fuel flow and time, DR navigation, and a lot of luck.

"As we pushed down through the clouds we were amazed to see the entrance to Subic Bay just ahead as we came out under a 2000-foot overcast. We turned toward Cubi and tried calling the tower to no avail. We leveled at 1500-feet, started our approach, and slowed to something just above vibration (stall) speed. Willard lowered the flap handle to 30 degrees; the flaps came down, but the slats remained up. He then lowered the gear OK, but as he hit the electric switch to get the slats out it caused the flaps to go to forty degrees ... and the bottom fell out. I, for the first time, reached for the face curtain, but noticed the VSI at 3700-feet-per-minute down and then decided we were not going to make it anyway. Somehow Willard had full power on and blowing rooster tails below runway level, and as we touched down the rescue helo was over us. As we rolled past the tower we heard a voice ask if we were ok. Answer: 'We are *now*, thank you.'"

The beach detachment repaired the A-6, and the crewmen eventually rejoined their squadron mates back in the Tonkin Gulf. One more line period followed, ending on 9 November 1966; notably, VA-65 once again found itself carrying the load in the face of inclement weather, flying 77 percent of CTF-77's missions over the North. As Rupe Owens puts it:

"Flying at night and in the heavy rain and clouds, the enemy couldn't see us, although we could see our targets. Using DIANE and the excellent search radar we were able to close in on the North Vietnamese targets and get meaningful damage or destruction night after night despite the weather. The A-6 was the only aircraft to do that; the F-105s were not in the ballgame."

CAG-15, Cmdr. J.D. Ward, concurred. In the 26 October 1966 "Fabulous Fifteen Yankee Bulletin," published as a daily summary of air wing operations, Ward stated:

"CVW-15 again operated on a limited scale today due to foul weather, with the 'Tigers' of VA-65 continuing to carry the ball. The A-6As twice hit a thermal power plant and transhipment point at Vinh, damaged a ferry complex near Thanh Hoa, hit a railroad yard at Nam Dinh, destroyed 2 barges, and got a large secondary explosion from a radar site near Ha Tinh and damaged eight barges in Package IV."

On the 30th CAG Ward reported:

"The winter monsoon continued to cover NVN with a low overcast, reducing strikes to those carried out by the A-6As. VA-65, flying through low clouds, hit a highway pass south of Ninh Binh, a storage area and cave complex on Ile Cac Ba, a radar site near Thanh Hoa, a naval port facility near Hon Gay, another port facility at Vinh, a SAM site near Haiphong, and damage to barges in PKG.IV."

CAG would add in a later report:

"Really a good show! You perfidious air pirates are really pelting the DRVs."

Constellation outchopped on 24 November 1966 and headed for the barn in San Diego via Yokosuka. In 111 days on line the *Tigers* had lost only two aircraft, with one pilot recovered, one KIA, and two POWs. The losses notwithstanding, the squadron *had* proven the viability of the A-6 *Intruder* in its toughest test to date. Rupe Owens sums it up thusly:

"The squadron lost no aircraft during night attacks and dropped a record 10,599,877 pounds of ordnance in 1239 combat sorties. We accounted for four strategic bridges, 16 PT boats, and 52 barges, plus barracks, trains, roads and trails interdiction, ferries and choke points damaged, plus a large number of KIA in actions conducted in support of ground troops in South Vietnam."

Notably, Owens and skipper Bob Mandeville also became the first inductees into the *Intruder* "1,000-hour club."

Finally, just prior to VA-65's return to Oceana, Mandeville summed up his pride in the squadron and its accomplishments in the last issue of *Tiger Rag*:

> "Looking back at the cruise, I could not help but go even further back to Oceana when we were handed the challenge of 'the A-6 program will live or die as a result of VA-65's performance.' It was pretty frightening for me, as I'm sure it was for you. However, at the same time, I knew that we would hack it, because we had the number one requirement in spades – and I mean people. Not numbers of people, but a group of individuals who represented the best available in the Navy, or anywhere else. Talented people, strong, dedicated, loyal and intelligent people – when you have a squadron made up of this type of person you really can't lose.
>
> "I don't think there's a doubt in anyone's mind that we cut the mustard – in fact, our performance was outstanding, far better than anyone (excluding ourselves) expected. We made a few mistakes, but not many, missed a few flights, but not many. We weren't perfect – but close enough! I'm sure that Uncle Ho and his boys are happy to see us go, which is a pretty good measure of our success – we sure tore up a lot of his stuff!
>
> "It's very difficult for me to find the proper words to express my pride and thankfulness for being privileged to be your commanding officer while you made history. Somehow, the traditional Navy 'well done' seems to say it best, 'well done and thanks.'
>
> "One last word, remember this fact: The most complicated airborne weapons system in the world is now a *proven* capability of Naval Aviation because 330 *men* made it work."

Finally, in his last familygram dated 14 November 1966, the skipper also credited another source of support the squadron had received: letters from the family members back in the United States:

> "I was pleased to read of your interest and continued backing of the squadron throughout this cruise. Probably the most delightful compositions we received were the 150 letters from fifth graders in Pennsylvania. Mrs. Sandra Lee Rodenbaugh, the wife of Steve Rodenbaugh ADR3, one of our fine plane captains, is their teacher, and when the children expressed a desire to correspond with military personnel in Vietnam, I guess the 'Tigers' were a logical and grateful choice. One hundred and fifty eleven and twelve year olds are now reading about the air war in Vietnam as told by an equal number of VA-65 combat veterans.
>
> "My pride in our people is equaled only by my pride and gratefulness for those parents, wives, and children who remained behind supporting us with your letters, thoughts, and prayers."

For Craig "Nellie" Dye, the first Marine A-6 B/N to see combat in Vietnam, the *Tigers'* successful deployment was also personally fulfilling:

> "I have to give a lot of credit to the ordnance and aircraft maintenance men who did a superior job throughout the grueling months, as they loaded and maintained aircraft, in the most difficult conditions, in order to keep those A-6s in the air. VA-65 was extremely fortunate not to lose any aircraft due to premature weapons explosions or maintenance problems. As a Marine on exchange duty with VA-65, it was an education. I was fortunate to be associated with the finest bunch of Naval Officers I have ever served with during my Marine Corps career. And my thanks to great commanding officers, who gave me the chance of a lifetime."

Says Rubery, who later went on to a CVW-17 intelligence slot:

> "When we finished the cruise, on the last day of flight operations after the last debriefing we felt very lucky, fortunate to be done, and proud of our contribution. We passed out – as a joke to all pilots and B/Ns – FUBIT buttons (F*** you buddy I'm Through). Unfortunately we had to leave three squadronmates behind. Two would come back seven years later and one would not."

The other squadrons in the young medium attack community took VA-65's lessons to heart. The *Intruder* was viable, the mission was doable, and the future now looked very bright.

More *Intruders* to Cherry Point

Training continued in Virginia Beach; by the summer of 1966 the VAH-123 det was just about ready to return to Whidbey Island, while the first west coast *Intruder* squadron, VA-196, was wrapping up its transition. AirLant's VA-35 prepared for its initial deployment in the type, as did VMA(AW)-242 and -533 down at Cherry Point. On 1 April 1966 a third squadron joined the Marine lineup when VMA-225 turned in its A-4Cs at Chu Lai and moved on paper to the station in North Carolina. The *Vagabonds* existed in an unmanned status under the auspices of VMA(AW)-533 for several months before receiving new personnel and aircraft.

Another long-time attack outfit, the *Vagabonds*, was first organized as VMF-225 on 1 January 1943 at MCAS Mojave, flying the F4U *Corsair*. The unit saw combat in World War II in Guam and in operations against several Japanese strongholds prior to moving to Cherry Point in May 1945. Through 1952 the squadron made five Med deployments in *Siboney* (CVE-112), *Leyte* (CV-32), *Midway* (CVB-41, two cruises) and *Franklin D. Roosevelt* (CVB-42) while still flying F4U-4s.

On 17 June 1952 the outfit redesignated as VMA-225, and within short order moved to MCAS Edenton, NC. Operations from Edenton continued through 1958, with the *Corsairs* giving way to *Skyraiders* in 1954. In May 1958 the *Vagabonds* moved back to Cherry Point; the following October the squadron transitioned to the A4D, beginning eight years of *Skyhawk* operations. This period included one Med cruise with CVG-10 in *Essex* (CVA-9) from August 1959 to February 1960 (flying A4D-2Ns) and a short deployment in *Enterprise* during the Cuban Missile Crisis.

Prior to joining the A-6 community VMA-225 made two more deployments, including a trip to Japan in September 1964 and to Chu Lai for combat in May 1965. After the squadron shifted in name to Cherry Point it redesignated as marine attack-all weather in June 1966 and separated from VMA(AW)-533 on 22 October 1966. Now fully manned and equipped with the A-6, the *Vagabonds* pursued a rigorous training schedule in preparation for a return visit to Southeast Asia. On 1 November the fourth Marine Intruder squadron, VMA-224, checked in for their conversion cycle. Again the Marines shifted the squadron on paper; the *Bengals* left their A-4Cs at Chu Lai, redesignated as VMA(AW)-224, and gained their first A-6s on 5 November.

The *Bengals* also possessed a long proud lineage, having first organized at MCAS Ewa, Territory of Hawaii, as VMF-224 on 1 May 1942. Operating F4F-4 *Wildcats*, the squadron quickly went into combat, arriving at Henderson Field on Guadalcanal in August. In October 1942, following two months of desperate uninterrupted combat, -224 returned to the United States, having shot down 60 Japanese aircraft. The squadron remanned and reequipped with F4U *Corsairs* at San Diego and El Toro prior to returning to combat in early 1944. Further campaigns in the Marshalls, Solomons, and Okinawa brought -224 two Presidential Unit Citations and one Medal of Honor, awarded to skipper Maj. Robert Galer for leading the squadron "repeatedly in daring and aggressive raids against Japanese aerial forces." The *Bengals* concluded their corner of the war stationed at Yontan airfield on Okinawa; following VJ Day, they relocated to Yokohama, Japan, and remained there on occupation duty until June 1946.

VMF-224 moved to MCAS El Toro and remained there for about two years before transferring to MCAS Cherry Point in June 1948. In the early 1950s the squadron received F2H-2 *Banshees*, followed by a five-month Med cruise in *Franklin D. Roosevelt* with CVG-6. Following their return to Cherry Point in May 1952 the *Bengals* traded in their F2H-2s for F9F-5 *Panthers* and continued training. On 10 September 1953 the squadron moved to NAS Atsugi, Japan, as a component of U.S. forces monitoring the Korean Armistice. It returned to ConUS at El Toro in November 1954, and on 1 December redesignated as Marine Attack Squadron 224. Three years later the *Bengals* transitioned to the A4D-1 and pulled a 14-month tour in Iwakuni.

Following their transfer to Cherry Point in early 1959, VMA-224 entered into a six-year period of deployments, training exercises, and other operations. These included a six-month Mediterranean cruise in USS *Independence* with A4D-2s as a component of *Carrier Air Group 7*, and another 14-month trip to Iwakuni. The year 1965 found the *Bengals* making their first deployment to Vietnam; by now operating A-4Cs, the squadron checked in with *Marine Aircraft Group 12* at Chu Lai in October and immediately entered combat. During the course of its 13-month tour, VMA-242 did the standard CAS mission while participating in several operations, including *Colorado*, *Prairie*, and *Shawnee*. With their year in Vietnam concluded the *Bengals* turned in their trusty *Scooters* and made the move back to Cherry Point.

According to pilot Capt. Jim Perso, training for the new all-weather attack -242 crews had its moments, particularly when it came to horsing the big A-6 in behind the tanker:

> "The Marine Corps used the KC-130 for aerial refueling. It employs the 'probe and drogue' method of refueling. The C-130 became a KC-130 when two refueling pods were added to the wings, outboard of the engines. Each pod contains a hose, reel, and a refueling drogue. The drogue stabilized the hose and coupling; it is the target for the probe. The drogue collapses when stored and opens as it is reeled out. The receiver aircraft has a probe fixed to the aircraft, which the pilot flies into a drogue streamed from the tanker. The drogue has a coupling with spring-loaded latches to hang onto the probe, once the pilot has flown it into the drogue. The hose had to move into the pod six feet to start fuel flow.
>
> "The KC-130 has four turboprop engines and is not very fast. Jet aircraft must slow to fly formation and refuel from the tanker. The technique is to fly formation on the tanker and move forward, flying the probe into the drogue. On the F-4 *Phantom* and F-8 *Crusader*, the refueling probe was behind the pilot's head, making refueling a challenge for good pilots. The A-4 *Skyhawk* and A-6 refueling probe was visible to the pilot. Even so, inflight refueling requires good airmanship. Refueling at night was, is, and shall remain more difficult. The drogue has three 'iso lights' that glow dimly at night. The three dots of light, in an equilateral triangle, form a visual target for the probe."

Perso says one particular tanker hop took place on the classic "dark and dreary night" with a VMGR-252 *Hercules* doing the honors. He and B/N 1Lt. Steve Paul were in one A-6, flying wing on the *Bengals'* operations officer:

> "The Major had a lot of time in the F-8 *Crusader*; he was a skilled aviator and had done this before. Steve Paul was my B/N.

Steve was new to the squadron; we had not flown together until this night. After takeoff we joined up and proceeded to find the tanker. Since we were in formation, lead turned off his rotating beacon (the anti-collision light, a.k.a. 'anti-smash' used to alert other aircraft). It is SOP for the wingman to have the anti-collision light on for the formation and for lead to turn his off, so as not to distract the wingman. Position lights (like a boat's running lights) are normally set to bright, but may also be set to dim or off.

"Our squadron mate 'Igor' (pronounced EE-gore) was on the tanker (in another A-6). We came up on a perch and checked in with the tanker. They cleared one of us on to the other side of Igor, and the other was to stay on the perch. Hearing this Igor transmitted, 'That's OK, send them both, I am leaving. IGOR - NOT - DO - TOO - GOOD!' His transmission did not bode well. Igor and I were just out of flight school; if Igor found this difficult, I was in for a hard night. I was not worried about the Major; he had refueled F-8s, a difficult task. I moved into position on the left side of the tanker and tried to see the drogue. The sky was moonless, with a high, thin overcast. The stars shone faintly and blurry through the overcast. On this night, the iso lights were indistinguishable from the stars. It was very difficult that night. I flailed around, trying to plug into several stars, only to find the drogue unexpectedly coming out of the dark.

"My 'anti-smash' light was bothering the major; he asked me to turn it off, which I did. However, I also turned my position lights to OFF rather than DIM in the process of trying to fly the bird, at slow speed, in formation with the drogue, while feeling around the panel for the light switches, low on my left side. I think the Major or the KC-130 stated my lights were off (which I deemed confirmation of switching off the anti-smash). Steve recognized what had happened and tried to tell me. I am really busy trying to fly. I interpreted his comment as a statement reflecting adversely on my ability to fly. True, I was not at my best; even so, I was offended and growled, fiercely, to SHUT UP. Steve didn't say anything more.

"I persisted on the tanker, trying to plug. It was bad that night. It was impossible to see the drogue. It would appear from nowhere, always in the wrong place. Sometimes the probe would hit the rim of the drogue, then the drogue would tip off and jump away or tip off and hit the aircraft. It even went all the way around the nose. It hit the windscreen. It tapped the canopy over Steve's head. All this time Steve did not say anything. The major left and I stayed. Finally I plugged. I moved forward, flying formation on the tanker, and stabilized in the refueling position. I wondered to myself, 'What would happen if I was low fuel?' I backed out, and the drogue disengaged and disappeared into the night. Then, I tried again ... and again ... and again. I could not plug again. I was really depressed at my lack of skill. Eventually I gave up and headed back to MCAS Cherry Point.

"I called Approach Control for Radar Vectors to a GCA. That may have been the smoothest GCA I ever flew. I was right on the vectors and altitudes. The call to 'Begin normal rate of descent' was followed only by 'you are on glide slope, you are on centerline,' until the final controller said, 'the Tower does not have you in sight.' I landed and they still could not see me. Duh! Having been quiet throughout this ordeal Steve said, 'I think the lights are out.' I turned the lights on and taxied in. After postflight and signing the yellow sheets we climbed the stairs to the ready room. I dreaded to admit my failings as a pilot. As I feared, the Major and 'Igor' were waiting for me. They asked 'how many times did you plug?' I was embarrassed, ashamed, and dejected. I responded 'God it was awful, I could only plug once.' They said, 'You *did*? Neither of us could plug!' The experience, awful

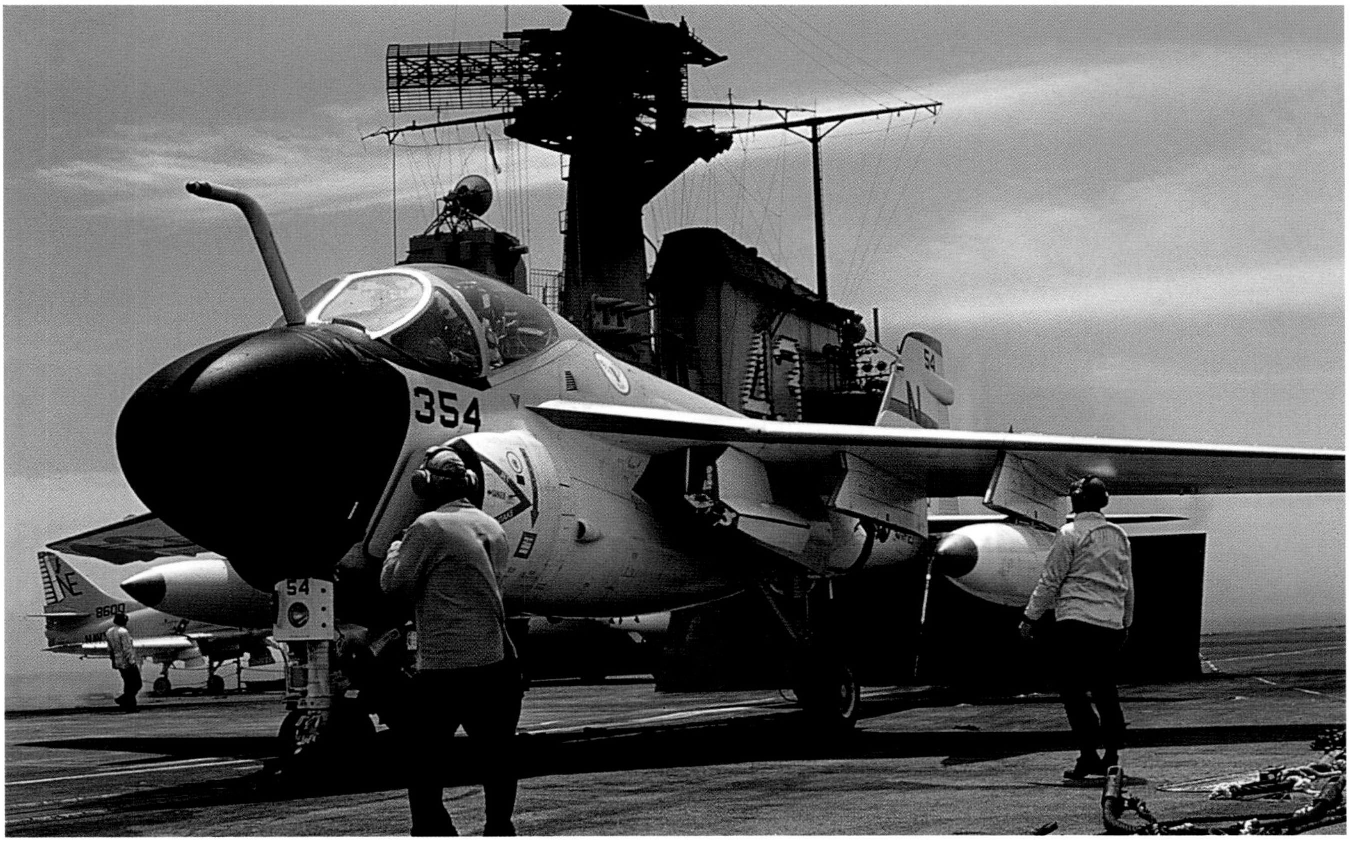

The West Coast *Intruder* training squadron was formed in 1966 as a part of VAH-123, the A-3 *Skywarrior* RAG. In February 1967 the det was spun off as a separate squadron, VA-128. Heavy-123's NJ354 is shown here conducting CQ on *Coral Sea* (CVA-43). (U.S. Navy)

as it was, boosted my ego. That night was a downer, until they said they did not plug. It was thirty years ago. I remember it well."

Fun with tankers and bad weather notwithstanding, the *Bengals* continued to prepare for deployment to Southeast Asia. Two more Marine squadrons would subsequently join them in the FMF *Intruder* lineup, but it would take a couple of years.

... and "The Rock"

On 17 August 1966 the VAH-123 A-6 Det left VA-42 and the friendly confines of Oceana and returned to NAS Whidbey Island with the first three AirPac A-6As, BuNos 152616, 152617, and 152618. Young Lt. Lyle Bull, one of the *HATRON 123* cadre, found the reception to be one of "curiousity; no one thought the plane was particularly attractive."

Led by OinC Cmdr. Jerry Patterson, the detachment adopted the nickname *Golden Intruders*, utilizing the "Winged No.1" emblem of First National Bank of Norfolk. According to Patterson the initial group had seen the bank's logo during their stay in Tidewater and approached FNB's president concerning its use by the future A-6 RAG. He not only agreed, he apparently "adopted' the squadron, and later attended VA-128's establishment ceremony at Whidbey.

For Whidbey, the sight of the bulbous *Intruders* could very well have been termed "curious," particularly in contrast to the huge – but sleek and powerful – *Skywarrior*. At the time of the A-6 det's arrival at the NAS the *Whale* had held forth as the Navy's northwesternmost station in the continental United States since 1957, when the heavy attack community relocated from NAS North Island. Five squadrons operated A-3Bs under assignment to *Commander, Fleet Air Whidbey*: VAH-123 (the RAG with A-3Bs and TA-3Bs); and four fleet squadrons (VAH-2, VAH-4, VAH-8, and VAH-10), all equipped with the A-3B.

One of the Navy's primary west coast jet bases, NAS Whidbey Island dated to 1942 and the establishment of a seaplane patrol facility on a spit of land immediately south of the small town of Oak Harbor. The Navy subsequently built a second facility for land planes and support operations a few miles to the northwest at Clover Valley, commissioned it on 21 September 1942, and named the field for Cmdr. William Bowen Ault, a Naval Aviator who died in the Battle of the Coral Sea. Through World War II Ault Field expanded, hosting multi-engine patrol squadrons operating types such as the PV-1 *Ventura*, PV-2 *Harpoon*, and PB4Y *Privateer*, while providing training for units operating out of NAS Seattle. Water-borne patrol activities with PBY *Catalinas* and PBM *Mariners* continued at the original installation, appropriately nicknamed the "Seaplane Base."

Postwar the Navy reduced both installations to standby operational status while studying their future; the service made the decision to expand Ault Field's operations for jet aircraft in 1949. The first P5M *Marlins* arrived at the Seaplane Base in 1950, followed by P2V *Neptunes* at Ault Field, which by 1953 had two new 6000-foot runways. The *Skywarriors* started arriving under *Heavy Attack Wing 2* in 1957, and the runways were extended another 2000 feet to handle the new "Heavy Jets." *HATWING 2* eventually gave way to *Commander Fleet Air Whidbey*, and on 30 June 1959 *Heavy Attack Training Unit Pacific* became VAH-123, under assignment to RCVW-12. Initially equipped with P2V-3Bs and A3D-1s, the *Professionals* would spend slightly over a decade producing trained aviators, bombardiers, and maintainers for the A-3 community. With the August 1964 redesignation of the heavy attack community at NAS Sanford, FL, to *Reconnaissance Heavy Attack Wing 1*, Whidbey became the Navy's center for *Whale* operations.

Following their arrival at Whidbey the personnel of the Heavy 123's A-6 det started developing the numerous course syllabi and operating instructions necessary to get the west coast *Intruder* RAG going. Jim McKenzie, now assigned to the det following his combat tour with VA-85, remembers the FNGs fit in well, partly due to their background in heavy attack:

> "We had a couple of desks over in the corner of the ops office. I always thought we did things better out here (at Whidbey Island). The A-3 types were accustomed to flying in airborne weapon systems, and were accustomed to dealing with NFOs. Back on the east coast they were used to single-seat, and it took time to overcome antipathy to the B/Ns. Out here, from day one, the NFO and the system were paramount to what we were training in. It was a notably smoother operation here."

On 15 November 1966, Whidbey welcomed its first fleet A-6 squadron with the arrival of VA-196 from Oceana. Five weeks later, on 22 December, former VA-115 *Spad* pilot and Korean War vet Cmdr. Leo Profilet relieved Cmdr. James Donovan as skipper of the *Main Battery*. The ceremony marked the first A-6 squadron change of command at Whidbey Island; according to Profilet, from this point things would pick up rapidly for medium attack at the station:

> "I transferred to Oceana from NAS Lemoore and went through the RAG with the rest of the squadron," Profilet says. "It took me a while to 'train' my B/N, but once the crews got used to each other, it worked out great. I relieved Jim Donovan about a month after our move to the northwest; three or four months later we went off on our first WestPac."

Bats and *Black Panthers* Go to Vietnam

While the Heavy 123 Det and VA-196 were setting up the A-6 program in Washington state, two other squadrons wrapped their preparations and headed to Southeast Asia. On 19 October 1966 VMA(AW)-242 departed Cherry Point under the command of Lt.Col. Howard Wolfe and headed to Da Nang Airbase, Republic of Vietnam. The transport of squadron personnel and equipment was made by U.S. Air Force C-141A *Starlifters*, while the squadron's *Intruders* made an equally lengthy transcontinental flight, followed by the first trans-Pac by Marine A-6s.

The *Bats* had had a busy time getting ready for their first big deployment overseas, with occasional difficulties. On 20 June 1966 the squadron suffered a mid-air; Capts. J.K. Engstrom and J.L. Anderson collided with 1Lts. A.C. Clark and W.D. Lewellyn after coming off Tangier target in Chesapeake Bay. During join-up one bird hit the other in the wing root area, and all four quickly ejected safely; one A-6 went down in the Bay, but the other one landed among some housing in Buckroe Beach, injuring several on the ground.

Unfortunately, the accident started one of those brief periods where *Intruders* seemed to stop flying for whatever reason. The Naval Air Test Center at Patuxent River had its own sequence starting on 4 Octo-

CVAN-65: *Enterprise* plows through the South China Sea during its 1966-67 deployment to Vietnam with CVW-9. A black-nosed VA-35 A-6A sits in the center of the bow – the squadron was making its first cruise with the type. A-4 *Skyhawks* from VA-113 and VA-56 flank the line of *Intruders*. A variety of A-3 *Skywarriors* are also visible, from VQ-1 (the "PR" marked EA-3B), VAP-61 (black RA-3B), and VAH-2 (A-3B). Two F-4B *Phantom* II fighter squadrons (VF-92, 96), RA-5C *Vigilantes* (RVAH-7), and an E-2 *Hawkeye* det (VAW-11) make up the rest of this typical mid to late 1960s "big deck" air wing. (U.S. Navy)

ber 1966. Lt.Cmdrs. P.E. Ericksen and Vandermullen were doing a J52-P-8A evaluation flight about 20 south of Pax, near Adams Island, when their aircraft stumbled into a low altitude-high sink rate situation. The crew ejected at 100-feet; Ericksen suffered a broken right leg with multiple lacerations, while Vandermullen received a broken vertebra and lacerations to the right foot. It got worse: on 10 November NATC's Lt. F.M. Hammond III and ADJ3 A. Adams launched from Trapnell Field with an unsymmetrical load of mines onboard. After lifting the plane settled back onto the runway, hit an embankment, exploded, and burned. Hammond ejected, but his chute never fully opened and he sustained a fractured ankle and compressed vertebra. Petty Officer Adams failed to eject and died in the crash.

Nine days earlier, on 1 November 1966 – and alongside with the recently arrived VMCJ-1, operating EF-10B *Skynights* and RF-8A *Crusaders* – the *Bats* entered combat over Vietnam. The initial tasking was CAS support for ground forces in the south, including U.S. Marine, U.S. Army, Army of the Republic of Vietnam (ARVN), and Republic of Korea troops. During its first month in theater the squadron dropped a total of 1321 tons of bombs, including 215 Mk.83s, 673 Mk.81s, and 3,236 Mk.82s. The crews then moved "Up North" as participants in yet another phase of *Rolling Thunder*.

While the Marines settled in at Da Nang, VA-35 also headed for combat. On 19 November 1966 the *Black Panthers* loaded up with CVW-9 in *Enterprise* at Norfolk and departed for their first combat deployment in the *Intruder*, the fourth for the type. Led by Cmdr. Arthur H. "Art" Barie, VA-35 had formally transferred to *Carrier Air Wing 9* the previous September. Skipper Art Barie was another one of the strong old hands of attack aviation who had made the transition to *Intruders* early in the type's career and, as with most of his counterparts, he had taken the scenic route to his penultimate assignment, starting his career as an *Avenger* driver with VC-33 at NAS Atlantic City:

> "We moved to Atlantic City and took over the night attack/special weapons business," he comments. "I spent 42 months there, got out of the Navy – it was 1952 – and stayed out just short of four years. I flew in the reserves at NAS Birmingham, a reserve fighter squadron with F8Fs and *Corsairs*. Then in 1956 I got a letter from the Navy, 'Hey, we'd like to have you back.' I thought well, what the hell, and I came back. They sent me to the NATTU (Naval Air Technical Training Unit) at Olathe as a maintenance officer, a real career enhancing position. We had to maintain a bunch of SNBs, but I only stayed there for six months, went through the school system, and ran the GCA controller school for two years; *that* was a very interesting job. I got selected for regular Navy, TAR, and lieutenant commander all within one week. I went to Washington and found out I was going to Sigonella, Sicily, as the maintenance officer. 'G.C.,' one of the biggest SOBs to ever drive people out of the Navy, was my detailer."

Barie survived Sig and even qualified as a plane captain in the R4D-8, having logged some R4D time at Olathe along with his first jet time in the Lockheed TV-1. Afterwards he received orders for postgraduate education and attended Harvard, studying in the international relations program with a gentleman named Henry Kissinger. Then, having been out of squadrons for something like 10 years, he received orders to A-1 training with VA-42 and within short order was with VA-85:

> "Four of us went through," Art continues. "The last of the *Spad* drivers. It was Whitey Gooding, Jim McKenzie, Tom Porter, and yours truly. I went to VA-85 as the ops officer after being a maintenance officer all my life, but I enjoyed it. We went to the Med in '63; about a month after I got there the skipper, Clint Mundt – a squadronmate of mine in Atlantic City – spun in and killed himself at the target south of Sicily. John McKee became the skipper, and I fleeted up to acting XO."

The *Falcon's* Med cruise in *Forrestal* on 2 March 1963 marked its final deployment with *Skyraiders*, and the Navy stopped ordering people in, in preparation for the squadron's transition to A-6s. Hence, Barie remained as the XO for quite a while and then fleeted up to commanding officer – as a lieutenant commander – when John McKee moved on. He recalls:

> "We got to go to sea for a couple of months, we won the Battle E and then – of course – we started dribbling people out left and right.
>
> "During that period the Great God of A-6s was Stu Nelson, the skipper of VA-42. He said he wanted to see me one day; I went over to talk to him, and he says, 'You know, you've been up there to Harvard and all that stuff, I just want you to know I'm going to do everything I can to keep you out of A-6s. Nothing against you personally, we just have too many people.' I said, 'That's okay, I'm just letting you know that I'm doing everything I can to get in them,' and we left it at that. We went on about our business. Eventually four of us *Spad* drivers (myself, Jim McKenzie, Bud Roemish, and one fellow who didn't make it through A-6s), I think we were the only four who transitioned while assigned to VA-85. Before we started they ordered in the A-6 skipper, Billie Jack Cartwright, and he relieved me as CO while I stayed on as XO. I made commander in there somewhere. Then Jay Keller – who was an old friend of mine from VC-33 – came in as XO, so I moved over to VA-42 as ops officer and relieved Bob Mandeville."

Barie rated his brief tour with VA-42 as:

> "Very hard working, but tremendously rewarding. Obviously, the war was really going on. We were working like hell to get people through the syllabus, but also to keep them from killing themselves and use the plane the way it was supposed to be used. We had great people in the ops department; Sweetpea Allen was the naval weapons phase head. Stu Nelson really gave me hell on the little stuff, but he pretty much gave a free hand; he *was* a tough son of a bitch as a skipper."

When VA-35 started its transition to *Intruders* Nelson called Barie into his office and offered him the XO's job, with the understanding that he would have to also hold on to the ops officer slot for a while:

> "So I did two jobs for a while," adds Barie, "But it was helpful, and I got to know all the guys who were going to my squadron. The skipper during that transition period was Don Ross; he was a 1948 Academy graduate, a former *Corsair* pilot, A-4 pilot, and a great guy. I relieved him in July 1966, we went through CarQuals, things like that, and then we went to join *Enterprise* in Alameda."

Enterprise's departure from the Bay Area marked the commencement of VA-35's first combat deployment since 1951. Notably, the squadron had been to Vietnam before; in January 1945 while embarked in *Yorktown* the *Panthers* had participated in strikes near Saigon and along the Vietnamese coast.

Fourteen days prior to VA-35's departure VA-85 had pulled out in *Kitty Hawk* with CVW-11 for its second *Intruder* deployment, making it the first squadron to make a second combat cruise in the type. *Kitty Hawk* checked into Yankee Station first, arriving for its initial line period on 4 December. *Enterprise* followed on 18 December 1966; its arrival marked the first occasion in the war where two A-6 squadrons were available to *Task Force 77* at one time. However, initial combat operations for both CVW-9 and CVW-11 were brief, as the United States once again declared a holiday bombing halt.

The War Picks Up

As 1967 opened, the Navy had two squadrons on line on Yankee Station (VA-35 and VA-85), while the Marines had one of theirs – VMA(AW)-242 – down south operating out of Da Nang. Other *Intruder* squadrons were coming, but for a brief period during the Christmas and New Year bombing halts the theater was relatively "quiet."

However, the traffic on the Ho Chi Minh trail and other supply routes to South Vietnam during the stand down was anything but light. During the 48-hour New Year's halt, pilots reported the Vietnamese coastal road looked, "like the New Jersey Turnpike," as the North Vietnamese rushed materials south. The Johnson Administration, taking this as evidence the North was not quite ready for peace, ordered the resumption of *Rolling Thunder IV* on 2 January 1967. The *Black Falcons*, embarked in *Kitty Hawk* and *Black Panthers* in *Enterprise* immediately responded.

Things got off to a bad start for VA-85. On 19 January, while on a day mission over North Vietnam, XO Cmdr. Allen Brady and his B/N, Lt.Cmdr. Bill Yarbrough, were tagged by AAA and shot down. Other aircraft noted two ejections from the burning aircraft, but only one parachute opened. Commander Brady survived and was taken prisoner; Yarbrough, veteran of the successful April 1966 Uong Bi powerplant mission, died. As a replacement for Brady the Navy selected Cmdr. Jerry Patterson, the OinC of the VAH-123 A-6 detachment. Patterson arrived in *Kitty Hawk* in February to assume the duties of *Black Falcon* XO.

Rolling Thunder IV concluded on 7 February 1967 with a seven-day truce for the Vietnamese New Year; but round five started on 14 February, as the Johnson Administration continued its program of "careful and measured" responses. However, at the end of the month CTF-77 was finally authorized to execute an operation it had long recommended. For the crews and personnel of VA-35, *Rolling Thunder* was the first big test. It was also CVW-9 and *Enterprise's* first opportunity to work with the still new *Intruders*. Fortunately, the Navy's only nuclear-powered aircraft carrier was the centerpiece of a great team effort, with Capt. James "Triple Sticks" Holloway III in command and Cmdr. Jim Shipman serving as CAG-9. Along with Art Barie and his *Black Panther* personnel, they had studied the previous combat reports and knew what to prepare for:

> "We had obviously read the reports and talked to the guys who'd come back," says Barie. "We had tremendous support from the ship. The A-6 was a maintenance nightmare in the early days, so we started out by saying we'd like to get one good hop every day from each plane, and we had nine airplanes. That would give us time to keep them up. I think Mickey Weisner was the flag officer at the time, later CINCPACFLT. He said, 'How about 12?' What can you say? But we successfully made the 12 fly every day. We had very good support from the ship maintenance-wise, and in every other way, and had great Grumman people with us. Bill Schultz was the weapons guy, and Charlie Meyer was the airframes guy.
>
> "What did we learn from the other squadrons? We knew the loss rates were horrendous, and in many ways we had no expectations we would do any better. At that time several A-6s had blown themselves out of the sky; we knew that, and we had some idea of the type of targets they went for. My philosophy was ... I had the feeling that a large percentage of the combat losses – the guys who were really shot down – they would have been considered operational accidents during normal times. My philosophy was and I tried to preach it to my guys, the ground always wins. You can wrestle with a missile, it might get you, you go through the heavy AAA. At places like Nam Dinh it was almost a matter of luck. But if you touch the surface, the game's over right there. I really liked to push that. I also pushed a 200-foot minimum over the Red River Delta; the A-6 radar could see the karsts, you knew where they were. It also helped in having experienced people; they were all really rock solid, steady people, which really helped the youngsters.
>
> "By the time we got over there we were pretty well restricted to working in Route Packages Five and Six. We and VA-85 – our deployments overlapped – were all working up north. I read and gathered all I could from the other squadrons, and you still kind of expect to lose a bunch, including yourself. As it turned out I was the first A-6 skipper to go the whole cruise."

On 26 February 1967 VA-35 mined the Song Ca and Song Giang Rivers, marking the first combat aerial mining operation since World War II. The squadron's crews saw some antiaircraft fire in the vicinity of Vinh but no SAMs, and the mission came off without a hitch. The following month, VA-85 put down more mines in the Song Ma, Kien Gang, and Cua Sot Rivers, again with good results. While the mining ops were not as extensive as *CINCPAC* and *CTF77* desired – by presidential order, North Vietnam's major ports of Haiphong, Hon Gai, and Cam Pha were still excluded – they were successful in ending North Vietnamese water-borne operations on the affected rivers.

According to Barie, the mining was a joint operation all along and employed real, no-kidding mines, and not the modified Mk.82 *Destructors* then coming into production. He recalls it was a pretty substantial evolution:

> "They brought out the mines to the *Enterprise* and they brought out a team to handle them. If I recall correctly, they brought four VA-85 crews over and we got the mines on, five per airplane, one for each station. One thing stood out: it was like flying with five 55-gallon drums on the airplane. It was a real 'Drut' as compared to our standard load of Mk. 82s. Hell, dropping mines isn't a lot different from the pilot and B/N's point of view, but the airplane flies differently. We did our own planning for these rivers, and some time during the night we went in. It was a beautiful night, and I violated my own rule and made two passes. The moon was full and I could see we were on the wrong river, so I went in again. The opposition was minimal."

Further north business continued to pick up, as more targets were authorized. On 24 March 1967 VA-85 crews thoroughly pounded the Bac Giang thermal power plant, but the bird of Lt.Cmdr. Buzz Ellison and Lt.j.g. Jim Plowman failed to rendezvous afterwards. They were never seen or heard from again. On 24 April Jerry Patterson led the first approved strike on Kep Airfield, another long sought-after target. The daylight raid used six A-6s – four configured for the strike, with two others doing the flak/SAM suppression mission – with ten *Phantoms* from VFs -114 and -213 serving as a mix of bombers and TARCAP. On the run in the attackers got word there were MiGs in the air but none attacked. Unfortunately, at the roll-in point one bomber – flown by Lt.j.gs. "Scurvy" Irv Williams and Mike Christian – got shot up and turned to turn back to the beach. They did not make it.

After they blasted the airfield, the XO and his B/N, Lt.Cmdr. Richard G. "Dick" McKee, turned back and settled into a protective orbit over Williams and Christian. Patterson recalls he was about two-thirds of the way around his first circle over the downed airmen when Williams called, "XO, you've got a MiG on your tail!" Says Patterson:

> "I dipped my wings in salute, then dropped down into the paddies and hauled ass out of there. Interestingly enough, none of the *Phantoms* ever saw the MiG in question."

Both Williams and Christian went into captivity.

According to Ted Been – making his second combat deployment – the *Falcons* quickly decided they were not done with the target quite yet and set up another strike for that night. The plan called for using

A pair of VA-85 *Intruders* trundle to *Kitty Hawk's* cat one during the squadron's second combat deployment, in March 1967. The *Black Falcons* made four deployments to the war zone in five years, losing 13 aircraft over that period, as well as 11 aircrew killed or missing and 3 POWs. (U.S. Navy, JOC R.D. Moeser)

four bombers and two *Iron Hand* aircraft; Ron Hays and Been would lead the first section, with Lt.j.gs. Roger Brodt and Erv Stahel, while the crews of Lt.Cmdr. Ron Waters/Lt.j.g. John Schalde and Lts. Byron Hodge and Fred Schrupp made up the second section. Lt.Cmdr. Fred Metz and Lt.j.g. Dante Kolipano were tasked with the suppression of the SAMs to the north of the target, while Lt.Cmdr. Sam Sayer and Lt.j.g. Mike Anderson handled the missiles to the south, employing AGM-45 *Shrike* anti-radiation missiles. Four *Phantoms* served as TARCAP.

"The tactic planned for the strike was to have the four strike aircraft stream the target at 30 second intervals using the new *Snakeye* (SE) retarded Mk.82 500-pound bombs," Been writes:

> "The 30 second interval was used so each aircraft could fly its own flight profile and conduct its own target acquisition without having to worry about a mid-air collision with its wingman. To our knowledge, the SE had not yet been used in North Vietnam. Every time we went on night or all weather low level strikes with normal or 'slick' bombs we would fly at less than 500-feet, but would have to pull up to 1800-feet for release to escape the bombs' frag envelope. It did not take the North Vietnamese very long to realize this, and they would cut the AAA fuses for detonation at that altitude. With the SE retarded fins – which opened at release and delayed the bomb's downrange travel – we could release at 500 feet. We had never used the SEs before. Each aircraft was loaded with 28 SEs, and the specific targets at Kep were the open revetments housing the MiG fighters. These revetments paralleled the runway orientation.
>
> "I don't remember the specific coast-in point, but I do know it was northeast of the Cam Pha area. The plan was to fly over the terrain to a point about 30 miles northeast of Kep and then fly to the southwest down a valley that roughly paralleled the main runway orientation. The over-water portion of the flight was about 130 miles, and the overland portion was about 80 miles. The target area had very few radar significant aimpoints, but runways showed up fine on the A-6A radar. The plan was for the B/N to pick up the runway, and then the pilot would swing to the runway orientation and visually drop on the revetments. There was a bright moon that night, but as it turned out it wasn't needed. All this was to be done under strict Emission Control (EMCON) conditions, i.e no radio transmissions. The aircraft would not rendezvous, but fly their own profile. The A-6A Digital Attack and Navigation Equipment (DIANE) allowed us to conduct strikes which required complex timing requirements with no outside help or visual contact with other aircraft in the flight. Retirement was to be along the reciprocal course of the run-in."

The crews executed a normal launch under EMCON conditions, but aircraft number three – with Waters and Schalde – developed mechanical problems and had to return to the boat. Hodge and Schrupp closed up and continued as the number three aircraft. Been recalls the ingress to the point where they turned down the valley about 30 miles was uneventful, then things started happening:

> "The Radar Homing and Warning (RHAW) gear started beeping and flashing. Obviously this was not going to be 'a piece

of cake.' We had stirred up a hornets' nest on the daylight raid, and they were on the alert. Off to our right we could see the muzzle flashes of 85mm heavy AAA, and about 30 seconds later we could hear the shrapnel rattling off the aircraft. It sounded like being in a tin hut during a hailstorm. At about eight miles I picked up the runways, and we swung into position to run down the revetments. The AAA was already heavy, and ahead we could see them firing a barrage fire around the base. There was enough AAA fire to make utilization of the moonlight unnecessary. We could see perfectly well just from the flashes of the AAA. It looked like a big Fourth of July celebration. However, our low level SE tactic surprised them, and all the AAA was exploding about 1,000 feet over our head. The RHAW gear was going off in fine style.

"I set up the ordnance switches and turned on the Master Arm switch. As we swung down the revetments, Commander Hays took over visually and manually released the bombs. The intervalometer was set to drop a 2,000-foot string of bombs. As we pulled to the right off the target I called 'Number one off target' (EMCON was moot at this point), looked over my right shoulder, and saw a couple of large secondary explosions. Since Roger and Erv were on their run at this time, and I just could not see how anyone could possibly make it through all this mess, I made the assumption that one of the explosions was Roger's aircraft. This fear was intensified when Roger did not call off target. As it turned out, Roger and Erv were rather busy with other things at this time. When they finally made the 'off target' call we both breathed a great sigh of relief."

Right about the time Roger Brodt and Erv Stahel started their bomb run, the North Vietnamese locked up their *Intruder* with radar-controlled searchlights, lighting up the cockpit of their aircraft like daylight at high noon. The VA-85 crews had never run into this phenomenon before; Brodt looked directly into the searchlight and developed night blindness to the point that he could not see his instrument panel. He pulled into the start of a snap roll and then pulled back, successfully breaking the searchlight lock. However, it was now dark and he was completely blind, as Been continues:

> "Erv coolly got on the intercom and coached him back to level attitude and correct bombing heading. Erv then directed Roger to the revetments and gave him the signal to pickle the bombs. The secondaries I saw were probably from the results of their bombs. Roger and Erv's aircraft was severely shot up. They had numerous shrapnel holes running from the cockpit back to the tail and in the wings. The number two weapons pylon was completely damaged. It is a tribute to the Grumman Iron Works that this aircraft survived the damage and made it back to the ship, with Roger pulling 5-8Gs to escape from the target area. It took a month to repair it."

As for the number three aircraft, it got locked up by radar-directed guns and sustained serious damage. Pilot Byron Hodge elected to abort the run. During the festivities the two *Iron Hand* aircraft did get some indications of SAM activity to the south, but neither had to fire any *Shrikes*. All of the strike aircraft returned safely to the ship, and the crews received a round of medicinal brandy, which was greatly appreciated. As a result of the attack, Brodt and Stahel received nominations for the Silver Star, Hays and Been were put in for the DFC, and Sayers, Metz, Kolipano, and Andersen were recommended for the Air Medal. Hell of a night.

The following month VA-85 and a large chunk of CVW-11 went after a big steel fabrication plant south of Hanoi. According to Patterson the strike involved all ten of the squadron's A-6s, all of the wing's A-4s from VAs -112 and -144, the *Black Lion* and *Aardvark* F-4s, and VAW-114's brand new E-2As.

> "On the way into the target the *Hawkeyes* called out that MiGs were forming up," says Patterson. "I thought, 'oh, *this* is going to be fun.' Fortunately, by the time the strike made it to the IP the MiGs were landing. There was no opposition over the target, above and beyond the 'usual' flak and missiles."

The wing hammered the target, but Patterson recalls there were a *lot* of "telephone poles" in the air, and one VF-114 F-4B was knocked down on the far side of Thud Ridge:

> "The strike group set up RESCAP; we got word the Air Force *Jolly Green Giants* were enroute, and we concentrated on bombing the gooks who were trying to get up the karst to the crew. We had perfect air cover for those crewmen, then some Air Force guy at Tan Son Nhut called the *Jolly Greens* back, saying the mission was unsafe. We couldn't believe it! We had those guys covered!"

With the rescue helos canceled the RESCAP pulled back; a day and a half later the North Vietnamese captured the unfortunate *Phantom Phlyers*.

Occasionally other missions were equally confused, with somewhat less disastrous results. Veteran B/N Ted Been recalls one:

> "On a fine spring day in 1967 Lt.Cmdr. Fred Metz and Lt.j.g. Dante Kolipano were scheduled for an *Iron Hand* mission over Hanoi to cover an Alpha Strike on a truck park, or maybe it was a rice field, or maybe a herd of water buffalo. At any rate, it was one of those well worthwhile targets that McNamara, in his infinite wisdom, used to enjoy assigning to military pilots.
>
> "Dante was a native Filipino, a graduate of Manila University in engineering, and a fine Naval Officer and B/N. However, he had one disconcerting trait, and that was to lapse back into his native tongue, Tagalog, whenever he became highly excited.
>
> "So, there they were at 20,000 feet over Hanoi trolling for SAMs. Suddenly, Dante became excited and started gesticulating wildly, pointing aft and starboard of the aircraft and rattling along in Tagalog. Of course, in the A-6 this is a blind spot for the pilot and he can't see anything in that direction.
>
> "Now Fred is getting excited and is trying to get Dante to tell him what the problem is.
>
> "'Dante, what's going on?'
>
> "Response in highly excited Tagalog.
>
> "'DAMMIT Dante, what the hell is going on? Speak English, dammit!'
>
> "Again more frantic waving, and a stream of Tagalog in response. Just then a missile exploded under the aircraft, throwing it into a series of violent uncontrolled maneuvers. Fred valiantly fought the airplane and was finally successful in getting it back on an even keel and in controlled flight, at which point Dante keyed the ICS and said in perfect, blasé English: 'Forget it.'"

VA-85 and CVW-11 returned home from their second combat cruise on 12 June 1967. Overall, the cruise was a major improvement over their first deployment, with "only" two aircraft lost, plenty of good targets, and regular employment of the *Intruder* in its designed role. Still, squadron members Buzz Ellison and Jim Plowman did not survive to make the return trip. Lieutenants Irv Williams and Mike Christian would come home, but not for over six years.

The *Black Panthers* of VA-35 also had what could be termed a successful cruise, completing five line periods through the end of June 1967. Their single loss came on 19 May 1967 when the popular Lt.Cmdr. Eugene B. "Red" McDaniel and Lt. Kelly Patterson were shot down by a SAM on a day mission against the Van Dien repair facility, five miles south of Hanoi. The flight received word of a missile launch and then observed an explosion near McDaniel and Patterson's aircraft; they successfully ejected, and the other crews observed two good chutes. Both checked in once they were on the ground – Patterson reported he had broken his leg – but rescue units were not able to penetrate into the crash site. On 22 May the SAR forces lost contact with both men and they were presumed captured; Red McDaniel would surface again during the 1973 release of the POWs, but Kelly Patterson disappeared off the face of the planet:

> "That happened during an alpha strike," says skipper Barie. "We never lost an airplane on A-6 single-mission business, which is remarkable. This was an alpha strike on the southwest of Hanoi, a group grope situation. All the alpha strikes at that time had to be led by a commander; I think it was my ops officer, Herm Turk."

When the word got back to the *Panther* ready room Barie was in the ready room, and he rushed right up to operations to keep track of the rescue attempt:

> "They did talk to somebody; there was some question in our minds as to who we were talking to, was it Red or his B/N Kelly? We later sent planes in there to make contact again, but I'm not sure we ever heard from them again. We also – there was an effort put on by the people in Saigon – sent one or two officers down there, and they got a C-130 and went up there, that was pretty hairy – but it was unsuccessful. That went on for several days, but we never got them out. The effect on the squadron was remarkably muted. Kelly was a great guy and a good B/N, was engaged to a girl in Virginia Beach. Still, they – as well as I – were aware of the loss records of the various units over there, and this wasn't *Enterprise's* first loss during the cruise. This is the real world; you get up and go back to work the next morning. Some of the guys I really, really admire – and we had several in my squadron – could've done anything else in the world if they wanted to. Instead they made back to back combat cruises without a whimper, they never flinched."

He adds a few other things that stood out from his squadron's first combat cruise in A-6s, such as "unique" ordnance loads like 10 1000-pound Mk.83s.

"We also shot *Shrike*," Barie comments:

> "I was a *Shrike* shooter during the first strike on Kep. That was kind of interesting; we got to see some great dogfights out there. I saw an F-8 cut out an F-4 and bag a MiG in front of him. Most of us went in as bombers on the subsequent strike. I got to nosing around, checked to see if we had any 'daisy cutters' (fuze extenders) in the bowels of the ship. Well, we did, so I armed our planes with them. What the hell, those North Vietnamese planes are revetted, we got some pretty good hits. I told Lefty Schwartz, skipper of one of the *Phantom* squadrons, 'Lefty, today I got more MiGs than you have in the entire war.'
>
> "One thing else I want to mention: VA-35's relationships with the flag and the ship, but I'm sure this applied to other squadrons too. Many people thought the A-6 was a miracle machine – which it was – but it was also 1958 technology, an absolute maintenance headache. One thing that was difficult to get across at times to people was that they'd send you to targets that were not really radar significant. About the best you could do was take a sectional chart, look up some coordinates, and take a look. Then you look down at the corner and note it's only accurate to 800-feet. They'd send you up sometimes to look at a railroad and it wasn't there, it didn't show up. That caused some consternation."

Enterprise completed its second war cruise on 30 June 1967 and headed back to Alameda; Attack Squadron 35 returned home to Oceana on 6 July, rejoining VA-85 and the other Navy *Intruder* units. On 14 July 1967 Art Barie turned the *Panthers* over to XO Cmdr. Glenn Kollman and moved over to VA-42 as its new skipper:

> "We came home, I got a week's leave – I think – and then I went over to VA-42 and relieved Bill Small, because he had to be somewhere else. He went to be the XO of *Forrestal* – went right to deep draft – and later commanded USS *Independence*. When I talked to my detailer he asked me what I wanted to do, and I said I wanted to be just like Bill Small ... it didn't take him long to tell me *I* was no Bill Small!"

By this time two other squadrons had assumed the duty in the Tonkin Gulf: VA-196, which deployed in *Constellation* with CVW-14 on 29 April 1967, and VA-65, which departed Norfolk with CVW-17 in *Forrestal* on 6 June. The pace of the war during this period continued to increase. For this and other reasons, both of these squadrons would end up having "eventful" cruises, as would their Marine counterparts further south.

EA-6A: The Electric *Intruder*

The A-6's most significant derivative carved out its own distinct niche in Naval Aviation history and rightly deserves a separate book to tell its story. The EA-6A was developed from a Marine requirement for a new tactical jamming platform to replace its portly Douglas F3D-2Q *Skyknights*. Grumman came up with an aircraft based on the *Intruder*, and the first EA-6A flew on 26 April, 1963.

The "new" aircraft (which was actually a modified bomber) had an eight-inch extension to its nose to house electronic warfare gear and featured an odd housing on top of the tail that would in time be called a "football" due to its shape. The aircraft could carry a variety of radar jamming and chaff laying pods, as well as external fuel tanks. The "Electric Intruder" would also retain its fuselage speed brakes and never received the wingtip brakes of the bomber version. Twenty-eight would be produced by Grumman, 13 being modified from A-6A airframes, and the remainder being built from the "keel up" as EA-6As.

The type was initially assigned to the three Marine Composite squadrons (VMCJ-1, -2, and -3), where they replaced the EF-10B, which was the post-1962 designation of the F3D-2Q. These units carried out both electronic jamming and photoreconnaissance missions for the Marine Air Wings, with the latter assignment being carried out by RF-8A *Crusaders* and later, RF-4B *Phantom IIs*.

The Electric *Intruder* was deployed to Vietnam by VMCJ-1 in October 1966, where they quickly proved to be the most capable jamming platforms in theater. For the next four years Marine crews provided critical EW support throughout South East Asia. The *Cottonpickers* were pulled out of Da Nang in 1970, only to return along with a VMCJ-2 detachment in 1972 to take part in *Operations Linebacker I and II*. Two aircraft would be lost while flying combat that year, along with two aircrew.

In 1975 the three Marine Composite squadrons were reorganized into distinct EW and recon units, as VMAQ-2 and VMFP-3. The *Playboys* of "Q-2" now supplied EA-6A detachments out of its MCAS Cherry Point, NC, base, supporting Marines world wide, as well as making carrier deployments with several Navy Carrier Air Wings.

VMAQ-2 started transition to the EA-6B *Prowler* in February 1977, completely replacing the "A" model by the end of 1979. The type continued in use with Navy VAQ-33 as a fleet EW adversary aircraft, as well as with reserve squadrons VAQ-209, -309, and VMAQ-4 until the last was retired in 1993.

For a largely unheralded airframe the EA-6A had at least one everlasting claim to fame, that being as the sire to the incomparable EA-6B *Prowler*, which continues as the world's leading electronic attack platform into the 21st Century.

The EA-6A "Electric *Intruder*" established an enviable, and little appreciated, combat record in Vietnam with the Marines. In time the type was replaced by the EA-6B, and it ended up with the Reserves and VAQ-33. *Firebird* GD14 is seen flying over Southern California with a full load of jammers and chaff pods, as well as a centerline fuel tank in 1980. The type never got the bomber's wingtip speed brakes and retained the fuselage boards up to the end. (Bob Lawson)

CHAPTER THREE

Vietnam, 1967-1968: Marine Intruders in Country

In March 1967 the *Nighthawks* of VMA(AW)-533 deployed to Chu Lai Airbase, adding a second *Intruder* squadron to Fleet Marine Force units in the Republic of Vietnam. Upon arrival, the squadron joined MAG-12 – in and around several *Skyhawk* operators – and prepared for war. Among those who made the trip was 1st Lt. Charlie Carr, the former enlisted man who remembers the field and operations the *Hawks* were involved in:

> "Chu Lai had a north-south runway, 9,000-ft long including overrun. This was the prime runway for the A-6s and F-4s; the A-4s utilized the 'tin' runways and SATS (catapult). During some monsoon wind conditions, A-6s and F-4s trapped on the tin runways but did not utilize the cat because of weight. We flew *Rolling Thunder* and I Corps CAS missions utilizing VMCJ-1 EW support for the former. The *Rolling Thunder* missions included targets from Route Pack 1 through Route Pack 6 (Dong Hoi across the DMZ through the Chinese border/buffer zone)."

During this period of growth in the theater there was a major turf battle between the services over who owned and controlled what, and the Marine *Intruders* were right in the middle of it. One particular point of conflict was over which service should control aircraft assets inside South Vietnam; the Air Force felt it should have operational oversight for *everything* through *Seventh Air Force*. The Marines reasonably felt they should control their own aircraft, and continue their primary mission of supporting grunts in the field.

"Our folding into the USAF/USN/USMC missions was not a problem," Carr comments:

> "The fight was over who would 'own' or have opcon/adcon over the two USMC A-6 squadrons. The Navy wanted us to bounce at Cubi and go aboard a boat, *Seventh Air Force* wanted us to fly night *Rolling Thunders* as part of their frag, and the USMC wanted us to fly day/night CAS in *I Corps*. The Air Force and *III Marine Expeditionary Force* won the fight; we dedicated our resources to *Rolling Thunder* at night, and any excess sorties were fragged by III MEF for close air support. The only time CAS took priority was during the siege at Khe Sanh."

Over the next few years, the *Batmen* and *Nighthawks* literally "wrote the book" on *Intruder* operations in country. According to Carr, the crews quickly determined what worked ... and what positively *did not* work:

> "Both -533 and -242 wrote new training and readiness manuals in regard to night ingress, strike, and egress tactics. Generally,

While the Marines didn't have to put up with night recoveries at the boat like their Navy brethren, they did have to deal with things like Vietcong sappers and mortar attacks. VMA(AW)-242 deployed to Da Nang and, in time, had their *Intruders* placed in revetments for protection. (Wolfe, via Steve Dumovich)

> we validated the requirement for low and as fast as possible maneuvering in a high threat area; when stuck in a AAA barrage, fly wings level as fast as possible through it; jink out of tracking fire; and both pilot and B/N fly the aircraft when evading SAMs, especially at night. On the other hand, using sophisticated maneuvers such as a barrel roll starting at 500-feet AGL in order to defeat a SAM usually put the aircraft in a good SAM envelope ... or the aircraft probably hit the ground on recovery."

Again, this was training, tactics, and evaluation under combat conditions, but the two Marine A-6 squadrons pressed on. On 23 March 1967, -242 suffered its first loss, albeit under unusual circumstances: during an early morning takeoff at Da Nang: one of the squadron's aircraft collided with an Air Force C-141A. During the takeoff roll the crew noticed the C-141 crossing in front of them, and pilot J Fred Cone attempted to abort, but they plowed into the right side of the big transport's cockpit area. Both aircraft burned; the A-6 came to rest inverted, but fortunately the crew was able to escape with minor burns and, in Cone's case, a broken kneecap. The *Intruder* was a strike.

According to former Cpl. Steve Dumovich, who served with the *Bats* during this period as a fire control technician, ops at Da Nang (aka "Walnut Hill") were reasonably "routine":

> "The VC/NVA had rocketed and mortared us a few times, we moved into our new hangar, and we were into a fairly stable routine in the squadron. By this time the maintenance shops were pretty comfortable, working twelve on and twelve off with an occasional 'day off' about every ten days or so. Most of our launches were after dark, so things around the flight line were pretty active all night long. It was summer, and working nights wasn't really so bad. During the daytime, the sun would heat the aluminum skin of the aircraft to the point that you almost needed gloves to touch the hot metal.
>
> "The other advantage to working nights was that most of the 'staff' (Staff NCOs) worked the noon to midnight shift, so you could get some work done without some Maintenance Control Gunny asking you, 'Is the bird up yet?' every five minutes. It was also during this period that the squadron was experiencing some fairly good system availability on several aircraft."

Unfortunately, VMA(AW)-242 suffered their first combat loss on a systems hop the night of 17 April 1967. The crew of Maj. James McGarvey and Cpt. James Carlton were listed as MIA after hitting their planned penetration point 10 miles south of Vinh at 500-feet. A following aircraft saw a flash, and there was no further communication with the crew. Vinh was notorious throughout the war for the intensity of its air defenses; it effectively served as the "gateway" for further adventure in Route Packages IV, V, and VI. Receiving an assignment for missions north of Vinh usually brought about shortened breaths and white knuckles, as related by former *Bats* pilot Cpt. Charlie Clark:

> "The 17th of April started as a great day. I had just received word that I was going to Hong Kong for R&R on the 19th of April. I was elated, as the 19th was my 25th birthday, and the thoughts of ushering in a birthday at Da Nang's 'O' Club left me cold! The day could not have been better, that is, until I got my first look at the flight schedule: 'Clark/Springfield, launch 11:30, RT.' A night flight anywhere in North Vietnam was not what I considered fun! When crews saw RT, or Rolling Thunder, it had a way of creating anxiety!
>
> "Captain Ray Springfield was my regular bombardier/navigator. Ray had the nickname of Magellan, for his uncanny ability to navigate with the system. I felt that he could put a bomb on a flea sitting on a dog's back while it was hiding under a bridge. The ability, however, did not carry over to out of the (radar) boot or visual hops. Ray was reputed to have a brain/finger coordination problem that caused his finger to move on the map about 10 knots faster or slower than the aircraft. But then, I had a reputation of having a hard time talking on the radio and keeping the plane in the air at the same time! It was this match that allowed Ray and me to see some interesting country, off the beaten track.
>
> "Once inside the briefing hut, Ray and I were pleasantly surprised with our mission draw. The butterflies caused by route package 4, 5, and 6 flew away. We had drawn an interdiction point north of Vinh and south of Thanh Hoa, with a secondary mission of road reconnaissance in route packages 1 and 2. When Ray and I looked at the target pack, the aerial photograph revealed truck tracks on both sides of the river. Being good Marines, we could only assume Washington had us bombing an underwater bridge. With a target of this quality, gone were the usual intense negotiations of the pilot and bombardier/navigator. Ray always started out with the best radar look angle. I always started out with a run-in line over the least amount of guns! Through negotiations and compromise we would arrive at a mutually acceptable run-in direction.
>
> "In fact, it was such a piece of cake target, Ray talked me into a fake run at Dien Chou, which was a quiet little town just north of Vinh, then breaking off to the north and proceeding to our primary target. This was a maneuver I had been taught in flight school and later, as a flight student, I was able to practice with low level passes on migrant farm workers in the Rio Grande Valley. From the classroom to real life – hot damn!
>
> "The pre-flight start and system alignment was uneventful. With 18 500-pound bombs and six 19-shot packs of 2.75 rockets armed and ready, we hit our launch time perfectly. With the gear sucked up in the wheel well, we climbed to flight level 200 (20,000 feet) and leveled. Ray checked in with our Air Force controller and updated the aircraft's system on Fire Island. Ray gave me the pushover point, and I reached down and flipped the external master switch, killing my navigation lights and rotating beacon. I was ready to make this mission history, as I upped the RPMs to 100 percent. My object was to get going as fast as I could; in fact, if I could have gotten one more RPM or knot out of the aircraft, I would have put my foot on the throttle. We leveled at 1500-feet. I had long ago stopped the 100-foot run-ins for a simple fear for my life.
>
> "As I crossed the beach I gave a 'Feet Dry' call, and all hell started breaking loose. You would have thought I had sent an invitation asking them to stay up and shoot at us. For a moment, I thought we were number two in a coordinated attack! I knew it was going to be a bad night when I received heavy weapon fire from the kindergarten/playschool in the area! I followed my first impulse, which was to get the hell out of there, as it was evident the North Vietnamese on the ground were not impressed with my tricky little move. While proceeding to our primary target we were receiving ground fire from areas that we didn't know had people. With the target acquired, we did a quick salvo of the 18 500-lb bombs and retired to the ocean to let the people on the ground quiet down. In fact, we decided we'd had enough of that area, and headed directly for Route Package I for a quiet evening of truck hunting.
>
> "It was a beautiful night with a quarter moon, a few puffy clouds, and a little ground fog forming. We eased down the highway and, as we approached the split in the road north of Dong Hoi, Ray picked up a mover. I pushed over and followed the pathway on my VDI. When in range I committed, and two 19-shot packs of rockets left the aircraft with a blinding flash. As I followed the rockets down it appeared as if they had done a U-turn and were heading straight for our aircraft. The ground fire streaking by our canopy scared me so badly I jerked the plane hard enough

to get Ray to look out of his boot. His only words were 'J-E-S-U-S CHRIST!' He went for the boot and I went for the water.

"I told Ray I'd had enough excitement for one night and we were heading home. As we climbed to 10,000-ft heading for Da Nang we popped our oxygen masks. It was about that time the aircraft went through a small puffy cloud. The aircraft was immediately bathed in a white light. My first thought was FIRE! A moment later, Ray suggested I turn off my taxi light.

"As it turned out, I had flown the entire mission with the taxi light – which is located on the nose landing gear door – shining straight down. The next day we wrote up a suggestion that the taxi light be added to the external master light switch. This was done on later models and retrofitted to our A-6s. Now, when Ray and I talk about our A-6 days with our grandkids, we say we were an integral part of the A-6 *Intruder's* development!"

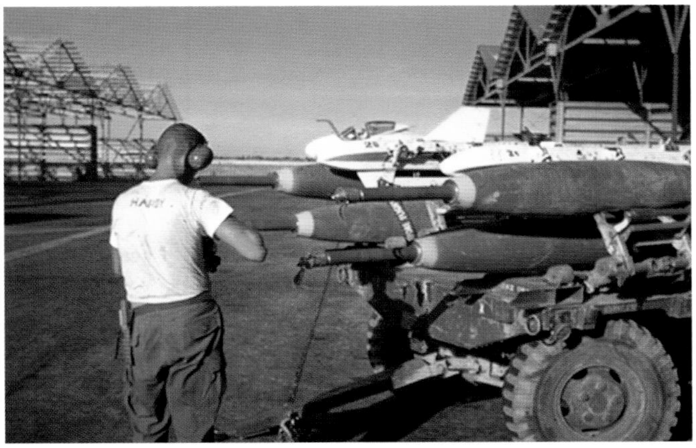

The ground crews rarely got the credit they deserved for keeping the A-6 in the air. A Marine ordie tends to the fuses of a rack of 500-pound bombs at Da Nang. (Steve Dumovich)

Good Days, Bad Days

Operations by both squadrons continued through the late spring of 1967 and on into the fall. Over time the Marines continued to get additional tasking Up North as their *Intruders* became an integral part *of Rolling Thunder*. However, the Air Force still retained operational control of *Rolling Thunder* strikes launched from Vietnam and Thailand, to the point of dictating tactics and numbers of aircraft. Many Marines questioned this arrangement, particularly as it did not take into account the A-6's particular capabilities. Pilot Joe Hahn recalls these circumstances in discussing one particular mission; notably, their target that night was near the POW camp at Son Tay:

> "The tale begins like all good *Intruder* tales, on a dark and stormy night when all the *Skyhawks*, *Phantoms*, and *Crusaders* were tied down on the line. (Captain) Dick Tinsley and I were fragged for a two-plane mission up the Laotian track to the Red River Valley to strike a 'Suspected Ammo Storage Area.' The target was on the south side of a reservoir approximately 30 miles west of Hanoi. Each *Intruder* carried 18 Mk.82s and two 300-gallon drop tanks, and the flight was supported by a VMCJ-1 EA-6A electronic countermeasures *Intruder*. We were also supported with SAM and flak suppression from two Air Force *Wild Weasels*, which at the time were the two-seat variant of the F-105 *Thunderchief*.
>
> "The flight lead was Maj. Andy Martin, and his bombardier/navigator was Cpt. Jack Butchko. The plan was a high-low-high section up the mountains at 20,000-plus, with a let down into the Red River Valley and delivery at 1,500-AGL on a west-to-east run-in heading (103 degrees magnetic). Because of the independent weapon systems in the two A-6s, our plane (Dash 2) would take a one minute separation on the flight leader as we began our let down over the mountain.
>
> "It was common wisdom within the Marine A-6 community that one plane had a far better surprise capability than two planes in tandem, especially at night, and therefore had a better probability of survival. The reality was that Dash-2 on a system attack routinely encountered far more defenses (AAA, SAMs, and small arms fire) than the flight leader. A single *Intruder* could usually get in and out of the target without much opposition. Yet, even though this grim fact was well known to the planners, they continued to launch two planes on Route Package VI strikes because of the misperceived need to coordinate with other support groups and prior experiences with visual bombers. At the pilot level we believed that a single A-6 with a *full system* and working ECM gear had a better chance of surviving a low-level attack in the Red River Valley, even without EA-6As, and especially without *Wild Weasel* support. However, certain political pressures prevailed, and total bombs on target per strike seemed to be an important objective. This inane practice continued until Captains Hugh Fanning and Steve Kott became its inevitable victims during a strike in October 1967."

On 31 October Fanning and Kott had disappeared at about 0200 while flying as the second plane in a two-ship mission to sew mines on the Red River near Hanoi. The first A-6 took heavy antiaircraft fire and SAMs, described as a "solid wall" by the time the second plane rolled in. A supporting *Cottonpicker* EA-6A saw a large explosion in the approximate area where Fanning and Kott would have been; no trace of them was ever found.

"As Dick and I joined up on Andy, climbing out over Khe Sahn, it became clear that our ECM gear was 'down,'" Hahn continues:

> "We had no S-Band or X-Band alerting, and were unsure of our jamming capability. Although it was a problem, it was not big enough for us to cancel our part of the mission. We concluded that Andy could advise us of active SAM sites as we got closer to the target, and we would attempt to saturate the run-in with chaff and hope the EA-6A would be effective in its role. We also had second thoughts about the *Wild Weasels* being active in the area, only because they tended to alert the neighborhood that something was up."

Dick Tinsley agrees, adding:

> "The requirement to work with the *Wild Weasels* did not sit well with us. Generally, we could be sure that coordinating with the *Wild Weasel* F-105s would mean that after the *Wild Weasels* entered the area, everyone on the ground would be wide awake with a whetted appetite. That's exactly what happened."

"In any event," Hahn continues:

> "It was perfect A-6 weather. There were substantial low clouds on the east side of the mountains and a right, moonlit night above. The attack plan had Andy running in at 500-ft altitude separation and breaking to the left after release, while we would start our run-in one minute later and break to the right after releasing our ordnance. The true importance of this mission was not in the 'suspected ammo dump,' but the fact that the Son Tay POW barracks were on the north side of the reservoir, and we had hoped to let our guys know that we were still thinking of them. The plan was simple. We would make a straight path attack with a long, narrow, 18-bomb footprint about a mile south of the prison.

"The mission was routine until we were approximately 20 miles from the target (at 350 KIAS, that's about 5.9 nautical miles per minute, or 3.5 minutes) and beginning our run-in, when Andy informed us the SAM sites seemed to be up and searching. Visibility was extremely limited because the run-in kept us in and out of the lower clouds. Andy pickled and broke left. Now it was our turn. We began our final one-minute run-in, and the SAMs began to fly. I could see the glow of what was clearly a missile in the air through the cloud layer, and I was convinced they were on us and not Andy's *Intruder*. Dick was doing all he could do – with his head in the boot, trying to keep the radar cursors on the target – because of our great concern that a miss could be disastrous on Son Tay. Nonetheless, it became clear, with less than a minute to go, that serious jinking and chaff were key to our immediate survival. Our purpose was to boost the POW's morale, not necessarily join them.

"At one point, I broke hard to the right and a SAM exploded directly below our *Intruder*, spraying the bottom of the aircraft with missile fragments. At least two other missiles undershot us and exploded a safe distance away. We thought we might have lost a 'lock' on the target; however, leveling out Dick advised me he still had a 'radar lock' on the target and a proper run-in heading, which allowed me to continue the planned straight path attack."

Tinsley comments:

"I had my head in the boot as we came over the mountains and easily picked up the target area near the reservoir. The radar predictions had been extremely accurate because of the size and configuration of the body of water. Also, there was a significant radar target on the south side of the reservoir that allowed me to get a lock earlier than expected. The high resolution of the radar aim point on this target made us comfortable that we would be able to accomplish the mission, even with the proximity of the Son Tay Prison. As we closed on the target and switched to the attack mode, Joe began violently maneuvering the aircraft. I never lost the radar lock, but was concerned that I would. Suddenly there were bright flashes all around the aircraft that lit up the inside of the boot and were followed by some buffeting. When Joe rolled out, we were within two miles of the release point, the lock ws still good, and we were in attack and the bombs came off as advertised.

"I can remember Andy or Jack calling as they came off the target that there was antiaircraft and missiles in the air, but 'It's all behind us.' I remember thinking at the time, 'Well, no shit, guess where *we* are.'"

Pilot Hahn lined up, pressed the attack, and completed their target run:

"After what seemed like the longest fifteen seconds (mile and a half) in the history of the world, the bombs released and we made our quick, hard break to the right to start our climb out over the mountains, only to experience a rough running left engine. The missile had obviously done some damage; I presumed there was a good probability it would quit, so I retarded the throttle, punched off the racks and drop tanks, advised Andy of our trouble, and began a slow climb over the mountains. We exited the target area and got as far from the SAM ring as possible. Because of the lack of power, we had to make a 360 in order to clear the 7000-foot mountains and climb out to 15,000-feet for the ride home.

"During the climb out, after my pulse rate slowed below 200 and my dry mouth began to permit conversation, I asked Dick for a drink of his survival water. He handed me his baby bottle. I started to take a slug, smelled something funny, and realized, as it poured in my mouth, that he had handed me a baby bottle full of gin. When I asked what was in the bottle, Dick simply announced, 'If we go down and I am going to get captured, at least I'm going to have one last good buzz.' He also had the bottle of water. He just grabbed the wrong one.

"All the way back to Da Nang we watched the damaged engine for any sign it might suddenly come apart. We made a precautionary single-engine approach (retarded left throttle), landed, and refueled. After shutting down, we inspected the bottom of the aircraft and were surprised to see how few holes were actually in the airplane (good armor plate). The next day, the BDA showed we had come as close to the 'suspected ammo dump' as anyone had expected. More importantly, it was well within earshot of the Son Tay POW barracks and the American prisoners. Over the years, Dick and I have often wondered whether the mission had its intended morale boost for the POWs."

During this period the *Bats* and *Hawks* were regularly tasked with flying specialized hops under the direction of ground controllers. Considered by many to be a miracle of technology, this method employed the AN/TPQ-10, with the A-6 crews flying a set course and altitude under the guidance of the ground personnel. Lt.Col. Earle Jacobson, the *Bats'* third CO, flew plenty of them with his B/N Cpt. Ed Weber:

"Ed Weber and I were scheduled for one of the ground radar controlled bomb runs (TPQ) directed to VMA(AW)-242 each night. These missions were an unexpected economy of assets utilization, and certainly not done with purposeful intent. The 'economy of utilization' was the ability of the squadron to use the non-system birds (aircraft) while providing a long time on station and superb ordnance carrying capability for the TPQ controllers. The favored bomb load was 28 Mk.82 bombs.

"The TPQ missions were rotated through the air crews, just like the 'Tally-ho' and 'Rolling Thunder' missions. The 'Tally-ho' (TH) was armed reconnaissance flights over the southern tier of North Vietnam. 'Rolling Thunder' was for the more heavily defended areas in the northern tiers. The 'in country' TPQ missions presented the mind numbing hazard of bombing controlled from the ground, ie., 'maintain heading of 270, airspeed 250 KIAS and 20,000-feet.. .5 miles to target ... standby, standby ... mark mark' ... 'turn right to 350 ... maintain 250 KIAS and 20,000-feet ... your target is now 15 miles ...' etc, etc. This went on for however many runs were required or desired in most combinations of 1 to 28 bombs. At the conclusion of the run we were given the map coordinates, target description, and unit supported. 'Coordinates 12345678, suspected harboring sites, in support of III MAF' was a typical communication. The target descriptions were creative if not humorous, ie, 'fighting holes,' 'elephant tracks,' 'suspected vehicle storage sites'; in other words, no one really knew what was there. Kind of like the kid's game 'Battleship,' where you might get lucky and do some damage. I knew with certainty that I hit the ground."

On 7 July 1965 Jacobson and Weber had their own "dark and stormy night" TPQ mission. They hit the target all right, but lost their A-6 *after* the fact while sitting on the ground in the fuel pits. Jacobson writes:

"The preflight briefing was to a word 'brief.' Ed and I confirmed the mission number, controller frequencies and call sign, bomb load and fuse settings, fuel load, switch and knobology, emergency procedures, launch times, anticipated recovery times, current and forecast weather, and divert fields. Our assigned aircraft was DT-5 (this was 242's second DT-5. The first was lost in a mid-air collision over Virginia prior to deployment).

"The aircraft preflight took longer, but checking 28 fuses, arming wires, and bomb racks takes time. The preflight start, taxi, and takeoff were comfortably routine. Hand-offs among the various agencies was routine, as was the final requirement of positive radar contact with the TPQ controller. The bombing runs were again routine, including the 10% that did not clear the racks. I can say with pride, we destroyed more 'elephant tracks' and 'fighting holes,' sometimes complete with descriptions as 3'x3'x5'. I guess the 3-ft depth addressed the needs of the 'vertically challenged' VC. The return to Da Nang was routine, with a TACAN/GCA recovery. After clearing the runway DT-5 was taxied up to the refueling area.

"The refueling area was actually a Marine Corps innovation for refueling aircraft at forward sites without 5,000-gal tank trucks or permanent fuel storage and transfer facilities. The TAFDS (Tactical Air Fuel Dispensing System) consisted of six 10,000-gal interconnected rubberized bladders, protected by shallow earthen berms and sandbags. There were several refueling points adjacent to the flightline of each squadron in the MAG. Modern jet fighter and attack aircraft have single point refueling systems that require aircraft to have electrical power to sequence the various internal valves. Therefore, the use of a limited inventory external power unit was required. The alternative and preferred method was 'hot refueling,' ie, refueling was done with the engines running and aircrews on board, and when refueling is completed the aircraft is taxied to the parking area. Towing the aircraft becomes unnecessary, which relieved the workload on the always short-handed GSE shop.

"Ed and I were directed to the refueling spot, stopped, chocked, and the routine of refueling started. It was about 2200. For the aircrew, it meant cleaning up the cockpit, unbuckling, and starting to unwind. For the ground crew it meant hooking up and fueling the aircraft, and insuring that all the necessary safety pins were placed in the landing gear and bomb racks. The routine ended with what we assumed was the first rocket or mortar round impacting in the fuel pit area. The announcement came as one hell of an explosion and flames. There was no time to think, just react. My immediate reaction was to secure the engines and abandon the aircraft, but I couldn't go down the pilot's boarding ladder because of the burning fuel beneath the aircraft. Ed's reaction was, of course, as immediate, but he could not go down his side either because of the flames below. The result was a mid-air, or rather a mid-cockpit collision, because I was going out Ed's side and Ed was equally intent upon coming out my side. Size prevailed; Ed was bigger, and we opted for my side.

"Ed went out and over the left wing. I started to follow, but paused because the flames seemed less intense. In fact, I felt I could and did descend the port-boarding ladder, hastily I might add. Ed went off the left wing tip that was some 6-7 feet above the tarmac. I sprinted through puddled burning fuel to a safe dry place well beyond the aircraft with my heels afire, but fading fast. There was then the realization that maybe only one round had hit. But why? Leaving that thought as quickly as it came, our attention was now focused on several of the ground crew fighting the fire, and more courageously, removing fuses from at least three 500-pound bombs to preclude setting off a high-order detonation. Very brave Marines, doing what Marines do best – selfless devotion to duty."

DT-5 burned, but the crew of Jacobson and Weber survived their unplanned ground evacuation without any undue injury. A subsequent investigation determined the refueling hose coupling was not properly seated to the aircraft receptacle, and when the fuel was turned on, the hose broke loose. Raw fuel went into the port engine, resulting in the explosion and fire. As Colonel Jacobson recalls:

152951/DT5 caught fire while hot refueling in the Da Nang pits on 7 July 1965. Both crew egressed successfully, and the ground crew put the fire out. Amazingly, the aircraft would fly again. (Steve Dumovich)

"The one lucky round theory was shot to hell. However, the real rocket attack occurred a few days later when the hangar and the flightline took several direct hits from Soviet-built 122mm rockets. Those magnificent young Marines shut down the TAFDS fuel pumps, successfully fought the fire, defused the bombs, and watched Ed's aerial escape and my 'winged mercury' dash out of the flames ... the most troubling thing for me has been the fact that I cannot recall the names of those Marines on the ground who acted so heroically."

Remarkably, the aircraft (BuNo 152591) was eventually repaired and returned to service, although for some time it sat forlornly, covered in what the colonel describes as "the foulest-smelling protein firefighting foam imaginable." Steve Dumovich saw the accident, and later participated in the aircraft's "recovery":

"On the night that DT-5 got flame broiled in the pits I happened to be working and saw some of the excitement. The fire did not last very long, but it did substantial damage to the area around the radome and the nose wheel well. The plane captain who was refueling suffered some second and third degree burns and was medevaced to the *Repose*. After the crash crew secured the scene Sgt. Dave Earnes and I were given the task of removing as much of the fire control equipment from DT-5 as was possible. The aircraft was still sitting on the spot where the fire occurred, and the entire front of the aircraft was covered with foul-smelling, tan-colored, protein-based firefighting foam. The foam is made as a byproduct of the meat packing industry, and although it works well as an extinguishing agent, it does maintain a strong odor and eventually dries into a sticky black residue.

"We eventually finished up our assignment around 0200 or 0300. Because of all the foam we had gotten on ourselves in the process, we were sent back to the living area to get cleaned up. Later that morning DT-5 was towed into the hangar and other shops removed their gear and equipment. The following day DT-5 was towed to the H&MS-11 flight line awaiting disposition. Because airframes were in such demand at that time, DT-5 was scheduled to be taken back to the states for a final decision as to keeping or striking the aircraft. It looked pretty pathetic, sitting there all alone on the open flightline and unprotected by revetments. Lots of MAG 11 people just walked around the aircraft during the days immediately after the incident."

A week later Da Nang came under a heavy rocket attack; fortunately, it was not followed by an actual attack on the field and the Marine and Air Force positions by VC sappers. Air Force AC-47 gunships went into action and started hosing down the suspected launch sites, and the following morning the Marines took stock of the damage. Fortunately, it was relatively limited; several casualties, some buildings knocked off their foundations or blown over, and a lot of shrapnel and fragments all over the place. Aircraft losses were two VMF(AW)-232 F-8Es destroyed and two VMA(AW)-242 A-6As damaged. The *Intruders* were parked in a hangar that took a direct hit; one aircraft suffered a destroyed left wing, while the other ended up with a heavily damaged *right* wing.

"So how does this relate to DT-5?" Dumovich continues:

> "Well, here sits two aircraft with non-repairable wings that will be (down) for several weeks. But there sits DT-5 with its melted aluminum and fiberglass and two perfectly good wings. In a cooperative effort between H&MS and our own metal, hydraulic, and electric shops, the wings were pulled and replaced on the damaged aircraft. It was only a few days, and 242 was back in business with only one aircraft short vs. three. Because of a previous loss of another DT-5 prior to deployment to Vietnam by 242, it was decided to number the replacement aircraft DT-00 and skip DT-5 for a while.
>
> "A final word on this story is this. I was going through the list of all the A-6 *Intruders* and where they are resting, and to my surprise, 152591 – the 'Crispy Critter' – is out there in the dry desert near Tucson, Arizona, standing in 'War Reserve.' Sad but fitting place for an old warrior."

Score One for the Good Guys

Over at Chu Lai the crews of VMA(AW)-533 were flying a similar mix of missions, including TPQ strikes and participation in the ongoing *Rolling Thunder* campaign. As with their sister squadron over at Da Nang, the victories were regularly balanced by losses; on the night of 27 August 1967 a *Nighthawk* A-6A went down over North Vietnam. The crew of Cpt. Paschal G. Boggs and Maj. Vladimir H. Bacik were heading for their target north of Haiphong and had just passed over the beach when communications ended. Both apparently ejected, but were listed as MIA.

Other missions were more uneventful, although the unexpected could and would occur at any time. Cpts. Bill Kretschmar and Don Hiltbrunner found this to be true early one morning. As the pilot Kretschmar relates, it would become, "the most gratifying single mission I would ever fly in two 'Nam tours":

> "We launched from Chu Lai in the middle of an early July 1967 on a no sweat mission into an area just north of the DMZ in North Vietnam. Time on target was 0330. It was supposed to be a simple TPQ mission ... the target was a 'suspected SAM site' several klicks north of the DMZ. The only thing that might get our adrenaline flowing was the fact that we would be making our pass at a level 20,000-ft, frozen on course and speed, no jinking. The only ECM protection we would have was our own internal black boxes. If there was more than a suspected site down there, we would be making the North Vietnamese shooters' job that much easier.
>
> "This wasn't your typical 'dark and stormy night' of which hairy tales are made. It was a clear night, and as a matter of fact, this isn't a hairy tale, at least from our perspective. The TPQ controller set us up on a run in heading and gave us occasional small heading corrections. Our ECM gear was quiet, indicating an apparent lack of radar activity by the enemy gunners. Then, as we approached that point in the sky where the TPQ controller wanted our 28 500-pounders to release he gave us a 'steady ... standby ... standby ... Mark-Mark.' I pressed the bomb release button on the stick, and the A-6 rocked and rolled a little and tried to climb, unburdened of the seven tons of ordnance it had moments ago carried. Now free to do my own thing, I banked the *Intruder* toward our egress heading. We saw the flashes of the bombs exploding in the impact area on or near the target area.
>
> "Moments later, we heard the high pitched wailing of an emergency beeper. This was followed by a voice calling 'MAYDAY MAYDAY MAYDAY.' We answered up and were asked over the emergency radio frequency if we had just dropped some bombs. 'Affirmative. What's up? Who are you? What's going on? Where are you at?' I asked. A nervous but somewhat reassured voice explained that he was a pilot and had ejected from an A-4 several hours earlier, 'Hit by a SAM missile ... got a broken leg and other injuries ... smoke from your bombs passing over right now ... been unconscious ... bombs, not far away, woke me ... can you get me out of here?' I replied. 'You bet, buddy. We'll go out over the Gulf and call for a *Jolly Green* (USAF SAR HH-3E). Be right back. Save your radio. Don't call me, I'll call you.'
>
> "We got a good fix on his position and passed it along to Da Nang. They advised the SAR would be there at first light, maybe a couple of hours away. I told them I would hang around until then to assist in the location. I did not know if I would have enough fuel to hang around until the SAR arrived, but the guy on the ground was scared and we couldn't just leave him there. We had dropped all our bombs, so the only thing we had to give at this point was our moral support and the sound of a pair of J-52s."

Kretzschmar and Hiltbrunner spent the next few minutes moving back and forth between the Gulf of Tonkin and the downed *Scooter* pilot's position, monitoring the rescue effort while reassuring their downed counterpart. Several times the A-4 driver attempted radio contact with the *Intruder* crew, primarily to reassure himself that they were still there. The A-6 pilot would move back and in a firm – but friendly – manner, remind him to save his radio for the rescue.

"The minutes went by for our newly found best friend," Kretzschmar continues:

> "And again and again the poor bastard didn't want to let go of us, his only contact with the real world. He did not really have many options at this point, and that reality was becoming very apparent as time passed. We couldn't blame him, and we really couldn't do anything more for him than just be there. Several times we flew back in to reassure him and to admonish him about his radio discipline, or I suppose his lack of radio discipline.
>
> "Our fuel was getting low, actually very low. Finally, from our altitude we could detect the sky lightening in the east. We began asking ourselves where are these guys (*Jolly Greens*)? Then came the call. The SAR was inbound for the rescue. I asked if we could be of any help in locating the guy. 'Not yet,' came the reply. 'Let's see what we got here.' 'God, these guys are cool,' I thought. We got word they were in contact with him. And then came the really good news: they had him visual. From this point we could only get in the way of these courageous professionals. We wanted so badly to stay in the area until he was in the helo and on his way home, but we were almost running fumes now. If we hung around any longer the *Jolly Green* might have to pick up two more downed aviators on his way back. We had held this guy's hand for a couple of hours and didn't want to let go until we were certain he was safe. But we had to, so we headed our A-6 for Da Nang Air Base. Prior to landing we got word that they had him. Rescue successful! *Jolly Green* and *Sandys*, 'Sierra Hotel ... Shit Hot.'"

The tired, but elated, crew recovered at Da Nang, but:

"Now a new emotion erupted. With the rescue complete we became infuriated. Infuriated that this A-4 driver, from a squadron at our very own base at Chu Lai (VMA-311), had been downed by a surface-to-air missile at about 1900, eight hours before our launch. And nobody knows about it, not us, not the TPQ controller, not our own group G-2, nobody. We didn't know that by virtue of the shootdown of the *Skyhawk*, the SAM site was no longer a suspected site but a hot active site manned by some pretty good shooters. Even more maddening was that no one saw fit to tell us to keep an eye and ear out for the downed pilot. We damn near killed him. That would have been the ultimate irony; poor guy gets nailed by an enemy SAM, survives the shootdown, and then the good guys from his own base come along and finish him off. It could very well have turned out much different than it did.

"Where were the intelligence guys, or ops guys, or anyone from his squadron, to let the *Intruder* guys know about this fellow? They were well aware that we A-6 pukes were always flying around up north at night. We raised hell about it later, but didn't get any satisfactory answers as to who dropped the ball. Fortunately, the story had a happy ending, but it could have had a tragic ending for a *Skyhawk* pilot for sure, and perhaps for a couple of *Intruder* types, as well. I was able to talk briefly by phone with the rescued pilot before he was medivaced out of country. The A-4 pilot was Maj. Ralph Brubaker, and he was happy to be alive and back in the hands of the friendlies. We were just happy to have been in the right place at the right time to help bring him back alive, even if he was a little damaged by the events."

A few months later Kretschmar rang up another evening for the *Nighthawks*, this time while flying with Cpt. Larry Friese as his B/N. He later wrote about the mission, describing it as:

"Take five dumb bombs, one dumb (non-system) A-6, one dumb target, and one dumb way to fight a war. Add a large portion of luck. What's the result? An averted disaster. But, leaving a crew who will be able to fly another day and an *Intruder* still around to fly in harms way.

"Back in September '67 'someone' found some old World War II 2000-pounders lying around somewhere. What good were they? What could they be used on? What airplane could carry them? 'Someone' had the answer. 'Someone' might have been a McNamara 'Whiz-kid' at the Puzzle Palace, or maybe just a lowly Wing staffer. 'Someone' said to strap those heavies onto an A-6 and go tunnel and cave busting where the Viet Cong take shelter. The 500-pounders don't make much of an impression; they merely bounce off. These behemoths will do some real penetrating, some real damage, 'someone' thought. That was the plan."

According to Kretschmar, he and his B/N were the "honorees" for the experience. They were given an *Intruder* with a single 2000-pound bomb on each of the five stations and would dive bomb the pernicious enemy into submission. Never mind there was not accurate mil setting data for the old weapons, nor that no one knew of a proper release speed or altitude, just go forth and be fruitful. Kretschmar continues:

"So off we went. Getting airborne was no problem. This was a mere five tons of steel and high explosives. The A-6 can easily slip the surly bonds of earth with seven tons (28 500-pounders). Our FAC turns out to be an Air Force O-6 (colonel), pretty high ranking for this job (could *this* be the 'someone?'). He was out of DaNang and was flying a little Cessna O-1 *Bird Dog*. He marked our target (tunnels) with smoke. We set up for our first run. Since this was 'Indian country' I wasn't too concerned about missing the target on the first drop. From the first impact spot I could easily adjust something (airspeed, release point, mill setting, dive angle) so as to be more accurate on subsequent runs. We pulled up after the first run, and the FAC was ecstatic, initially. 'Right on target, Armorplate. Right on target' he shouted over the radio. Then this: 'But it didn't go off. It was a dud. Check your armament panel. Put the next one in the same place.' The armament switches were all as they were supposed to be. But the second run had the same results – a dud. Our FAC was getting pissed, and so were we.

"Third run. 'Right down the pipe, Armorplate. Good job! No dud that time. That got their attention. Put your next one right on top of that.' I managed to do just that, but it too failed to explode, as well. That's awful. One out of four. One bomb left. Hope it's got some life in it. I set up for the dive. Banked hard left, kicked left rudder, and let the nose drop into a 30-degree dive. The airspeed is building as I walk the pipper on the gun sight up to the target. I press the pickle to release the last bomb, then pull the nose up and start a left bank. Something is not quite right. The airplane feels a little sluggish, out of trim. There's some excess airstream noise. Then there's banging. I glance out to my left. I couldn't believe what I saw. The front lug on the bomb rack had released but the rear lug hadn't. The 2000-pound bomb was nose down about 45 degrees and flopping in the breeze! It was oscillating and banging against the fuselage right where the left engine bay access door is located. I leveled off, reduced airspeed, and zeroed the bank so as to try to eliminate the banging. The question of the moment was: would this thing detonate?"

Kretschmar and Friese pulled some Gs, but nothing happened. The pilot then tried fishtailing the rudder, and all he got for his trouble was more bomb-to-fuselage contact. At that point the crew decided it was time to call home:

"We called Da Nang and told them to launch the rescue helo, the *Jolly Green*," he continued. "It appeared we might have to leave this machine if the bomb didn't leave it first. We certainly couldn't land with it like that; we'd end up a ball of fire on the runway. So, as the late afternoon turned to dusk we headed for the South China Sea in the vicinity of Hue. We'd continue to try to muscle this thing off, and although we had plenty of fuel to work at it for a while, we were now limited by the approaching darkness.

"Our plan was to eject just about over the coastline so as to parachute into the offshore waters. We didn't want to come down on land, which may or may not be Indian country, especially with night approaching. We also wanted to have *Jolly Green* get a visual on us as we left the aircraft. Our options are limited to either the bomb leaves or we leave. Time to review the ejection procedure. Cinch down those leg restraints. Tighten up that chinstrap on the hardhat. This is a plus; given the nature of the beast, an ejection is usually a spontaneous thing. You're on fire, you're out of control, pieces are falling off your airplane, and the time to get out is now. But we have some time to talk about it.

"We can now make out the coast up ahead. But there's a cloud deck below, which increases in solidity out over the water. Great. Wonder what the base of the clouds is? We get word that the rescue chopper is inbound to the area. The light is fading fast. We're just minutes away from crashing through the canopy of this perfectly airworthy A6. Airworthy, but not landworthy. What a shame.

"We reach the coast and plan on making one or two orbits while coordinating with *Jolly Green*. Then we hear a 'clunk' and the *Intruder* lurches a little. I look out and it's gone! Bombs away! Hoot and I are elated. What timing! What luck! I bank around to see if the bomb impact can be observed. Maybe the explosive flash.

Maximum effort. A *Batman Intruder* prepares to taxi from alert at Da Nang with 28 *Snakeyes*. With targets sometimes within sight of the base, external tanks could frequently be dispensed with. (Wolfe, via Steve Dumovich)

> We see nothing. Good. It's probably another dud. Hopefully it went into the Gulf."

The crew cancelled the rescue effort and headed back to Chu Lai. Fortunately, the flailing bomb didn't do any damage to the landing gear door or anything else and the gear came down as advertised prior to their successful recovery at home plate. Afterwards, they nervously checked the aircraft for BDA – on the airplane, not the target, as Kretschmar put it – and learned there was nothing wrong with the plane that:

> "A good body and fender guy from the Airframes Shop couldn't patch up. Later at the club, where we were all on our way to oblivion, we learn that 'someone' has decreed that our A-6s won't be carrying around World War II 2000-pounders anymore. Hoot, let's drink to that!"

It was that kind of war, and the *Nighthawks* and the *Bats* carried on, proving the viability of Marine *Intruder*s on a nightly basis in several roles with the odd Combat OpEval. The high value of the dedicated Marine crews would come into demand again, particularly as the year drew to a close.

The *Main Battery* and *Tigers* Deploy

By the spring of 1967 the carrier forces in the Tonkin Gulf were employing two squadrons' worth of A-6As at all times. The current lineup incorporated the *Main Battery* of VA-196 – marking its first combat cruise in *Intruders* and the first A-6 deployment from NAS Whidbey Island – and the *Tigers* of VA-65, making its second run to Vietnam.

VA-196 was first out of the barn, departing with CVW-14 in *Constellation* on 29 April 1967. Led by former *Spad* driver Cmdr. Leo Profilet, the squadron started initial operations from Dixie Station on 18 May. Within a few days they were on Yankee Station and checking out the target-rich environment of North Vietnam:

"The morale was fine," says Profilet:

> "It was new for all of us. We were the first A-6 squadron in *Air Wing 14* and the first in *PacFlt*. The Airlant guys had already been over there, but we had to win acceptance from the rest of wing. We did it, albeit initially by flying more like *Spads* than *Intruders*. That summer there were a lot of Alpha Strikes going on, daylight dive bombing, that sort of thing. This was interspersed with the classic night stuff; we quickly established ourselves."

The Alpha Strikes VA-196 participated in included attacks on Hanoi's main electric power plant in June, done in concert with aircraft from *Enterprise* (VA-35 with CVW-9) and *Bon Homme Richard* (with CVW-21), as well as additional strikes on the powerplant at Vinh. In the latter operation, heavy-cruiser *St. Paul* (CA-73) battered the facility with accurate 8-inch fire before the air wing moved in to demolish the site.

VA-65 arrived on station in *Forrestal* with CVW-17 on 8 July 1967, relieving the *Enterprise*/CVW-9/VA-35 team. Barely five months had passed since the *Tigers* completed their successful "*Intruder* field evaluation cruise" in *Connie*, thereby saving the program, and now they were back with a new carrier and air wing. For both the 12-year old *FID* (for "First In Defense," the carrier's motto) and *Carrier Air Wing 17* – which established on 1 November 1966 – it was the first combat deployment. Fortunately, VA-65 returned with several old hands. Among them was B/N John Diselrod, again paired with Bob "Willard" Williams. According to John, it was a unique situation: this time around the experienced *Intruder* squadron was helping prepare a new air wing for combat:

"Back at NAS Oceana, picking up a bunch of new pilots and B/Ns, and its introduction time to the real world," he comments. "The saving grace is that one gets in a lot of flying, but it's frustrating not being with your own pilot. Things worked out, but one wonders how. We were now assigned to a new air wing and ship; VA-65 was now a member of CVW-17, assigned to USS *Forrestal* (funny; I was in VF-174 back in 1953 as a member of the old CVG-17 at NAS Jax). It was something new for me, the first ship I've been assigned to that I could *drive* to and take my family aboard for a tour.

"The squadron workups went pretty well, and then we got to air wing workups. The two F-4 squadrons were from Oceana (VFs -11 and -74, F-4Bs), the two A-4 squadrons were from Cecil Field (VAs -46 and -106, both flying A-4Es), and the splinter dets from everywhere in between. VA-65 was the only one with combat experience, which was a blessing (or a shortcoming). The savior of the whole situation was CAG Jack DeWinter; he pulled us all together and made us work."

Cmdr. Bob Mandeville, who was relieved as commanding officer of the *Tigers* by Cmdr. Frank Cramblet in June 1967, concurs:

> "Jack DeWinter used VA-65 as instructors for the air wing in all aspects of wing tactics and strike training."

Under DeWinter's leadership and VA-65's guidance, CVW-17 successfully completed Fleet Training at Guantanamo and weapons training at Puerto Rico, and departed for the Gulf of Tonkin. Diselrod continues:

> "We survived the Fleet Training tour at Gitmo, weapons training at Puerto Rico, and a flight deck picnic. I learned one thing during weapons training: when an A-6 with 2.75 rockets fires at a low glide angle at a rock in calm seas, you have to expect the rockets to ricochet back up and just miss the nose of the A-6 at pullout. Surprise, surprise. The next stop was Rio for a 'typical Navy visit.' The Brazilian people wanted to host the Americans; the consulate asked for 250 officers in dress white uniforms to attend. They stated the affair was to be at 2000.
>
> "The American Consulate advised the admiral and requested the officers be there by 1730. The admiral advised the commanding officer of the request and requested the officers be there around 1700. The CO of the *Forrestal* told the CAG to select officers from each squadron and have them ashore around 1630. The CAG told the squadron COs to notify the appropriate personnel and have them at the party by 1600. We all got fitted out in starched whites, caught the launch around 1500 to 1530, and prepared to party. Upon arrival at the designated party place, we found the rooms empty and no one around. There you go, 250 officers, in choker whites, staring at each other. About the time we were ready to cut out some waiters showed up and started serving drinks – soda and sweet vermouth – the food came out, the bar opened, and people started showing up. Brazil is very formal; all of the young ladies were chaperoned by their parents and none of us spoke Portuguese. Shortly afterwards a multi-stringed orchestra showed up and started playing music suited for an afternoon tea, and we started wondering what we had gotten ourselves into.
>
> "Just about panic time; a gentleman from the Brazilian consul walked out, motioned for the orchestra to depart, and out came a jazz band. THE PARTY WAS ON!"

After the fun and games in Brazil everyone crawled back aboard FID, and the ship wended its long way around the tip of Africa and across the Indian Ocean. *Forrestal* started its first line period on 25 July; as John puts it, "we were ready to pick up where we left off." Notably, VA-196 and VA-65 went to war with only eight and nine aircraft, respectively. For this and other reasons, it would turn out to be a memorable cruise for all concerned.

Barely five days into the first line period, on the morning of 29 July 1967, disaster struck *Forrestal* while the carrier was preparing for its second launch. A *Zuni* rocket ignited and flew across the aft flight deck into the pack, starting a major conflagration. The fire – fed by burning fuel and exploding ordnance – quickly expanded, destroying a multitude of aircraft, holing the flight deck, and killing 134, while wounding an additional 62. Several *Tigers* were among both groups:

"Willard and I had three combat flights and were spotted abeam the island for our fourth sortie on 29 July when the flight deck exploded," recalls B/N John Diselrod:

> "It is difficult to mention all the young sailors seen throwing 500-pound bombs over the side or pulling burning bodies out of the mess of wreckage.
>
> "We had a survival equipment man striker in the squadron. When we were outfitting in Norfolk, his workstation was just forward of the port sponson in the rigger's compartment. He was hanging hooks on the bulkhead for his life vest, flashlight, flares, etc, because he couldn't swim, but he wanted to be ready to go over the side in case of emergency. Would you believe, since he was missing from the all hands muster during the fire, and since that area was devastated, we figured he was a goner. He showed up on the destroyer that was following us as plane guard. Like he said, he couldn't swim, but he could walk on water if necessary."

With the assistance of other CTF-77 ships and aircraft – particularly helicopters, which delivered firefighting assistance and flew off the wounded – *Forrestal's* crew got the fires out after 17 hours. After the disaster a total of 21 aircraft were stricken, with most if not all going over the side; the number included seven F-4B *Phantoms*, eleven A-4E *Skyhawks*, and three RVAH-11 RA-5C *Vigilantes*. Another 43 planes received varying amounts of damage. The carrier buried its dead at sea, medivaced the wounded, and retired from the line. After initial repairs at Subic Bay she went home, leaving the theater on 22 August 1967. *FID* returned to Naval Station Norfolk on 14 September, battered but still functional. She would return to making routine deployments to the Med; however *Forrestal*, the first of the "Super Carriers," would never return to Vietnam or make another sustained combat deployment.

As both VA-65 and VA-196 had started the cruise with a reduced number of aircraft the Navy decided to build up the *Main Battery* with several men and planes from the *Tigers*. John and Willard were among those who crossdecked:

> "When VA-65 was in *Connie* in '66 we were a 12-plane squadron, but with VA-196 aboard *Connie* and VA-65 in *Forrestal*, both squadrons were reduced," says John. "Following the fire the Navy decided to up VA-196 to 12 A-6s again. Three flight crews and 49 enlisted men from VA-65 transferred to the *Main Battery* to beef up their complement; I don't know the rhyme or reason, but my roommate Jim Joyner and I were sent as a crew, as were two other pilots and B/Ns. We knew some of the VA-196 flight crews and a lot of the *Connie* personnel, including the admiral's staff, so there was not a lot of discontent over the transfer; everyone had a job and knew how to do it."

The merger worked out fine, according to Jerry Menard, by that time a maintenance officer assigned to VA-196:

> "We had started with eight aircraft, and with VA-65's people we grew to 12. It worked real well. They melded right into the squadron."

The crews remained with VA-196 and CVW-14 through the end of *Constellation's* combat tour in December.

Boomers, *Knightriders*, and *Golden Intruders* Check In

When squadrons lost aircraft and aircrew it fell to the two RAGs back in the United States to provide replacements. Oftentimes the A-6s came right off the Grumman production line on Long Island, but producing trained pilots and bombardier/navigators took time. By mid-1967 VA-42 at Oceana had trained a total of six squadrons (VAs -35, -65, -75, -85, and -196, and VMA(AW)242), and was preparing to receive a seventh. At Cherry Point, VMA(AW)-533 had gone through the mill and on to Vietnam with the *Vagabonds* of -225 and *Bengals* of -224 in loose trail. While the Marines did not have their own stand-alone RAG yet, efforts were underway to provide one.

Which left NAS Whidbey Island and the A-6 detachment of VAH-123. The det was already turning out qualified personnel and had supported the bedding down and deployment of VA-196 once it had relocated to the Rock from Virginia Beach. Now, with the *Intruder* replacement program running smoothly within the *Whale* RAG, it was time for some serious transitioning of AirPac *Skyraider* squadrons. The *Boomers* of VA-165 were the first to check in, arriving on 1 January 1967 following their last *Spad* cruise with the "all attack" CVW-10 in the nominal "anti-submarine" carrier *Intrepid* (CVS-11).

Dating to their establishment at NAS Jacksonville on 1 September 1960 as the fifth squadron in *Carrier Air Group 16*, the *Boomers* had flown nothing but ADs for six years. Initially equipped with AD-6/A-1Hs, they gained several A-1Js prior to their second deployment, changed homeports to NAS Moffett Field in September 1961, and further moved to NAS Alameda in March 1964. While a component of an AirLant-designated air group, the squadron made four cruises to WestPac before changing over to *Intruders*. The first two deployments were in *Oriskany* and ran from June 1962 to December 1962 and August 1963 to March 1964. It was during the second cruise – with the November overthrow of Vietnamese President Ngo Dinh Diem – that the first hints of combat in Vietnam started rumbling. During the crisis *Oriskany* maintained station in the South China Sea and prepared for contingency operations.

Following a 1964-1965 combat cruise in *Coral Sea* with CVW-15, VA-165 found itself right back in Vietnam, this time deploying with CVW-10 in *Intrepid*. The *Boomers* had returned to Alameda from their CVW-15 deployment on 1 November 1965; barely six months later they pulled out again, this time as part of a cobbled-together air wing in the ASW carrier. The ship deployed with four attack squadrons (VA-15 and VA-95 with A-4Bs and VA-165 and VA-176 with A-1Hs) and a UH-2 detachment from HC-2. During the course of the six-month cruise VA-165 lost two *Spads* to AAA, fortunately with both pilots recovered. One was the XO, Cmdr. William S. Jett, III, who was the skipper in May 1967 when VA-165 started receiving A-6As.

Among his young B/Ns was Terry Anderson, a private pilot and proud graduate of the NROTC program at Georgia Tech who had originally planned on becoming a Phantom RIO:

> "I found out about the school program; ROTC was a requirement," he laughs. "The Navy had about 100 in its unit, and the Air Force seemed to have about 3000, so I decided to go to the smaller group. I wanted to be an F-4 RIO out of Pensacola, that was the hot deal. They had us go into the CO's office and choose a seat; the higher up you were (in your class), the better your chances. I saw they had an A-6 seat up at NAS Whidbey Island, and one of my instructors had flown A-6s so I chose that."

When Anderson got to Whidbey Heavy 123 was still serving as the Intruder RAG; he remembers Lyle Bull, Charlie Hunter, and Brian Westin as his instructors, among others:

> "I was one of the first five to graduate from VAH-123. We flew about three hops in the A-3 and then went to the A-6. I then went to VA-165, which was standing up. When the squadron stood up the guy I was supposed to fly with was killed. At a party out at the golf course he got drunk – he had a Corvette – and hit a tree. I ended up with Bill Boissenin, as he didn't have someone to fly with. Mild-mannered guy, smoked like a fiend ... excellent pilot, I enjoyed flying with him and we made two tours."

The squadron completed transition training, and on 16 June officially transferred to *Carrier Air Wing 2*.

While the *Boomers* prepared for their first A-6 combat deployment, VA-52 took their place in the pipeline at Whidbey. The *Knightriders* first saw the light of day at NAS Olathe, KS, in the late 1940s as naval air reserve fighter squadron VF-884. Nicknamed the *Bitter Birds* – and with the squadron emblem displaying a rampant Kansas Jayhawk bearing a spiked club – the squadron was called to active duty for Korea on 20 July 1950. It reported to NAS San Diego a week later, and on 2 March 1951 departed in *Boxer* with CVG-101.

That first combat deployment was a rough one and saw the death of squadron CO Lt.Cmdr. G.F. Carmichael on 24 May after he bailed out of his stricken F4U-4 *Corsair*; according to the *Dictionary of American Naval Aviation Squadrons* the men of VF-884 were not too happy about being in Korea in the first place, hence the "Bitter Birds" nickname. The loss of another skipper during the second Korean War cruise – again with CVG-101, but this time in *Kearsarge* – undoubtedly did not help morale either. On the plus side, during the second tour Lt. E.F. Johnson turned the tables on a MiG-15 and shot it down, marking the first kill for both squadron and air group.

On 4 February 1953 while still deployed in *Kearsarge,* CVG-101 redesignated as CVG-14; all of the squadrons also redesignated with VF-884, becoming the regular Navy's VF-144. Upon its return from the second deployment the squadron settled at NAS Miramar, transitioned to F9F-5 *Cougars*, and started a regular schedule of WestPac deployments, with one Med cruise thrown in for good measure. In April 1956 VF-144 transitioned to the F9F-8B *Cougar*, followed by AD-5s in late 1958, a transfer to *Carrier Air Group 5*, and redesignation as VA-52 on 23 February 1959. Deployments to WestPac continued in *Ticonderoga*, and on the day of the Tonkin Gulf incident VA-52 was in the thick of things.

By now known as the *Knightriders*, the squadron responded to both the initial attack on *Maddox* on 2 August 1964 and flew again on the 4th when a second round of attacks were reported. On 5 August four VA-52 *Spads* participated in the *Pierce Arrow* strikes against the Vinh POL facilities; all four returned. The *Knightriders* made two more Vietnam deployments in *Tico* – one each with CVW-5 and CVW-19 – prior to their final return to NAS Alameda. On 1 July 1967, after turning in its A-1Hs and A-1Js, the squadron transferred to COMFAIRWHIDBEY and started gathering B/Ns and other personnel for the long transition cycle.

With one squadron through the mill and the second in training, VAH-123's ramp was beginning to look decidedly thick with *Intruders* and thin on *Skywarriors*. Indeed, with three fleet A-6 squadrons assigned to Whidbey and four A-3 squadrons the *Intruders* were on the verge of taking over the ramp, a development not viewed with much joy in some portions of the *Whale* community. The A-6 det now functioned as a full-fledged RAG, in purpose if not in name. Hence, on 1 September 1967 the Navy formalized the det's status by creating VA-128. At the ceremonies at Whidbey Island Cmdr. Clinton B. Warwick stood up *The Golden Intruders* as its first CO. Warwick, the officer who had fleeted up to command mid-cruise during VA-75's first combat deployment, inherited a large, well-trained staff that immediately returned to work, training both squadrons and replacement personnel.

For their part the *Professionals* of VAH-123 would continue in the A-3 fleet replacement role, albeit supporting a rapidly dwindling pod of *Whales*. *HATRON* 8 would disestablish at Whidbey on 17 January 1968, and later that year two of the three surviving fleet heavy attack squadrons (VAHs -2 and -4) would transfer to NAS Alameda to redesignate as tactical electronic warfare (VAQ) squadrons -132 and -131. VAH-10 remained at Whidbey, but would also redesignate, becoming VAQ-129 on 15 August 1970. That left VAH-123 as the last heavy attack *Skywarrior* squadron; it continued to serve as the Navy's A-3 RAG until its own disestablishment on 1 February 1971.

Thus, in the fall of 1967 the newly established VA-128 was bringing VA-52 up to speed in *Intruders*, and VA-165 was getting ready for its first war cruise in the type. The *Boomers* were cocked and ready to go, and took great pride in their status as the first Whidbey-trained A-6 squadron. They apparently also had a tendency to claim they were the first *fleet* A-6 squadron from Whidbey, a statement that conveniently ignored VA-196. Admittedly, the *Main Battery* was absent at the time, what with the war in Vietnam; according to one former *Milestone*:

> "They (VA-165) were the first squadron to *complete* the *RAG* at Whidbey Island; *we* were the first fleet squadron at Whidbey Island. They were the candy asses of the (medium attack) wing ... we were always in trouble, and for some reason they could never do anything wrong."

Whatever the subject of discussion, about this time the augmented VA-196 found its plate full. Due to the unplanned departure of *Forrestal* and CVW-17 it was the only *Intruder* squadron on station with CTF-77, and there was heavy weather right ahead.

Four Go Out, One Comes Back

Onboard *Constellation* the war continued to pick up, with regular strikes against targets that most aviators had for years only viewed wistfully. According to John Diselrod:

> "1967 had opened up better targets than the previous year, and we had CBUs (Cluster Bomb Units) for road work. The A-6's system was top notch, and the maintenance crews kept them that way. I never had a system that I couldn't use to its fullest capabilities; lots of fun doing AMTI in the triangle with CBUs. Things were pretty routine until 22 August, when Joyner and I got to run the ferry slip next to the Paul Doumer bridge in Hanoi. The Air Force announced they had knocked out the bridge and power plants a day or so before, so we figured it would be completely dark and a piece of cake. HA! All we had to do was follow the street lights and the tracers."

While the level of combat was increasing, there was still no sign of any give on the part of the North Vietnamese government, or in its efforts to defend itself. Accordingly, the carriers assigned to CTF-77 continued to run a full slate of strike operations. Unfortunately Monday, 21 August 1967 proved to be a particularly tragic day for the men of VA-196.

On the 21st *Connie*, *Intrepid* (CVW-10), and *Oriskany* (CVW-16) launched scheduled strikes against multiple targets in North Vietnam. The aviators from CVW-14 were tasked with hitting the Duc Noi railroad yards and Kep Airfield; VA-196's contribution was four A-6As led by skipper Leo Profilet against the rail targets. The rail facility was located about eight miles north of Hanoi and 72 south of the Chinese border and, as with all targets in Route Package VI, was heavily defended:

> "The strike plans were up to the flag," Profilet recalls. "The general idea was we'd fly four planes against four individual targets at night and four on one target during the day. I was strike lead that night with (Lt.Cmdr.) Bill Hardman as my B/N. We were coordinating with some Air Force F-105s out of Thailand, and I remember being quite pleased, because we hit our TOT at the roll-in within ten seconds."

From this point things took a major downturn, as Profilet continues:

> "We came in over the 'armpit' (an inlet between Thanh Hoa and Nam Dinh) and dodged a bunch of SAMs – that was real exciting – and my number four called and said he'd been hit. There wasn't anything we could do because we were about five seconds from roll in, so I told him to make a 180 and get out of there. He said he'd rather go in with the rest of us. We saw the SA-2 that got us coming up as we rolled in. I had it electronically, actually saw it – I'd seen a lot of SAMs, so I wasn't too impressed – and we were in a standard dive, about forty degrees, going like hell – and it went off right outside the cockpit. Bill yelled 'JESUS CHRIST!' and punched out.
>
> "Obviously we'd been hit. I had pickled, but the aircraft was all over the place doing all sorts of violent things; I found out years later that the right wing had been shot off. In fact, someone later came up with a photo, taken by one of our escorting F-4s. The plane looked like something out of Hollywood, a great big ball of fire. I got out at about 3000-feet, right into the middle of the North Vietnamese, and didn't get a chance to do any of that fancy evasion stuff; it was just not a good place to get shot down. I wasn't injured in the ejection, but they beat me up somewhat, hit me with things."

The crews of the other three A-6s saw a ball of fire and two good chutes, and then heard five minutes of beeper signals as they departed the maelstrom. However, on the return leg a large cluster of thunderstorms split the group. One, flown by Lt.j.g.s Phil Bloomer and Kenneth E. "Denny" Berman, turned south and successfully made it back to *Connie*. The other two *Intruders* – crewed by Lt.Cmdr. Jim Buckley/Lt. Bob Flynn and Lt.j.g. Dwayne V. Scott/Lt.j.g. Jay Trembley – turned north, were

The *Main Battery* of VA-196 was the first Whidbey-based squadron to deploy. NK401 is shown in May 1967. (U.S. Navy)

jumped by Chinese MiGs, and shot down. Bombardier/navigator Bob Flynn was the only one to survive, and was taken prisoner by the Chinese. He subsequently spent six and one-half years as a "guest" of China, of which all but 13 days were in solitary confinement. The other three crewmen were initially listed as Missing in Action, but later changed to KIA. As for Leo Profilet and Bill Hardman, the North Vietnamese held them through March 1973. According to the skipper:

> "I ended up doing my first 27 months in solitary confinement. Guess I pissed them off or something."

Back at the ship, they knew the CO's plane had gone down, but there was no word on the other two missing aircraft. Maintenance officer Jerry Menard still remembers the day:

> "We were up on the flight deck, waiting ... and waiting ... and waiting. We found out later that the other two were shot down by the Chinese. It turned up in *Stars & Stripes*, with the Chinese claiming that the U.S. Navy had invaded their country. Of course, they failed to point out that both aircraft were unarmed."

On 28 August 1968 XO Cmdr. Edward C. Bauer formally assumed command of the *Main Battery*. Maintenance Officer Lt.Cmdr. Robert G. Blackwood – who had jumped out of a VA-85 *Intruder* during the *Black Falcon's* first combat cruise – fleeted up to executive officer, in the process becoming the first NFO to become an XO, albeit under unusual – and not yet legal – circumstances.

To Jerry Menard's recollection, Blackwood was the next senior man in the squadron and a *very* direct individual. "Basically, he announced 'I'm the XO,'" says Menard. "And it took."

While 196 regrouped and went back into combat, the Whidbey RAG sent several replacement crews. Among them was the *HATRON 123* pilot-B/N team of Charlie Hunter and Lyle Bull. According to Menard, the two made quite an appearance when they joined the squadron on Yankee Station:

> "Bull looked to be something like 9'2", and Hunter was – oh – 5'1". But they worked good together."

Lt.Cmdr. Charlie Hunter was a former F7U *Cutlass* and A4D-1 pilot with VA-83, and had served an additional tour flying A-4Es with VA-81 prior to his assignment to the A-6 RAG. Lt. Bull, a former B/N with the *Fourrunners* of VAH-4, was a veteran of a 1962-1963 WestPac cruise in *Bon Homme Richard* with CVG-19. Both were charter members of the VAH-123 A-6 Det. According to Bull, Hunter came up to him one day and said, "VA-196 just lost a bunch of aircraft; do you want to go to war?" He responded in the affirmative, and off they went:

> "VA-196 had just lost their commanding officer when we were sent over;" Hunter recalls. "The powers that be assured me that I'd make commander and fleet up as the CO. Well, I got out there, and they pulled both Blackwood and me and put (CDR Louis C.) Ditmer in there. I was told, 'You're just too junior!' Anyway, Lou Ditmar came over to be the new XO, and I immediately went to the admiral and threatened to turn in my wings, resign, etc. He laughed and said, 'Go back to work.' Things worked out in the end. At the end of the cruise we were down on Midway Island with a busted Constant Speed Drive (CSD) when I got the word that I had been promoted to Commander. All they had to do now was find a squadron for me."

Renewed Concerns Over the *Intruder*

With the loss of three out of four aircraft on a single hop – never mind the circumstances – some voices in higher circles again started questioning the future of the *Intruder*. While VA-65 had proven the plane barely a year before, there were still nagging doubts. Several critics thought cheaper aircraft – and a lot more of them – was the solution.

A big part of the difficulty remained with the A-6A *Intruder* itself. According to Lyle Bull, the system had to be constantly "massaged." Adding to the situation, there were still those on the operations side of the house who wanted to use the plane as day/VFR bomb truck, leading to situations similar to those faced by the *Sunday Punchers* and *Black Falcons* on their initial combat deployments. The aircraft were getting sent out in the day, which resulted in degraded systems accuracy at night. In addition, VA-196's pilots had not received much training in using the iron sight for daytime, visual drops. It was, "embarrassing, we were laughed at," adds Bull. "When you miss with 28 Mk.82s everyone can see the results."

Other A-6 crews had similar feelings. Career attack pilot Capt. Richard J. "Dick" Toft, who had served with VAH-2, VA-112, and VA-34 on its last A-4 cruise, says:

> "It came down to the old philosophical question of whether you wanted cheap aircraft and lots of them or fewer, high-tech planes. The A-6 was complex and always at the leading edge of technology. Over the years different skippers would do different things with the aircraft. A lot of times former A-4 skippers would try to operate the plane like a big, twin-engine A-4, i.e., 'We'll load it up with lots of bombs and bomb the target with the iron gun sight only.' Others said, 'We'll fly the plane and mission as designed: if the system isn't up, the aircraft isn't up.'"

Occasionally problems with the plane's systems status got unwanted attention. Toft recalls one instance some years later during workups. One VA-196 *Intruder* crew was having a particularly difficult time getting their A-6A's systems to come on line for a night hop. What they did not realize was that with the carrier's SINS (inertial navigation) cable plugged into the plane the ship could hear the ICS communications from the cockpit. The B/N informed his pilot – and the ship – that he couldn't get the system to work and the plane wasn't ready for flight. The pilot's response was, "Goddamit, I want to go flying tonight. Let's go!" The conversation bounced back and forth for several seconds before the Air Boss called down with a friendly, "Now, let's not have any marital discord. Either you're up or you're not." Toft felt it was a nice way of handling the situation.

Still, the maintenance crews kept at it while the old hands among the pilots and B/Ns came up with "alternative means" to complete the missions. Bull recalls one strike against the San Song barracks when they used a secondary infrared photo for BDA. The A-6 would go through and drop the bombs, and then a VAP-61 RA-3B *Black Whale* would come through to take the IR photos. The photos were used to tell the difference between the old (cold) and new (hot) craters. It was very effective and, as it turned out, on this particular hop they scored a bullseye:

> "The gouge from the skipper was to not drop on a night mission if you did not have a ground lock on the target," he continues. "You had the angle and the distance to the target using the radar; the computer took the altimeter reading to figure vertical separation and then determined a release solution. When the first A-6A squadron came back from cruise, they stated that – among other things – they never achieved ground lock, hence poor night time accuracy. It was a farce."

Bull says he dropped on the inertial navigation vertical loop system, updating the system's altitude over water by flying an altitude using the radar altimeter and updating the inertial altimeter prior to "feet dry." Using this method they got good accuracy, as demonstrated at San Song. He adds the skipper never found out and finished the combat cruise assuming his crews were using the mandated track radar "ground lock" method of delivery. Charlie and Lyle used the same technique later on a strike at Cat Bai Airfield. On their one pass they dumped 13 1000-pound bombs, which hit the corner of the first hangar, flattened the second, and wiped out most of the surrounding buildings. According to Bull, the blast also blew the camo net off a parked AN-2 *Colt*, riddling the aircraft in the process. They got credit for the kill.

Navy A-6 crews were not the only ones having systems problems; the Marines at Chu Lai and Da Nang reported similar difficulties. For the month of October 1967 VMA(AW)-533 reported 333 sorties for 435 combat flight hours total. According to the squadron's report only *49* missions were accomplished with full systems, reflecting "serious maintenance challenges." Still, they managed to dump 1,557.5 tons of ordnance on the enemy, and like their afloat brethren pressed on, occasionally in spectacular fashion.

On the night of 24-25 October 1967, three groups of *Bats* A-6As went Up North to hit Phuc Yen Airfield, the site of the North Vietnamese introduction of MiG-15s and MiG-17s in 1964. Led by squadron CO Lt.Col. Lewis H. Abrams, the aircraft executed a coordinated attack while successfully coping with a hail of SA-2s and intermittent jamming support and devastated the field. According to the official record of the strike, the crews flew over 200 miles at 500 to 1000-feet AGL to hit their target. Maj. Fred Cone and his B/N – leading the second wave – had a particularly intense time of it over the target. Alerted to an SA-2 launch, Cone turned into it then down, then dropped to 300-feet to avoid a second missile. While proceeding at low altitude he managed to dodge a third SAM, but a fourth holed his *Intruder*. Cone recovered, climbed for ordnance separation, jettisoned his drop tanks to avoid the *fifth* missile, then got the hell out of Dodge.

The raid marked the first use of Marine aircraft against a facility of the North Vietnamese Air Force. For their roles in planning and executing the mission flight leads Lt.Col. Abrams, Maj. Kent C. Bateman, and Maj. Fred Cone received the Navy Cross, marking the only award of that medal during the Vietnam War to Marine *Intruder* aviators. Sadly, Colonel Abrams would receive his award posthumously; on the night of 25 November, while flying against a target near Haiphong, Abrams and B/N Cpt. Robert Holderman were shot down and lost.

There were victories. The following night Bill Kretschmar and Don Hiltbrunner flew a solo low-level mission into North Vietnam. After traveling 190 miles at 400 to 600-feet AGL they successfully laid a string of mines into the waterways near the notorious Paul Doumer Bridge. They recovered safely, having put a crimp in North Vietnamese waterborne transport on that part of the Red River.

Charlie Hunter, Lyle Bull, and the Navy Cross

On the afternoon of 30 October 1967, Charlie Hunter and Lyle Bull were busy onboard *Constellation* planning the night's excursion. They had worked up a "routine" hop in which they would go after trucks using the *Intruder's* AMTI; by this time, AMTI was one system that worked well most of the time, leading Bull to comment, "Finding moving targets in complete darkness was no trick." As per their usual drill, the B/N planned the mission while the pilot looked over the work. Also per usual, Hunter made very few changes.

Their final weather briefing was scheduled for 1800, but at 1630 the Squadron Duty Officer called and told them to report to the Integrated Operational Intelligence Center (IOIC) for a change of plans. Air Intel Officer Lt.j.g. Pete Barrick delivered the news: they were now going after the Hanoi railroad ferry slip – a *very* high value target – which meant plenty of guns and SAMs. Three nights earlier Bob Blackwood had led a six-plane strike against the target. The A-6s were spaced ten minutes apart, and each successive aircraft – not surprisingly – took an increasing amount of fire. The last *Intruder* in was forced to dodge six SA-2s. However, Blackwood felt a single plane might be able to get in and bomb the ferry with "some ease."

Charlie and Lyle were going to find out. In Hunter's words, there was no "best" way to get in or out; all of the options were poor. Bull recalls there was something like two-dozen SAM sites, plus 597 known gun emplacements. Adding to the problems was the *Intruder's* own ECM gear:

> "It was embarrassing, because we didn't know much about it. One guy recommended you turn on the ALQ gear in the repeat position so that you didn't have to worry about forgetting to turn it on in the target area. What we didn't know was that the equipment was beaconing us; it would break (radar) locks okay, but it was also beaconing us. Another problem was with our terrain avoidance system ... the two radars in the nose of the aircraft were not very reliable."

The crew finally came to an agreement on how best to approach this hornet's nest and launched a few hours later. They went feet dry at the "armpit," making 350-knots at 500-feet altitude. Their planned ingress route used the karst hills southwest of Hanoi in order to take advantage of radar masking, but 18 miles from the ferry slip the first *Fan Song* lit them up. Bull talked Hunter down to 300-feet and then guided the pilot to the initial point, an island in the Red River:

> "I took Charlie down a valley between two karst ridges – used A-3 BN radar shadow tactics – and then headed us for an opening in the ridge line where we were going to come out. As soon as we broke out the returns showed the valley in front of us with shadows on both sides. When it opened up a bit I put the crosshairs on the next opening, told Charlie 'Steering's good,' and added, 'You'd better put some Gs on this thing, or we're gonna run right into the goddamn mountain.'"

At the IP Hunter turned in towards the target, ten miles ahead. Again they got a SAM warning, followed by an actual launch. They dropped down to 200-feet while observing an SA-2 launch that guided perfectly to the A-6. Bull says things then picked up a bit:

> "Much to our chagrin they had us right away; the cockpit lit up. The wisdom was that the SA-2 was no good below 1500-feet – the *Fan Song* couldn't discriminate from ground clutter – so we were fat, dumb, and happy at 400 to 500-feet. We saw the missile come up at 12:00 ... and it came right at us. I yelled at Charlie just as he started pulling Gs."

Hunter waited until the last second, then yanked back on the stick, pulling the A-6 into a steep climb. With the nose of the *Intruder* pointed almost straight up the SA-2 exploded directly underneath, rattling the aircraft. At the peak of the high-G maneuver Hunter rolled the A-6 onto its back at 2,500-feet altitude and then continued back onto the target heading. Bull stepped the system into attack ... and here the situation *really* got sporting, as Charlie recalls:

> "At this time the AAA fire was so heavy that it lit up the countryside, and I could see the details on the ground pretty well."

Lyle looked out and said, "I've got two missiles at two o'clock." His pilot responded with "I have three at ten o'clock," and dropped the aircraft down even further; Bull remembers seeing the radar altimeter bottom out at 50-feet. The missiles guided but did not drop down; Bull sensed they exploded above the canopy from the glow and rattling, but he was a tad preoccupied with ID'ing the target. About this time the searchlights came on and illuminated the aircraft intermittently. On signal, Hunter pulled the aircraft up to 200-feet. The ordnance – 18 500-pound Mk.82 *Snakeyes* – sprang from the aircraft, and Hunter pulled away to the right in a 7G turn. Lyle gave steering to the southeast, while Charlie jinked to throw off the gunners. Four more missiles were fired at the *Intruder* on the way out but failed to connect; however, an 85mm flak site outside of Hai Duong almost tagged them, as Bull says due to "complacency on our part."

Following their recovery aboard *Constellation* Hunter and Bull decided they had cheated death several times during the single mission. Bull says he later tracked down the wing EW officer and said something to the effect of, "You know those SA-2s that won't track down below 1500-feet? The first one would've hit us right in the f'ing canopy if Charlie hadn't pulled up!" Sources later confirmed the North Vietnamese had modified their *Fan Songs* to eliminate the ground clutter, allowing the missile to receive guidance at 500-ft.

For the B/N, the mission finally answered the personal question of how he would face combat: once the shooting started; all he could concentrate on was putting bombs on target. Hunter and Bull received the Navy Cross for their mission, the fifth and sixth for *Intruder* crews in Vietnam. A few nights later, on 2 November, VA-196 lost its last aircraft for the cruise. Bull's roommate, bombardier/navigator Lt. Jim Wright, and pilot Lt.Cmdr. Richard D. Morrow were shot down by AAA on their way to Hanoi. Before the mission Wright had told Bull he was "a dead man" and had written a last letter to his wife. Unfortunately, he was correct in his premonition; both radar and radio contact were lost just south of Hanoi during the aircraft's run-in at low level. The rest of the squadron carried on; in Bull's words, "We didn't have time to mourn yet because we had to fly the next day."

The *Tiger*-augmented *Main Battery* lost a total of four *Intruders* in combat, resulting in three POWs, four KIAs, and one MIA. However, no aircraft were lost for maintenance reasons. According to Jerry Menard:

> "We did have a lot of maintenance write ups early in the cruise ... so many we couldn't read them all. However, towards the end of the cruise my roommate – one of the B/Ns – came up and said, 'Gee, the planes are really getting better now.' I said, 'No, you dumb Polock, *you crews* are getting better now.' Once they got comfortable in the airplane and had it figured out, and once we started flying the system missions we were designed for, we had a lot fewer problems.
>
> "We had good PCs (Plane Captains), good maintenance, and had no maintenance incidents – not even a Fod'ed engine the entire cruise. In a lot of squadrons, if you screwed up in the shops you were sent to the line. We did the opposite; if you screwed up on the line you were sent to the shops. That helped us keep the best people as plane captains and flight deck maintainers. I had problems with one maintainer. Our guys could fix planes on the flight deck rapidly. One day this guy dropped the birdcage, fixed the radios and some other components; the aircraft went back 'up' and launched ... and never came back. This poor guy took it personally; I could never convince him that it wasn't his fault."

While tough, the cruise had been memorable, particularly for the VA-65 crews and maintenance personnel. When *Connie* headed back to North Island the *Tiger* personnel flew back to Oceana:

> "We finished the cruise before Thanksgiving and were on our way home again," comments John Diselrod. "The only problem with our *Magic Carpet* flight was the flight attendants. They recognized us from the previous year and cut us off."

Finally, the cruise was memorable for one other reason: Lt.Cmdr. Bob Blackwood, the first NFO to serve – however briefly – as a squadron executive officer. Blackwood's legend would grow over time, partly through incidents such as the following, as relayed by Lyle Bull:

> "After the last line period Connie pulled into Subic Bay to offload Yankee Station assets before going home," Lyle recalls. "A bunch of us were at the Calleyanne Club at Subic, and there were all sorts of guests that night. Bob was at the bar, holding a cigarette with his famous long cigarette holder, waiting for a drink when a young female came up and stood next to him. Bob struck up a conversation with the young woman. He asked her what she did and she said 'I'm a school teacher.'
>
> "'Oh, we have a lot in common.'
>
> "'Oh? What's that?'
>
> "'You train 'em, I kill 'em.'"

The teacher's response was not recorded, but her exit from the scene was immediate. Bull adds, "In those days, if you let life get too serious it would get you down." He also recalls they had some fun with a reporter during the same deployment:

> "It was during the Bob Hope show, first cruise just prior to Christmas, and there were reporters aboard. Phil Bloomer was the player. The reporters were not supposed to cover anything but the Bob Hope show. Max Spiegel was the escort, and he brought him into a room where several of us were having 'Kool-Aid.' Phil was talking about being on an AMTI hop that night, and the reporter asks him, 'what's that?' Phil pointed out that through radar we could pick up moving targets.
>
> "The mission was to bomb trucks carrying supplies south. He said this night, going up the prescribed route – 1A between

The introduction of the A-6B in 1967 gave the Navy a superb aircraft with which to carry out the *Iron Hand* mission. Prototype 155628 is shown with a Standard ARM at Grumman's Peconic Field. (Grumman History Center via Tom Kaminski)

Navy Cross

The Navy Cross is the naval services' highest combat decoration, and is awarded for extraordinary heroism not justifying the award of the Medal of Honor. More men have been awarded the Navy Cross flying the A-6 *Intruder* than for any other fixed-wing jet aircraft in history. Over a two-year period in America's longest war 14 *Intruder* aircrew earned the award, six more than pilots in the next closest type, the A-4 *Skyhawk*.

A total of 485 Navy Crosses were awarded during the war, with 79 going to aviators for "extreme heroism" while airborne. Of the A-6 crews, eight went to Naval Aviators (NA) and six to Naval Flight Officers (NFO). Befitting the *Intruder's* crew concept, five of the Navy pairs of awards went to Naval Aviator/NFO teams for the same mission. Three Marine pilots won the medal for the same mission – a night, low-level attack on Phuc Yen airfield. Brian Westin was awarded for saving his wounded pilot's life at grave risk to himself.

Curiously, all but one was awarded for actions during a relatively short period of the war – October 1967 through March 1968. Since Vietnam, no fixed-wing Navy aircrew has received the Navy Cross, although individual awards have gone to Marine helicopter pilots in both Grenada and *Desert Storm*.

LTJG Brian Westin	NFO	VA-85	27 Apr 1966	
LCOL Lewis Abrams	NA	CO, VMA(AW)-242	25 Oct 1967	KIA 25 Nov 1968
MAJ Kent Bateman	NA	VMA(AW)-533	25 Oct 1967	
MAJ Fred Cone	NA	VMA(AW)-242	25 Oct 1967	
LCDR Charles Hunter	NA	VA-196	30 Oct 1967	
LT Lyle Bull	NFO	VA-196	30 Oct 1967	
CDR Jerrold Zacharias	NA	CO, VA-75	24 Feb 1968	
LCDR Michael Hall	NFO	VA-75	24 Feb 1968	
CDR Glenn Kollman	NA	CO, VA-35	24 Feb 1968	KIFA 12 Mar 1968
LT John Griffith	NFO	VA-35	24 Feb 1968	KIFA 12 Mar 1968
LT James Pate	NA	VA-165	24 Mar 1968	
LT Roger Krueger	NFO	VA-165	24 Feb 1968	
LCDR Gerald Rogers	NA	VA-165	30 Mar 1968	
LCDR Robert McEwen	NFO	VA-165	30 Mar 1968	

NA: Naval Aviator, NFO: Naval Flight Officer, KIA: Killed in Action, KIFA: Killed in Flying Accident

No A-6 crews won the Medal of Honor; the two Navy awards in Vietnam for inflight heroism went to UH-2 Combat SAR pilot Clyde Lassen and A-4E driver and *Iron Hand* specialist Mike Estocin. A single Marine pilot, Steven Pless, won the award while flying UH-1 *Hueys*.

Presidential Unit Citation

The Presidential Unit Citation (PUC) is the highest award a squadron can receive, ranking in precedence above the Navy and Meritorious Unit Citations (NUC and MUC). Established in 1942, the PUC was typically given to units during WWII that excelled in combat, frequently while also taking severe losses. Several ships were awarded the ribbon in Vietnam, but over the following thirty some odd years since no aviation squadron has been granted the award, although it has gone to numerous submarines involved in their own silent business during the Cold War.

One Persian Gulf-based carrier put itself in for the PUC at the end of *Desert Storm*, but the award was downgraded to a Navy Unit Citation by NavCent, the NUC being given to most, if not all, units participating in the war on a blanket basis. Nonetheless, the ship's mess deck awards experts passed the word, and sailors from that ship were seen for some time proudly wearing the ribbon in blissful ignorance.

According to the Navy's official award reference (OPNAVNOTE 1650), four carriers, along with their embarked CVWs and all of the squadrons and detachments, were awarded the PUC during Vietnam. Three of these had *Intruder* squadrons assigned, which are listed below. The fourth air wing was CVW-21, which was embarked in *Bon Homme Richard* for their 1968 deployment with a standard "small deck" two F-8 squadron/three A-4 squadron makeup.

Criteria for unit awards is frequently a highly contentious issue. Why some units get awards and others do not, remains grist for hangar flyers Navy-wide. Nonetheless, these three squadrons can take pride in the unique achievements represented by the award of the PUC to their ship and Air Wings.

A-6 unit	*Ship*	*Air Wing*	*Cruise*	*Other squadrons*
VA-75	*Kitty Hawk*	CVW-11	1967-68	VF-114, 213; VA-112, 144; RVAH-11; VAW-114; VAH-4 det 63, VAW-13 det 63, HC-1 det 63
VA-165	*Constellation*	CVW-9	1971-72	VF-92, 96; VA-146, 147; RVAH-11; VAW-116; VAQ-130 det 1, HC-1 det 3
VA-115	*Midway*	CVW-5	1972-73	VF-151, 161; VA-56, 93; VAW-115; VFP-63 det 3, VAQ-130 det 2, HC-1 det 2

Haiphong and Hanoi – they picked a mover, went down to investigate, and it was a school bus. When he said *school bus*, the reporter swallowed and asked:

"'How did you *know* it was a school bus?'

"'We went down, could see a real big sign, 'Yeah Haiphong! Beat Hanoi!' We broke off, came around, dropped bombs on the school bus, and the pom-poms went everywhere.'

"By this time the reporter knew he'd been had, and we were rolling on the floor. It didn't happen very often where you could get one. When you have the chance, you have to take it."

Hunting SAMs: The A-6B

By the time VA-196 and the VA-65 detachment returned to their respective home stations two other squadrons had shipped out to take their place. VA-165 departed ConUs with CVW-2 in *Ranger* on 4 November 1967, followed on 17 November by the *Sunday Punchers* of VA-75 with CVW-11 in *Kitty Hawk*. It was the first *Intruder* combat cruise for the *Boomers*, while VA-75 was coming in for their second go-around; their arrival would once again give CTF-77 two A-6 squadrons on a full-time basis. It also brought the forces on Yankee Station something extra: an A-6 modified specifically for the *Iron Hand* SAM-suppression mission.

Earlier in 1967 Grumman stripped ten A-6As of their ground attack equipment and fitted them with the ER-142 receiver system, Bendix AN/APS-107B radar homing and warning equipment, AS-2Q50 homing antennas on the radome and intakes, and AN/ALR-55 warning antennas in the wing tips. In addition, these aircraft were equipped with specially developed LAU-77 racks and wired to carry the new AGM-78 *Standard Anti-Radiation Missile (ARM)*. The missile, a modification of the Navy's *Standard* surface-to-air missile program, offered increased range, speed, a larger warhead, and improved homing capabilities over the AGM-45 *Shrike*. It was also much larger than the *Shrike* – measuring 15-feet long and weighing 1,356-pounds.

Grumman followed the first ten aircraft – designated A-6Bs – with three additional conversions incorporating the improved Passive Angle-Tracking/Anti-Radiation Missile (PAT/ARM) equipment, which retained the A-model's bombing avionics. The three aircraft fitted with the Johns Hopkins-developed equipment were BuNos 155628 to 155630, with the first example flying on 26 August 1968. A subsequent six A-6As were modified in 1970 with the IBM AN/TPS-118 Target Identification Acquisition System (TIAS); the first of the lot, BuNo 151820, flew on 1 October 1968. These 19 airframes constituted the entire production of A-6Bs, as the Navy subsequently cancelled the planned $22 million acquisition of 54 new construction aircraft (BuNos 154046-154099). Instead, each squadron heading to Vietnam received two to four B-models giving their air wing an organic *Iron Hand* capability. The aircraft rotated to another squadron at the end of the deployment. VA-75 was first out, flying a mix of A-6As and Bs on their second combat cruise. The A-6B would eventually serve in combat with a number of units and would achieve some success.

While the *Sunday Punchers* had the new aircraft, both they and VA-165 would see more of the usual during the last month of 1967, ie intense air combat in the face of a determined enemy. Despite the violence and frustration brought by the war, the occasional inflight problems led to unexpected situations, sometimes bordering on the hilarious. Terry Anderson, at the time a B/N lieutenant with the *Boomers*, writes about one combat hop that took a whole 'nother turn:

> "We had begun flying combat flights about the first of December. On the 15th Bill (pilot Lt.Cmdr. Bill Boissenin) and I were slated to drop DSTs on a coastal target. DSTs were Mk.82s with special fuses; the bomb didn't blow up on contact with the ground, but armed during flight and would be set off with the metal in a shoe eyelet within 50-feet. As usual, we had planned the primary target, but also a secondary target was ready in case we couldn't hit the primary. It was a night launch, and the weather, as I recall, was none too good. We pointed the nose to the beach, dropped to about 500-feet, and headed for the target. It was going to be the standard drill; find the target, step the system into attack, accelerate to 480-knots, watch for flak, and keep everything under control until release. It was going to be a straight path delivery. Nothing fancy.
>
> "Well, it was all looking really good, but at the release point – zip, nada. The bombs didn't release. So there we were, heading inland with the same 18 bombs 'cruising' at 500-feet and 480 KIAS, and now we had to horse around getting feet wet again. With the max trap weight of about 34,000-pounds – leaving about 5000-pounds for fuel – there was no way to get back aboard the ship with those 10,000-pounds of bombs. At the time MERs were so valuable that jettisoning the whole shebang would have been severely frowned upon, if you get my drift. We decided to go for the secondary target to see if we could get the bombs off, not knowing why they didn't release the first time.
>
> "OK, here we are approaching the secondary target. 'Now ... Now ... DROP!'Zip, *nada*, nothing again. Man, we can't understand why those things won't drop. Anyway, we managed to get back out over the Gulf without getting shot down. We called Strike Operations on the USS *Ranger* and told them we were heading to Da Nang, some one hundred miles or so to the south, I think. I don't remember that fuel was a problem, but we were facing a maximum weight landing at a relatively hot airport. Field length was not going to be a problem. The approach was IFR all the way until we cleared whatever mountain (Monkey Mountain) one flew over on the approach to the south. I can still remember the runway shining at us in the night from about two miles out. Looked LOTS longer than *Ranger* did that night, ya sure, ya betcha!
>
> "Ever make a max weight landing? Pretty high airspeeds, eh? We noticed that right away! Took the entire 10,000-feet to get her slowed down enough to taxi off the end. About halfway to 242 – which, as I recall, was about the last outfit on the left taxiing north on the west taxiway – we felt the airplane slow to a stop and more power didn't help. I got out to see what the problem was and noticed that both wheel brakes were glowing a nice cherry red and burning nicely. Both fuse plugs had blown (releasing all of the air from the tires). We called the tower and asked them to send help. As I recall, some 242 guys showed up with a tractor to tow the airplane, and I'm sure they put out the wheel fire."

Once towed to the *Bats* line, the crew disembarked and worked with their Marine brethren to get the plane fixed ... or at least, bedded down. Anderson continues:

> "At the 242 line, it was pretty obvious the bombs didn't drop – duh! The ordnance guys said they'd take them off for us (and arrange to have them delivered up north). We put our flight gear in the cockpit and headed for the 'BOQ.' Even had mosquito netting, as I recall. I think we were briefed as to where to go if the field came under rocket or mortar attack, which they suspected was going to happen that night. Since we never went to the shelter that night I assume nothing happened.
>
> "The next morning we went to the airplane to preflight for the flight back to *Ranger*. I think we had an overhead time of 1300. We didn't have to get too close to the airplane to observe that the paint scheme had changed somewhat overnight. There were bats painted everywhere! Even the birdcage was painted with bats. They looked especially good on the inside of the wing-tip speed brakes. When we opened the canopy to get our flight gear, we found that truly nothing had escaped the bats. Even our helmets were painted

– fortunately, not the visors. As I recall, Bill and I were both tickled by the whole thing. I know I was, as were obviously the dozens of grinning Marines who were there to see us depart. My recollection is that I kept my *Bat* helmet clear through the end of my next cruise, at which time I departed the squadron. It was a great memento.

"Anyway, we launched back to the boat and landed. Naturally, everyone in the tower saw the bats on our airplane, and word soon filtered down to our CO, who was not pleased with the honors bestowed upon one of his A-6s. I believe he might have written a letter to the CO of -242 expressing his displeasure about his airplane being all marked up, but I'm not totally sure about that."

According to a Marine who was with VMA(AW)-242, at the time the *Bats* were always more than happy to help out their wayward Navy counterparts, any time in any way. He writes:

"This may have been the first of many such coups that were scored by the *Batmen* of 242 on our Navy brothers. It got so bad that the Navy started posting a guard on the aircraft, which only meant that we had to paint the bats on the aircraft as it taxied from the flightline. It really became great sport. Besides, what were they going to do with us? Send us to Vietnam?

"In addition to unloading the bombs from this Navy *Intruder*, the Marine maintenance guys of -242 worked all night swapping out parts from the Navy A-6 with parts – shall we say less than RFI – from our inventory. We wanted to make sure those two Navy guys didn't get hurt on the way back home, but we didn't have the heart to send them back with all that working gear aboard. Besides, it really wasn't stealing, was it?"

Interservice cooperation aside, the *Intruders'* war took a major downturn on the last day of the year. On 31 December 1967, VA-75's Lt.Cmdr. John D. Peace and B/N Lt. Gordon S. Perisho were shot out of the sky by an SA-2 while bombing storage caves near Vinh, and both men were killed. That same day the Johnson Administration formally ended *Rolling Thunder V* and declared the now traditional New Year's bombing halt. By all standards 1967 had been one of the most hectic, violent, and ultimately frustrating years in the U.S.'s short time in Vietnam. The Navy had lost a total of 133 aircraft in combat, with an additional 48 lost to operational accidents during the year. Despite optimistic pronouncements from some military and government leaders, the end was most definitely not in sight.

Yet, no one could possibly foresee what would take place in 1968, during which a total of seven Navy and two Marine *Intruder* squadrons would go through the crucible.

Year of Decision, Year of Frustration

The New Year showed no promise of peace talks between the warring parties and, on 3 January 1968, President Johnson ordered the commencement of *Rolling Thunder VI*. What ensued – above and beyond ongoing combat in Vietnam – was one of the most turbulent twelve-month periods in the history of the nation. Over the next twelve months there would be a crisis in Korea, a massive offensive to fight off in Vietnam, and riots at home. These were events that shook the nation and changed history.

The same day *Rolling Thunder VI* kicked off VA-35 departed ConUS in *Enterprise* with CVW-9 for its second combat deployment. While the plan was to continue bombing the north as before, the weather over North Vietnam didn't cooperate, limiting CTF77 operations through January into February. This, of course, meant perfect operating conditions for the two *Intruder* squadrons on Yankee Station (VAs -75 and -165), which became the only units that could regularly hit their assigned targets. However, the situation quickly turned sour for the *Boomers*, who lost two aircraft in short order. In an operational accident on 23 January 1968, an A-6A flown by squadron XO Cmdr. Leland S. Kollmorgen and Lt.Cmdr. Gerald L. Ramsden impacted the water on ingress for unknown reasons about 25 miles from the target. The XO was able to eject prior to the impact and was recovered with cuts and bruises, but Ramsden died. Three days later Lt.Cmdr. Buzz Eidsmoe – one of the original VAH-123 cadre – and Lt. Michael E. Dunn went down due to unknown causes on a night mission near Vinh Airfield. Their aircraft dropped off of radar coverage about six miles north of the field, and apparently both men ejected but were not recovered.

The news rattled the wives back at Whidbey Island. As was usually the case during the long, hard war, it affected the wives in different ways; one was young Betty Anderson, who was expecting her and Terry's first child:

"I met him (Terry) in France while he was on cruise with the *Forrestal*," she recalls. "He was with his Georgia Tech buddies, and I saw him on the beach. Two years later we were married – after he got his wings – and we moved to Whidbey Island via Memphis. We had our first child here, Christy; she was our 'war baby.' I was expecting when we lost two planes within a couple of days of each other. Gerry Ramsden went down; I think the whole time he (Terry) was gone that was the worst feeling. I wasn't the youngest wife in the squadron, but I was probably the most naïve. I was afraid my daughter wouldn't have a daddy. It was peaceful at Whidbey, but Terry wasn't the world's best letter writer. The best part was the support of the military wives; we were a tight, supportive group."

It was a rocky start for VA-165's cruise, but the nation's attention was diverted elsewhere. On 27 January *Ranger* was directed to the Sea of Japan in response to another crisis; on the 23rd North Korean naval units had surrounded and captured *Pueblo* (AGER-2) while the American surveillance ship was operating in international waters off Wonsan. The *Ranger* battle group sped to the Sea of Japan as *Task Force 71* and established Defender Station, while the U.S. government tried to decide what it was going to do next. *Enterprise* – about seven hours out of Subic and enroute to its first line period in the Tonkin Gulf – was also ordered to take station off Korea. Both air wings and their squadrons started planning possible combat operations against the North Koreans; however, all contingencies went by the boards when the North Koreans towed *Pueblo* into Wonson harbor and imprisoned her crew.

The *Pueblo* operation left VA-75 in *Kitty Hawk* as the only *Intruder* squadron in the Gulf when the Tet Offensive erupted on 30 January 1968. Within 72 hours communist forces captured Hue and briefly penetrated the grounds of the U.S. embassy in Saigon. Over the following weeks VA-75 concentrated on targets in the "Iron Triangle," including bridges, barge construction facilities, railroad yards, and communications sites. The two Marine squadrons – VMA(AW)-242 at Da Nang and -533 at Chu Lai – flew similar sorties, many of which described as of the classic "take off, gear up, drop bombs, gear down, land" variety.

Both squadrons suffered several losses while flying against both local and distant targets. On the night of 17/18 January 1968 a -533 *Intruder* disappeared while on a system strike against a Western Track *Rolling Thunder* target, the Ven Bay ammo dump. The weather was 300-foot scattered, with three miles visibility in fog; two minutes before release contact ceased with the A-6. Maj. Hobart Wallace and Cpt. Pat Murray were listed as KIA. On 23 February a second *Nighthawk* bird was lost when it suffered a mid-air with a VMF(AW)-235 F-8E while on a GCI approach to Chu Lai. The controllers notified the crew there was a *Crusader* in the area, the B/N looked up – crunch! – and out they stepped. The *Intruder* crew was recovered while the fighter pilot was lost. The following day the squadron lost yet another A-6 on a night strike to Hanoi, when it checked in two minutes before hitting the target ... and then disappeared. Maj. Jerry Marvel and Cpt. Larry Friese were believed hit by a SAM and became POWs, both being released in 1973.

VMA(AW)-242 lost a plane on 1 May 1968. After making an 0530 IFR departure, pilot Cpt. E.J. Fickler noticed he did not have an "up" indication on his slats. He slowed to 220-knots to recycle, slowed again to 180-knots, raised the slats, and then the A-6 made an uncontrolled roll to the left. Fickler blew the canopy, ordered his B/N, Cpt. G.H. Christensend, out, and then followed. Two days later -533 lost a bird and crew over Route Pack 1. 1st Lts. Robert Avery and Thomas Clem were on an armed reconnaissance mission when they dropped off the radar screen approximately six miles northwest of the coastal town of Dong Hoi in Quang Binh Province. The Marines called in an immediate SAR effort, but the rescue crews did not locate any wreckage or evidence of the two crewmen, and they were eventually declared killed in action.

While Tet served as a major defeat for the Viet Cong, effectively decimating VC leadership and ground forces for the remainder of the war, the perception in the United States was one of futility and rage at both the Johnson Administration and the military leadership. For President Johnson and his advisors it was a major political and public relations disaster. No one could know it at the time, but the administration was deeply divided over the conduct of the war; the aftereffects of this disastrous month of January would continue to hound both the president and his party through the elections in November.

Enterprise did not arrive on Yankee Station until 22 February 1968, while *Ranger* remained away from the Tonkin Gulf through 9 March. That left *Kitty Hawk* – and VA-75 – to hold the line until the end of February. That month the National Command Authority authorized another new target: Hanoi's Red River port facility, on the southeast side of town. The *Punchers* responded on the 24th with three aircraft led by skipper Cmdr. Jerrold M. Zacharias and B/N Lt.Cmdr. Michael L. Hall. The *Intruders* flew through a gauntlet of SA-2s and heavy antiaircraft fire to get to the release point. Commander Zacharias later described the final few seconds of the attack in the May 1972 issue of *Proceedings*:

> "At about six miles, every gun in town opened up, and they were really awake by now. Mike picked up the target. As we got near it, I popped up and delivered 24 bombs ... at bomb release I broke down and left and crossed the downtown part of Hanoi at 530 knots, just to wake up the heavy sleepers."

For their planning, leadership, and execution of the mission, Zacharias and Hall received the Navy Cross.

On 21 February, with the impending arrival of *Enterprise* and VA-35, *Kitty Hawk* departed for R&R and upkeep, ending nearly 60 days of sustained combat operations. The ship returned on 4 March, and two days later VA-75 suffered its second and final *Intruder* loss of the cruise. AAA batted down Lts. Richard C. Nelson and Lt. Gilbert L. Mitchell while they were pressing an attack on a railroad yard outside of Haiphong; the Navy subsequently declared Nelson MIA and Mitchell KIA.

Kitty Hawk and CVW-11 departed the line for the last time on 1 June 1968 and returned to the states on 20 June. For the *Sunday Punchers* the third combat deployment was their best yet, with "only" two aircraft lost, albeit resulting in one MIA and three KIA. On the plus side, the squadron had completed the first A-6B cruise, which included the first combat *Standard ARM* shot by an *Intruder* squadron in March. VA-75 returned to Oceana on 28 June to commence a period of recuperation, training, and preparation for several Mediterranean deployments. The *Punchers* would eventually return to Vietnam, but not for four years.

Black Panthers Return

When *Enterprise* arrived for its first line period on *Yankee Station* on 22 February 1968, VA-35 already had one A-6 cruise under its belt. During the first go around combat losses were limited to one aircraft and crew; this time, however, it was quickly and tragically different. Four *Black Panther* aircraft went down in two weeks, a situation reminiscent of the initial A-6 cruises two years prior.

The first occurred on the 28th when Lt.Cmdr. Henry A. Coons and Lt. Thomas Stegman hit the water during a night strike, killing both. They were enroute to Bai Thoung Airfield, and their last call was "execute," indicating they were on their way in to the target, but their plane did not rejoin after the strike. After studying their planned route *Jouett* (DLG-29) and *Southerland* (DD-743) went in to a point off the coast to search for wreckage and found an oil slick and debris. Coons and Stegman were pronounced MIA and presumed dead. On 1 March Lt.Cmdr. Thomas E. Scheurich and Lt.jg. Richard C. Lannom also went out and did not come back. They launched at 1800 local as part of a three-plane strike on the barracks at Cam Pha, and again, were heard to call "excute." Only two *Panthers* rendezvoused after bombing the target; they immediately alerted the SAR forces, but no one ever heard any beepers or found any wreckage. This circumstance was quickly becoming known as the "A-6 peculiar" loss, as the first indication the ship had of any problem was when the plane didn't report back to marshal on time.

On 12 March disaster struck the crew of commanding officer Cmdr. Glenn Kollman and first tour B/N Lt.j.g. David H. Griffith. Following an 0430 cat shot their A-6 impacted the water about three miles ahead of the ship, with no ejections. The only indication of problems was the last call heard from the aircraft: "Rolling, Rolling, *Eject! Eject! Eject!*" On 15 March XO Cmdr. Herman L. Turk fleeted up to the command slot. The following night, Lt.Cmdrs. Edwin A. Shuman III and Dale W. Doss launched on a single-ship strike into North Vietnam and disappeared. The Navy later received word the crew had been shot down by AAA; fortunately, they managed to eject, but were taken prisoner. The two men joined the ever-growing number of Navy, Marine, and Air Force crews held by the North Vietnamese.

On 19 March 1968 *Ranger* returned to Yankee Station with CVW-2. While the unfortunate crew of *Pueblo* underwent a series of public show trials, courtesy of the North Koreans – they would not be released until 22 December – the Navy disbanded *Task Force 71* and closed Defender Station. There would be no strikes on North Korea, but at least now CTF-77 had its second *Intruder* squadron back on line. However, for all concerned, surprising political decisions back home in the states would drastically change the prosecution of the war in Southeast Asia.

The Bombing Halt

March 1968 was a particularly rough month for Lyndon Baines Johnson. The power and prestige of the office of President – which he had sought so strongly during the 1960 primaries and had gained through the death of President John F. Kennedy – was battered by domestic unrest, the seemingly endless war in Vietnam, and a revolt within his own party.

By now his own Secretary of Defense, Robert Strange McNamara, had been making noises about the futility of the war, of which he was the chief architect. Accordingly, on 1 March McNamara resigned and was replaced by Clark Clifford. The resignation ended McNamara's seven-year stint as SecDef, dating to his early 1961 appointment by President Kennedy. Yet, it got worse for the president. Senator Eugene McCarthy of Minnesota challenged Johnson for the Democratic party's nomination for president, and remarkably nearly beat Johnson in the 12 March New Hampshire primary. Several party members subsequently started pressing Senator Robert F. Kennedy to run for president.

On 31 March 1968 at 9:00 p.m. eastern time, the president appeared before a national audience on television to address these and other subjects. In his speech – viewed by approximately 70 million

Americans – Johnson made two announcements that floored the nation. In the first, he announced he was "taking the first step to de-escalate the conflict. I have ordered our aircraft ... to make no attacks on North Vietnam, except in the area immediately north of the demilitarized zone." The president further stated his action was intended to indicate American willingness to make concessions towards opening peace talks with the North Vietnamese.

His second announcement was even more startling.

"There is a division in the American house now," the president stated:

> "I have concluded that I should not permit the Presidency to become involved in partisan divisions. Accordingly, I shall not seek, and I will not accept, the nomination of my party for another term as your president."

The following day the United States suspended all combat operations north of 20 degrees; the line subsequently shifted south to the 19th parallel. The peace talks that the Johnson Administration had long sought finally started with the North Vietnamese government on 13 May in Paris.

According to *Boomer* Terry Anderson, he and his pilot Bill Boissenin had already flown 47 missions Up North when the President's surprise declaration shifted operations elsewhere.

"By that time we were the only squadron flying over North Vietnam," he comments:

> "The A-7s and A-4s were sent south. One mission – one of those things that you find out later how bad it was – we were supposed to do a road recce near Ha Tinh. We went down this highway, we're IFR, and I see all these 'flashbulbs' and can't figure out what they are. We later found out we'd flown over their antiaircraft training center. Then there was the time near Vinh. We thought we'd be stealthy, over two ridges, pop in, seed a river, and we thought we'd fool them. We popped up over that ridge, and the whole bloody world broke loose. It was as bright as day. We got in, dropped the mines, and hauled butt out of there. We never did that again.
>
> "One thing I absolutely remember, the 31st of March 1968 we were working up for a midnight launch. The Air Boss comes up, scrubs the mission, and we download the bombs. That night, I was sure were going to die; it was a single mission into downtown Hanoi."

This sudden switch in the direction of the war did not impact the Navy's deployment schedules; the carrier air wings would still go over and fight a war, primarily over South Vietnam and Laos. Laos was another former French colony that had suffered through centuries of bloodshed and warfare. France granted Laos independence in 1953, but the communist Pathet Lao – formed by Prince Souphanouvang and Kaysone Phomvihan – immediately sprang up to overthrow the government of Prince Souvanna Phouma. The United States started sending assistance to Laos during the early 1960s – primarily through the Central Intelligence Agency – as a balance to North Vietnamese activities, but it was a classic "secret war," denied by all the participants. In late 1962 the Air Force started regular armed reconnaissance flights over Laos, operating from bases in Thailand, which subsequently supported Marine A-4s and Royal Australian Air Force *Sabres*. A 1963 Pathet Lao offensive against the Laotian Army and organized Hmong tribesmen led to an increased Navy presence with RF-8As and RA-3Bs. By 1964 these operations grew into regularly scheduled air strikes and close air support missions under the title Operation *Barrel Roll*. The U.S. Air Force's famous *Raven* FACs – pilots officially loaned to the American attaché in Vientiene, but more often tasked by the CIA – called in and spotted the strikes.

By 1968 the North Vietnamese Army was flooding northern Laos' *Plaine des Jarres* – the Plain of Jars – and mountainous areas with regular troops, forcing the U.S. to become more directly involved. Airspace which had seen Air America C-46s, C-123s, C-130s, and a multitude of helicopters and STOL aircraft, as well as the *Ravens'* O-1Es and T-28s, suddenly saw large numbers of jets with the star and bars on the wings. Due to presidential directive they could not bomb North Vietnam anymore, but they did prove useful in Laos. As VA-165 and VA-35 in *Enterprise* moved the war further south and west their reliefs departed the United States. On 10 April VA-85 headed out of Oceana for its third *Intruder* combat tour; and VA-196 followed, departing Whidbey Island on 29 May for its second trip. Both squadrons wondered what kind of war they would find when they got there.

For the *Boomers*, the remainder of the cruise was relatively unremarked. They suffered no more combat or operational losses through two more line periods, and *Ranger* departed the theater on 18 May 1968. By the time of VA-165's homecoming at Whidbey Island on 25 May other A-6 squadrons had assumed the mantle in the Tonkin Gulf. After the squadron's return home there was the usual period of adjustments for all concerned, including B/N Terry Anderson, who came home to a new daughter, Christy:

> "Before each cruise, particularly the war cruises, the CO would meet the wives," says his wife, Betty. "One thing I remember him say was 'handle as much as you can here, but don't tell your husbands.' I took that to the extreme. Christy was born while he was overseas. Six days before the bombing halt I sent Terry a picture of his daughter, and he didn't like it at all. He came home, Christy was seven months old, and he decided she was the most beautiful thing in the world. He was really gone a good part of her first year, but she just adjusted to him. I got home, and she could say things like 'mommy' and 'puppy,'" adds Terry. "She kept calling me mommy until she was almost two years old. By the end of the fourth cruise I'd settled into it."

While the *Boomers* were finally home from their first combat deployment, the crews of *Carrier Air Wing 9* in *Enterprise* were still engaged in a major shooting war, and the wing was battered through the remainder of its remaining line periods. On 13 May 1968 North Vietnamese AAA bagged its fifth *Black Panther* A-6; fortunately, rescue forces pulled out both pilot Lt. Bruce B. Bremner and B/N Lt. John T. Fardy. The sixth and last *Intruder* shootdown for the deployment came on 24 June, near Vinh, again due to antiaircraft fire. The crew, Lt. Nicholas M. Carpenter and bombardier Lt.j.g. Joe Mobley, were initially listed as missing in action, but Mobley later turned up as a prisoner of war.

Two days later *Enterprise* departed the theater. VA-35 had lost a total of seven aircraft, with six crewmen killed, one MIA, two rescued, and two POWs. The air wing's total also included one RA-5C, three A-4s, and four F-4Bs, three of which were lost in operational accidents. The *Panthers* finally made it home to NAS Oceana on 18 July 1968. They returned to find several changes at the station, including a new aircraft type and yet another squadron undergoing transition to *Intruders*.

Community Development: *Double Eagles*, "Ticks," and *Thunderbolts*

During that rough year of 1968 several actions took place in the now four-year-old medium attack community that helped further define the *Intruder*, both in operations and in training. The first occurred on 15 January 1968, when the Marines stood up VMAT(AW)-202 at MCAS Cherry Point. Led by Lt.Col. J.K. Davis and under assignment to *Marine Aircraft Group 24*, the squadron's establishment finally gave the Marines their own A-6 RAG.

The following month HQMC approved -202's emblem of a deep red disc with black shield and golden double-headed eagle – not unlike

The Marines formed their own *Intruder* replacement squadron at Cherry Point in January 1968. The *Double Eagles* of VMAT (AW)-202 operated for 22 years before being deactivated. KC04, an early A-6E, is seen at NAS Memphis, TN, in November 1977. (Mark Morgan)

the Romanov-period Russian crest – giving rise to the squadron's unofficial nickname, the *Double Eagles*. The squadron set to work training Marine aircrews with an obvious emphasis on close air support; it graduated its first class of five B/Ns on 22 March 1968, followed by the first four pilots on 10 May. Over time the squadron would also provide introductory training for EA-6A crews in the VMCJ/VMAQ units, although it would never operate the "Electric" version of the aircraft.

Approximately two weeks later, on 28 January, the *Green Pawns* took possession of their first Grumman TC-4C. The appearance of the aircraft was rather startling, even among a flight line full of *Intruders*. The basic airframe was the G.159 *Gulfstream I* corporate transport powered by two Rolls Royce *Dart* Mk.529-8X turboprops. The basic design was sleek and very popular with corporations, but Grumman's modification grafted on the nose of an A-6, complete with search and track radars and ancillary equipment. The interior was anything but "corporate," with exposed insulation, a complete A-6 cockpit, and additional radar repeaters. While replacement pilots would on occasion get fam hops in the type, its main purpose was the training of replacement bombardier/navigators. Once they completed initial ground and academic training at the RAG they would load up their nav bags and, under the guidance of an experienced fleet B/N, sally forth for actual radar and systems training at a somewhat sedate speed. All B/Ns received this training, which included navigation and bombing practice, prior to their first flight in the right seat of an *Intruder.*

The initial airframe, BuNo 155722, made its first flight from Calverton on 14 June 1967. While officially nicknamed the *Academe*, the aircraft quickly adopted the popular sobriquet of "Tick," and Grumman eventually delivered a total of nine *Ticks* (BuNos 155722-155730) to the three RAG squadrons at Oceana, Whidbey Island, and Cherry Point. Within short order the sight of "attack" TC-4Cs plying the airways and making slow – but steady and educational – runs on the bombing ranges at Dare County and Boardman became commonplace. One of the authors recalls an incident that was decidedly *un*-commonplace; it occurred in 1978 at Oceana:

> "I don't exactly remember when we made this particular flight; probably early spring 1978 with VA-42 at NAS Oceana. The intent was to navigate down to the Dare County range in North Carolina with my amigo, John Stahura, getting the first few passes as B/N. I occupied the pilot's seat in the cockpit at the rear of the plane; our instructor was LT Kent Horne, who later went on to be *Blue Angel #7* and commanding officer of VA-205 at NAS Atlanta. I don't recall who the pilots were; I do recall that they both had beards and were E-2 drivers in another life (the only *Green Pawn* pilots with beards were the *Tick* pilots).
>
> "It was pretty much standard until we took our place at the end of runway 5R and the pilots ran up the engines. At that point the cabin started to fill with smoke, and all three of us started yelling to the pilots that we had a fire in the back. Kent had us unbuckle and start up toward the front, while the pilots reduced power and parked us in the grass alongside the runway. We were convinced the fire was in the tail of the plane, and so told the two guys up front. However, they maintained it was in the *front*; seems the air conditioning system had sucked up the smoke and was pumping it into the cockpit. After a very brief period of "out the front – no, out the back!" they won, and we lowered our heads, ran back the length of the cabin – still a lot of smoke back here – opened the emergency door at the right rear of the *Tick*, and jumped down into the grass, about, oh, eight feet or so. We then proceeded to haul ass in a military fashion while observing the arrival of the fire trucks and other emergency vehicles.
>
> "Upon inspection it was determined that the APU in the tail of the plane had caught fire; on the ground, no big thing and not too much damage, but if we'd taken off it might've become a whole 'nother thing. The squadron towed the plane back to our hangar, and the five of us "survivors" caught a ride back on some yellow gear, where we were greeted with the usual irreverent frivolity. John and I eventually made up the hop in the other *Tick*."

During 1968 four more squadrons joined the still growing *Intruder* community, one each at Oceana and Whidbey Island, and two at Cherry Point. The first in the barrel were the *Thunderbolts* of VA-176, who moved to Oceana from NAS Jacksonville in May. VA-176 was established on 1 June 1955 at NAS Cecil Field, FL, and equipped with AD-6s; the squadron moved to Jacksonville in February 1956 prior to their first deployment. That first cruise, as part of *Air Task Group 202* in *Randolph* (CVA-15), saw the *Thunderbolts* operate as a component of *Sixth Fleet* forces monitoring the 1956 Suez Crisis. Otherwise, the squadron made a total of seven Med and one Northern Atlantic deployments, including seemingly regular trips to the Caribbean in response to regional unrest.

The year 1966, however, brought something different. Assigned to the stripped down CVW-10, the *Thunderbolts* pulled out in USS *Intrepid* on 4 April as part of the ASW carrier's famous "all attack" combat cruise to Vietnam. Notably, it flew A-1Hs alongside VA-165, which was making its last *Spad* cruise prior to transition to *Intruders*.

The *Thunderbolts* of VA-176 traded in their *Skyraiders* for *Intruders* in 1968. Never deployed to Vietnam, the squadron instead made multiple trips to the Med in *Franklin D. Roosevelt* with CVG-6 during the war's last years. AE510, shown here on *FDR's* cat no. 1, didn't last long with the outfit, being lost on 26 Sept. 1969 due to inflight fire near the Bahamas. Both crew successfully ejected. (U.S. Navy)

More notably, on 9 October 1966 VA-176's Lt.j.g. William Patton gunned down an attacking MiG-17 during a RESCAP operation. It marked the second such shootdown of a North Vietnamese MiG by a *Spad* – following a VA-25 victory on 20 June 1965 – and resulted in the awarding of the Silver Star to Patton.

Following their return to NAS Jacksonville on 21 November 1966, the *Thunderbolts* made one more "near-combat" deployment to the Med in mid-1967. With word of the attack on *Liberty* (AGTR-5) on 8 June four VA-176 *Skyraiders* quickly sortied from *Saratoga* as part of the protective force for the stricken ship. However, the strike was recalled in mid-flight by Washington, and diplomacy resumed its course. Upon its return to Florida the squadron turned in its last A-1Hs, and in May 1968 moved to NAS Oceana for assignment to *ComFAirNorfolk*. Now led by Cmdr. Charles L. Cook, VA-176 received its first A-6As on 5 February 1969. Following completion of training the squadron went to CVW-6 for a series of Med cruises; the *Thunderbolts* would never see Southeast Asia as an *Intruder* squadron.

Community Development: *Swordsmen* and *Polka Dots*

At the other end of the nation, VA-145 also dumped its trusty *Spads* and got in line for A-6s. Like their counterparts at VA-52, the *Swordsmen* were another old-line naval air reserve squadron that got called up for the Korean War and effectively never went home.

Originally known as the *Rustlers* of VA-702, the squadron was activated at NAS Dallas on 1 December 1949 and went on active duty for the Korean War on 20 July 1950. It departed ConUS with CVG-101 on 2 March 1951, operating AD-2s and AD-4Qs from the deck of *Boxer* alongside the *Bitter Birds* – and future *Knightriders* – of VF-884. The squadron made two combat deployments to Korea, redesignating as VA-145 on 4 February 1953 with the changeover from CVG-101 to CVG-14.

Post-Korea VA-145 returned to NAS Miramar, transitioned to AD-6 *Skyraiders* in 1956, and made a series of WestPac deployments from the decks of *Hornet, Ranger, Oriskany, Lexington*, and *Constellation*. When Vietnam erupted in the summer of 1964 the *Swordsmen* were a component of CVW-14 in *Connie* and participated in the *Pierce Arrow* strikes. Sadly, during the attacks on the North Vietnamese naval facilities at Loc Choi, NVA gunners shot down VA-145's Lt.j.g. Richard Sather; he crashed into the sea with his *Spad* off target and became the first Naval Aviator killed in the conflict. There were victories; on a subsequent cruise while operating with CVW-14 in *Ranger* skipper Cmdr. H.F. Griffith received the Silver Star for his combat leadership against North Vietnamese targets and during a successful SAR operation. On 1 February 1966, during another strike against targets in the *Steel Tiger* region of Laos, *Swordsman* Lt.j.g. Dieter Dengler got shot down. The Pathet Lao captured the young pilot, but on 29 June he escaped; he was finally retrieved on 20 July after 21 days of evasion. For his escape and his successful efforts to rejoin his squadron Lt. Dengler received the Navy Cross.

VA-145 made a total of three Vietnam deployments, including one in USS *Intrepid* with CVW-10 during 1967, before turning in its A-1Hs. The *Swordsmen* moved to NAS Whidbey Island on 28 January 1968 for training under VA-128 and received their first *Intruder* on 4 June; they would eventually report *to Carrier Air Wing 9* for a return visit to Vietnam.

Next came one more Marine squadron at MCAS Cherry Point, the first to report to VMAT(AW)-202 for transition. The initial unit to check in was the *Polka Dots* – aka *Moonlighters* – of VMA-332, on 20 August 1968. The *Dots* dated to Marine Scout Bombing Squadron 332, activated on 1 June 1943 at Cherry Point. They equipped with SBD *Dauntlesses* and, after a period of training at MCAAS Bogue Field, NC, headed to the west coast. In February 1944 VMSB-332 moved to MCAS Ewa on Oahu, and subsequently to Midway Island for a few weeks. The *Dots'* combat operations were limited to patrol and escort missions with SB2Cs with no contact with the enemy, and in July 1944 the squadron moved back to Ewa. It shifted to TBM *Avengers*, redesignated as VMTB-332 on 1 March 1945, sailed for San Diego at the conclusion of the war, and deactivated on 13 November 1945.

As with many of its sister Marine aviation units, the next call to arms for -332 came with the Korean War. Now designated VMA-332, the squadron reactivated at MCAS Miami, FL, on 23 April 1952. It subsequently moved to Itami Airbase, Japan, and from there deployed in *Point Cruz* (CVE-119) from 11 April to 18 December 1953, incorporating a brief cross deck to *Bairoko* (CVE-15). Apparently Korea is also where the squadron picked up its "Polka Dot" name; reportedly, when it acquired VMF-312's F4U-4s and -4Bs the -332 personnel repainted the *Corsair's* checkerboard nose markings into dots. In and around operations from the escort carriers the squadron rotated through Itami and continued supporting the "mud Marines" on the Korean Peninsula in their fight against North Korean and Chinese "volunteer" forces. Finally, in December 1953 VMA-332 returned to MCAS Miami and transitioned to the AD *Skyraider*.

In 1957 the squadron began a five-year period of annual rotations to MCAS Iwakuni. A year later the *Dots* relocated to Cherry Point and

The TC-4C, based on Grumman's *Gulfstream I* business aircraft, proved to be a critical airframe for training aircrew throughout the *Intruder's* years. Nine were bought, with all three RAGs operating the type. (VA-42, -128: Rick Morgan, VMAT (AW)-202: Mark Morgan).

The A-7 *Corsair II* started replacing the A-4 *Skyhawk* in the Navy's light attack squadrons in 1966. The type's first deployment was in *Ranger* with VA-147, shown here conducting CQ prior to its 1967 deployment. The A-7 continued as the *Intruder's* stablemate until its final replacement by the FA-18 *Hornet* in 1991. (U.S. Navy)

exchanged their *Skyraiders* for the A4D-2 *Skyhawk*. They continued their overseas rotation, scheduled through May 1962, when rising tensions in Thailand resulted in their transfer to Udorn Royal Thai Air Force Base as part of a Marine Expeditionary Force dispatched in response to a communist threat to the nation. VMA-332 remained in country until 2 July, departed Iwakuni for MCAS Cherry Point in October 1962, and eventually transitioned to the A-4C and A-4E variants of the *Scooter* before starting A-6A transition in 1968.

For their next medium attack squadron the Marine Corps selected the *Green Knights* of VMA-121. However, at the end of 1968 that outfit was still stationed at Chu Lai with MAG-12 and would not report in with the *Double Eagles* until early 1969. Before then the complexion of the war in Southeast Asia would once again change, as the nation prepared to inaugurate a new president.

The End of One Hell of a Year

On 10 April 1968 the combat-proven *Black Falcons* of VA-85 pulled out of Tidewater for their third *Intruder* deployment to SEA. Having completed two previous trips with AirPac's *Kitty Hawk*, they were now going back as part of the AirLant team of *America* and CVW-6.

This time the squadron headed into combat with a mix of A-6As and A-6Bs. *Buckeye One* was still Cmdr. Jerry Patterson; notably, one of his first-tour JOs was an ensign named James B. Dadson, reputedly the youngest A-6 pilot in the Navy and known to one and all as "JB." In discussing his status Dadson recalls:

> "I managed to get A-6s directly out of TraCom and was one of the first pilots assigned to VA-42 as an ensign. When I showed up everyone said, 'What the hell are *you* doing here?' They thought I was there strictly to stand the duty. Everyone said I was the first ensign A-6 pilot to fly combat; I was 22 when we got there, and turned 23 during the cruise. This was *America's* first WestPac. We had a normal workup and everything, and got into combat operations pretty good."

The *America/Air Wing 6* team chopped to CTF-77 on 12 May 1968 and started the first of four line periods on 31 May. Seven weeks later, on 20 June, Cmdr. Ken L. Coskey relieved Jerry Patterson as commanding officer. Patterson, the *Whale* veteran who had been told in late 1965 he may never get another chance at command, finished the war with one and a half combat cruises as the *Black Falcons'* executive and commanding officer. In looking back on his tours in Vietnam, Patterson says that despite everything he saw, he only got hit once. It was on a night interdiction mission; one of his own flares failed to ignite, and instead impacted his right horizontal stab, putting a "tree trunk-sized dent" in the leading edge. Unable to get the plane slowed enough for safe flight at the boat, Patterson and B/N Dick McKee diverted to Da Nang. He found that base to be a singularly interesting place to fly, what with getting shot at in the approach pattern.

Constellation was also inbound to the theater for its fourth combat go-around, once again with VA-196, which was making its second appearance. The carrier departed NAS North Island on 29 May 1968 and arrived in theater on 14 June, and as with their Oceana-based counterparts, the *Main Battery* entered the combat zone with several B-model *Intruders* among the mix.

Cmdr. Lou Ditmar was still the CO of VA-196. Among his officers was Bud White, who was now making his first A-6 combat cruise following two in A-3s. He said that as a lieutenant commander he now had his choice of pilots:

> "During the previous cruise VA-196 had lost a *lot* of senior people, and as the incoming Ops Officer I adopted a philosophy: while in the RAG, I looked around for the best, sharpest, youngest pilot and latched onto him. I eventually identified a young guy – and future astronaut – named Dan Brandenstein, and away we went. By the end of the cruise Dan and I took pride in the fact that we could fly an entire combat mission and not say a word to each other. We knew what each other was thinking, and were able to communicate strictly by hand and through gestures. It was a good cruise."

Lyle Bull was still with the *Main Battery*; by now he too was a lieutenant commander, and the senior bombardier/navigator in the squadron. His pilot for the cruise was old friend and *Whale* buddy Lt. Michael L. "Mike" Bouchard. According to Bull, back in their RAG days, on occasion Bouchard would turn over the wheel of their TA-3B to his friend, then head to the hold to show the student B/Ns how he used to do it when *he* was a B/N. That way Bull managed to get some yoke time in the A-3.

Lyle recalls that the restrictions on bombing north of the 19th parallel caused CVW-14 to fly most of its missions over Laos and South Vietnam. He in fact spent a period of the deployment at Nakon Phanom Royal Thai Naval Base working with the Air Force on coordinating A-6 operations along the Ho Chi Minh trail.

Despite the reduced tempo Up North the bad guys were still on the ground shooting at aircraft. VA-196 and VA-85 both lost their first aircraft during the last two weeks of August, and both were the *Iron Hand* A-6B variant. The *Battery* dumped the first plane on 20 August when B-model 151560 went out of control while flying a tanker hop. Five minutes after takeoff the wing slats malfunctioned, and the pilot attempted a zero-G maneuver to aid in their retraction. Instead, the plane rolled 90-degrees right wing down and 30-degrees nose up; the driver managed to recover at 45-degrees, but then the aircraft rolled again 90-degrees to the right. Both of the crewmen safely ejected and were recovered. The pilot was Bud White's usual driver, Dan Brandenstein; flight surgeon Lt. William A. "Bill" Neal, Jr. – later the head of the pediatric cardiology department at the University of West Virginia – occupied the right seat during the unexpectedly short flight, getting his monthly flight time in one of those "good deals."

Nine nights later, early on the 29th, Lt.j.gs. Robert. R. Duncan and Alan .F. Ashall went out in their A-6B and failed to come back. An E-2A last painted them about 10 miles southwest of Vinh, at which point an A-7A several miles to the south called out the launch of three SA-2s. The *Hawkeye* alerted the A-6 crew and got no response; the *Intruder* went off radar about 90 seconds later, presumably a victim of the SAM site they had been hunting. The Navy declared both men KIA.

On 6 September 1968 the *Black Falcons* got the word *Buckeye One* was down. The skipper and his B/N, Lt.Cmdr. Dick McKee, ejected and came down on opposite sides of a small island; McKee was quickly rescued, but Cmdr. Coskey was nowhere near as fortunate:

> "Coskey and McKee went down south of Vinh," JB Dadson recalls. "We got McKee out okay, but Coskey landed on the little island, and there was a village at the other end. He could hear the North Vietnamese coming – the plane was burning on the island – and his ankle or leg was hurt, so he couldn't run much. Other than that, he was fine. The SAR helo went in, tried to get Coskey, but it took so much small arms fire they had to back out. The skipper called on the radio a couple of minutes later and said, 'I can see the torches ... they got me, pull out.'"

Cmdr. Coskey survived his internment at Hanoi and was released on 14 March 1973. VA-85's XO, Cmdr. Charlie Hunter – who a few months earlier had been sitting on Midway Island with a busted airplane, wondering when he would get his turn at command – took over:

> "I came back from the VA-196 cruise as a brand new commander and was still waiting for word on a command tour," Hunter says, "when I got a call saying, 'Hurry, sell your car, fly your family back to Virginia Beach, and buy a house. You've got five weeks to report to VA-85 as executive officer.' We did all that; I made a quick sweep through the RAG, and became XO of VA-85. Five weeks later, the skipper was shot down and captured ... I'm now the CO."

Two weeks later, on 30 September 1968, VA-196 lost its second aircraft of the cruise and the first in combat. An SA-2 caught the A-6A of Lt.j.g. Larry Van Renselaar and LDO-Aviation Maintenance Officer Lt. Domenick A. Spinelli over North Vietnam while they were prosecuting a night armed reconnaissance mission southwest of Phu Dien Chau. Another A-6 in the area observed two SAMs detonate, and approximately twenty seconds later saw a substantial explosion on the ground; an E-2A received one sweep of IFF emergency and nothing more. The following night Radio Hanoi announced it had shot down an *Intruder* over Nghe An Province, and the Navy subsequently declared both men MIA.

The *Buckeyes* and *Milestones* carried on, and within short order a third squadron joined them on *Yankee Station*. The arrival of VA-52 in *Coral Sea* on 10 October marked the first time CTF-77 had three A-6 operators available. The cruise, which began at NAS Alameda on 23 September, also marked the first deployment of the *Intruder* in a *Midway*-class carrier.

The *Knightriders* set to work with the other units of CVW-15. Three nights after their arrival, they suffered their first and only loss of the cruise when XO Cmdr. Quinlan "Quin" R. Orell and B/N Lt. James D. Hunt went down due to unknown circumstances. They had launched at 1831, and some time later Orell called out a "Singer," i.e., a *Fan Song* lock. A few minutes later he reported they had lost the lock and called in base altitude minus 1000-feet, followed by their disappearance. Hunt was killed, while the XO – who was scheduled to relieve *Knightrider* CO Cmdr. Lester W. Berglund, Jr., in January – was placed in MIA status. As a result Cmdr. James A. McKenzie detached from VA-128 and headed to *Coral Sea* to assume the duties of executive officer. He relieved Cmdr. Berglund on 20 January 1969, with Robert H. Kobler moving into the XO slot.

Out With A Bang ... and a Whimper

Marine Captains James Perso and Capt. Don Diederich had an entertaining evening during this period. A lot of what transpired came down to basic switchology and crew coordination:

> "The human engineering of the A-6 was good," Perso writes. "Items that were the pilot's concern were the left side of the instrument panel, or on the left console. Items that were the B/N's concern were on the right side of the instrument panel, or on the right console. Those things that both the pilot and the B/N needed were in the center of the instrument panel, or on the center console. Radios, for example, were on the center console. All of the armament panels were in the center of the instrument panel, except for the Multiple Release Switch, which was located adjacent to the B/N's right knee. It allowed the B/N to select the number of bombs to release. It would select OFF, 2, 3, 4, 5, 6, 9, 12, 18, 24, or 30 pulses to release the selected number of bombs. We called it the 'Dial-a-Bomb' switch.
>
> "Don and I met in VMA(AW)-224, where we trained together. We went overseas about the same time and joined VMA (AW)-242. We teamed up, flew 116 missions together, and were roommates (a good combination; when you return from the same night mission, you won't stumble around in the dark and wake your roommate). On several missions both of us got scared, but on one mission, I scared Don.
>
> "In the fall of 1968 we had a night mission in Route Package One (the southern 60 miles of North Vietnam). Our ordinance load was eighteen 500-pound bombs and ten 500-pound mines. We laid the mines, as planned, in a river, and were about to start an armed reconnaissance to hunt trucks when we were told to contact a high speed FAC with a priority target."

Perso and Diederich checked in with the FAC and received their briefing on the way to the target. The FAC had observed trucks hidden in hooches with tire tracks leading in but not out. He added the crew could expect pretty heavy ground fire, and then added he would drop a parachute flare to illuminate the target.

"I thought, 'Oh boy, night dive-bombing with the tail illuminated by a flare,' something I had not done since weapons deployment to Yuma, Arizona, six months previously," Perso continues:

> "We got to the location of the truck park in the foothills of the Annamite Mountains. The pilot radioed that he wanted six of our 18 bombs on the first run, and that he was dropping a flare. I pulled up to establish a dive bomb run. We popped through a cloud layer that was thin enough that I could see the glow of the flare. I let the nose fall through to start the dive when it seemed right. We punched through the cloud and found the flare way out front of where we thought it would be located. I converted a planned 30-degree dive to a 15-degree dive because I also knew that my target release height would be lower because of the shallow dive angle. I could not see the target yet, so I pointed the refueling probe at the flare to guide me to the target. I closed one eye, so the brilliance of the flare would not night-blind both eyes.
>
> "Eventually I made out the hootches and the target. I was so busy trying to line up that I never saw any ground fire, but Don recalls that we took some fire. I got the pipper lined up on a hooch and mashed the bomb pickle. I held the dive as the bombs came off, expecting six thunders, but there were a lot more. I pulled up and the FAC said, 'Great hits! On the next run get the hut to the south.' I said to Don, 'How many bombs did we drop?' Don replied, 'All of them. Let's get the hell out of here!'
>
> "I radioed the FAC: 'We experienced an intervalometer malfunction. We are ammo minus and headed to homeplate.' I knew that Don had revised the armament switches during the run. I knew how professional he is; I knew that I must have upset him. After we climbed to a safer altitude and got wet feet Don explained that he kept watching that ridge coming while simultaneously reaching down to the 'Dial-a-Bomb' switch and increasing it to nine; watching longer, cranking to 12, then to 15, and finally to 30. The lower we got, the more he twisted the switch. I *did* get a bit low."

On Thursday, 31 October 1968, President Lyndon Johnson ordered the complete cessation of all bombing of North Vietnam. He made his announcement just five days before the presidential election between Richard M. Nixon and Hubert H. Humphrey. Thus, aerial combat over North Vietnam ended, for now; the war itself would continue for another five years. The announcement was a kick in the teeth for the crews who had been faithfully prosecuting the war in Southeast Asia. Former VA-85 CO Jerry Patterson remembers "a lot of angry young JOs" forced to deal with the frustrating ROE, poor guidance, and now this.

Capt. John Peiguss, a former VA-145 A-1 pilot who later commanded VA-196 and VA-42 feels the anger was understandable:

> "It was a combination of Vietnam – not being allowed to fight – and a lot of these JOs who wanted to get in the fight were being plowed back to training command. When they got over here they couldn't believe how things were being run. It was rough on everybody, from the leadership on down through the JOs to the maintenance troops. Even some of the things we did in the Med during this period seemed incongruous."

Through these first 40 months of the conflict a total of seven *Intruder* squadrons made 12 deployments, in the process losing 36 aircraft to all causes. The personnel toll was 31 killed in action or in operational accidents, 8 missing in action, 16 prisoners of war, and 17 rescued or recovered. The two Marine squadrons operating from Da Nang and Chu Lai racked up similar numbers with 11 aircraft lost – including two in mortar attacks – 1 KIA, 14 MIA, and 7 aircrew recovered. Still, the A-6 *Intruder* – and its pilots, bombardier/navigators, and maintainers – had overcome its early difficulties and made a lasting contribution in Vietnam. They had taken out the targets that had frustrated entire air wings, regularly operated in weather conditions that grounded their counterparts, and had rained untold millions of tons of ordnance on the heads of the enemy. Vice Adm. Ralph J. Cousins, outgoing *Commander Task Force 77*, would later express his appreciation and deep encompassing pride for his men during his farewell speech to the fleet:

> "In these years, we have seen Task Force 77 take the war to the enemy – into the very heart of North Vietnam – 'downtown' Hanoi and Haiphong, as the pilots say – into an area where by general consensus the flak and the surface-to-air missiles presented our aircrewmen with the most hostile environment in the history of warfare. In late 1967, in a single day, as many as 80 SAMs were fired at our air wings over Haiphong. We lost aircraft and aircrewmen – and several hundred of the finest men in the world are now in prison in Hanoi.
>
> "But in all that time, the morale of our aircrewmen never wavered. There was never any doubt in anyone's mind but that we could continue to dish it out – and take it – as long as necessary. If there were ever a force – a fleet that came to stay – this is it.
>
> "The fact is that we hit North Vietnam so hard during the fall of 1967 – and during the first few months of 1968 with A-6s – that Hanoi decided they had better go to the conference at Paris and see what relief – and concessions – they could win by negotiation.
>
> "I am certain that the United States has never fought a war in which our young men have been as courageous – as competent – as they have in this one."

On 20 January 1969, Richard Milhous Nixon took the oath of office as the 37th President of the United States of America. In his inaugural address he said, in part:

> "The peace we seek to win is not victory over any other people, but the peace that comes 'with healing in its wings;' with compassion for those who have suffered; with understanding for those who have opposed us; with the opportunity for all the peoples of this earth to choose their own destiny."

Nixon's victory marked the culmination of a remarkable political comeback. The 56-year old former Naval Officer, congressman, senator, and vice president had himself gone down in defeat in the 1960 presidential election; a subsequent humiliating loss in the California gubernatorial election had made him a pariah in his own party. He had overcome these setbacks and – running under the slogan of "Bring Us Together Again" – had acceded to the presidency.

In one real sense, Nixon's election was a repudiation of the Johnson Administration's handling of the War of Vietnam. In reality, the new President faced a daunting task. The nation was torn by racial and political violence, including massive protests against the war in Vietnam. His margin of victory over Vice President Hubert Humphrey was only 510,314 votes out of a total of more than 63 million cast nationwide. In addition, President Nixon would have to deal with a decidedly Democratic 91st Congress ... and that Congress was *not* happy with the conflict in Vietnam. During his campaign, Nixon had spoken of various plans he was formulating for bringing peace, both to the nation and to Southeast Asia. He announced "new leadership" would solve the war, but declined to produce any specific plans or proposals. Officially, he did not want to jeopardize the negotiations in Paris.

However, upon election, Nixon did reveal two initial steps. One was the continuation of the Paris Peace Talks with representatives of the People's Democratic Republic of Vietnam. Johnson had sacrificed his presidency, his place in history, and a large number of American fighting men to achieve these talks, and the new administration intended to continue them. The other was a program whereby the South Vietnamese government would assume more responsibility for its own defense. Nicknamed "Vietnamization" – and directed by the new Secretary of Defense, former Wisconsin Representative Melvin R. Laird – the program would result in a drastic change of America's force laydown in Southeast Asia over the coming two years. Indeed, within the year he would announce the withdrawal of 25,000 Americans from the over 543,000 engaged in the war. Further withdrawals would reduce the remaining forces by half by the end of 1971.

Less was said of the plight of the POWs held by Hanoi, now numbering in the hundreds. As for those crews still fighting the war from Yankee and Dixie Stations, combat operations would continue over Laos and the Republic of Vietnam. What kind of war they would actually fight, no one really seemed to know. Still, the *Intruder* had arrived. Time would tell what part it would play in the resolution of the war.

CHAPTER FOUR

Vietnam, 1968-1971

At the time of the 31 October 1968 total bombing halt over North Vietnam three of the carriers assigned to CTF-77 had *Intruder* squadrons onboard: *America*, with VA-85 assigned to CVW-6; *Constellation*, with CVW-14 and VA-196; and Coral *Sea*, with VA-52 as part of CVW-15. *Ranger*, with VA-165 under CVW-2, was inbound and checked into the theater on 12 November.[1] The *Black Falcons* had been on line the longest, dating to the start of their third combat cruise in April. The *Main Battery* followed in May for their second SEA deployment with A-6s. The *Boomers* of -165 were at the other end of the spectrum; *Ranger's* first line period did not start until 29 November, nearly a month after the suspension of bombing up North.

By that date *America* had departed the line and started back for Norfolk. VA-85 returned to Oceana on 16 December 1968, completing its most successful *Intruder* deployment to date and earning the squadron, airwing, and ship the Navy Unit Commendation. Still it was ultimately frustrating, what with the bombing halt. According to young pilot J.B. Dadson:

> "You could go up the other side of Vinh, but you couldn't go up to Hanoi or Haiphong. It was really frustrating, because we knew our guys were up there and we were giving the North Vietnamese time to build back up. We knew it was only a matter of time before we'd have to go back up there."

As for *Buckeye One*, Charlie Hunter would turn the squadron over to Cmdr. Herb Hope, Jr., on 6 June 1969 after only eleven months at the helm:

> "I'd sent in some reports on the *Intruder* during the cruise," he recalls. "Someone must have liked them, because after I was relieved I went to DC as A-6 Program Coordinator. I had a great tour with VA-85. The amazing thing is that after what happened back with -196, I got to do a combat tour as CO and Lou Dittmar was *still* an XO."

About this time VA-196 also headed for the barn, returning in *Connie* on 31 January 1969. For most of the cruise Lt. Lyle Bull had paired with friend and former A-3 B/N and pilot Lt. Mike Bouchard. According to Bull, the two men's friendly predilection from their A-6 days of swapping seats had continued through this cruise:

> "We came off a bombing hop and went over the ship on autopilot. We safed the seats, undid the straps, I crossed behind him, we hooked up, and we did fine. Steve Richmond and Larry Roberts – an LDO B/N – came up in close, and we were flying formation. Mike took his helmet off – while sitting in the right seat – and looked over at Steve and smiled. Steve *immediately* opened up. We did a loop, and Steve was on my wing; he claims I rolled into him. I got several landings one time – about a dozen – at Kizarazu, a field across Tokyo Bay. My comment was to Mike – pulling his leg – 'All the time I thought you guys (pilots) were special, but this doesn't require shit.'"

The end of the deployment also marked the conclusion of the *Milestones'* first with A-6Bs. According to Bud White, operations with the *Iron Hand* birds had been fairly standard, with one B-model normally escorted by an A-6A carrying either bombs or the AGM-45 *Shrike*.

"The North Vietnamese knew we'd go by cycle," he commented:

> "Funny thing; when they knew the A-6Bs were in the area all of the radars would shut down. Once the cycle ended they'd turn them all on again. Well, at one point we were flying double cycles, so when the radars came back up we had plenty of good targets.
>
> "We did the mission differently than the Air Force *Wild Weasels* and never escorted strike groups into the target area; we'd just sit up there trolling for SAMs. Towards the end, when the A-7s started providing *Iron Hand*, there were places we'd go where they'd refuse to follow, so we went back to the two plane – one bomb, one *Iron Hand* – configuration.
>
> "Dan Brandenstein and I had a good cruise, albeit with one event that left us looking at each other," Bud adds. "On one hop we had a transfer failure from the centerline drop tank; the tank was about half full, and of course you can't recover with a half full drop, so we were directed to jettison it. However, when we punched the button only one hook opened, so now we had a half full drop hanging on by the rear attachment. Now we *really* couldn't recover, so we were ordered to Da Nang. I called back and said we didn't have the fuel; they sent an A-3 tanker up with another A-6 to provide escort service.
>
> "Dan proceeded to put the probe *through* the basket, so now we were *really* in trouble. When we got to Da Nang we were on fumes. This heavy had taken off in front of us, stirring up a lot of turbulence, but we didn't have the fuel to go around. About 100, 150-ft off the deck we hit the turbulence and the plane stalled. The

[1] The other two carriers were *Hancock*, with CVW-21 and its typical "27C" wing of F-8s and A-4s and *Intrepid*, with its modified "all-attack" CVW-10.

normal reaction is to pull back on the stick, but Dan correctly pushed forward, regained airspeed, and we hit the runway. The hung tank hit first and immediately burst into flames, so we rolled out, opened the canopy, and ran like hell. The Da Nang fire crews managed to save the plane."

Events like Da Nang notwithstanding, overall the cruise had seen an improvement over the 1967 deployment, but there were still rough moments. Bull recalls that with the onset of the Johnson's bombing halt the VA-196 crews were briefed that after combat over North Vietnam, missions in Laos would be "a piece of cake." Sure enough, the squadron lost two aircraft in quick succession over Laos towards the end of the cruise. On 18 December one of the -196's *Intruders* was shot down over Laos on a day strike in excellent weather. The aircraft was on its eighth run and was rolling in to drop its last two Mk.82s when it fell out of the sky, killing the crew of Lt.j.g. John Babcock and Lt. Gary Meyer. The escorting planes did not see any flak or tracers, but suspected Babcock and Meyer had taken a burst of ground fire.

The following night pilot Mike Bouchard went down in Laos while on a mission with B/N Lt. Bob Colyar. Their A-6 was on its first run and had dumped 16 Mk.82s at 6500-feet; they were apparently hit by ground fire at approximately 5500-feet during the pullout. Again, the other (A-6 in the flight) didn't observe any visible flak or tracers, but the plane was observed to auger in with multiple explosions some 30 miles southeast of Tchepone. Other aircraft made initial contact with Bouchard – who indicated he'd been injured and burned – and Colyar later checked in.

Word of the two losses quickly got back to the squadron wives in Oak Harbor. Among them was Diana Bull, who was not aware her husband was not flying with Bouchard that day. On the day of the loss, Lyle Bull was sent by his CO to Nakon Phanom RTAFB to work with the Air Force on incorporating the Air Force into their projects. Once he heard his friend was down he attempted to call his wife to let her know he was okay, but when he made contact, the CarGru ops officer – who was with Lyle – had to do the talking. After some delay, CO Ed Bauer's wife finally got word to Diana Bull that her husband was safe. Colyer, the B/N, was recovered after twelve hours, but Bouchard was never seen again. According to Bull, the word was the Pathet Lao did not take prisoners. Both Mike and Lyle had heard about an RA-5C guy who had successfully shot his way out of trouble by carrying a second pistol, so they made it a practice to carry the issue pistol on their vest and a second weapon inside. In Bull's words, he and Bouchard had also agreed they were going to "resist to the end;" and he feels that's what Mike could have done once he hit the ground.

The loss was the third and last for the cruise. VA-196 returned to the Rock on 31 January 1969 and, like its sister squadrons, immediately began preparations for another trip. Still, some memories of the cruise would be perpetuated; the *Main Battery* would later establish the Bouchard Leadership Award for outstanding junior officers in the squadron in memory of their fallen comrade.

The *Main Battery's* departure left VA-52 and VA-165 to hold the line in Vietnam, in whatever form that might take. Back in the state – while the Nixon Administration still considered its options for the war – two more squadrons headed east: VA-65, which deployed in *Kitty Hawk* with CVW-11 on 30 December 1968; and VA-145, embarked with CVW-9 in *Enterprise*. While the *Tigers* and *Swordsmen* worked their way east the *Knightriders* and *Boomers* continued operations. For VA-52 in particular it was a unique situation: they were halfway through their first combat cruise in *Intruders*, while concurrently proving the ability of *Midway*-class carriers to handle the big Ironworks product.

One of the "Turtle Herder" pilots of VA-52 was a rather senior lieutenant commander making his first A-6 deployment following transition training. Among his counterparts, most of whom had come to *Intruders* with light attack backgrounds, Daryl Kerr stood out: he was a former *Stoof* driver, coming from Grumman's twin piston-engine anti-submarine aircraft (occasionally described as "Two T-28s flying in formation with a garbage truck"). But he had fiercely *tried* for attack aviation in order to get to Vietnam, and that still counted for something:

"I was commissioned through AOCS following graduation from Portland State College (Ore) and was designated a Naval Aviator in 1960," he states. "I was then assigned to VS-21 at NAS North Island, flying the S2F-2 and S2F-1S1. After that I became part of the original cadre from the *Fighting Redtails* to form the first *Stoof* RAG, VS-41, and served for two years as an instructor and ASW phase head."

By 1966 Kerr wanted a transfer to a combat unit going to Vietnam. He says that after bugging the detailers for the appropriate amount of time he received verbal assurances he would go to VA-127 at NAS Lemoore for transition to A-4s. However, the "needs of the service" raised their ugly head, and in August 1966 he instead found himself with orders to the *Bonhomme Richard* as Assistant CATTC Officer. Upon querying about the change of orders the detailer said something to the effect of, "Well, you didn't have any jet time, so we couldn't send you to jets."

While at Naval Technical Training Center Glynco, GA, Kerr corrected the omission, gaining some fifty hours in the T-39 over two and one-half weeks, as well as some time in the Basic Naval Aviation Observer school's *Whales* and TF-9J *Cougars*. After completing his tour in *Bonnie Dick* Kerr finally reported to the Jet Transition Training Unit at NAS Kingsville, and then moved on to VA-128 at Whidbey Island. He picked up O-4 along the way:

"When I arrived at VA-52 after completing the RAG I was the senior lieutenant commander in the squadron ... and one of the least experienced *Intruder* pilots in the squadron."

For *Coral Sea* and *Carrier Air Wing 15* it was a curious period. With the initial suspension of strikes north of the 20th, and then 19th parallels, the squadrons went against targets in Route Packages II and III. President Johnson's subsequent ban on all bombing in North Vietnam sent wing aircraft against targets in South Vietnam and Laos. During the Tet period of 1969 uniformly poor weather reduced operations; once again, the *Intruders* got through, although the target environment was nowhere near as profitable as in early 1968. There was one advantage to the curtailment of operations Up North: the frequency of R&R trips went up substantially. In and around five line periods ship's company and air wing personnel from *Coral Sea* managed liberty in Subic Bay, Yokosuka, Singapore, and Hong Kong. They completed their last stint on the line on 30 March 1969 and headed back to Alameda. For the *Knightriders* – with one combat loss – it had been a reasonably good first tour with *Intruders*.

As for VA-165 in *Ranger*, its second combat deployment with A-6s also passed relatively uneventfully. This cruise was unique in that there were five different varieties of "attack" aircraft in Air Wing Two, as well as two squadrons of *Phantoms* (VF-21, 154). Along with the *Boomers* and their *Intruders*, VA-155 had A-4F *Skyhawks*, VA-147 was taking the new A-7A *Corsair II* to sea for the first time, and there were also RA-5Cs (RVAH-6), as well as KA-3Bs (VAH-2 det 61) onboard. Maintenance and parts supply issues aside, the ship went through five line periods totaling 91 days, and finally outchopped from CTF-77 on 10 May 1969. Remarkably, there were no combat losses among *Carrier Air Wing 2*; however, four aircraft – one A-4F assigned to VA-155, one A-7A from VA-147, and two VF-21 F-4Js – were lost in operational accidents with four aircrew recovered.

Ranger returned to NAS Alameda on 17 May 1969, and the *Boomers* rejoined their community at NAS Whidbey Island. They too started preparing for the next trip back to Southeast Asia, but would not deploy for 11 months; by that time the squadron would operate three variants of the A-6 *Intruder*.

Disaster On *Enterprise*

Now led by Cmdr. St. Clair Smith, VA-65 had departed Oceana at the end of December for their third combat deployment in *Intruders*; the squadron was also making its second cruise in an AirPac carrier, in this case *Kitty Hawk* with CVW-11. In keeping with now standard practice, the *Tigers* numbered a few A-6Bs among their compliment. On 6 January 1969, one week after VA-65's departure, the *Swordsmen* of VA-145 hauled out as a component of *Carrier Air Wing 9*. Under the leadership of skipper Cmdr. Niles R. Gooding, Jr., they too deployed with a mix of A and B-model *Intruders*. While the *Tigers* had seen combat before with A-6s, this was VA-145's first time out; for this and other reasons, it would turn out to be an auspicious debut.

On 14 January 1969, while *Enterprise* was conducting an ORI off Hawaii, an explosion resounded among the aft pack on the flight deck. The exhaust from a "huffer" (portable starting cart) ignited a *Zuni* rocket pod mounted underneath the wing of one of the Phantoms. The resulting conflagration destroyed 15 aircraft, killed 27, and injured 344, and it took the crew and air wing personnel over three hours to control the fire. According to *Swordsman* Lt.Cmdr. John Peiguss his squadron managed to dodge the bullet for the most part: `

> "Most of our aircraft had gone off on the first launch, and the remainder were spotted in the pack forward. Fortunately, no one from the squadron was injured or killed in the subsequent fire."

This was the third such disaster to occur during the Vietnam War, following the tragedies onboard *Forrestal* and *Oriskany*. Above and beyond the loss in personnel and aircraft, the big carrier took several hits. Its fresnel lens mirror landing system was destroyed, and several pieces of fragmentation went into the island, one destroying the captain's elevator. The biggest damage, however, was to the flight deck aft, where multiple Mk.82s cooked off. Fortunately relatively few of the bomb detonations opened up spaces below the deck.

Big E limped into the Pearl Harbor Naval Shipyard and remarkably was ready to go again in just two months. VA-145 and the other squadrons of *Air Wing 9* went ashore at NAS Barbers Point and remained there through 11 March, when the carrier loaded up and resumed its journey. She checked in with CTF-77 on 17 March 1969 and began her first line period 14 days later.

Yet "things" continued to happen. On 14 April 1969 North Korean interceptors shot down a VQ-1 EC-121M engaged in surveillance operations over the Sea of Japan, *90 miles* from the Korean coast. The aircraft launched from NAF Atsugi on a routine mission, but at 1350 – about seven hours after takeoff – an Air Force radar site monitoring the flight noted two hostile blips rapidly closing on the *Warning Star*. Pilot Lt.Cmdr. James Howard took the alert warning and immediately started turning away, but the MiGs blasted the aircraft out of the sky. All 30 crewmen onboard the stricken "Willy Victor" died with their aircraft, with only two bodies recovered.

The Navy responded by re-establishing *Task Force 71* in the SOJ to enforce freedom of the sea operations. *Enterprise* departed Yankee Station on 17 April to join CTF-77, where she joined *Ranger* – with VA-165 onboard – *Ticonderoga, Hornet* (CVS-12), and their escorts. Reconnaissance flights over the Sea of Japan resumed in April, and during May *Kitty Hawk* relieved *Enterprise*. The size of the force slowly decreased, along with the threat of renewed action with North Korea. By the time *Enterprise* returned to the firing line on 31 May 1969 its cruise was already almost half over.

During this period the U.S. was reducing its presence in the Tonkin Gulf to four attack carriers at any one time; in fact, *Kearsarge* (CVS-33) completed the last combat deployment by an ASW carrier on 4 September 1969. According to VA-145's Jack Peiguss, despite the disruptions and abrupt break mid-cruise the remainder of the deployment went smoothly:

> "I was used to flying with other people in my aircraft, dating to my initial service in AD-5Qs with VA(AW)-33, so I didn't have much trouble adjusting. The air wing was also used to having A-6s in the complement. Niles Gooding was the senior squadron CO on board, which helped a lot; that, and the *Intruder* had been around for a while now. We were very busy. When you go into combat most of the petty stuff gets pushed away pretty quick."

Enterprise departed the Tonkin Gulf on 2 July 1969, having spent a grand total of 35 days on the line with CTF-77. Not counting the fire off Hawaii, air wing losses could be considered "minimal": one RVAH-6 *Vigilante* shot down with both crew killed, and a single VA-215 A-7B lost in an operational accident. With the reduced tempo in Vietnam, it would be over a year before the *Swordsmen* reappeared. On the other hand VA-65 – with CVW-11 in *Kitty Hawk* – had what could be termed a more "traditional" cruise. The ship underwent five line periods before departing for a total of 111 days on Yankee Station. Total losses included three aircraft in combat and another four in operational accidents. The *Tigers* lost one aircraft on 3 April 1969 when Lt.Cmdr. Edward G. Redden, and B/N Lt. John F. Ricci were shot down over Laos by AAA. Thankfully, both men were rescued.

Retired Rear Adm. George Strohsahl recalls that, except for the one loss the cruise proved to be "benign." At the time of the deployment he was a junior lieutenant commander fresh out of A-4s:

> "There was a lot of heroic action done in the mid-60s, the early days of the war," he comments. "During the period of our cruise we were in the middle of one of several bombing pauses, so our activity was limited to South Vietnam and Laos. The interesting thing – from an A-6 perspective – is we went with three *Standard ARM* aircraft in addition to our bombers, so we were an augment squadron with 15 aircraft.
>
> "We were pumped up; we had 52 officers in the squadron. I was number 13 in a seniority of 13 lieutenant commanders, and the department heads were full commanders. My job was weapons and tactics training, and it started out pretty good, because every day we did something really different. The flexibility of the aircraft in flying with different loads and tactics was really the hallmark; others didn't have the legs to fly double cycles over Northern Laos, so we got the mission. The one crew got bagged by a lucky shot, but they survived, so we ended up bringing back everyone home and all of our aircraft, except for that one. I wasn't flying the A-6B, but those that were had some interesting missions in support of the *Blue Tree* missions over the north (photo-recce missions). In any case, I ended up with 72 missions in nine months."

"There was one interlude in the middle of that deployment," Strohsahl adds:

> "That was when the EC-121 was shot down. *Kitty Hawk* was in port in Hong Kong, and we'd just arrived – had a plane load of wives coming over for mid-cruise – got a whole eight hours of liberty, and then we steamed out and joined *Ranger* and *Enterprise*. We prepared for a possible retaliatory strike against North Korea, but it never happened. We went back on the line in the Gulf of Tonkin and sent six planes and crews to *Enterprise* – the three A-6Bs – and I led the other three to *Ranger*, where we augmented VA-165 for a couple of weeks until everything went away."

As indicated, there were no further incidents involving *Kitty Hawk* for the remainder of the cruise, and on 4 September 1969 VA-65 returned to NAS Oceana. The arrival marked the successful completion of its third and last combat deployment of the Vietnam War.

Green Knights to Cherry Point, *Vikings* to Da Nang

In mid-1969 President Nixon announced the first reduction in U.S. ground forces in South Vietnam, totaling 25,000 troops and support personnel. All of the services started selecting the units that would go home, with an eye towards even greater reductions within two years. Until those reductions – which would serve to cut the forces in Southeast Asia by another 50 percent – took effect, there were still over 500,000 involved in combat on the ground, and the level of violence did not subside. Adding to the moment, the North Vietnamese continued to shift supplies southward, taking advantage of the lengthiest "let's catch up" period they had yet seen. Thus, while the air war for those stationed in the Tonkin Gulf "lightened" somewhat, the in-country units still had plenty of work to keep them occupied. It was under these circumstances that the U.S. Marine Corps established its sixth – and last – *Intruder* squadron while dispatching a third to Vietnam.

On 14 February 1969 the *Green Knights* of VMA(AW)-121 stood up at MCAS Cherry Point. Previously assigned to MAG-12 at Chu Lai, the squadron had turned in its A-4Es in mid-1968 and became a paper outfit, with personnel scattered to other units. Once it re-manned with pilots and bombardier/navigators the *Knights* entered training with VMAT(AW)-202, receiving their first A-6A on 24 February.

Another veteran unit, VMF-121, was initially established at MCAS Quantico, VA, on 24 June 1941. They would see notable combat in World War II, albeit via the "scenic route," disbanding at Camp Kearney, CA, on 28 February 1942, and reorganizing the following day. Equipped with F4F-4 *Wildcats*, -121 departed the states on 30 August 1942, and after a brief stop in Noumea, New Caledonia, moved on to Guadalcanal. The squadron participated in the defense of that island from October 1942 through January 1943, running up a magnificent record against overwhelming Japanese forces. One member, Capt. Joe Foss, scored 26 kills, making him the all-time number two ace among Marine aviators. In recognition of his record and combat leadership Foss received the Medal of Honor. The following eight months saw VMF-121 operating from Espiritu Santo and the Russell Islands, with participation in the invasion of Rendova and New Georgia in the Solomons. In August 1943 the Marines broke up the squadron, with its flight personnel assigned to other units and ground crews sent back to San Diego.

VMF-121 subsequently reformed at MCAD Miramar, reequipped with F4U *Corsairs*, and prepared to reenter combat. On 18 July 1944 the squadron went aboard *Kwajalein* (CVE-98) at NAS Terminal Island, CA, and headed west again. Through the end of the War in the Pacific the outfit engaged in combat operations in the Marshall Islands and other locations before ending up at Peleliu; it returned to San Francisco on 6 September 1945 and disestablished at Miramar three days later. VMF-121 finished the war with 208 enemy kills, the most by any Marine squadron in World War II, and the Presidential Unit Citation for their service at Guadalcanal.

The squadron returned to service as a reserve fighter outfit on 1 July 1946 at NAS Glenview, IL. Called to active duty for Korea in 1951, VMF-121 transferred to MCAS El Toro, reequipped with AD-2 Skyraiders, and redesignated as an attack squadron 15 May 1951. The next departure for the newly renamed *Wolf Raiders* came on 2 October when they went aboard Sitkoh Bay (CVE-86) for a short jaunt to Yokosuka. VMA-121's arrival at Pohang/K-3 airfield on 22 October marked the commencement of two years' aerial combat in Korea; the *Raiders* flew their first combat sortie on the 27th, led by CO Lt.Col. Alfred N. Gordon. The squadron ended its part of the "police action" at Pyongtaeng-ni/K-6 and remained in South Korea for several years following the armistice.

After moving back to El Toro, VMA-121 went through a period with F9Fs before transitioning to the A4D-2 *Skyhawk*. February saw the squadron – now nicknamed the *Green Knights* – arrive in Japan for a year's tour; the squadron went aboard *Coral Sea* on 5 October for a six month WestPac deployment with CVG-15. With the conclusion of the cruise on 10 April 1961, -121 returned to El Toro, where it upgraded to the A4D-2N. The Cuban Missile Crisis the following year sent the squadron to NAS Jacksonville, where it remained until 2 December. The *Knights* made yet another trip to Japan in March 1964, remaining for one year, at which time they moved without personnel and equipment back to MAG-33 at El Toro.

On 25 August 1966 the first squadron A-4Cs flew across the Pacific via MCAS Kaneohe Bay and Wake Island for operations from MCAS Iwakuni. On 14 September eight pilots and planes reported to Chu Lai for initial familiarization operations; the entire squadron fol-

VMA (AW)-225 was the shortest lived Marine *Intruder* squadron, operating the type for about six years, half of which was in combat from Da Nang . The *Vikings* returned to CONUS in June 1972, only to be deactivated. They would return as an FA-18D squadron in 1991. CE04 sits on alert with a load of 12 Mk.81 250-pound bombs. (Steve Dumovich)

lowed in December. Operations from that austere base continued through 1968, with occasional breaks at Iwakuni and Naha Air Base on Okinawa. The *Green Knights* continued supporting Marine ground forces and those of other services and nations until their stand-down and subsequent transfer to Cherry Point.

Concurrent with the arrival of VMA(AW)-121 in North Carolina, VMA(AW)-225 moved to Da Nang AB to join the *Bats*. By now known as the *Vikings* – the Corps approved the name and emblem change on 7 October 1968 – the squadron immediately went into combat in support of I Corps. 1st Lt. R. James "Diamond Jim" Garing was one of the Viking pilots during their first year in Southeast Asia. He writes:

> "I arrived at Da Nang on 17 July 1969 and departed 19 June 1970. Commanding officer at the time of my arrival was Lt.Col. Don Harvey, followed by Lt.Col. John Metzko. When I left Da Nang, Maj. Pete Busch was the commanding officer.
>
> "During most of my tour the squadron operated its 24-hour a day schedule guided by a team of six 'PRODOs' – Professional Duty Officers ... if you were a 'PRODO,' you were on six hours a day, seven days a week, unless otherwise relieved by one of the two spares. This system worked well, because in fact, we 'PRODOs' knew everything there was to know about the availability of airplanes, systems, ordnance, maintenance needs, and who could fly what with airplane! The above was important, because out of our total number of squadron airplanes (12 or 13), typically three, maybe four would be complete, all-up systems airplanes. The rest of the airplanes would have varying degrees of system availability, being relegated to various levels of combat need. 'Iron bombers' could be used for close air support during day VFR. All of the rest of our missions required some degree of systems availability. For IFR operations, a search radar was the minimum system required, because of the penchant for controllers to turn you on a base leg towards the mountains outside Da Nang and forget you on that course."

According to Garing, the squadron flew the following types of missions, in order of increasing systems demand:

> *Close Air Support (CAS)* – Self-explanatory; 30º dive, ordnance delivery by pilot-controlled release. Absolutely forbidden under flares at night; also, A-6 drivers did not get much practice at this and were reputed to be not as accurate as the A-4 drivers.
>
> *Ground-based radar control delivery (TPQ)* - Ground-based radar directed your course and altitude to release. Very close control of altitude and heading was required during the last seconds, awaiting the ground command – "mark mark" – designating the release point. typically, one half to one degree heading corrections were given during the last seconds before release.
>
> *Direct Air Support Alert (DAS Alert)* - Kind of a standby "Close Air Support mission," although occasionally on these missions the "bore sight mode" was utilized, wherein the pipper on the gunsite was used to tell the computer where the target was. In bore sight mode, delivery was by aircraft systems utilizing the computer and commit switch on the stick. Release occurred during a 4g pull-up, when the computer saw the proper ordnance release window.
>
> *Beacon (BCN)* - This mission required complete aircraft systems, including track radar. Delivery was by straight path (unaccelerated), computer-released. Target acquisition was through the utilization of a ground-based radar beacon that displayed brightly on the radarscope. The Beacon was typically operated by a Forward Air Controller, and the target was given as an offset from the Beacon. Because the Beacon was co-located with friendly troops, the first run on a Beacon run was always cold, except under emergency circumstances.
>
> *Armed Reconnaissance (AR)* - the armed reconnaissance was an out-of-country mission, to "other." From mid-1960 to mid-1970, "other" was Laos. Since Laos was a free fire zone, Armed Reconnaissance mission allowed us to pick targets of opportunity through the use of the AMTI (Airborne Moving Target Identifier) mode of the search radar. In AMTI, a moving target on the ground showed up as a bright blip. In the event that we were unable to locate a "mover," we always had a secondary target identified by coordinates during the mission brief, upon which we expended the four Delta 4s (Mk83 1000 pounders). The Delta 8s (5" Zuni rockets) were too expensive to expend unless we located a "mover."
>
> These missions were often frustrating, because the "movers" knew the sound of an A-6, and as soon as they heard us, they would stop. A good B/N could get the cursor on the target before he stopped, or under the best of circumstances, get a track lock-on before the mover stopped.
>
> *Commando Bolt (CB)* - The Commando Bolt was a modification to the Armed Reconnaissance, wherein airborne and ground-based command centers utilized sensors placed along the heavily traveled highways in Laos, principally the Ho Chi Minh Trail. Through the use of these sensors the airborne and/or ground base command centers would sense the presence of a "mover," or group of movers. The typical mission would also include use of an F-4 as an AAA CAP. The F-4 would typically carry a load of CBUs for suppression of groundfire.
>
> Both the A-6 and the F-4 would proceed to the general target area and begin to loiter at maximum endurance airspeeds until a target was identified, or else bingo fuel was reached. In this case, bingo fuel was sufficient fuel for the A-6 to deliver ordnance to an alternate target and then return to base. If the airborne or ground-based command center picked up a group of "movers," coordinates were given for the movers, and we would begin a run-in to the target with the F-4 providing cover.
>
> *Barrel Roll (BR)* - We very infrequently would fly into far northern Laos, far enough north to be opposite Hanoi, but still in Laos. These missions required reduced ordnance and the addition of two drop tanks for additional fuel. Depending upon fuel states, we might bingo to Udorn or Ubon, Thailand. Other than the length of mission, the Barrel Roll mission was similar in character to an Armed Reconnaissance.

"(The) ordnance load varied," Diamond Jim continues:

> "We considered 28 Delta 2s (500-pound bombs) as the standard load (varied fusing). This load was typically used for Direct Air Support, Close Air Support, Beacon, TPQ, and Commando Bolt missions ... typical Armed Reconnaissance ordnance loads (consisted) of four Delta 4s (1,000-pound bombs) and four Delta 8s (rockets). When we carried CBUs it was generally for AAA suppression, and the individual bomblets had a mix of fusing to allow both instantaneous and varying levels of delay before detonation. The lowest drag index load we carried was the five 2,000-pound bomb load with standard bomb fins. The highest drag index load carried was the 28 Delta 2 load with banded *Snakeye* fins.
>
> "My log book shows a total of 138 combat missions, of which 56 were out of country to 'other,' 73 were night, and 21 resulted in arrested landings at Da Nang using the MoRest gear. Hung ordnance/wet runway day landings were required to be arrested, and

wet runway night landings were required to be arrested. Since the A-6 had no indication as to whether or not ordnance had in fact been released from all rack stations, all missions carrying ordnance were assumed to be hung ordnance upon return to Da Nang. In fact, it was frequently the case that we did have hung ordnance, but were unable to detect it because we could not see the centerline station. This did lead to bombs skittering along the runway after touchdown once in a while.

"With regard to my total of 138 combat missions, the average in the squadron after a year of uninterrupted flying was close to 180 to 200. In my case, I spent three months in the field with 2d Battalion/1st Marines as a Forward Air Controller learning much, not least of which was to step into the footprint of the man ahead of you. During this period I was located just south of Da Nang. The neighboring battalion's Forward Air Controller at the time was a friend of mine, Bruce Cruikshank, an A-4 driver out of Chu Lai.

"Being out in the field as a Forward Air Controller was serious business, particularly in this area of Vietnam. 2d Battalion/1st Marines lost a man a day (approximately) to booby traps during the period I was with them. In addition, Bruce Cruikshank unfortunately ran afoul of a mine, losing both legs in the process."

While all Marines train as Marine *infantry* first and foremost, most of the crews felt in the air they at least had a hand in their survival. Still, the *Vikings* lost two aircraft during their first year at Da Nang; fortunately, all four crewmen survived and were rescued.

The first *Intruder* loss came on 21 September 1969, when Maj. Pete Busch and B/N 1st. Lt. R. "Kip" Hardgrave were blasted out of the sky by AAA. In Jim Garing's words, the crew was:

> "Accompanied by another aircraft in section on a close air support mission near the DMZ. While in a 30-degree dive on their second pass Pete Busch noticed the gooners had shot off one of the wings. That made the airplane difficult to control, and therefore both punched out. Busch lost one large toe in the ejection, and Kip Hardgrave – unfortunately – was more seriously injured, severing a nerve in his right arm in the process. My belief is that Kip did not return to duty as a result of these injuries. Pete Busch, on the other hand, came back to us several months later as Commanding Officer, being known as the Marine Corps' only nine-toed Major. As a footnote to this, Pete Busch much later lost his life as a civilian while on a flight in a civilian airplane, campaigning for Congress."

The second -225 bird went down on 16 November 1969 when two *Viking* first lieutenants were forced to step out at night. Garing recalls:

> "An aircraft piloted by Lt. Jess Jensen, accompanied by Lt. Dean Tutor as B/N, suffered a ground fire hit shortly after takeoff from Da Nang. Incidentally, for this reason, the protocol was a sharp left turn after takeoff out of Da Nang, insofar as one can accomplish that in a loaded A-6 (60,000+ pounds). Pursuing this one thought, takeoff roll with 28 Delta 2s and full fuel typically ran 7,000+ feet. Rotation speed was 180-kts. Because of marginal maneuverability, flaps were left at takeoff setting until the aircraft passed through 2,000-ft.
>
> Continuing on with Jensen and Tutor, upon being hit with ground fire, the aircraft immediately lost one engine, whereupon Jess rightfully punched off the ordnance. To add to the thrill, this was a night IMC hop. Continuing around the GCA pattern to try to regain Da Nang, the fuel gages were observed to rapidly decline, and the remaining engine flamed out on long final. Both pilot and B/N were rescued. There were some colorful moments captured on tape when Dean Tutor, on the way down in his chute, yanked out a survival radio to broadcast on guard as to where he was and how to pick him up, in no uncertain terms!"

The *Vikings* finished 1969 with 9,661 sorties, in the process unloading over 51,000-tons of bombs on the collective heads of the enemy. In doing so, -225 managed an overall 75.5-percent "up" rate on its jets, while managing somewhat less reliability with the A-6A's cranky systems. Thirty years after the fact Garing – now the president of a small engineering company in Southern California – recalls his squadron's first year at Da Nang with satisfaction. At least they knew how to blow off steam, as he relates:

> "We naturally were a wild bunch while in Vietnam, routinely blasting away at parties with tennis ball guns, pencil flares, and – after all else failed – 'arc lighting' squadron members during the wee hours. 'Arc Light' was the code word for multiple B-52 strikes. A typical arc light by our definition consisted of a dozen or more squadron members appearing at your doorstep after 0200 to regale you with war stories and drink your liquor supply."

Somehow, one feels the Air Force *Buff* crews – safely ensconced at Anderson AFB on Guam – didn't have near as much fun between missions. Then again, unlike their Marine counterparts, they *did* have to fly a long way to get into combat.

Funny ... They're Still Shooting at Us

The two other Marine squadrons in Vietnam – VMA(AW)-242 at Da Nang and -533 at Chu Lai – also had their hands full during the bombing pause. The *Batmen* hit a milestone on 16 January 1969 when CO Lt.Col. Adnah K. Frain recorded the squadron's 10,000th combat mission in Southeast Asia.

In the *Bats'* history for the period, recorded by the Marine Corps' Historical Division, the squadron also reported:

> "November of 1968 brought an end of the bombing of North Vietnam, and A-6 operations were contained to the Steel Tiger Laos missions and TPQ, CAS, and experimental-type 'beacon' missions in South Vietnam. These missions continued through the early months of 1969. Later in that year, the A-6 beacon missions were promoted to the extent that most Marine Corps units in the Da Nang area possessed an operating Motorola Beacon and '242' began flying six Beacon missions a day. By the end of 1969, the ground units had learned to depend on the A-6 all-weather beacon bombing. A total of fifty beacons were being actively used in the field by Marine and Army units as well. Beacons were operated with the 101st Airborne, 1/5 Mechanized Infantry, the Americal Division, and Special Forces units and all Marine Corps battalions.
>
> "In December 1969, the '242' A-6s began flying in the *Commando Bolt* area of Laos. Simultaneously, *Barrel Roll* operations began and A-6s were sending fifty percent of their missions into Laos for interdiction missions."

Jim Perso and Don Diederich, still with -242, had another one of their interesting flights during this period. They launched one fine day on a TPQ mission, but this one was during daylight, and Perso writes the ordnance strapped to their trusty *Intruder* was "unusual":

> "We had ... five 2000-pound bombs of World War II manufacture with a fat, blunt, high drag profile and box fins. Since Easter was approaching, our artistic squadron mates had repainted the bombs as psychedelic colored 'Easter Eggs.' We got airborne and checked in with the DASC, who diverted us from the planned TPQ and told us to contact a FAC flying around the Ashau Valley. We checked in with the FAC, and a flight of A-4s checked in be-

hind us. The FAC briefed us on a target he wanted us to hit with our 2000-pounders, a bunker in the middle of Ashau Valley. That presented a small problem. We were under direct orders from the Group CO to not perform any single aircraft dive-bombing. I radioed the FAC a terse version of the order. The trailing A-4 leader radioed that he would be our wingman. That seemed to make it legal, neat, and tidy.

"The next problem was that we did not have any ballistic information on the WWII vintage – mils of sight depression. Kentucky windage is OK for drift, but this was really pulling numbers out of the air. The FAC described the target as a bunker. An earthen bunker, in the middle of a field not easy to see up close and from a slant range of 20,000-feet is impossible. I rolled out looking for the bunker; Don was calling the altitude, airspeed, and dive angle. About halfway down I thought I saw something, corrected the run, and watched. I was lucky, as I approached the release altitude, I was sure that it was the target. The pipper moved through the bunker at the release altitude, at the proper airspeed and the correct dive angle. I pushed the pickle, releasing three of the bombs, got the aircraft coming up, and then radioed the FAC, 'Three Easter Eggs – on the way.' The FAC said, 'What did you say? Oh wow, that's neat.' The FAC followed shortly in an excited voice, 'Bull's eye! You demolished it. Great hit!'

"As we turned downwind I inquired, 'Where do you want the last two?' He replied, 'Hit your smoke.' The next run was easy to start, with the huge column of smoke and dust. I pickled the last two bombs with the pipper centered on the base of the dust cloud. I'm not sure what we hit, but the FAC was ecstatic. We checked out with the FAC and thanked our A-4 wingmen. They replied with, 'Enjoyed your show!' We returned to Danang, and Don and I commented that we had never heard a FAC get so excited. We did not know then and to this day we have no idea what was in the bunker or why FAC was so excited. Maybe it was the bombs painted like Easter Eggs."

On the down side, both squadrons continued to suffer loses. On 17 January 1969, a -242 *Intruder* went down while supporting operations in the same Ashau Valley. A *Red Eye* Army OV-1 lost contact with the plane, and it was never heard from again; the pilot, Capt. Ed Fickler, and B/N 1st LT Robert Kuhlman were declared missing in action. On another mission the night of 17 March a VMA(AW)-533 A-6A flew into the ground – probably due to AAA fire – while on an armed recce mission. The FAC saw an aircraft crash and burn at 2130; other aircraft in the vicinity saw one explosion in the vicinity of the target and then a second explosion. No one saw any parachutes and no emergency beepers went off, and after a few days the military terminated search and rescue efforts. The Marines subsequently declared Maj. Charles Finney and his B/N 1st Lt. Steve Armistead Missing in Action. Later in the year, on 29 September 1969, another *Bat* disappeared. This time it was squadron executive officer Maj. Luther J. Lono. Once again, the plane went out and did not come back, adding Lono and his B/N, 1st Lt. Pat Curran, to the list of MIAs.

Still, for all good *Batmen* and *Night Hawks* there continued to be plenty of trade. Former -242 pilot Jim Perso writes:

"The A-6 *Intruder* possessed a unique capability to hunt trucks by using the 'system.' Trucks were attacked at night, or in bad weather, or both. The search radar had an AMTI capability. When the AMTI was enabled and the vehicle exceeded a minimum speed, the doppler shift of the return radar echo was sensed and displayed on the B/N's scope.

"Quite often the roads we were interested in searching were in valleys, surrounded by mountains. Good B/Ns could enable AMTI, canceling just enough ground return to pop out the 'movers,' but still see the mountains to navigate clear of a sudden and

Even if the *Intruder* was considered an "all-weather" aircraft, tropical conditions could wreck havoc with the A-model's DIANE system. The Marines had to put up with heavy rain at Da Nang on a frequent basis. (Wolfe, via Steve Dumovich)

final stop. We usually cruised through the valleys around 400-kts and 3,000-ft AGL. The speed was fast enough to maneuver, allowing just enough time for the B/N to identify a target and set up an attack, and for the pilot to fly the attack commands on the VDI. The altitude was above small arms fire, but right in the AAA zone. These missions were termed an Armed Reconnaissance, usually referred to as an armed recce (pronounced *wreck KEY*). After President Johnson canceled the *Rolling Thunder* missions to deep North Vietnam armed recces in Route Package One were the typical A-6 direct air support missions. When all missions to North Vietnam were suspended, we flew armed recces in Laos (along the Ho Chi Minh trail supply route).

"During the spring and summer of 1968 trucks were attacked with CBUs. The preferred ordnance to attack trucks varied over the course of the war. CBUs were a clamshell canister containing many softball-sized bomblets. The CBU was fused to open the canister in the air and spread a large pattern of the bomblets, giving good coverage on the ground. CBUs generated a lot of fine shrapnel. Truck tires were punctured and gas tanks were pierced, but all too often the trucks were only damaged. They were recovered, repaired, and returned to service. Later we used 500-pound bombs with daisy cutter fuse extenders. A daisy cutter is an 18- or 36-inch 'pipe' filled with high explosive that screws into the bomb fuse well and accepts the fuse in the other end. The bomb explodes when the fuse hits the ground, increasing the effectiveness of the shrapnel for targets at ground level. These weapons were more effective in destroying trucks, with the tradeoff of a sparser coverage over the ground."

All three Marine *Intruder* squadrons were still battling systems reliability with the A-6A. However, on occasion, the crews were graced with full, up systems that worked perfectly. On those strikes they were rewarded with the instantaneous, gratifying view of a tough target going ka-boom in the night. Perso discusses one such mission that took place on 19 May 1969. His B/N for the hop was Capt. Don E. Diederich:

"In May 1969, both Don and I were getting short. We both had received orders to the 2nd MAW at MCAS Cherry Point, NC. This night we were assigned an armed reconnaissance mission along Route 9 in Laos. We had been there before, and knew that it was well defended by 23mm and 37mm AAA.

"This would be my 197th mission and the 273rd for Don. The squadron had more pilots than B/Ns, so the B/Ns flew more. We adhered to the typical A-6 aircrew routine. We attended the evening briefing for A-6 aircrews at 1700 and then went to supper. After sunset we went to the Tun Tavern (the officer's club), watched the movie on the patio, and had a soda. The 'pilot van' (a 6x6, 2 1/2 ton truck) took us to the flight line. We were airborne a little after midnight. It was a dark and virtually moonless night.

"We headed northwest into Laos. As we crossed the mountains in western Vietnam and eastern Laos we let down towards the valley containing Route 9. Don set up the radar AMTI to look for 'movers.' Shortly after we entered Laos, Don spotted a mover (a single truck) on Route 9. He set up the armament panel, selecting two *Rockeyes*. Don locked on and stepped the computer into attack. The computer generated steering to the release point and I followed it. As we were executing the attack we took AAA fire. The NVA gunners were good; we had a number of airbursts very close to the aircraft. I began to jink, and Don came out of the boot to see what was happening. The flak was on the right side and getting closer. Don said, 'Break left!' I broke left, hard. The fire continued to follow us until we were well south of Route 9.

"We wondered about how the gunners were able to deliver such accurate fire. There were no indications of any search or fire control radar. I double-checked that the external lights were off, and it was a moonless night. We even wondered if the gunners were using a 'star-light' scope to spot and track us. By this time I had continued the turn so the aircraft was heading southeast, back towards Vietnam. We decided simultaneously that 'this is bullshit, and we need to get that bastard.' We set up the attack again, heading northwest."

Once Person and Diederich turned back to make another run, things got even more interesting:

"Don reacquired the mover, stepped the computer into attack, and I followed the steering. We were fired upon again. I was jinking the aircraft, and the AAA tracked us. But this time we completed the run. Two *Rockeyes* were released. I broke hard left and jinked out of the area. Don saw several secondary fires in the target area. We brought home four of the six *Rockeyes*. We also brought home 22 Mk.82s we had on the wing stations. This was one of the few times I dropped *Rockeyes*. The next day an RF-4 took photos of the burned-out truck located where we dropped. This was the only time I was told of BDA confirmed by photo-reconnaissance. I assume the intelligence section crosschecked debrief information with the immense amount of daily aerial photos. I presume that we were (usually) not told about BDA so that if we were captured, we could honestly claim ignorance of the results of our missions."

Late in the year, the VMA(AW)-225 crew of Capt. Jim Henshaw and B/N 1st.Lt. Fred Amend had their own late night festivities. Again, the mission involved nocturnal prowling for moving vehicles, only this time, their efforts were coordinated with an Air Force controlling agency. The hop took place at 0100 on 17 December 1969. As background, pilot Henshaw writes:

"At the time we were having very little success in locating truck convoys on the AMTI. Since this was the one feature that made the A-6 attractive to the Air Force during that period, the heavies were just as frustrated as the aircrews with our lack of BDA.

"We finally figured out, after analyzing the daily reports from the listening stations for a 'gazillion' or so acoustical sensors which the Air Force had air-dropped along the Ho Chi Minh Trail, that the average north Vietnamese convoy moved at about 1.8 nautical miles per hour. The A-6 search radar system, however, required a minimum of 4-knots of target speed to show up as a 'mover' (technology defeated again by dirt roads). Late in the year, somebody (rumor has it was Maj. Dick Skelton, the A-6 liaison to Seventh Air Force in Saigon) came up with the idea of combining the capabilities of the acoustic sensors and the A-6 Intruder. Certain terrain features along the trail that were visible on the radar and distinctive enough to be easily identified were cataloged, and offset bearings and distances developed from these points to various 'choke points' in the Ban Karai and Mu Gia areas. Now, the choke point might be a stretch where the road ran across the face of a steep hillside, or maybe through a particularly marshy area. In all cases, it was somewhere that the trucks had to travel single file and could not pull off the road and hide under the trees.

"An A-6 would launch out of Da Nang with 28 Mk.82s and proceed to a designated holding point west of the DMZ in Laotian airspace. They would check in with the sensor-controlling agency up in Nakon Phanom, Thailand, whose call sign was 'Copperhead,' and settle down to wait for some traffic. Copperhead monitored all those gazillion sensors and would let the A-6 know when they heard trucks. Copperhead controllers would then work up a direction of movement and speed on the convoy and DR them ahead to the next choke point. Once they had developed an ETA for the convoy at the choke point, they would query the A-6 if they could drop on, '... point Alfa-Charlie at time '44?' The B/N would then enter the coordinates of the terrain feature associated with Alfa-Charlie into the computer, along with the offset bearing and distance, cycle steering, and let the computer come up with seconds to release. If that number of seconds worked out to time '44 or earlier, they would give Copperhead an affirmative and start positioning the aircraft for the run. They would push over from their holding point and regulate their descent so as to be 'bombs away' over point Alfa-Charlie at exactly '44, with the airplane going as fast as its blunt profile and the drag of 28 Mk.82s would let it go. The whole procedure had the code name of '*Commando Bolt*.'

"The results were spectacular! A-6s had more confirmed BDA in Laos in the next two weeks than we had over the rest of the armed recce program combined. At one night flyer's brief I attended at the MAG-11 CP, the S2 presented Lt. Jim King of -225 an 8x10 glossy photo of 14 burned-out truck chassis taken by an RF-4B from VMCJ-1 at his *Commando Bolt* drop coordinates from the night before. It was a 'turkey shoot.' Unfortunately, as with most successes in Vietnam, it wasn't long before the 'turkeys' began shooting back.

"After about two weeks of basically unopposed *Commando Bolt* ops, the ridge lines which defined the Ban Karai and Mu Gia passes began to sprout guns. For a couple of nights things got pretty colorful. Wing (1st MAW) finally got the idea of sending a 'flak suppression' bird along with the strike A-6. So, an F-4 from VMFA-542 would load with wall-to-wall CBUs and fly in a three-mile radar trail (the F-4's radar, that is) behind the A-6. When the A-6 started their bomb run, the F-4 would parallel track at a higher altitude, from which he could initiate a visual bomb run. As the blacked-out A-6 down in the valley went by the first NVA gunner, he would start shooting at the noise. As his buddies saw his tracers go up, they would start to shoot, laying down a barrage fire that looked just like the now famous CNN video clip of downtown Baghdad on the first night of Desert Storm. The A-6 would transmit via a pre-briefed code word for 'bombs away,' and pull up and out of the pass in a pre-briefed direction so that the F-4 – who's been watching the light show from altitude – would know roughly where the A-6 was. The F-4 was now center stage, and the last set of muzzle flashes got ten CBUs down the throat. Advantage MARINES! The next escalation would prove to be a SAM site, but that came later."

Henshaw and Amend's turn in the barrel came on the afternoon of the 16th, when they learned they had been assigned a *Commando Bolt* mission. Henshaw recalls:

> "I was getting kind of short and thought this might be my last out of country mission (it was), and I wanted to experience this latest tactical innovation before I went home. Fred and I went to the Ringneck 1600 night-fliers brief, then went over to the squadron area to brief with Capt. DaHart and his RIO, who were going to fly our escort. Adrenaline was running high as we manned our aircraft and departed. The F-4's radar was working 4.0, so the join-up and transit to the orbit point was uneventful.
>
> "With 28 Mk.82s aboard the A-6 had a (very high) drag index of 195, which is somewhat like flying with the (landing) gear down. Consequently, we had about 20 minutes of loiter time at the orbit point before we had to either drop or leave. As we hit the holding fix and settled into our racetrack pattern, we asked Copperhead if he'd been busy that evening, and he said no, it had been very quiet. About four laps later there was still no activity from the sensors, and it looked like we were going home empty-handed. We told *Copperhead* that we were approaching 'bingo' fuel and asked him if he had a secondary target. He gave us the coded point for a suspected truck park and a drop time that was easily makeable. Fred typed all the data into the computer, and once he had the radar cursors dressed up on the offset aimpoint, I pushed the nose over and we started down.
>
> "Descending at full power, you can almost get 500-knots out of an A-6 – even with that bomb load – and that's about what we were doing as we leveled off at 3,000-feet and steadied into the run-in heading. The airspeed began to drop off smartly once we were level, and about that same time the ridgelines started to light up. I was genuinely looking forward to getting the hell out of there when the aircraft began its familiar shudder and five six-packs of 500-pounders departed. The speed had bled off to about 420-kts by this time, but being suddenly 15,000-pounds lighter, the bird was ready to go. As I took a peek out of the top of the canopy it seemed that our pre-briefed left pull off was not going to be a good idea, due to the large number of airbursts currently popping on that side. I told Fred, 'I'm going right,' and transmitted 'Reverse,' so Reid would know that I was going the other way. I pulled back hard, and once the nose was up about 20 degrees, I rolled into about 75 degrees of bank to the right and kept pulling. Through 7,000-feet we seemed to be above the airbursts, so I relaxed the back pressure and just let her climb to about 20-30 degrees nose up.
>
> "About this time I heard Reid transmit, 'Son of a Bitch!' and I looked back in the mirrors to see what was up. All I could see was a bright orange glow. I reversed my turn back to the left and honked on the nose to let me look back down into the valley. There was the biggest orange fireball I'd ever seen, and it seemed to just hang in space for several seconds. I thought DaHart and Reid had gone in, but just about that time he said, 'Boy, you sure hit *something* down there,' and I knew our escort was okay. I was still honking the nose around to see if there were going to be any more explosions when Fred said over the intercom in an utterly calm voice: 'There's 80 knots.'
>
> "Numerous thoughts were racing across my mind as I looked back inside the cockpit at the instruments, mostly relating to how I could have been so stupid, along with a brief review of ejection and E&E procedures. I found us on our back at about 14,000-feet with the nose about 60 degrees above the horizon and about 20 degrees left wing down, with the airspeed indicator stubbornly stuck on zero. On the plus side, both J-52-P8s were running at full power, and the airplane was now nearly clean with only five empty MERs for external drag."

Faced with an *Intruder* that had effectively stopped flying, Henshaw recalled a recovery method he had learned from another Marine pilot during his RAG days:

> "One of the many fine qualities of the A-6 was that the airplane really wanted to fly, and if you didn't really screw around with it very much (like initiate spin recovery when you really weren't spinning), it WOULD fly. This fact had been impressed on me during my training in the airplane by Capt. Steve Palmason (who is still the best A-6 stick I ever saw). Steve formulated the 'Three-Step Palmason Zero-Airspeed Recovery Method.' I began reciting his method that night as we sort of hung in space. It goes like this:
>
> 1. With the throttles full forward, let go of everything
> 2. Now with your right thumb, push gently forward on the back of the stick
> 3. Resist the urge to maneuver at 150-kts
>
> "The forward stick pressure took all the aerodynamics load off the aircraft and let the engines devote their full effort into producing speed. As the nose fell through the horizon, the airspeed needle wiggled off the peg, and I mumbled a thank you to Palmason. We were still above 10,000 as we hit 200-knots, and I rolled the aircraft upright. 'Whatdayasay, let's go home?' I asked Fred. 'Good idea,' he answered. We were pretty quiet the rest of the way back to Da Nang.
>
> "We never did find out what blew up on the ground."

The three Marine *Intruder* squadrons continued to take the war to the enemy through the end of 1969. In January 1970, as part of the ongoing Vietnamization of the war, VMA(AW)-533 would depart for MCAS Iwakuni; they were followed in September by -224, leaving the *Vikings* of -225 as the sole remaining Marine *Intruder* squadron in Southeast Asia. For the *Batmen*, the departure would mark the end of a remarkable 58 month tour in continuous combat, a mark unequaled by any other *Intruder* squadron and only exceeded by VMA-311, VMFA-542, and Detachment 110 of HC-7, the Navy's afloat Combat SAR squadron. CO Maj. Patrick J. McCarthy flew the *Bats'* last mission in theater; it brought the total number of combat hops for -242 to an incredible 16,783.

After leaving SEA, -242 checked in at MCAS El Toro, making it the first west coast-based A-6 squadron in the Fleet Marine Force. Others would join them over time.

Back to the Gulf: VA-85, VA-35, and VA-196 Deploy

While the Marines were prosecuting the war from Chu Lai and Da Nang – while also preparing to send several squadrons home – the Navy maintained its deployment schedule for the Gulf of Tonkin. By the fall of 1969 the *Tigers* of VA-65 had concluded their third and last combat cruise in the *Intruder* and were back at NAS Oceana preparing for their first Med cruise in the type. At the other end of the country, VA-145's *Swordsmen* had wrapped their first combat deployment and were preparing for another go-around.

On 11 August 1969 VA-85 departed Virginia Beach for its fourth – and last – Vietnam combat cruise with the A-6. Now led by Cmdr. Herb Hope, the *Black Falcons* went to war with a mix of A-6As and Bs as a component of the proven *Carrier Air Wing 14* in *Constellation*. Regrettably, they lost one aircraft prior to the deployment. On 18 April 1969 one of the squadron's A-6As refueled at NAS Olathe, KS, while enroute to NAS Fallon. The port engine failed on takeoff; the pilot aborted the launch and dropped the hook, but they missed the long field gear. The *Intruder* went off the end of the runway, flipped over, and caught fire; fortunately the station's crash crew was able to get to the wreck in good time, put out the flames, and tunnel underneath to save the crew.

"One Month in Combat": July 1969 VMA (AW)-242

Sorties: 317 (151 day, 166 night)
Mission tasking: for Marines: 157, for 7th Air Force 102, for Army: 45, for Korean Marines: 6
Missions, by type: TPQ: 119, Armed Recce: 86, Beacon: 79, Direct Air Support (DAS): 19, Close Air Support (CAS): 6, TEST: 5, "other"-3
Combat hours: 378.8, non combat hours: 6.0

Ordnance: MK-82 500 pound bombs: 5867, MK-83 1000 pound bombs: 219, MK-20 Rockeye: 9, Zuni 5 inch rockets: 48, MK-75 500lb mines: 186.

Targets destroyed:

Structures destroyed: 20	Bunkers destroyed: 70
Bunkers damaged: 2	Secondary explosions: 27
Secondary fires: 37	Trails cut: 6
Meters trench uncovered: 20	Meters trench destroyed: 167
Meters trench cut: 25	Meters tree line destroyed: 350
Harbor sites destroyed: 1	Fighting holes destroyed : 7
Rice storage bins destroyed: 1	Infiltration routes uncovered: 1

Source: *Command Combat Chronology*

On 23 September, sister squadron VA-35 followed the *Black Falcons* out of the blocks, departing on its third combat deployment. The *Panthers'* departure also marked their first *Intruder* cruise in a *Midway*-class carrier; in this instance they replaced VA-52 in *Coral Sea*. At NAS Whidbey Island VA-196 was also preparing for their third war cruise in A-6s. With their slot in CVW-14 filled by VA-85, the *Main Battery* shifted to *Carrier Air Wing 2* in *Ranger*. Like VA-35, this time around they would solely operate the A-6A, leaving VA-85 to haul around the big old *Standard ARM* under their A-6Bs.

For VA-196, getting to their next deployment had not been without incident. The squadron had transferred from CVW-14 to CVW-2 about eight months after their return from the last deployment. Prior to the move they and several other wing aircraft were selected for participation in an underway air show, one of those standard "good deals" designed to demonstrate the power of the attack carrier for families and VIPs. Unfortunately, during a practice session one *Milestone* crew ran into a wee bit of difficulty, almost losing their plane in the process.

The crew of the subject aircraft were Lt.j.gs. Steve Richmond and Dick Littke; Richmond writes they were:

> "At 1,000-feet over the *Connie* during a practice air show. We were supposed to have two A-3 tankers; an F-4 was to plug into one, and we were to hit the other one just as we overflew the boat. Unfortunately, there was only one A-3 up this day, and CAG briefed the F-4 to plug in first, then unplug and let us plug in overhead. The F-4 unplugged with about a mile to go and – in my haste to get plugged in and look good – I had to use full power to get in and then idle/speed brakes to stop the closure rate. Unfortunately, I shut both engines off by mistake. Realizing what I had done I unplugged, started a glide, pulled the RAT (Ram Air Turbine, to provide emergency electrical power), and started a relight attempt. I told Dick to 'get ready to eject' (fortunately I had my boom mike on so he could hear me – it was real quiet in that cockpit)."
>
> "At 400-MSL, Dick ejected. Boom! Kneeboard cards went everywhere. He went through the canopy. I had idle power at 100-feet. I started leveling off at approximately 50-feet, got it trimmed up, and reached for the face curtain over my head. Just then, the old A-6 engines started really thundering, and it zoomed back up. Dick landed just aft of the destroyer (we were making our pass bow to stern). The helo came in to pick him up but wouldn't go into a hover until I left the area. I was circling him at 500-feet, 60-degree angle of bank, full power, and the adrenaline was pumping big time. My buddy Dan Brandenstein and his B/N Bud White came down and joined on my right wing, coaxing me away from the crime scene. The captain came up on the radio and asked me if I had any other problems, except for a hole in the canopy. I said '*NO SIR*' ... so he just told me to hold until the rehearsal was over and recover aboard. I dropped my 10 Mk.82 bombs one at a time in a safe area and watched them blow up (we were supposed to be in a bombing demo later in the show ... but I was excused).
>
> "I landed solo and received a grade of 'OK-3 wire (little not enuf B/N).' Of course, the helo had already picked up my B/N and deposited him safely aboard the boat, where he told everyone that '*his stupid pilot shut the engines off.*' I was mad because I had my story already worked out – this might have been the first A-6 double flameout caused by fuel ingestion from a defective air-refueling drogue. My skipper, Cmdr. Lou Dittmar, asked me if I thought that would ever happen to me again – I said, '*No Sir!*' So he let me fly in the actual airshow, but the B/Ns had to draw straws to see who would fly with me, because Dick was down for a few days with a compression fracture. He forgave me eventually, and we flew together for one whole cruise. I never did that again!"

There were no recorded incidents during the remainder of workups, and on 14 October 1969 *Ranger* pulled away from the pier at NAS Alameda ... with *all* of its aircraft. As a postscript, 28 years later – during VA-196's disestablishment ceremony at NAS Whidbey Island – one of the featured speakers would repeat the story of Richmond and Littke's unplanned "flight demo" of low-altitude restart capabilities. In closing, the speaker advised the crowd that former Lt. Richmond was now flying for a major airline and suggested everyone, "... take that into account when you make your next travel plans."

All three squadrons had completed their previous cruises under some aspect of the Johnson bombing halt, so no one really expected any surprises; missions over Laos and South Vietnam remained the norm. *Constellation* arrived first, reporting to CTF-77 on 1 September 1969 and taking its place on the line on 12 September. Notably, while at Naval Station Subic Bay COMTF 77 Rear Adm. Maurice R. "Micky"

Weisner moved his flag to *Connie*; later in the deployment, Rear Adm. Frederick A. Barshar, the carrier's third skipper, would relieve Weisner in ceremonies held onboard.

The *Falcons* deployed with a total of 14 aircraft – 12 A-6As and 2 A-6B PAT/ARMs – and put their resources to good use, with emphasis on night interdiction and truck hunting along the Ho Chi Minh trail, *ala* their Marine brethren. During the course of six line periods in the Tonkin Gulf totaling 128 days VA-85 succeeded in unloading some 11 million pounds of weapons over the span of slightly less than 1,500 sorties. It was a good cruise, more so for the fact the squadron did not lose a single aircraft in combat or operational accidents; in fact, the *Falcons* were the first *Intruder* squadron to return from Vietnam with all of the aircraft it started with. CVW-14 did suffer a total of five combat losses, consisting of one VF-143 F-4J and four A-7As from VAs -27 and -97, with two more *SLUFs* lost in operational mishaps.

Connie turned east on 29 April 1970 and returned to North Island on 8 May. VA-85 returned to their home station the same day, in the process retiring the record for the number of consecutive Vietnam deployments by an AirLant *Intruder* squadron. After the traditional post-cruise stand down, VA-85 transferred to *Carrier Air Wing 17* and prepared for its first Mediterranean cruise with *Sixth Fleet*. The squadron would see combat again, but not for 20 years and in another part of the world.

The *Main Battery's* cruise in *Ranger* was more painful. The previous time out they had lost four aircraft, with the majority of the crews either KIA or MIA. Unfortunately, this deployment brought more of the same, even though the war was ostensibly winding down, the peace talks were continuing, and combat operations were "limited" to runs in Laos and the Republic of Vietnam. The United States had been fighting a "limited" war in the theater for five years now, without adequate explanation or results, and -196 paid the price again.

The carrier rolled in for its first line period on 17 November 1969. Three days later, the *Milestones* lost their first aircraft when XO Cmdr. Lloyd W. Richards and Lt.j.g. Richard C. "Dick" Deuter went down over Laos. Rescue forces successfully recovered Richards, but the fate of Deuter was never determined. That same day the squadron suffered an "A-6 peculiar loss" when Lt.Cmdr. Richard F. "Dick" Collins and Lt. Michael E. "Mike" Quinn launched and never came back from a mission over Laos.

Milestone One Lou Dittmar recalls it being an exceedingly rough start for the cruise:

> "Dick Collins was my best friend. He, Lloyd, and I went back a long ways. Lloyd and Pee-Wee Reese had bought these orange curtains and stuffed them in their survival vests, just in case they got shot down. When Lloyd hit the ground he was a real mess – two broken legs, a broken back, and other injuries – and was just laying on the ground. The bad guys were coming to get him, and he couldn't move, but he was able to get the curtains out and waved the helo in. They pulled him out just in time. I did a tanker hop that night, the late launch, then went down to Da Nang to see him in the hospital. When I got back to the ship I learned Dick Collins was down. They eventually shipped Lloyd back to Oak Knoll (the naval hospital in Oakland), and I guess his wife saw him all beat up, in traction and all that, and she got very, very upset. Shortly afterwards they learned she had Lou Gehrig's disease, and she died within the year. It was not a good time."

Back at Whidbey, VA-128 grabbed several replacement crews and sent them east to join VA-196. The group included Cmdr. John R. Wunsch and young Lt.j.g. Dave Williams; the latter would in short order get his "graduation exercise" in combat. As for Cmdr. Richards, he would overcome his injuries and family tragedy and assume command of VA-196 some 19 months later. In his stead the Navy assigned Cmdr. Wunsch as the *Milestones'* new executive officer, and the squadron moved on.

The squadron's third loss was particularly tragic, coming right after the New Year. On 2 January 1970, Lts Bruce C. Fryar and Nick Brooks had just rolled in for a 40 degree visual dive on a target near Mu Gia Pass when ground fire struck their aircraft. The *Intruder's* right wing separated, followed immediately by a severe yaw and two rapid rolls to the right. Two objects believed to be ejection seats were seen to depart the aircraft immediately before the wrecked A-6 burst into flames. As Fryar and Brooks' squadron mates came in to provide top cover for a possible rescue one crewman was observed on the ground, lying immobile in his parachute. Additionally, an intermittent beeper was heard on Guard, which may have been from the second crewman. A helo lowered a pararescueman to the ground, where he found Lt. Fryar dead, apparently from a broken neck; incoming fire then forced the SAR forces

While the war continued, the RAGs continued pumping out students. VAH-123's *Intruder* det was formally established as VA-128 in September 1967. Here a buddy store equipped A-6A practices tanking with a *Golden Intruders* squadron mate during 1971. (U.S. Navy)

to depart, and when they returned the following day the body and parachute were gone. Efforts to find the two crewmen continued through the 7th, but neither man turned up, and both were eventually listed as Missing in Action.

The fourth and final shootdown for the deployment had a happier ending. On 6 February – once again over Laos – Lt.Cmdr. Evan P. "Pee Wee" Reese and his B/N, Lt.j.g. E.R. "Don" Frazer, were forced to step out after their A-6A caught a burst of AAA. Both men were recovered, as were the two crewmen of a squadron *Intruder* dumped in an operational accident on 20 February.

Ranger and CVW-2 wrapped up and turned for home on 23 May 1970, having completed five line periods for 103 days; with everything else that was going on, they had pulled an additional six-day tour with *Task Force 71* on Defender Station off Korea in early April. Regrettably, VA-196 had suffered the only combat losses for the entire air wing, with four aircraft down, three *Milestones* recovered, four missing in action, and one killed in action. *Air Wing 2* totaled seven aircraft lost under operational circumstances, including the single A-6A, four A-7B *Corsairs* from VAs -56 and -93, and individual F-4Js from VFs -21 and -154. Yet, morale remained good among the *Intruder* crews and maintainers. Veteran B/N Bud White would remember some years later:

> "Our squadron lost five aircraft on each of my two combat cruises. Still, the morale was really high; we had people fighting for flights, even taking other guy's flights. A very aggressive, *gung ho* squadron.
>
> "I ended up flying 205 combat missions over two tours, most with Dan Brandenstein. We only got hit once, on a day, visual dive-bombing run. My windscreen was shattered but remained intact, and we made it back to the ship okay. Now, Dan and I had an agreement, something of a 'you break the plane, you get to keep the part.' After they pulled my windscreen the maintenance guys offered it to me as a souvenir. I declined, but told Dan he could have it; his father, a woodworker, turned it into a coffee table for his living room. He still has it."

The *Main Battery* completed this, their third combat cruise with *Intruders*, on 1 June 1970. After their return Bud White departed the squadron for a shore tour with VA-128 as Aircrew Training Officer, where he could join others in passing on the lessons learned in hard combat. As for Bud's pilot Dan Brandenstein, some years hence he would join NASA's *Space Shuttle* program and eventually fleet up to the position of Chief of Astronauts, relieving fellow Naval Aviator John Young. Dan ended up flying four *Space Shuttle* missions, one where he was forced to make a night landing and two where he was Mission Commander.

Thankfully, VA-35's cruise in *Coral Sea* was more similar to that of their sister Oceana-based squadron than that experienced by VA-196. Again, the targets were limited to the usual places, occasionally leavened by reconnaissance sorties into North Vietnam; in fact, the only real major drawbacks to the cruise were its length – which stretched into a full nine months; the Navy was taking advantage of the bombing halt to overhaul several carriers – and the high number of operational accidents.

Coral Sea arrived on Yankee Station for the first of five line periods on 27 October 1969. Two months later, on 26 December, the *Black Panthers* suffered their sole loss for the deployment. While making a case I approach to the carrier a squadron A-6A hit the water. The two crewmen ejected late and were not recovered. The cruise was a rough one for CVW-15, primarily because of the multiple operational losses. The only aircraft downed in combat was a VA-86 A-7A, which was hit by AAA over Laos on 7 January 1970. Otherwise, in non-combat circumstances the record was the single VA-35 *Intruder*, four more A-7As, one E-2A from VAW-116 with five killed, and an EKA-3B assigned to VAQ-135.

Having completed 125 days on Yankee Station, *Coral Sea* departed the theater on 18 June 1970. The *Panthers* arrived home in Tidewater on 1 July and started preparations for a Med cruise.

Blue Blasters and A-*Rabs*

The arrival of the New Year traditionally signifies a time of hope and renewal, but most Americans could probably be excused for looking upon New Year's Day 1970 as anything but upbeat. With one year on the record books for the Nixon Administration no real progress had been seen in the Paris Peace Talks. The war, now entering its sixth year, truly seemed endless; protests were continuing at home, and a large number of POWs were still held in North Vietnam.

On the plus side, units and personnel were starting to move home from Southeast Asia. Indeed, the next 18 months would see the greatest movement of U.S. forces from the theater, as more and more equipment and responsibility shifted to the Republic of Vietnam. By mid-1971, the majority of Marine and U.S. Air Force units stationed in Southeast Asia were on their way home, along with a large portion of the Marine and Army's ground forces. In the meantime, the medium attack community continued to grow and gain experience. By this

VA-34 was established at Oceana in April 1970 and began making deployments to the Med with the new carrier *John F. Kennedy* (CVA-67). This *Blue Blaster* A-6B is using a D-704 buddy store to tank an RA-5C *Vigilante* near NAS Albany, GA, in a shot dated July 1971. 151563 was lost only three months later, when the pilot inadvertently ejected from the aircraft while on a hop with VA-42. Without a pilot, the B/N wisely followed. (U.S. Navy)

date, the Navy was regularly stationing two or more *Intruder* squadrons in the Tonkin Gulf, as well as one squadron with Sixth Fleet in the Mediterranean. In order to maintain this level of operations the service stood up two new squadrons on 1 January 1970.

VA-34 was the first Navy squadron established specifically to operate the A-6. In ceremonies held at NAS Oceana on 17 April the squadron adopted the famous *Blue Blasters* nickname and emblem of its predecessor, an A-4C operator out of NAS Cecil Field. Led by Cmdr. Robert W. Miles and assigned to *Carrier Air Wing 1*, the *Blasters* equipped with A-6As and Bs and prepared for a fall Med deployment in *John F. Kennedy*.

Also on 1 January, in ceremonies at NAS Whidbey Island, the Navy returned the *Arabs* of VA-115 to the roll of active squadrons. The unit dated to its World War II establishment as *Torpedo Squadron 11* (VT-11) and possessed a proud record, but had been held in an inactive status since August 1967.

VT-11 stood up at NAS San Diego on 10 October 1942; equipped with TBF-1 Avengers, the squadron conducted initial patrol and anti-submarine operations from Hawaii and Fiji before shipping out to Guadalcanal in April 1943. In September 1944, VT-11 went aboard *Hornet* (CV-12) as part of *Task Force 38* for strikes against Okinawa. Its embarkation with the other squadrons of *Carrier Air Group 11* marked the commencement of six months of non-stop combat operations against Japanese forces in the Southwest Pacific. During the Battle of Leyte Gulf on 25 October 1944 squadron TBM-1s launched against the Japanese Central Fleet while still 340 miles away from their targets. Racing in to assist with the defense of *Taffy 3* – a group of escort carriers under fire from a hugely superior Japanese force off Samar – VT-11's aircraft managed to hit one battleship and two cruisers. They then navigated the 300 miles back to *Hornet* and successfully recovered; seven squadron pilots received the Navy Cross for their actions that day. The following day another VT-11 pilot was awarded the Navy Cross for his successful torpedo attack against a Japanese light cruiser. The squadron's combat performance directly contributed to the final tally of 26 major Japanese surface combatants sunk during the battle.

The squadron continued operations in the vicinity of the Philippines through December, in the process riding out the infamous typhoon that sank three destroyers assigned to Adm. William F. Halsey's task force. January 1945 found *Hornet* executing strikes against targets on Formosa and Luzon before moving to the South China Sea. VT-11 ended the month participating in raids on Hong Kong and Cam Ranh Bay. On 1 February 1945 the squadron transferred from *Hornet* to *Kasaan Bay* (CVE-69) for transport out of the theater. Upon arrival in Hawaii the squadron moved to *Curtiss* (AV-4) and returned to the states. The final honors for their six months of combat with the fast carrier task forces included three Presidential Unit Citations.

The squadron would remain a component of CVG-11 for the following 21 years, during which time the unit redesignated to VA-12A (on 15 November 1946) and VA-115 (15 July 1948). Initially based at NAS Alameda, VA-115 changed home stations three times before settling in at NAS San Diego. In late 1948 the squadron exchanged its TBM-3s for AD-1 *Skyraiders*, beginning an association with the Douglas product that would last 18 years.

During the inter-war period VA-115 trained for deployment with CVG-11 and participated in an around-the-world cruise in *Valley Forge*. However, the squadron was onboard *Philippine Sea* in 1950 when the Korean War broke out. By now flying AD-4s, VA-115 made two combat cruises to the Sea of Japan and Yellow Sea, including participation in the Inchon landings of September 1950 and the evacuation from the Chosin Reservoir in December 1950. During their second war cruise (December 1951 to August 1952), the squadron concentrated on interdiction missions against North Korean and Chinese forces and raids on hydroelectric power plants. Postwar, the squadron settled in at NAS Miramar and upgraded to the AD-6 variant of the *Spad*; it was during this period -115 adopted its famous – and allowable, this being the 1950s – *Arabs* nickname. Still assigned to CVG-11, the *Arabs* made eight WestPac cruises in carriers *Kearsarge*, *Essex*, *Shangri-La*, and *Hancock* prior to the Tonkin Gulf incident. Operations included assignment to the *Seventh Fleet* covering force during the Quemoy-Matsu Crisis of 1958, and patrol activities during the Laotian Crisis of 1964.

On 19 October 1965 – and by now assigned at NAS Lemoore – the squadron ventured forth for the first of two Vietnam combat cruises. The *Arabs* made their initial deployment in *Kitty Hawk* with CVW-11, losing two *Spads* prior to their return on 13 June 1966. One pilot was recovered from Laos, but the other, Lt.j.g. William L. Tromp, died in North Vietnamese captivity. The second cruise, made as a component of *Carrier Air Wing 5* in *Hancock*, ran through mid-1967. Flying alongside A-4 operators VA-93 and VA-94 and *Crusader* squadrons VFs -51 and -53 the *Arabs* lost two aircraft in quick succession, fortunately with both men recovered. However, another three VA-115 *Skyraiders* went down in operational accidents with two fatalities.

Battered but proud, the *Arabs* returned to beautiful Hanford on 22 July 1967 and almost immediately stood down. The following month the Navy placed the squadron in inactive status and assigned it to VA-125, the A-4 RAG at NAS Lemoore; the situation was highly unusual in Naval Aviation with the *Rough Raiders'* CO dual-hatted as the skipper of VA-115. Yet, the arrangement kept the squadron's record and combat honors alive. Finally, on 1 January 1970 the *Arabs* returned to duty, with Cmdr. Conrad J. Ward assuming command in the formal ceremonies held at Whidbey on 16 January 1970. In April the squadron received its first A-6, came under assignment to *Carrier Air Wing 16*, and commenced preparations for another trip to Vietnam.

Intruders to the Med

While the stalemate in Vietnam continued, several *Intruder* squadrons labored in relative obscurity on the other side of the world. These were the A-6 units that started deploying to the Mediterranean under *Commander, Sixth United States Fleet*. For years attack aviation in the Med meant A-1s, A-3s, A-4s, and – very briefly – A-5s. By 1970, however, the Oceana medium attack community had six squadrons on line, and with the force reductions in the Tonkin Gulf A-6s could now regularly go to the Med. Their assignment substantially increased the carrier air power available in the region.

Once again, the *Sunday Punchers* took the lead. On 13 June 1966 – barely a year after they made the A-6A's combat debut in Southeast Asia – VA-75 headed east with *Carrier Air Wing 7* in *Independence*. The Indy was originally scheduled to head back to South East Asia, but a theater bombing halt allowed the Navy to send the ship to the 6th Fleet instead. The *Puncher*'s attack duties in CVW-7 would be shared with two A-4 *Skyhawk* squadrons, one of which was Marine.

Their first Med cruise in the type concluded on 1 February 1967 and was quiet; in fact, following their return to Oceana they went back to Vietnam for their second go-around and did not return to the Med for over two years. The *Punchers*, under Cdr. Deke Bordone, sortied to the Mediterranean again in July 1969 with Cdr. Bob Mandeville's CVW-3 in *Saratoga*. Once again the deployment was fairly quiet, but this time there was a switch; when VA-75 outchopped in January 1970, VA-176 in *Franklin D. Roosevelt* took its place as the duty A-6 squadron. The *Thunderbolts* also had a standard Med cruise with CVW-6, which concluded on 27 July. Their return to Oceana marked the squadron's first successful deployment with *Intruders*, but there would be plenty more, and several other squadrons would get in line to pull liberty in places like Naples and Palma instead of Cubi and Hong Kong.

The cycle picked up in mid-1970 when both *Saratoga* and *Independence* returned to the Mediterranean with A-6 squadrons, in this case VA-75 and VA-65 respectively. The arrival of the carriers marked the first occasion where two *Intruder* units were assigned to Sixth Fleet at the same time; the deployment also featured the Atlantic Fleet debut of the A-6B, assigned to the *Punchers*.

At the time John Diselrod was the *Standard ARM* training officer at VA-42. He later wrote he was right in the middle of enjoying his shore tour at the RAG when he got a phone call from VA-75's skipper:

> "One day Deke Bordone called and said, 'What are you doing?' I told him 'not much' and he said, 'I'm calling BuPers to get you over here.' I told him I had a year left on my shore tour, and in 1966 when my son graduated from high school I was in WestPac; I didn't intend to miss my daughter's graduation. He responded with, 'If we can guarantee that, will you come?' Okay. Two days later I got my orders, and I made her graduation, although they complained about my missing RefTra at Gitmo.
>
> "They deployed with something like 100 *Standard ARMs* onboard, and we wanted to pull them all and test them in flight, particularly the seeker heads. I talked to the admiral (Pete Harmisch) and said here's what we have to do. We worked with the EWs on every hop and started categorizing every radar, using the APS-107, the ER-142, and the *Standard ARM*. The system was aligned towards the threat in Southeast Asia, but we got enough ER-142s for every A-6B. We shared the ready room with VMCJ-2, and when they saw the equipment they ordered it for their EA-6As."

As can be expected, the appearance of the B-Model *Intruders* in the Med drew the attention of the Soviets, particularly since the A-6s were now purposefully hanging around Soviet combatants. According to John, that made for some interesting situations:

> "We worked with the General Dynamics service rep and mounted the (ER-142) scope on the B/N's side by the left knee, but not where it would take his knee off in an ejection. It seemed to work; you dial in the missile, and it would snap right to the emitter. The Russians were very cooperative with emitting; we used to run low passes until they started sending missiles up the rails and started tracking us. Whoops! We kept at 1500 to 2000-feet after that. The first time we went up with a live round and shot it I think every Russian cruiser and destroyer in the Med was lined up behind us, taking pictures. One day I was flying with John McNabb, the maintenance officer, and we went out to investigate something weird. It turned out to be two whales chasing their tails. John saw a cruiser and said, 'Hey, let's buzz them!' We came in perpendicular to the bow, and you can imagine what they thought on the bridge, because there was a *Vigi* (RA-5C) coming up their starboard side at the same time! We crossed at the bow, low and fast; I don't recall how close we came to each other, but we heard the rumble."

Impromptu air shows aside, both the *Sunday Punchers* and the *Tigers* moved to the forefront in early September due to major unrest in the Middle East. Following a campaign by the government of Jordan to remove Palestinians from their nation Syria invaded Jordan, and terrorists hijacked three airliners. *Independence*, with CVW-7 and VA-65 embarked, and *Saratoga*, with VA-75 under assignment to CVW-3, moved to a position in the eastern Med and established Bravo Station (which, with subsequent contretemps in the region, would eventually gain the name "Bagel Station"). *John F. Kennedy* was also enroute with VA-34 onboard, making its first deployment. On 18 September the ship and *Air Wing 1* were on their way to the Caribbean for the pre-deployment ORI when they got the call to make a hard left turn. Once the *Blue Blasters* arrived on Bravo Station it briefly gave Sixth Fleet three on-call *Intruder* squadrons, two of which (VAs -34 and -75) were also operating the A-6B.

The crisis subsided in October and VA-65 and VA-75 headed back to the barn. VA-34 remained in the Mediterranean until the end of its initial cruise in *Kennedy*, returning to Oceana on 1 March 1971. By that time the *Black Falcons* of VA-85, led by Cmdr. Donald H. Westbrock and embarked in *Forrestal* with CVW-17, had assumed the duty, as well as the B-model *Intruders*.

The area would heat up again in a couple of years, but in the interim, the A-6 provided regional commanders with a proven "big stick," one that would be useful in any sort of contingency or combat operations. One of these contingencies included participation in the SIOP, or Single Integrated Operations Plan. SIOP was a euphemism for the employment of nuclear weapons against the Soviet Union. While the A-6's ability to carry and deliver nuclear weapons was incorporated in the design from the start, its capabilities in that area had been pushed to the rear through the mid-1960s due to the war in Vietnam.

According to veteran VA-52 pilot Daryl Kerr, the *Intruder's* development as a nuclear weapons platform was another result of the Navy's mid-1950s search for a large carrier aircraft to assume the SIOP mission. The AJ *Savage* was considered a disaster, so the mission fell to the A3D *Skywarrior*. Over time, as technology successfully reduced the size of air-deliverable nukes, ADs and A4Ds assumed a large portion of the mission. When the *Intruder* came along it too was wired for special weapons.

Kerr adds the targets that were assigned to the Navy squadrons were 'fourth or fifth level' of importance. As the SIOP was calculated, Strategic Air Command's ICBMs would go after the primary targets; the Navy's *Polaris* submarines would take the secondary; and SAC's bombers and Naval Aviation would take the tertiary. The general consensus was that any nuclear delivery strike would be an "absolute suicide mission," with crews and aircraft pressing on through areas that had already been hit and irradiated multiple times. Still the Navy persevered, maintaining an at-sea air deliverable nuclear capability from both the Mediterranean – against targets in the eastern and southeastern Soviet Union – and Western Pacific against Petropavlosk, Vladivostok, and other targets on the Kamchatcka Peninsula.

The Med-deployed *Intruder* crews shouldered the mission, flew practice strikes and studied their target folders, in and around their other tasking. Through the end of the Vietnam War, *Sixth Fleet* would continue to rotate medium attack squadrons through the region at the rate of one or two at a time. However, by the end of 1971 the situation in Southeast Asia would heat up again, and several AirLant attack outfits – notably the *Black Panthers* and *Sunday Punchers* – would find themselves back in a shooting war in the Tonkin Gulf. Within a few short years the Mediterranean and Middle East would assume their own primacy in defense planning, in the process becoming the focus of the nation's attention. As could be expected, *Intruders* would play a big role.

Arrival of the A-6C TRIM

On 10 April 1970, VA-165 departed with CVW-9 in *America* for its third Vietnam cruise. The carrier arrived for duty with CTF-77 on 12 May and started its first line period on 26 May. That same day *Boomer* skipper Cmdr. Fred Bachman flew the newest variant of the *Intruder*, the A-6C TRIM, into combat for the first time.

The C-Model was the result of the Navy's TRIM (Trails, Roads, Interdiction Multisensor) program, an attempt to improve night attack and targeting effectiveness. The project resulted in a podded system fitted with a Doppler Frequency (DF) receiver, a low-light-level television (LLLTV), and forward-looking infrared (FLIR). Four P-2 *Neptunes* with TRIM equipment installed in a centerline pod received the AP-2H designation and went to VAH-21 at NAS Sangley Point. The squadron sent the aircraft to Cam Ranh Bay for combat evaluation, which was inconclusive at best; the big, slow, and noisy *Neptune* did not lend itself to stealthy attacks on suspected targets, and the aircraft proved to be highly vulnerable in country.

At this point, program attention turned towards the A-6. Grumman modified NA-6A 147867 as a test bed, fitting it with two pods mounted to the outboard pylons. Further testing resulted in the development of a large single pod mounted on the centerline of the aircraft. In 1969 the

VA-165 introduced the A-6C TRIM to combat in 1970 while a part of CVW-9. The large sensor assembly took up the entire centerline store and was a continuous maintenance headache, but it pointed the way towards what would become the A-6E TRAM. (U.S. Navy)

Navy authorized $50 million for the conversion of up to fifteen aircraft, although only twelve A-6As received the modification, with BuNo 155667 serving as the production prototype. To handle the added weight of the pod during recovery each A-6C was fitted with an EA-6A tailhook.

Evaluation by the Naval Air Test Center at Patuxent River was continuing in February 1970 when VA-165 received its first examples. Skipper Bachman recalls the *Boomers* got the planes simply because:

> "We were the next in line. In general, the way the A-6s were dispersed really pissed off a lot of people, because during this period the new A-6As coming off the line would go directly to the Marines at Da Nang. They flew 'Lead Nose' *Intruders*, partly because they didn't maintain the systems. There were occasions where VA-165 would have to lend aircraft to VA-128 so the RAG could get the students through their systems hops."

With the arrival of the C-model *Intruders* – barely three months before its departure on cruise – VA-165 had to scramble to get in some training:

> "We got eight of the twelve A-6Cs built," Bachman says. "The only internal change was an auxiliary scope between the B/N and pilot. They could choose one of three displays: basic radar, LLTV, and IR. Grumman trained our crews, and then we trained a couple of B/Ns from each of the other squadrons.
>
> "We'd get one up and then attempt to run up the mileage. Our day started at 1800 and ended at 0600. We'd go to Boardman, drop bombs, then tank, fly up to Spokane, do a half dozen runs, then fly a low level back to Whidbey. Once while we were back east on a training deployment for our cruise in *America* another A-6C showed up at Whidbey. I wasn't there, so the wing commander, Yates, called my wife and asked her to come down to the station and sign for the squadron. She did."

Concerning the modified *Intruder's* flying characteristics, the skipper recalls:

> "It was like flying with a big blob in the belly. With 3 to 3.5K extra weight on the centerline it made handling different and really affected the fuel load coming back to the boat. We had to land with a lot less fuel than anyone else, due to the extra weight of the pod. Fortunately, no one from the squadron ever got into trouble. As an example, I launched from Oceana once in a C, with (XO) Dick Zick following in an A-6A. Once we got the gear up I throttled back for the climb out, and Zick called, 'Give me a couple! (percent of throttle)' I thought he meant slow down, so I pulled the throttle back a tad. Zick came back with, "No, give me the *other* couple!' as he was about to overrun us."

The *Boomers* ended up deploying with five A-6As, eight A-6Cs, three A-6B *PAT/ARMs*, and *fifteen* tech reps. Fred says the large number of Grumman and contractor personnel outraged the CAG, but they were necessary:

> "We needed the different people for the airframes, the avionics, PAT, TRIM, you name it. However, as the cruise proceeded they cross trained, and eventually we were able to send some home."

Carrier operations with the TRIM bird revealed another problem; according to Grumman, it took two days to do an engine change because the pod had to be dropped first, *very* carefully. That took up a lot of valuable time and space on the carrier's crowded hangar deck. Bachman says his squadron solved the problem when it sent several mechanics back to Grumman for some training. One of the Es came back and said, "Ah, no, I can change the engine without dropping the pod." The mech demonstrated the process, proving it could be done and reinforcing one of Bachman's basic leadership techniques.

"Get a good Navy mechanic and let him figure out how to do it. It works every time. This was definitely a 'Model T' system," Bachman concludes:

> "At Boardman they had this radio controlled jeep target. You'd almost have to be on top of it to pick it up with the IR. But the jeep had a flashing light that showed up at five miles on the LLTV. It'd wash out the screen it was so bright, so the LLTV wasn't too useful. However, using the LLTV over Yakima – 57 miles away – we could count the runway lights at the civilian field east of Boardman.
>
> "The system didn't really pan out over in Vietnam. On one mission over Laos with Jack Hawley as my B/N the TFA at Nakon Phanom put us on five trucks. Jack picked them up on AMTI, locked on with the IR at five miles – it was a cool night – and within five miles Jack was calling the system every name in the book, because the screen had blossomed to the point it was unusable. I said, 'You want to know what it looks like? Look outside.' Jack looked out and saw wall-to-wall flak. We decided to let them keep their trucks."

Twelve A-6As were modified by Grumman into A-6C TRIM birds. (Grumman History Center via Tom Kaminski)

Boomer Blue-Suiters

VA-165's 1970 deployment in *America* attracted interest for another reason: the presence of the first two U.S. Air Force exchange officers to serve with a fleet *Intruder* squadron. The men were pilot Maj. Larry Beasley and B/N Capt. Doyle Ballentine.

Beasley had a particularly colorful background. As an enlisted Marine in 1955 he was accepted for AVCAD training at NAS Pensacola. Following his graduation and commissioning he returned to the Corps and flew with the service through 1959. After leaving active duty Beasley went back to school and joined the Ohio Air National Guard. During the Cuban Missile Crisis the government called his F-84F *Thunderstreak* squadron to active duty; upon the conclusion of the facedown with the Soviet Union, Beasley elected to remain on active duty. He subsequently flew *Phantoms* out of MacDill Air Force Base, FL, and Naha, Okinawa, with the *12th Tactical Fighter Wing*; did a brief tour in special operations flying A-1s at Bien Hoa; did another tour in F-4s with the *555th Tactical Fighter Squadron* at Udorn Royal Thai Air Force Base, and was then sent to VA-128. Upon completing the RAG he checked in with VA-165, where he became the squadron TRIM officer for the deployment:

> "I didn't have much trouble adjusting to the Navy due to my Marine aviation background," Beasley would later comment. "There was a *big* difference between the Navy and the Air Force in standardization. They made me squadron NATOPS officer, and I later picked up the TRIM project officer job for the cruise. I picked up the first aircraft at Calverton on 25 February 1970, and we deployed on 9 April.
>
> "Flying the A-6C wasn't much fun at the start. With that pod it had the same drag coefficient as 28 Mk.82s; you got the high chevron right off the end of the catapult. The tough part was doing the Test & Eval *and* combat evaluation at the same time, in a deployed fleet squadron. It was a good cruise, on a good ship. Once we got the aircraft figured out it worked pretty well. In fact, the main problem was not with the *Intruder* but with the supply system. We were told, 'Don't worry about the parts, they'll be on station.' Naturally, when we got out there, there weren't any parts. The sensor's quartzite windows blistered once we got aboard the carrier, and they wanted something like $5000 to replace each one. We ended up replacing them with plexiglas, which seemed to work okay and reported it back to the states. There was no response. We were so pushed to get the concept out and into operation that the lack of feedback didn't bother us."

On 9 June 1970, during the first line period on Yankee Station, Cmdr. Richard A. Zick relieved Cmdr. Bachman as *Boomer One*. For Fred Bachman it had been great being at the "top of the heap;" as with most officers who made it to command, he rated his year at the helm as the highlight of his career ... although the first week was anything but auspicious:

> "I relieved Leland Kollmorgen on 27 June 1969," he says. "One week later we lost an airplane in the Cascades due to control problems."

The A-6A, flown by Lt.Cmdr. Dean Herman and Lt. Phil Soucek, was on a night hop in instrument flight conditions. According to Bachman:

> "They were over the Cascades and heading to Seattle when they lost elevator control and started oscillating up and down. The plane would descend in a level attitude, then climb and fall off on one wing. They eventually were forced to eject. One guy landed in a tree and used his 50-foot rappelling line to drop down. When he got to the end of the rope he figured he couldn't be too far above the ground, so he let go – and fell another fifty feet, breaking his ankle in the process. We couldn't get in there to rescue them until the next day due to fog. The plane came down about a half mile from the highway. They checked the hydraulic fluid and systems and couldn't find anything wrong, so this one was checked off as 'unexplained.'"

Bachman, a former *HATRON Four* pilot, has several other "fond" memories of his tours at Whidbey Island. Prior to assuming command of the *Boomers* he had helped stand up VA-128 as its first executive officer. He was later relieved by the legendary Cmdr. Bob Blackwood, the first B/N XO from VA-196. Fred recalls:

> "One night, on the way home from the Officers Club at Whidbey Blackwood ran his VW microbus into a ditch and flipped it. When the police showed up the van was on its back, wheels spinning. The cop asked, 'What seems to be the problem here?' Blackwood responded with, 'I don't know, officer, I can't seem to get any traction.'"

While he possessed certain status as a plankowner with the *Golden Intruders*, Bachman says the relationship changed somewhat when he got his own squadron. He admits to having fun with the RAG's skipper, Cmdr. Niles "Whitey" Gooding ... all in the spirit of wholesome competition, of course:

> "We used to steal Gooding's baseball cap from his desk," Bachman recalls with a laugh. "It got to the point where he'd post a guard outside the door. One time a couple of my guys snuck in through the ceiling, lowered a hook, and made off with the hat.

We later presented it to Gooding at the O'Club and he was absolutely livid. We also trained the wing commander, Buddy Yates, in the A-6. He was a rather outspoken individual. He'd regularly tell Gooding, 'Look at these -165 aircraft. Look how well *they're* maintained. *This* is what they're supposed to look and fly like.' It drove Gooding nuts! I had a great tour. I concentrated on keeping the men informed, from top to bottom, and it worked well."

Now led by Dick Zick, the *Boomers* continued their combat cruise with the mix of As, Bs, and Cs. Despite problems with the TRIM birds, the squadron had an excellent deployment, losing no aircraft over five line periods.

"I didn't fly it (the A-6B) much," Zick later commented:

"We set it up that only four or six crews in the squadron were qualified in the aircraft. It was a long, lengthy process to train people so we decided to limit it. The C was a standard A-6 with a lot of stuff hanging off of it, including that big tub underneath; it set the stage for the TRAM bird. The B/N had a scope while the pilot had a repeater. Most B/Ns would take the hood off so the pilot could see the display. We took the first ones out expecting good things. Hell, we didn't know what we were going to be looking for. Some targets we were told we'd look at never showed up. It was supposed to be used for trucks and other targets – supposed to pick up diesel engines or something to that effect – but at least on our cruise, it never did work properly. That, and I think the Vietnamese were a lot smarter than we were giving them credit for.

"It was an interesting cruise, operating three different types of aircraft. I think we had sixteen planes – a *lot* of A-6 – but we didn't lose any. The maintenance we understood; it was easy, nothing gave us a lot of problems. We had pretty good people in the squadron that had been there for several years, so overall we had a great maintenance crew. I don't know of anyone who didn't like the A-6; we could do just about anything with it. I once beat up on an F-4 on cruise while hanging around the pattern waiting for my Charlie time. A clean A-6 is a pretty smooth airplane – obviously, it doesn't look like it – but no one ever understood that, other than A-6 drivers. The F-4 people never understood that."

Otherwise the *Boomers* got along just fine with their mates in the other squadrons. According to Stan Richardson, one night a squadron A-6 even escorted an RA-5C from RVAH-12, which in itself was an unusual act: F-4s usually went along with the *Vigis* on their nocturnal sojourn. He heard afterwards the two planes were tooling along when they attracted a sudden burst of heavy groundfire, and both pilots immediately went to full military power to get out of Dodge. The A-6 crew then started furiously looking for the *Vigilante*, knowing it to be unarmed and undoubtedly in some peril. They finally found it: two small afterburner flames in the distance, moving off at very high speed. So much for the armed escort.

CTF-77 released *America* on 23 November 1970, and the carrier started back for Norfolk. VA-165 returned to Whidbey on 21 December, having turned over their A-6Cs to VA-145. In recognition of their excellent cruise and combat introduction of the TRIM Intruder the *Boomers* later received the AirPac Battle E.As for Cmdr. Zick, Cmdr. Thomas W. Conboy relieved him on 17 June 1971, and he prepared for a trip to the Naval War College, but the "needs of the service" intervened:

"The detailer called and said, 'Hey, have I got a good deal for you!' I didn't know what it was, but I knew it *wasn't* good. As it turned out, during a launch in *Independence* the Air Boss had called out, '*Phantom* off the bow! You're on fire! Eject! Eject! Eject!' Unfortunately, there were *two* F-4s off the bow; all four crewmen ejected, and the Boss was relieved."

Zick and his family changed their plans, and shortly afterwards he found himself on *Indy* as the unfortunate Boss's relief. He would return to the A-6 community on 31 August 1972 as the CO of the *Green Pawns* and *Medium Attack Wing One* at Oceana.

Now equipped with the A-6Cs, the *Swordsmen* arrived in the Tonkin Gulf on 21 November 1970. For this cruise – their second in combat with *Intruders* – VA-145 was assigned to *Carrier Air Wing 2* in *Ranger*. Among the pilots was former VA-25 *Spad* driver John Juan, who is still proud of the fact he'd made his first cruise in *Midway* while it still had a straight deck. He had already made a couple of combat deployments to Vietnam in A-1s, doing the usual jack-of-all-trades work from strikes to close air support to search and rescue:

"I flew a RESCAP once, got a call to hit some PT boats, found them and shot them up," John comments. 'Okay, wait there and keep an eye on them until the whole air wing can get there.' We started getting low on gas – which is hard to do in a *Spad* – and they finally showed up and we diverted to Da Nang. It was a 9.5-hour flight, and we were looking for the bottle of medicinal brandy after that one!"

When he left he figured the war would be over soon, and he ended up taking a tour at Meridian flying T-2s. After three years – Juan describes the TRACOM tour as a "good way to burn out" – he applied for a return to combat duty. When he called the detailer and said "what have you got?" the detailer replied with, "How about A-4s in Jacksonville?:

"I flew up to the detailer and said 'What's this about A-4s in Jax?' He said with his forked tongue, 'Whatdoyamean? You're going to be cat officer on the *Saratoga*.' 'WHAT?' It actually turned out to be a good tour, as I got to go to the Med. I finally reported to VA-128 in 1971 for A-6s, and for whatever reason they didn't have a spot for me. It took something like a month and a half to get my last TC-4C flight. (Later on) I ended up in VA-145 under George Matt and Rupe Owens. I remember one liberty (in 1972), we went down with the ship to San Francisco and Rupe went to Jack London Square. He was kind of pugnacious on liberty; when it came time for the cruise photos he wasn't available because he had two shiners."

Juan's B/N for the cruise was Lt. Ray Cinco. John recalls they were in the bar one night, and he turned to Ray and said:

"I bet you I'm the only Filipino/Lithuanian in the entire Navy!" He said wrong, he was too; turned out my dad knew his dad. We'd always have these minority meetings, we'd go off in the corner and have a minority meeting. Our lives ended up paralleling quite a bit: we both became teachers, and we both married nurses. Nice guy."

In a sense -145's cruise was quiet because they saw little ground fire and lost no aircraft on combat missions. However, over five line periods totaling 123 days the squadron did lose two in operational accidents. The first and only loss of an A-6C by any squadron occurred on 8 January 1971 when BuNo 155647 hit the water after a cat shot. B/N Lt.Cmdr. Gerry Smith ejected and was rescued, but sadly executive officer Cmdr. Keith R. Curry died in the mishap. The squadron lost a second *Intruder* on 24 February, but both crewmen were recovered. With no further incidents, the *Swordsmen* returned home to Whidbey on 17 June 1971.

VA-35 would make the third and last combat deployment with the A-6C one year later, but overall the TRIM design never panned out. Other squadrons (VAs -34, -75, and -176) took the plane to the Med for cruises, but it never saw combat again; the 11 surviving A-6Cs were taken out of service in late 1975 and modified to the A-6E standard. However, the experience Grumman, the systems contractors, and the Navy gained through the development and employment of the C-model would pay big dividends several years down the line.

Knightriders in the Gulf of Tonkin: *Iron Hand*

When VA-52 arrived in theater with *Kitty Hawk* and CVW-11 on 27 November 1970, it marked the squadron's second combat *Intruder* cruise and first with A-6Bs. As with other squadrons deployed during the extended bombing halt the *Knightriders* found themselves concentrating on targets in Laos and South Vietnam. However, the North Vietnamese continued to prove themselves a wily and intractable adversary; they had used the now two-year-long "vacation" to move large numbers of people and equipment south, including SA-2 sites that were also moved closer to the Laotian border. As a result, every now and then a highly trained team of aviator and NFO would have to sally forth to deal a deathblow to the dirty communist SAMs. On occasions though, things did not go as planned, as related by retired Capt. Paul S. "Raoul" Bloch, a *Knightrider* JO at the time:

> "In the spring of 1971, VA-52 was flying the A-6B *PAT/ARM*, which was capable of firing the General Dynamics *Standard ARM* (*STARM*) missile. This missile was effectively a *Tartar* missile strapped to the wing of an aircraft and was able to travel about 50 miles. The *Tartar* was a shipboard missile designed to shoot down incoming aircraft, but the *Standard ARM* version had a seeker head that acquired and homed in on enemy radar signals.
>
> "That spring most of our missions were against the Ho Chi Minh trail in Laos. Previously, it had been very quiet there, with only some AAA, but then the North Vietnamese started shooting SAMs at aircraft in the area. These SAMs were being launched from sites both in Laos and in North Vietnam along the border. To counter this, a *STARM* Patrol was established. The intent was to fly a long figure eight pattern over Laos along the North Vietnamese border. If signals received in the cockpit indicated that the North Vietnamese were about to launch a SAM, the A-6 was to launch a *Standard Arm* to prevent it.
>
> "One day a *Knightrider* crew who shall remain nameless (but one of them spoke with a strong southern accent) were flying the *STARM* Patrol when they got a strong signal coming from the direction of North Vietnam. The signal remained steady, so they maneuvered the aircraft to fire the missile. When all conditions were correct, and the missile had locked on, it was launched in an almost due easterly direction. It was then they realized that North Vietnam was about 30 miles wide at that point, and they had just fired a missile that could travel in excess of 50. A little simple arithmetic showed that a good-sized portion of the Gulf of Tonkin was now included in the lethal envelope of the missile, some of which contained presumably friendly ships from the United States Navy.
>
> "So, suddenly, everyone in Southeast Asia heard a voice with a strong southern accent issue the following transmission over Guard frequency: *'ALL YOU SHIPS IN THE GULF OF TONKIN, SHUT DOWN YOUR RADARS!'* Presumably, either they did shut them down or the missile was in fact locked on some enemy SAM, because no U.S. ships pulled alongside the *Kitty Hawk* complaining about missing a radar dish."

Lt.Cmdr. Daryl Kerr also made the cruise, and recalls the squadron in general made very limited strikes against North Vietnam, usually in response to attacks or the SAMs. One day a Navy aircraft collected an SA-2 near the Laotian-North Vietnamese border, followed by additional firings against Air Force aircraft. *Air Wing 11* personnel located the missile maintenance and assembly facility and sent in a classic Navy "protective reaction" strike. According to Kerr, VA-192 and VA-195 A-7s did the *Iron Hand*/flak suppression runs using *Shrikes*, guns, and bombs, while the first two *Intruders* went in with iron bombs and blasted the place.

Kerr came in 60 seconds later with 28 CBUs and proceeded to completely destroy the facility, with all sorts of secondaries, flames,

A division of *Knightriders* flies over the mouth of the Skagit River in Washington State, near the town of La Conner, in April 1970. Their Whidbey home sits through the haze, about seven miles beyond them. (U.S. Navy, PHC R.E. Halcomb)

and smoke into the sky, the works. Notably, it was his first strike lead mission; he was not the senior lieutenant commander in the squadron but he still got the job, which for him marked a major step up from trolling for snorkels in a *Stoof*:

> "There were a lot of SAMs and AAA in the air that day, but everyone made it back okay," he comments. "I was later awarded the Air Medal, although I heard it was originally submitted as a Distinguished Flying Cross. We learned later that the follow-up strike, led by the CAG, arrived to find the place destroyed. He still attacked it, then wrote himself up for the Navy Cross for bombing a destroyed target. He received the Silver Star; go figure."

The key remained bombs on target, and in this area the *Intruder* crews continued to excel. Some crews even showed themselves to be true *Knightrider* material before they left the RAG. Cat I (nugget) crew Larry Yarham and B/N Fred "Ferd" House were furiously working up at VA-128 during this period, with a planned mid-deployment assignment to VA-52. House writes they were:

> "Flying our graduation hop in the RAG, as we two students were to join VA-52 in mid-cruise. The profile was a high altitude leg, dropping down to fly the low level that ends up at Bravo 16 (target) near Fallon. For this simulated nuclear mission we carrier a 2,000-pound shape with a smoke charge on the centerline. All our pre-flight planning was done. Charts, fuel figures, divert fields, RSP, and kneeboard cards were cascading from my nav bag.
>
> "The high portion and the low level went fine. Entry into Bravo 16 was right on our target time with a high loft delivery planned. My 12-mile radar prediction didn't look like my scope or the RSP the squadron gave me ... expanded display was worse. We're accelerating to 500-knots now, and the ride is bumpy ... eight miles to go and I'm still not sure of the target ... 'Master Arm is ON ... the pickle is hot!' At two miles to go, I 'fess up and tell I don't have the target ... He said, 'I got the run-in line and the target visually.' I felt so relieved ... I said, 'Great, we'll do our LABS backup!' I reached up and selected LABS, but unfortunately the wrong one of two choices. We were doing a LABS TGT instead of what I selected, LABS IP. Well, as advertised, at 55.2 degrees a thud from the ejector foot and the shape was on its way. Larry was busy flying the half-Cuban eight as I looked at the armament panel (A-6A). I immediately knew what I had done ... and lowered my seat as low as I could.
>
> "'*Goldplate One*, no spot, sir ... oh, wait a minute ... holy smoke! *It's at the base of that mountain out there!*'
>
> "Switchology will get you every time. The way I saw it, a five-mile hit with a nuke could still be a bullseye."

Yarham and House successfully completed their training and, within short order, joined VA-52 in mid-cruise. Reportedly, they fit right in with the rest of the *Turtle Herders*.

... and "Tanker Posit"

It was during this cruise that VA-52 started finding itself flying an additional mission: aerial refueling. Daryl Kerr remembers the new role was a long time coming, but once it arrived, it quickly became very important to the air wing:

> "Back when the F-4 *Phantom* arrived in the fleet it quickly became the 'darling' of the air wing. The *Phantom* was better looking than most of the aircraft in the wing, and had all the bells and whistles, so it received much of the high-visibility tasking. During this period the funny looking A-6 was introduced and concentrated on bombing."

As Kerr recollects, the air wings quickly discovered that the F-4 required a *lot* of inflight refueling to be effective. One immediate result was to retain detachments of A-3Bs in the air wing with a primary tanker role. These were augmented by A-4s fitted with the D-704 inflight refueling pod. The *Skywarriors* spent most of their time passing gas, the *Skyhawks* pitched in on occasion, and everyone was happy.

When Kerr made his first A-6 combat cruise in *Kitty Hawk* as maintenance officer, the scheduled VAQ-133 EKA-3B tankers were briefly pulled from the boat. It seems the A-3s had wing crack problems, and all of the aircraft were grounded until they could be fixed. That left *Air Wing 11* with quite a problem; there weren't any A-4s in the wing, so who was going to provide the inflight refueling? The answer was VA-52, flying the A-6A. According to Kerr, the squadron quickly received fourteen D-704 pods and assumed the tanker mission with their *Intruders*. Kerr says the subsequent cruise was spent:

> "Trying to keep the damn things operating. Primarily, we spent the cruise dropping iron on Laos and then returning to the stack to refuel other aircraft. We went into combat with a standard load of D-704 on station three, drop tanks on stations two and four, and MERs on one and five. With proper planning, we would usually return to refueling point with 6,000-pounds available for transfer. If the squadron was called on to bomb up north, we'd download the D-704s and replace them with bombs."

Every A-6 on the cruise did double duty as a tanker, and VA-52 quickly became the most popular unit on the ship, at least where VFs -114 and -213 were concerned, as related by B/N Tom "Wacker" Wyckoff:

> "The Whale tanker can't fly. We take an A-6A with a buddy store and four drops and set out for *Red Crown*. We hear the fighter bubbas getting vectors, giving angles, etc. I look on the radar, and there they are about 20-nm out, so I sez to *Red Crown*, 'Got a Judy 10 left, 5 low at 20.' *Red Crown* wants to know if we need vectors, and I tell him 'Negative.' The 'Bubbas,' of course, also refuse assistance.

"As we close with the sun at our six, they haven't seen us on radar or picked us up visually. We see them and Deb says, 'Hold what you've got and we'll join on you.' The normally verbose 'Bubbas' tank with nary a word, while we just giggle a lot. Back at the boat over sliders, the 'Bubbas' want to know how we did that. We just smiled and chewed. From then on, the first words those two crews heard from any tanker were, 'Hey Bubba, you need vectors?'"

Lts. Debbenport and Wyckoff had several other tanker encounters with the *Phantom Phlyers* on their cruise, including this one, again reported by "Wacker" Wyckoff:

"Feet wet after chasing monkeys out of trees; carried a buddy store just in case somebody needed a sip. *Triple Sticks* (a VF-213 F-4J) joins up, and we slip him 1.5. As he departs, he lights the burners and pulls his nose to the vertical. Deb looks at me and sez, 'Whatta jerk!' We switch tower and hear, 'Boss, this is *Triple Sticks* for a low fly by.' Deb sez, 'Watch this.' I sez, 'O crap.'

"We tuck in underneath the F-4, and as we fly past the boat I get a nice view of the hangar deck overhead. We follow *Triple Sticks* through a victory roll and break off in the opposite direction. After our post-flight interview with CAG the Skipper (Cmdr. Doug McCrimmon) has his turn.' There's only one reason I'm not grounding both of you ... *the troops loved it!*'"

Another *Intruder* type commented:

"What always gave you a warm and fuzzy feeling (NOT!) was the fact that these F-4 'tactical fighter-bombers' (as dubbed by the media), who would carry six – count them, six Mk.82s – on a bombing hop and come back with two hung, also doubled as 'the eyes of the fleet.' Why then were these guys never able to find the A-6 tanker overhead the ship (high or low station)? 'Tanker Posit,' I have come to believe, was part of their call sign. Why did we, as A-6 crews, always seem to see them first and rendezvous with them? Inquiring minds want to know."

Fun with *Phantoms* aside, VA-52 had an exceptional cruise, with 138 days on the firing line over six line periods and no combat or operational losses. In fact, CVW-11 lost only one aircraft during the entire cruise; on 27 February 1971 a VA-195 *Dambuster* A-7E went down in non-combat circumstances. The pilot was successfully retrieved.

Perhaps the greatest excitement resulting in emotional duress occurred on the trip home. Again, the story is relayed by "Raoul" Bloch, and concerns *Knightrider* Lt. Bob Berg:

"On the return from the 1970-1971 cruise, *Kitty Hawk* departed Subic Bay enroute Pearl Harbor and home. The ship commenced an extensive housecleaning to insure that no drugs or other illegal materials were going to be smuggled into the States via the ship. The ship's XO, Cmdr. 'Mo' Peale, was directing a search with the Master-at-Arms force, and all hands were asked to assist. The U.S. Customs agents were going to come aboard in Pearl Harbor and search the ship prior to its arrival in San Diego.

"Now, Bob was a very regular fellow, and preferred to use the same stall in the same head every day. He also was very observant and, on departure from Subic, noticed that some of the screws from an access panel to a void located in his favorite stall had the paint stripped from them. It has to be understood that disturbing six or seven coats of thick Navy paint applied in the typical shipboard maintenance program could probably be spotted by a three-year-old. Sensing the drug bust of the century (it was a big void), Bob dutifully reported his suspicions to the Master-at-Arms force. They got some screwdrivers and entered the compartment.

"There they found, not the Mafia's drug stash, but instead case upon case of some very fine liquor. *Chivas Regal, Johnnie Walker, Wild Turkey* ... the really good stuff. It must be remembered that liquor like that was very cheap at the package store in Cubi ... perhaps half the stateside price, if that. Well, there was nothing to be done but take the contraband to the fantail and chuck it over. I heard 'Mo' Peale himself say there were tears in his eyes as he was deep-sixing the booze.

"Bob Berg almost followed the liquor, if cooler heads hadn't prevailed. It turned out that instead of the Mafia, some F-4 guys from the fighter outfits had gotten together to buy the booze, store it in the void, then unload the stash in San Diego for home consumption. As we used to say at Mom's in Fallon, *'you pay your money, and take your chance.'* They lost."

For the *Iron Hand* and tanker-mod Intruders, there would be other cruises. No one could know it at the time, but one year hence the squadrons deployed with the A-6B would find more trade than they could handle. As for the concept of using the A-6 as a tanker, Grumman eventually sent tech reps to the *Kitty Hawk* to look at the system and decided they could come up with a more permanent solution. The company took four older wing-stressed A-6As, returned them to the plant in Bethpage, and created the KA-6D.

Full-Time Tanker: the KA-6D

The steady conversion of light attack squadrons from the A-4 *Scooter* to the A-7 *Corsair*, and the "annual impending demise of the *Whale*" led the Navy to work with Grumman in developing the *Intruder* as a permanent tanker alternative. Grumman took the first four high-time A-6As to Bethpage, modified them and – on 16 April 1970 – flew KA-6D No. 1, BuNo 151582, for the first time.

The tanker differed from the standard A-6A in that its avionics and bombing systems, including the radar, were removed. In their place a hose-drum air-refueling unit was installed in the aft equipment bay, or "birdcage"; the aircraft retained its centerline buddy store option. The conversion program also included rewinging several aircraft, replacement of internal bulkheads, and installation of improved fuel cells. With five 300-gallon Aero 1D external tanks installed the KA-6D could carry approximately 3844-gallons/26,000-pounds of JP for the *Phantoms*, *Vigilantes*, and other aircraft in need of a drink. Over the subsequent two decades Grumman modified 78 A-6As into KA-6Ds at their St. Augustine and Stuart, FL, facilities, along with 12 A-6Es. Notably, the early variants retained a non-systems bombing capability, and reportedly some thought was given to mounting 20mm cannon. In any event, the tanker *Intruders* never dropped ordnance, and the later conversions had even this minimal bombing capability deleted.

The *Thunderbolts* of VA-176 became the first fleet squadron to operate the KA-6D, receiving their initial examples on 25 September 1970. However, VA-85 would score the first deployment with the tanker, departing Tidewater for a Med cruise on 5 January 1971; VA-176 followed in *Franklin D. Roosevelt* three weeks later, while VA-115 made the first combat deployment with the tanker variant in April 1971.

Although there would be exceptions over time, squadrons would typically deploy with four or five KAs alongside their bombers; Marine squadrons were not assigned the tanker version unless they were part of a Navy air wing. As for the B/N in the right seat, their jobs on tanker hops were reduced to operating the tanker package, serving as a safety monitor and communicator, and staring at the KA's right side stark, empty instrument panel.

End of the Year and the Last A-6A

As 1970 drew to a close there was no sign or indication that the war in Vietnam would be resolved any time soon. The peace talks continued in Paris, but in the interim squadrons were still deploying, both to the Tonkin Gulf and the Mediterranean. The A-6 had finally come into its

The KA-6D was introduced in 1970, with 90 bombers being converted to the dedicated tanker role through the years. Here a VA-196 KA fuels a squadron mate from a buddy store off their Whidbey Island home on an unusually snowy day in 1986. Saratoga Passage, the Skagit Valley, and Mount Baker are in the background. (Rick Morgan)

own, with seven AirLant squadrons, six in AirPac, and six assigned to the Marines. Concurrently, the three RAGs at Oceana, Whidbey Island, and Cherry Point continued to train and qualify pilots, B/Ns, and maintenance personnel. Their impact on their respective stations was startling, as recalled by Jerry Menard, now assigned to VA-128:

> "By this time I was one of the maintenance officers at VA-128. Every third person on the station belonged to the squadron. We were the largest squadron around – humongous – pumping out a *lot* of people. There was no FRAMP, just a lot of people working to get crews, maintenance people, and aircraft ready for the fleet squadrons.
>
> "We regularly prepared aircraft for TransPacs, gave them a lot of attention, and never lost one. We used our own crews, but occasionally other squadron's people. One Marine came in and had all sorts of questions, 'How do I do this? How do I do that?' etc. When he asked 'How do I do get the inflight refueling system to work?' I got *real* nervous. It turned out he was just verbally reviewing procedures, although those of us who were listening didn't realize it. One other pilot delivered an *Intruder* to us with the deck and canopy rails full of cigarette butts. Now, we knew that people smoked while flying, but this was blatant, and he'd trashed the cockpit. We made sure he never flew again."

There were other occasional problems, a result of the vibrancy of the A-6 program. At Oceana the two communities – fighter and attack – coexisted on a fairly peaceful basis, as both the *Phantom* and *Intruder* wings were growing. According to Menard, it was somewhat different at Whidbey Island, long the home of heavy attack. Now the A-6s were in ascendancy, and they needed a *lot* of space:

> "VA-128 was only a couple of years old when I rolled in for my shore tour, and we were still expanding, still looking for space for training and other activities. The A-3 squadrons at Whidbey had 'owned' the station for years and had problems adjusting to the rapidly growing A-6 community. They seemed to run on the idea of 'Get it done tomorrow'; in A-6s we couldn't wait until tomorrow, it had to be done today. We had a lot of resistance. Finally, the A-6 COs got together and went to the wing and said something like, 'We need these things and this space for our purposes or we'll be forced to close the Skywarrior (movie) Theater and use *it*.' They got the room they needed, and in the process remade the station to fit their needs."

Menard says despite this rapprochement, there were still the occasional clashes between the *HAT/ VAQRons*, the *Intruder* squadrons, and the station staff:

> "The A-6 guys started showing up at the Whidbey O'Club in flight suits and got kicked out. Their COs had to spend a lot of time explaining things to the admiral. This all changed when Jig Dog Ramage[2] got here; he used to go to the club in *shorts*.
>
> "I had particularly bad relations with the station 'chop,'[3] a captain. Here was this lieutenant commander telling him he needed things *right now* – I'd already had people complain that I didn't

[2] Then Captain, later Rear Admiral, James D. "Jig Dog" Ramage was a legendary attack pilot who spent a period as C.O. of the Whidbey wing. He had won the Navy Cross in 1944 as skipper of Bombing-Ten during the Battle of the Philippine Sea and became heavily involved in development of the Navy's nuclear strike mission in the 1950s and as Commodore of Heavy Attack Wing One in Florida where he is generally given credit for helping the A3D *Skywarrior* program survive its early problems.

[3] "Porkchop", or "chop" is a slang term for a Naval Supply Officer. The term comes from the odd shape of the oak leaves they wear on their uniform which distinguishes officers in the Supply Corps.

treat people nice – and he did not like my attitude. So we were always butting heads. However, I spent a lot of time with the JO 'porkchops' showing them how to acquire, prepare, and store parts and prepare for cruise. I did such a good job one of the JOs recommended me for the station's 'Supply Officer of the Month' award – and I won. This Captain had to make the presentation, and he did not like it one bit. At least I got a free lunch out of it. 'Service to the fleet,' what a concept."

For Grumman the year ended on a slightly different note. On 28 December 1970 the company rolled out the last of 488 A-6A *Intruders*. The A-model, along with its variants (the A-6B, A-6C, and KA-6D), would continue in service for some time to come, but a better *Intruder* was on the way. However, the new model would not arrive in time to play a role in the resolution of the Vietnam War.

1971: Steaming as Before

As 1971 opened the war continued at what could still be officially considered a reduced pace. In South Vietnam, two Marine Intruder squadrons (VMA(AW)s -242 and -533) had gone home, leaving the -225 *Vikings* at Da Nang. Also shipped home were several other *First Marine Aircraft Wing* units, including MAG-12, MAG-13, VMFA-542, VMAs -211, -223, and -311, and VMFAs -115 and -122.

VMA(AW)-225 had a particularly busy last month of the year. On 12 December 1970 – and in what was probably the first use of the type ordnance by an A-6 squadron – *Viking* aircraft expended 12 GBU-24 laser-guided bombs on targets in Laos. The targets were designated by personnel on the ground.

However, -225's time in theater was also drawing to a close. On 28 April 1971 the last squadron A-6A departed Da Nang for MCAS El Toro, bringing to conclusion four and one-half years of combat operations by 1st MAW *Intruders.* During their stay in South Vietnam, the *Vikings* had completed over 13,000 combat hours.

The Marines' Air Force counterparts were also shifting units out of the theater, as more and more equipment and responsibility were turned over to the South Vietnamese military. In 1971 alone, the *12th Tactical Fighter Wing* (F-4Ds, Phu Cat Air Base), *14th Special Operations Wing* (multiple types, Phan Rang AB), and *35th Tactical Fighter Wing* (F-100D/Fs, A-37Bs, Phan Rang AB) either inactivated in place or went back to ConUS.

As for the carriers of CTF-77, they remained on the line keeping an eye on things, also in reduced numbers. Still, on 10 March 1971 the two wings in *Ranger* and *Kitty Hawk* – CVWs -2 and -11 – set the record for one day of sustained operations, with 233 strike sorties over Laos and South Vietnam.

Ranger was first to leave, departing on 9 June 1971. VA-145's two operational accidents were the squadron's only losses for the cruise, although CVW-2 sustained additional combat or operational losses of four A-7Es, two F-4Js, and a C-2A. The latter was particularly tragic, with nine aircrew and passengers killed when cargo apparently shifted during a catapult shot, leading to an out-of-control condition. The *Swordsmen* returned home to Whidbey on 17 June; they were followed exactly one month later by VA-52.

Their places were taken by the *Arabs* of VA-115, making their first *Intruder* cruise with CVW-5 in *Midway*, and VA-196, coming around for their fourth Vietnam deployment. *Midway* had been absent from the war since November 1965, due to a massively lengthy and expensive overhaul at Hunters Point. Suitably refurbished, the carrier pulled away from NAS Alameda on 16 April 1971 for only her second combat deployment and arrived for duty with CTF-77 on 7 May. As for the *Arabs*, their initial assignment to CVW-16 had been brief; the wing disestablished on 30 June 1971, and the squadron shifted over to *Air Wing 5*.

The *Main Battery*, reunited with CVW-14 but now embarked in *Enterprise*, departed the states on 11 June and checked in on the 27th. While they could not possibly know it, in this third year of the bombing halt, both squadrons would have what could be termed "eventful" deployments.

NFOs in Command and Other Developments

It was during this period of the Vietnam bombing halt that several events took place that affected the A-6 community. While several had long-term implications, a couple turned out to be of the "gee, that's nice, what's the purpose?" variety.

Perhaps the biggest one was the resolution of the question of NFO command of squadrons. All along the bombardier/navigator had served as an equal partner in the cockpit of the *Intruder*, more so than their counterparts in the *Phantom* community. Even Navy F-4 RIOs had it better than their Air Force brethren; the Air Force initially tried putting rated junior pilots in the back seats of their F-4s before coming around to the idea of a non-pilot, Weapon Systems Officer or "Guy In Back." The A-6 pilot and B/N were teamed with the B/N assuming primary responsibility for mission planning, navigation, systems management, and setting up the weapons and delivery. Already in at least one case – the famous instance where Lt.Cmdr. Bob Blackwood assumed the XO duties with VA-196 while in combat – the B/Ns had proven themselves highly qualified for leadership rolls.

One of the leading proponents of NFO command was Lt.Cmdr. Michael R. "Mike" Hall, who had helped work up the A-6C TRIM program. He and others lobbied to bring the question to a resolution, but they faced heavy opposition. Reportedly the biggest enemy of the concept was COMNAVAIRPAC, Vice Adm. Allen M. Shinn. However, following Shinn's relief by Vice Adm. William F. "Bush" Bringle on 31 March 1970, a large part of the opposition dissipated. Vice Adm. Thomas F. Connolly, OP-05, then pushed through the necessary legislation authorizing command at sea for NFOs. Accordingly, in ceremonies at NAS Whidbey Island on 23 November 1971 A-6 Bombardier/Navigator Cmdr. Lennart R. "Lenny" Salo relieved Cmdr. Doug McCrimmon as CO of VA-52. NAVAIRLANT followed suit in May 1972 when Cmdr. Mike Hall relieved Cmdr. David W. Timberlake as *Black Falcon One* at NAS Oceana.

According to veteran B/N Lyle Bull – who himself would later command VA-196, VA-128, *Constellation*, and two carrier battle groups – once NFOs assumed command the opportunities quickly opened up others, with minimal problems. The RA-5 and A-3 communities quickly moved their RANs (Reconnaissance Attack Navigators) and B/Ns up to command slots, as did the fighter and airborne early warning units. These two men were the first of many, and once again, medium attack had shown the way.

There were other changes. On 1 October 1971 the Navy established COMMANDER MEDIUM ATTACK WING ONE (MATWINGONE) as the functional commander for the A-6s at NAS Oceana. Until this date all fleet *Intruder* squadrons had reported to COMFAIRNORFOLK; they had been joined by VA-42 on 1 June 1970, following the disestablishment of RCVW-4. Now all seven fleet squadrons at Oceana and the RAG were united under one command. The initial CO of MATWINGONE was Cmdr. Michael F. Andressy, who was serving at the time as commanding officer of the *Green Pawns*. This dual-hat arrangement continued through 5 January 1973, when former *Black Panther* CO Capt. Herman L. Turk became the first commodore of the wing.

Otherwise, the squadrons of MATWINGONE continued operations as before, rotating through the Med and otherwise training. On 25 September 1970 another crisis erupted in Jordan when Palestinian commandos attempted to oust King Hussein. In response, *John F. Kennedy* (VA-34, with CVW-1), *Saratoga* (VA-75/CVW-3), and *Independence* (VA-65/CVW-7), as well as their task forces hustled to the eastern Med to monitor the situation. Hussein managed to beat down the threat, and after a period the carriers resumed their deployments.

Occasionally a unique event came up that required the presence of an A-6. One example was Operation *Storm Fury*, a test program that involved the chemical seeding of a hurricane and observation of the results. VA-176, VA-85, and VMA(AW)-224 at Cherry Point were assigned the mission. The *Thunderbolts* participated in *Storm Fury 69*, which saw several squadron aircraft penetrate Hurricane Debbie, fly through the eye, and then return through the wall of the storm while dumping seeding chemicals. The storm was the fourth of the 1969 season and formed during the same period as the infamous Hurricane Camille that hammered the Louisiana/Mississippi Gulf Coast with 165-knott winds. Debbie was somewhat more of a lady at category 3, "only" reaching a sustained wind velocity of 105-knots at 951 millibars pressure and never threatening the coast of the U.S. At its closest point of approach on 29 August 1969 Debbie was approximately 200-miles south-southeast of Bermuda and 820-miles southeast of Oceana.

The *Bengals* took the assignment the following year, and in 1971 it was VA-85's turn. Over September and October 1971 the *Buckeyes* trained, studied, and prepared, culminating in a mission on 28 September. On that date three squadron A-6s navigated to Hurricane Ginger, located the storm at a point 340-miles east of Jacksonville and 480-miles south of Virginia Beach, and spent over an hour seeding it with silver iodide. The mission provided a unique change in the routine for the squadron; however, the hurricane did end up hitting the United States, going ashore at Cape Lookout, NC, on the morning of 30 September with winds of 80 to 85-knots. Ginger moved inland rapidly, lost strength, bounced off the Blue Ridge, and exited as a tropical depression on 3 October near Cape Charles, VA.

In October 1971 the *Sunday Punchers* had their turn with, "... and now, for something completely different," when they participated in an airborne ASW exercise. At the time the deployed *Saratoga* and CVW-3 were engaged in test activities to determine the suitability of big-deck carriers for VS and HS operations. By now the Navy was quickly discarding its designated antisubmarine warfare carriers. To date, *Essex* (CVS-9), *Hornet* (CVS-12), *Randolph* (CVS-15), *Bennington* (CVS-20), and *Kearsarge* (CVS-33) had decommissioned, and the surviving three carriers – *Intrepid* (CVS-11), *Ticonderoga* (CVS-14), and *Wasp* (CVS-18) – weren't long for the world. The faithful Grumman S-2 *Tracker's* days were equally numbered; its replacement, the turbofan-powered Lockheed S-3A *Viking*, was in development at Burbank. Concurrently, the numbers of Soviet nuclear-powered attack and cruise-missile submarines increased, and several of the new types were specifically tasked to take out U.S. carriers. Hence, the Navy worked to determine the best mix of aircraft and support systems that would give the attack carriers an organic fixed and rotary-winged ASW capability.

Operational test of the program – titled the "CV Concept" – saw the assignment of VS-28 S-2Es and HS-1 SH-3Ds to Cmdr. Paul Gillcrist's *Air Wing 3* for *Sara's* 1971 Med deployment. As a result *Saratoga* departed Naval Station Mayport on 7 June 1971 with a highly eclectic mix of aircraft on the flight deck, including *Stoofs, Sea Kings*, VAW-123's new E-2B *Hawkeyes*, and the EA-6As of VMCJ-2 Det 60. After participating in Exercise *Magic Sword II* in the North Sea, *Sara* moved to the Med and relieved *Forrestal*. The deployment proceeded apace, the more typical sound of J52s and J79s interspersed with the melodic (?) roar of twin R-1820s and distinctive whine of SH-3 rotors.

Not that everything went as planned, as while anchored off Athens the *Sara* started taking water through a massive failure of a gasket in the No.3 Main Machinery Room (MMR). The ship went to General Quarters (GQ) and was able to stop the sea's inrush, although the ship reportedly took up to 49ft of water before it was stopped. Ten aircraft were launched from anchor to lighten the ship, which had assumed a nasty list, and at least one F-4J almost went into the water after the non-standard, dead-in-the-water cat shot. Ten days later the No.4 MMR flooded in similar fashion, and for a period the ship was limited to only one screw ops out of four. Needless to say, this put a crimp in flight operations.

Despite the problems – particularly in gaining adequate wind over the deck – *Saratoga* continued the ASW operation on a limited basis through the remainder of the deployment. VA-75 played its part towards the end of the cruise, on 7 October 1971. Using the radars and anti-radiation sensors of their A-6As and Bs the *Punchers* positioned and released several strings of sonobouys. The Anti-Submarine Classification Analysis Center onboard the carrier then processed the signals from the sonobouys. The event marked the first use of the *Intruder* in such a role, and reportedly it was considered a success; by all accounts Cmdr. Everett W. "Hoot' Foote's VA-75 – aka "F-Troop" – did a sterling job with every role and mission that was tossed at them, while still managing to have fun in and around their extended stay at Athens' Congo Palace Hotel.

As a final note, Foote's regular B/N was the irrepressible John Diselrod, by this cruise a full-fledged member of the *Sunday Punchers*. Concerning the ASW evaluation, he wrote:

> "Hoot and I were both ex-*Mushmouths*, or *Savage Sons*, from VAH-5. We figured if we're going in with sonobouys they must be worth something. We did the usual stuff in the Med with *Sara* on one side and *Roosevelt* on the other.
>
> "Once the Admiral from the *Sara* announced he wanted to fly to *FDR* to discuss an upcoming war at sea exercise. The Hooter stuffs a sonobouy behind his seat. When the admiral's topside, Hooter takes the sonobouy into the island, slops saltwater from a toilet into a bucket, and throws the sonobouy into it (which activates the bouy). The next day the wargame starts, we launch the S-2s, they call up the sonobouy, and there's the *FDR*. Alpha Strike to follow."

Saratoga and CVW-3 finally made it home on 9 November 1970, following the two repair stops in Athens and five days in Naples:

> "It was a good cruise, and the CV concept worked well until the ship almost sank," said Diselrod. "I had the duty, was wearing my tropical white longs, stopped out of the ready room, and this fire hose went right across my chest, ruining my uniform. Then a chief ran by, absolutely soaked from head to foot."

However, other than with similar evaluations in the Pacific Fleet, the A-6 was never again called upon to exercise this unique new capability. The S-3 was on the way, and the Navy proceeded with the integration of ASW assets into the carriers and their air wings. On 30 June 1972 *Saratoga* redesignated as CV-60, making it the first of the multi-purpose carriers; notably, on that date the carrier, CVW-3, and VA-75 were in the middle of a major shooting war in the Tonkin Gulf.

Meanwhile, Back at the Karst

As 1971 progressed there was continued evidence of North Vietnamese movement of personnel and supplies down the Ho Chi Minh; that aspect of the war had not changed. However, during the course of the year CTF-77 regularly operated with a minimum of carriers on Yankee Station. The cycle normally saw only one carrier on the line, with one or two others elsewhere engaged in R&R and upkeep. By summer 1971 the two on-call *Intruder* squadrons were VA-115, with CVW-5 in *Midway*, and VA-196 with CVW-14 in *Enterprise*. For the *Arabs* the cruise constituted their graduation exercise with the A-6 following transition; the squadron had been to Vietnam before, but their last tour was in 1967, flying A-1Hs and Js off *Hancock*. Both squadrons deployed with KA-6Ds, marking the tanker version's arrival debut in Southeast Asia. Unfortunately, -115 and -196 also lost the first tankers within a one month span of each other.

On 12 August 1971, during *Midway's* third line period, a VA-115 KA-6D suffered a massive fuel leak, followed by an inflight fire. The crew of Lt. John McMahon and right-seater Lt.Cmdr. Bart Wade had an interesting time of it, according to McMahon:

The *Intruder's* cockpit changed through time as new versions and equipment were introduced. These shots are actually of an AWG-21 equipped A-6E TRAM. (U.S. Navy, via Mick Roth)

BOMBS/GUNS
ROCKETS/MISSILES
MASTER ARMT
MECH ARM
ROCKET
GUNS
THREAT INDICATOR
NUCLEAR
DATA SELECT
COMPUTER
CORRECT POS
RADAR
RANGE X 10
DISPLAYS
LASER
TEMP
STORE
CLEAR
CHAFF | FLARE | JAMR

"I was named the 'AirPac KA-6D Fleet Introduction Officer' for the tanker, responsible for working them into -115. On this flight Bart and I launched on a yo-yo[4] tanker op around 1700 and climbed up overhead. There was an Alpha Strike in progress; while we were in the pattern waiting we tanked off a VAQ-130 *Whale* to ensure our package was working, and then he pulled in behind us to take on our excess fuel. The A-3 pilot – I think his name was Bruce Hardin – said he looked up and saw this huge ball of flame come out of the back end of the *Intruder*. He started yelling, 'Arab! Arab! You're on fire! *Eject! Eject!*' We didn't have any fire indications in the cockpit, so we kept flying along. We learned later that Gary Wheatley's B/N – in one of the other A-6s in the vicinity – immediately went for the handle to eject, and Wheatley had to reach over and restrain him.

"About the third call the *Whale* driver finally got more specific: 'Arab 517! Arab 517! *You're* on fire! *Eject!*' Bart looked out the right side, saw a glow, and ejected, just like that with no warning to me. I still didn't see any indications – other than my missing B/N, the plane looked good, but I decided there was no sense in sticking around, so I ejected. On the way down Bart and I could see each other and were gesturing that we were okay. I watched the tanker circle in, streaming smoke. As soon as it hit the duty AGI rushed over and started pulling parts out of the water. They were heading for *us* when the helo came in and pulled us out. The best we were able to determine, the attachment point where the hose attached to the reel failed. The other tankers in the squadron showed obvious wear in that area."

One month later, on 10 September a VA-196 tanker – flown by Lt. Charlie J. Taylor and Lt.j.g. James A. Dickey – lost all of their pressure instruments following a night launch from *Enterprise*. Both men survived their ejection and were recovered.

[4] Yo-yo tankers launch first, pass all of their gas and then land at the end of the following recovery cycle.

Other than McMahon and Wade's unplanned swim call – and the loss of the first KA-6D navy-wide – and another operational loss of a VAW-115 E-2B, *Midway's* first deployment in over five years was "standard." She wrapped up 74 days on Yankee Station on 10 October 1971 and started back for NAS Alameda on 24 October 1971. In her stead *Constellation*, with CVW-9 and VA-165, arrived on 27 October. The *Arabs* returned to the Rock on 6 November and immediately started preparing for another deployment.

As for the *Main Battery* – now commanded by Lloyd Richards – their deployment in *Enterprise* was a little more auspicious. The cruise was the squadron's fourth in combat with the A-6, making them the high-time AirPac outfit, along with VA-165. As for *Enterprise*, this was her fifth deployment to SEA, albeit the first since July 1969. The big nuclear carrier had spent the intervening two years in the yard, refueling and completing repairs from the prior cruise's flight deck fire.

On 3 December 1971 *Enterprise* was three weeks into her fourth Tonkin Gulf line period when war broke out between India and Pakistan over East Pakistan. On the 10th she and her task force were ordered to the Indian Ocean to monitor the conflict, balance a growing number of Soviet Navy combatants, and prepare for evacuation operations. The carrier and her escorts – designated *Task Force 74* – entered the Indian Ocean on 15 December, the same day Pakistani forces in the east surrendered to India. By that time the Royal Air Force had successfully evacuated Western nationals from East Pakistan, obviating the requirement for a U.S. rescue mission. However, *Enterprise* and her units continued into the Bay of Bengal and remained on station to keep an eye on things.

As a result of the removal of the *Enterprise* battle group, *Constellation* extended on Yankee Station through the end of December and subsequently greeted *Coral Sea* with CVW-15 and VMA(AW)-224 embarked. For the Marines it marked the first deployment of one of their A-6 squadrons as a component of an attack carrier air wing; the *Bengals* flew their first strikes on 15 December. *Enterprise* and VA-196 returned to the Tonkin Gulf on 18 January 1972 following the settlement of hostilities in the Indian subcontinent and creation of Bangladesh, giving CTF-77 three carriers with A-6 squadrons. More would be needed ... and soon.

CHAPTER FIVE

Vietnam, 1971-1973

In the last month of 1971, while *Enterprise* and her escorts were meandering around the Bay of Bengal, carriers *Constellation* and *Coral Sea* continued operations "as before" in the Gulf of Tonkin. However, on 26 December – in a sudden and startling change in the Nixon Administration's prosecution of the war – aircraft from both air wings entered combat over North Vietnam again.

The concept of the protective reaction strike was nothing new. Since the onset of the Johnson bombing halt in the spring of 1968 and subsequent prohibition of all bombing of the PDRV, each service regularly responded to attacks on reconnaissance aircraft. The strikes still fell under the highly restrictive Rules of Engagement so familiar to the combatants in the theater, although the restrictions were loosened somewhat with Nixon's accession to the Presidency. Some old hands recall that by this stage of the war the Navy was, for all intents and purposes, dispatching Alfa Strikes with its "Vigis" and photo-*Crusaders*, all but daring the North Vietnamese to hose off a few rounds.

By this date the Paris Peace Talks were dragging into their fourth year, with National Security Advisor Dr. Henry Kissinger heading the American delegation. Unfortunately, the talks had also proven to be more of the same, with the North Vietnamese "negotiators" – led by Lee Duc Tho and Xuan Thuy – arguing every point. For one lengthy period the talks halted while the North Vietnamese representatives argued about the size and shape of the table. The North Vietnamese maddeningly and consistently reiterated the same demands, including the complete withdrawal of United States forces from Vietnam and the withdrawal of support from the government of the Republic of Vietnam. For its part the PDRV would not even admit they had forces in South Vietnam, maintaining there was only one nation and one legitimate Vietnamese government – in Hanoi – so there could be no discussion of an invasion.

On the home front some observers described the President's mood as "gloomy." At the end of 1969 the total U.S. force level in South Vietnam peaked at 543,400 troops and support personnel; two years later the number was down to 140,000. However, anti-war protests in America continued to escalate, including a massive turnout of 250,000 in Washington DC on 15 November 1969. When United States and ARVN troops went after NVA forces in Cambodia on 30 April 1970 there was another outbreak of protests and violence, culminating in the shooting of four students at Kent State University on 4 May. The 1970 Congressional elections – which saw the return of another Democratic majority – did not help. Privately, the President expressed frustration with the lack of progress in the peace talks and speculated he would not win renomination, let alone another election. Adding to his frustration was Dr. Kissinger, who advised the President the North Vietnamese were not interested in negotiation, just humiliation. It was becoming apparent that only overwhelming U.S. military action would force the enemy to bargain in good faith.

In the interim U.S. forces noted increased SAM activity near the DMZ and actual incursions by MiGs into Laos. By the end of the year even the lower Route Packs had become high threat areas. Finally, in December President Nixon started giving signals that he had had just about enough. Hence, on 26 December he initiated the single largest protective reaction strike to date, designated *Proud Deep*.

Operation *Proud Deep*

Officially termed another "protective reaction strike," *Proud Deep* ran five days and saw the largest number of attacks north of the 20th parallel since early 1968. While the remaining Air Force units in the theater went after targets in the western and northwestern portions of the PDRV, aircraft from CTF-77 flew a total of 423 strikes in the vicinity of Dong Hoi, Quang Khe, and Vinh.

The *Boomers* in *Constellation* and VMA(AW)-224's *Bengals* in *Coral Sea* carried the load as the only two *Intruder* squadrons on Yankee Station.VA-165 was on its fourth combat deployment in the A-6; led by Cmdr. Thomas W. Conboy, the squadron had pulled out on 1 October 1971 with a mix of A-6As and KA-6Ds as part of CVW-9. The cruise proved memorable – and intense – for both the *Boomers* and the rest of the air wing.

On 30 December, the last day of the operation, an SA-2 collected the A-6A of Lt.Cmdr. Fred Holmes and B/N Lt. Charles Burton. Holmes – another one of the original VAH-123 *Intruder* crews – was killed, while Burton apparently blew clear of the aircraft and went in the water off Hon Nieu Island. An HH-3A operated by the *Big Mothers* of HC-7 rushed in and successfully retrieved him. Other SAR forces located an ejection seat and life raft that may have belonged to Holmes, but were unable to locate him among a large number of North Vietnamese boats. It proved to be the only combat loss for VA-165 during the deployment.

Over on *Coral Sea* the crews of -224 also got the short course in aerial combat courtesy of *Proud Deep*. The fourth of six Marine *Intruder* squadrons, the *Bengals* had made history in April 1971 when they replaced VA-35 in CVW-15. The squadron departed ConUS on 12 November 1971. Its initial strikes in Southeast Asia on 15 December against warehouses and airfields near Vinh marked the first-ever combat sorties by a Marine A-6 squadron as part of an air wing.

In order to prepare the squadron for its deployment the Navy assigned several veteran *Intruder* types, including pilots and B/Ns. Among

Two VMA(AW)-224 A-6As hang out in low holding over *Coral Sea* in a familiar picture from 1972. The *Bengals* were the first Marine *Intruder* squadron to deploy aboard a Navy carrier, and spent 148 days on the line with CVW-15. (Grumman History Center via Tom Kaminski)

the group were Phil Bloomer, Samuel A. "Buck" Belcher, Phil Schuyler, and former VA-52 maintenance officer Daryl Kerr, who was briefly assigned prior to the deployment. He describes his six weeks with the Marines as "not to be believed":

> "For one thing, they kept their USMC structure. The Master Gunnery Sergeant – their equivalent of our command master chief – was invariably a senior maintenance type. Instead of concentrating on the maintenance aspect of keeping the A-6s flying he went around beating up on people to keep their rifle quals up to date. We'd have trouble keeping the planes flying, and he'd be found on the fantail with his Marines shooting at the water. The squadron also didn't believe in selective cannibalization. I'd worked up a method in VA-52 where we'd rotate aircraft as 'donors' to keep the other planes flying. We'd rob one aircraft (of parts) for fifteen days, then designate a second A-6 and spend the next fifteen days getting the original plane back into an up status using the new parts that were coming in. That way no single airframe exceeded the 30-day mandatory status change. The Marines never did pick up on this method; as a result, we had trouble keeping the aircraft up. Then again, they suffered from the same supply problems we did. Parts supply, particularly for the electronics, was gross."

There were other occasional problems, as Kerr continues:

> "One day during preps two Marines took an A-6A to Tinker AFB, OK. When it was time for the plane to come back we could not find the crew; they had apparently downed the plane and gone on extended liberty. I flew to Tinker with a nugget pilot in the right seat, left a note for the other crew saying, 'we took your plane to Alameda. You have until 0800 to get there yourselves,' and we flew the planes back to California single-seat. The errant Marines eventually made it to Alameda after laying out major personal funds and flying commercial."

Former M*ain Battery* and *Golden Intruder* maintenance-type Lt.Cmdr. Jerry Menard also made the deployment, this time as ship's company. As *Coral Sea's* AIMD officer – and later, IM-1 (Production Control) officer – for the cruise he too had close contact with the *Bengals*. Apparently his experiences were somewhat more positive:

> "I reported onboard the carrier while it was in overhaul at Hunters Point Naval Shipyard (on San Francisco's south side), one of the more 'interesting' locales I ever had the fortune to visit. We were told that after the third street down, if you hit somebody with your car keep going and don't stop. *Fun* area. All of the Marine AIMD people came out of an H&MS. The squadron didn't know these guys, but fortunately it worked well. The Marines were real good about boosting morale and building up the maintenance troops, particularly the avionics people. After each flight the crews would come down and tell the maintainers they were doing a great job and delivering great airplanes, and by the way there was a small problem here, a small problem there ... the Navy people like Buck Belcher also did it. It was a nice touch, morale was great, and everything worked out real well."

According to Kerr, one major problem continued to afflict -224 well into the cruise; unfortunately, it was at the top of the squadron's chain of command:

> "The Marine COs were assigned by their air wing, which rotated staff officers through; -224's was a staff type whose most recent experience was flying EF-10Bs. He had absolutely no carrier experience. The ship and squadrons did their initial boat work, workups, CQs, and deployed – all within a month – and unfortunately this guy was in completely over his head. He never did make a night CQ, as his planes kept mysteriously going down. I even taxied up with a good airplane and a ready B/N one night, said 'this one's up and up,' and the CO immediately downed the plane."

Kerr, Belcher, and a couple of others eventually approached CAG-15 and told him the Marines were willing, but this guy had to go. CAG, Cmdr.Thomas E Dunlop, said fine, he's gone. However, Dunlop subsequently received a message from the Commandant of the Marine Corps to the effect of, 'Who the hell do you think you are? *You* can't fire a Marine lieutenant colonel!' The Marine skipper remained in place. As a result the *Bengals* had a rough go of it well into 1972. According to *Bengal* B/N Charlie Carr, things did eventually pick up ... albeit under circumstances that could be rated unusual, even for combat:

> "It would take pages to fully portray the 1971-1972 VMA(AW)-224 cruise as part of *Air Wing 15* aboard USS *Coral Sea*. Suffice it to say we were just awful, with the CAG not trusting the squadron and the squadron not performing well around the ship. The squadron's potential was enormous, but the leadership was just shitty all around.
>
> "The first CAG (Dunlop) got bagged during *Linebacker 1* by an SA-2; it blew his airplane right out of the sky. The new CAG – Cmdr. Roger 'Blinky' Sheets, a fighter pilot – arrived and flew with -224 as his primary. I was lucky enough to fly with him. The squadron CO was sent home and CAG Sheets became, in fact, the CO and the CAG."

The original Marine commanding officer, relieved in place by the new CAG, eventually took emergency leave back to Cherry Point – under circumstances that were never adequately explained – and never returned. His former squadron pressed on, completing its *Proud Deep* tasking and then resuming operations over Laos and South Vietnam while maintaining an almost serious – yet not necessarily respectful – approach to "work." Apparently the primary ringleader was Capt. Charlie Carr, aided and abetted by several illustrious Naval Personnel, such as Lt.Cmdr. Phil Schuyler and CAG Sheets himself. Schuyler later wrote:

> "A master storyteller and one of the wildest members of the squadron in port, Carr rarely entered the ready room without several jokes up his sleeve.
>
> "One of his best took place in the context of a joint raid against the San Hoi bridge with A-6s from the *Constellation*. Before the joint briefing with the *Connie* crews, Carr went down to the parachute shop and had them make up a set of white headbands with a big rising sun in the middle. The whole squadron put on the headbands for the briefing. When the bridge called for them to man up, they ran up to the carrier deck, ran to the bridge, bowed three times, and then jumped in their aircraft, just like the scene in the popular 1970 film 'Tora! Tora! Tora!' When everyone had returned from the mission, Captain Harris paid a visit to Ready Room 5: 'Guys, that was great, it was one of the funniest things I've seen in my life, but you can't do it anymore. I'll surely have sailors writing home saying I'm launching *kamikaze* raids off my boat.'
>
> "Another common activity among the 'Vultures' was singing songs in the Ready Room. Roger Wilson, one of the pilots in the wing, saw himself as a future rock star and even brought a Fender guitar with him for the cruise. Apparently he was absolutely terrible, but his squadron mates still loved to hear him play."

Schuyler, nicknamed "Beer Barrel" – apparently after the character in Michener's "The Bridges at Toko-Ri" – built on his own substantial reputation during this period of the *Bengals'* hard playing and hard fighting. A native of Carpinteria, near Santa Barbara, Phil graduated from college in Los Angeles before entering Navy flight training in 1962. He got his wings and subsequently became a *Stoof* driver, notching a couple of deployments in the Tonkin Gulf. During one cruise in *Bennington* he flew some 120 gunfire support missions in lieu of tracking enemy submarines, as there weren't any enemy submarines in the area (at least, as best as anyone could tell). As he would later put it:

> "We would check in with the destroyer, we would set up a racetrack pattern, he would fire, we would spot the round and then adjust his fire accordingly. Occasionally we'd get fired on by small arms or 50-cal machine guns, but other than that the tour was relatively uneventful."

It was during that 1967 deployment with CVSG-59 in *Bennington* that Schuyler decided to go for A-6s. He went through jet transition at NAS Kingsville, arrived at NAS Oceana in December 1967, and made an earlier combat deployment with VA-65 in *Kitty Hawk* in December 1968. He found the cruise – primarily spent over Laos, due to the bombing halt – to be almost as uneventful as his previous S-2E tour in the theater, but decided to stay in the Navy and stay with *Intruders*:

> "I liked the people and the flying," he commented. "Besides, the war was going on, and I did not feel it was the time to get out."

Once with the Marines Schuyler proved to be a natural warrior, although he did express some initial qualms:

> "I had some misgivings about the Marines at first. We had always kidded the Marines about their maintenance and how dirty their airplanes were. Once I got to know the people, I felt better about it."

Otherwise, he apparently fit right in with the other personnel of VMA(AW)-224, as related by another squadron member:

> "Carr had gone to town and left Schuyler ('Beer Barrel') at the O'Club with someone else. Occasionally somebody would bust out the front window of the club. The window costs 50 bucks, so 'Barrel' went over and got the night manager of the club and said, 'How much is the window?' and the guy says, '50 bucks,' and Barrel says 'Here's 50 bucks' and promptly took a chair and busted the window out. Well shit, he paid for the window, that was the rule, so what? Well, the next day – you know – he now has to see Capt. Bill Harris for busting out the window at the O'Club. Barrel is pretty much shit-faced and hung over, but anyway, we kind of get him brushed up a little bit, spruced up, and got him in his uniform, got him in a jeep, and off to the goddam pier. He then went up to the bridge, where Capt. Harris made him stand and wait for a good amount of time before calling him over to his chair. Then he proceeded to read him the riot act.
>
> "'Goddamit, I would have expected this out of my Navy officers, but not from a Marine!'
>
> "Phil's standing there in a Navy uniform, looks down at his Navy belt buckle, then said, 'But Captain, I'm *in* the Navy.' Harris nearly fell out of his damn chair laughing."

While all this was going on there was still a war to fight. The armed reconnaissance flights over the North continued, generating a frenzied military and political response by the North Vietnamese government. Something was obviously up, and most observers took it to mean there was going to be a major offensive in the spring. They were proven right in short order.

Politics, Invasion, and a First Response: *Freedom Train*

With the reverberations from *Proud Deep* subsiding and peace negotiations continuing, 1972 opened pretty much as the previous three years had with continuing strikes on targets in South Vietnam and Laos. CTF-77's carrier assets received a brief boost on 19 January when *Enter-*

prise returned from its Indian Ocean tour. While operating with *Task Force 74* CVW-14's squadrons had spent their time flying surface surveillance and reconnaissance hops and keeping an eye on the Soviets. However, their return to combat in Southeast Asia was brief; *Enterprise* completed this, her fifth and last line period, barely a week after returning and headed back to NAS Alameda. The *Main Battery* returned to Whidbey on 12 February, eight months and one day after their departure.

February also saw something incongruous on the political front. On the 21st President Nixon – responding to an invitation by Premier Zhou Enlai – arrived in Beijing for an eight-day visit. Termed by the President as a "journey for peace," Nixon's visit marked the commencement of an extended period of rapprochement between the United States and China. The two nations did not establish formal diplomatic relations until January 1979, but in the interim they did set up liaison offices.

One thing that definitely did not change on the diplomatic side of the house was North Vietnamese intransigence. Despite the efforts of Dr. Kissinger and his aides their opposite numbers continued to attack the United States, both privately and publicly, while issuing the familiar demands. Le Duc Tho and his associates regularly referred to political unrest and antiwar protests in the United States while maintaining negotiations could not proceed without a complete U.S. pullout from South Vietnam.

On 23 March 1972 Kissinger determined the talks were completely stalled, and the U.S. government broke off negotiations. One week later, on Thursday the 30th, six North Vietnamese divisions – supported by tanks, heavy artillery, and brand new SA-7 *Strela* shoulder-fired SAMs – invaded South Vietnam. By Easter weekend over 120,000 NVA troops, constituting 12 divisions, were southbound and heading for An Loc, Hue, and Saigon. The long-held pretense of a "Guerilla War" was finally over.

This was the big one, the major spring offensive that American intelligence specialists had long expected. With the invasion – called the *Nguyen Hue Offensive* by the North Vietnamese, for a national hero who had defeated the Chinese in 1789 – the North hoped to end the war by conquering the south once and for all. Unfortunately for the South Vietnamese, three years of efforts by the Nixon administration to train and equip them for an adequate defense quickly came apart. In the face of the onslaught the Army of the Republic of Vietnamese fell back in disarray. Doubly unfortunate, there were now barely 95,000 U.S. troops and support personnel on the ground in the Republic of Vietnam. Most of the Air Force combat units had gone home, and all of the Marine aviation assets had long since departed. CTF-77 was in little better shape, with only *Hancock* and *Coral Sea* in the Tonkin Gulf for immediate response.

The U.S. scrambled to put adequate forces back in the theater. In the meantime, on 5 April 1972 President Nixon authorized the commencement of "demonstration" bombing of targets in the vicinity of Haiphong. The operation, designated *Freedom Train*, was prelude to a year-long campaign of aerial violence on a level not seen in Southeast Asia in nearly four years.

The initial strikes took place on the 5th, with the U.S. concentrating on the incoming North Vietnamese forces south of the 19th parallel and at several more strategic targets further north. For its part the Navy immediately launched strikes by *Carrier Air Wing 21* in *Hancock* and CVW-15 in *Coral Sea* while recalling *Constellation* and *Kitty Hawk* to the Gulf. The surface Navy also contributed, running its World War II-era destroyers and other units in towards the Vietnamese coast to blast NVA troop concentrations. Many of the ships, like cruiser *Newport News* (CA-148) and guided missile cruiser *Oklahoma City* (CLG-5), it was their last hurrah in combat, and they were going out swinging. These actions, combined with massive Air Force B-52 strikes from Guam, helped slow down the North Vietnamese and bought the South Vietnamese some breathing room; still, it was a violent, chaotic period. Back in the states the Navy, Marines, and Air Force scrambled to put forces back in the region, but it took time. In the interim the invitations were strictly "come as you are."

The onset of *Freedom Train* was particularly busy for CVW-15. The Marines of VMA(AW)-224 had received A-6Bs in March and were still trying to figure out the *Iron Hand* birds when the war heated up. In their official history from the period the *Bengals* reported:

> "Saddled with an additional mission, to supply *Standard ARM* support for *Air Wing Fifteen*. The squadron was assigned three A-6B aircraft, which further required augmentation of 35 inexperienced enlisted personnel and two likewise inexperienced aircrews from 1st MAW."

Still, three *Bengal* A-6Bs and a like number of A-6As sortied at 0300 on 5 April, enroute to the SA-2 sites ringing Haiphong. Reportedly, going in so low they needed to climb to clear ships in the harbor, the *Intruders* were highly successful. So too were an additional two Alpha strikes later that day by CVW-15.

The next day, however, an SA-2 shot CAG Dunlop out of the sky while he was leading a strike in a VA-22 A-7E. As a result the Navy dispatched Cmdr. Roger Sheets to *Coral Sea* to assume command of the air wing. The new CAG was definitely a combat leader; one VF-51 *Screaming Eagle* pilot later described him as resembling (the actor) "... Don Knotts, but (he) had more guts than Rambo":

> "I had been undergoing training to go to CAG-9 on *Connie*, and I was going to relieve John Snyder," Sheets comments. "I had gone through the F-4 refresher at Miramar, had jumped in the E-2s and helos at North Island, and had just finished the A-6 training at Whidbey and was down at Lemoore just starting the A-7. When I got the call it was Thursday night, and I was just home from Fam 1 in the A-7, which had turned out to be a night flight due to delays. At that time someone from AirPac said he'd (Dunlop) been shot down and they might want me to take CAG-15; I had a lot of experience with *Coral Sea*. They told me to get my shots in order and they'd let me know.
>
> "The next afternoon they decided they were going ahead to send me out to *Coral Sea*. I asked how soon, they said pretty quick. That was Friday. On Saturday at 0700 the duty officer was at my front door with Op Immediate orders, and at 1900 Saturday I was on a plane out of Travis that went to Hawaii to Guam to Clark Air Base, where they had a plane standing by to take me to Cubi. They put me aboard a C-118 and flew me to Da Nang, I spent the night there, and the next morning *Coral Sea's* C-1A was there to take me to the ship."

As previously mentioned, Sheets' background was in fighters: F9F-2s with VF-653/151 in *Boxer* and *Wasp* during the Korean War, followed by F8U-1E/F8U-2N *Crusaders* with VF-154 for two cruises in *Coral Sea*. He then transitioned to F-4 Phantoms and served as XO/CO of VF-161. Still, with his wholly VF background he gravitated to *Coral Sea's* all-weather attack Marines:

> "Tom Dunlop and I were about the same vintage, had known each other for years. He, for whatever reason, had no use for A-6s and even less use for Marines. Here you had kind of a sad situation: a Marine A-6 squadron been put together with all volunteers – I guess when they put out the call, something like 200 or 300 people volunteered – and this was a highly select group. Marines have a certain amount of pride, and there they were. When I was told I was taking CAG-15 I was told the air wing was considered number seven out of six air wings and the A-6 squadron was really bad.

"When I got to the ship, first off, I walked into what had to be a dream situation," Sheets continues. "The CO, Big Bill Harris – there were two carrier COs at the time, sometimes known as Big Bill and Little Bill, sometimes as Good Bill and Bad Bill – we had known each other for years off and on. When I got to the ship he said 'CAG, welcome aboard, your job is to run the air wing, my job is to run the ship. I'll stay out of the air wing business. If you need anything anytime, you let me know.' Adm. Howie Greer was the embarked flag, we had also known each other for a number of years. He said, 'Welcome aboard, my job is to get you whatever you need.'"

Sheets also recalls Dunlop had to brief the entire admiral's staff – for whatever reason – before a strike, but he never had to, add in:

"Everything I could ever want or ask for, I got it. For instance, the mining of Haiphong Harbor: because of the weight – and because we didn't have enough nose and tail cones – I wanted the ship to be way north so it was a short flight in. I said where I wanted to be and then assumed flag would get permission. We actually went into an area where they had to go to GQ and stay there for the mission. I told them that's where we had to be to get the best success for the mission, and that's where we went. It was that kind of support that I worked with. I have never encountered anyone who had as great a relationship as I had, just plain staying out of my hair and letting me do my job. It was everything I'd hoped a CAG tour would be."

As for the Marine component of his air wing, Sheets says they got everything turned around within a number of weeks; indeed, within short order he considered CVW-15 to be the equal of any air wing in the Navy:

"They got to be a very highly motivated group," he concludes. "The kind of professionals you'd expect them to be as volunteers and hand-picked people. They moved up and did well enough that at the end of the year they received the Commandant's trophy as the best Marine squadron, which is what they deserved."

Bengal Down

On 9 April 1972, the *Bengals* lost their first A-6A for the cruise when Maj. Clyde D. Smith and B/N 1st.Lt. Scott D. Ketchie were tagged by AAA while hitting the Ho Chi Minh Trail in the *Steel Tiger* region near Tchepone, Laos. Ketchie died, but Smith was able to eject; unfortunately, the bad guys saw him coming down in his chute and went after him. It took four days of evasion and the efforts of a large number of aircraft and personnel, but an Air Force HH-53 *Super Jolly* of the 40th Aerospace Rescue & Recovery Squadron was finally able to pull Smith out under circumstances that became legendary:

"Scott and I were fired off the catapult about 1800 on a road interdiction mission – call sign *Bengal 505* – near Tchepone, Laos, on Route 9, west of Khe Sanh," Smith writes. "We carried 12 Mk.82 500-pound bombs and 12 Mk.20 *Rockeyes*. Tchepone was a major transshipment location along the Ho Chi Minh Trail and had a reputation as a hot spot; an Air Force AC-130 was shot down near there just two weeks earlier. The Vietnamese had mounted a major offensive into South Vietnam in early April, but we knew little about the ground war.

"Soon after arriving on station we saw several trucks on the road trying to get a head start on the evening run down the trail. It was about 1900, and the day was turning to dusk; the weather was clear. We made two attacks, saw hits on two trucks, and rolled in on our third pass from about 16,000 feet, planning a 45-degree, 500-knot visual delivery. After we pulled off, I heard – more than felt – a thump like a door closing. I said something to Scott but then realized our intercom was not working and he couldn't hear me. The aircraft was doing strange things, and almost every warning light in the cockpit was flashing just before we lost all electrical power. The nose began to move up and down independently; I couldn't control it with the stick. I attempted a turn toward the mountains, and as I turned my head to look that way, I saw a huge ball of flame where the tail was, or had been. Shortly after that, the aircraft went into an inverted spin. When I looked up at Scott he was looking down and reaching for the lower ejection handle. I faced forward, reached up for the face curtain, and ejected."

Smith came down next to his *Intruder's* wreckage, which was fully involved; the remaining bombs were cooking off and shrapnel was flying all over the place. Another A-6 flew over – he assumed it was Capts. Roger Milton and Charlie Carr – and he attempted to contact them on Guard, but there was no reply and the plane departed. Shortly afterwards it got very dark, and Smith moved off into the woods, keeping an eye out for Ketchie. He recalls that about an hour later he heard shouting and several shots and assumed his B/N had been taken prisoner. Fortunately, Milton and Carr had in fact confirmed the crash site, and the rescue operation was already starting up. A couple of hours later Smith heard another aircraft and turned on his survival radio:

"A voice speaking perfect English came up on the rescue frequency. He came in clearly, sounded very close, and asked me where I was. 'I'm in the vicinity of the wreckage,' I answered. 'We'll be there in a few minutes,' the voice replied. It was totally dark by then, and we had been briefed that no rescues were ever attempted at night. I asked him his call sign, but there was no answer. Nothing like that happened again.

"About two hours after the bogus call I heard an aircraft fly over, and I immediately beeped on my survival radio. I transmitted my call sign as *Bengal 505 Alpha* (pilots used the Alpha suffix with the tactical call sign and B/Ns used Bravo) and received a response from a crusty fighter pilot who asked me how I was. He told me to stay hidden and said they would be back in the morning. I was confident that he, at least, was a friendly. Less than six hours after getting blown out of the sky and landing in the middle of Laos, my exact location had been confirmed and the search-and-rescue (SAR) group was organizing a rescue. As bad as things looked, at least someone knew that I was alive and where I was. I wondered if my family knew anything. Thinking of them strengthened my determination to make it through. I thought about our latest arrival; our son Tony was born the day before we deployed. My wife Jackie is a strong person, and I knew that she would hold the family together. She was being tested in a big way in 1972: her mother died in February, she had major surgery in March – and now this (she learned within 24 hours that I was alive in Laos and that a rescue attempt was underway; she told me later that the support from the families at Marine Corps Air Station Cherry Point, North Carolina, our home base, was overwhelming)."

What followed was a four-day ordeal. Smith heard trucks overnight, and the next morning there were plenty of people around, not surprising considering his proximity to the Ho Chi Minh Trail. At one point someone walked within a couple of yards of his hiding spot in the elephant grass, but fortunately the individual meandered away. At about 0900 an Air Force OV-10 (*Nail 17*) flew over and confirmed the major's identity by asking a few personal questions:

"'What's your mother's middle name?' asked the FAC.
"'I think it's Marie,' Smith answered.
"'What's your favorite family pet?'
"'Our dog Tootsie.'

"I immediately realized that I had written Tinker Bell, the name of our cat, on the card," Smith continues. 'Nail, you may not believe this, but I put our cat Tinker Bell on my card because we didn't have the dog at the time. I like the dog better, so that was my initial response.'

"Hard to believe, looking back. A life-or-death situation and I'm talking about liking the dog better than the cat, a very confused survivor trying to explain things. The Nail FAC just gave up. 'Okay, that's enough,' he said, and the radio got quiet. I thought I had blown it. Any bad guys monitoring the radio were probably more confused than I was.

"The on-scene SAR commander, call sign *Sandy 01* – an unforgettable Air Force officer named Jim Harding – arrived on station about 1500 in his A-1 to pinpoint my location and coordinate the FACs and F-4 fighter-bombers – call sign *Gunsmoke* – that were going to support the *Super Jolly Green Giant* HH-53 rescue helicopter. Ground fire was intense, and the aircraft took a tremendous amount of fire from a number of antiaircraft artillery sites. Listening to these professionals calmly going about their job under fire was something that will stay with me the rest of my life. At one point, *Sandy 01* asked one of the fighters about the location of a particular AAA site; *Gunsmoke 01's* response was, 'I don't know, I haven't been able to get him to shoot at me yet.'"

At that point the weather moved in and *King 21*, the SAR mission commander in an orbiting HC-130, called off the rescue for the day. The next morning a series of *Nail FACs* returned to the crash site and called in additional air strikes. Smith had moved into a hole at the base of an uprooted tree; at times he could not answer the SAR folks due to the proximity of North Vietnamese, and in any event, the rescue forces were fighting both accurate ground fire and intermittently bad weather:

"I heard some people talking in a whisper on the other side of the ridge. From the sounds they were chopping wood, but I assumed they were looking for me. At times like these, I could not respond to calls from the FACs; at other times I took a chance and whispered. I had a dilemma: if I didn't talk to the SAR people frequently they might assume that I had been killed or captured and terminate the SAR, but if I were careless the bad guys might be close enough to hear me. I fell asleep at one point and woke up to hear *Nail 46* telling *Nail 68* that I had not been talking lately. He expressed concern that it had been 30 minutes since he asked me to come up on the radio and he had not heard from me.

"It started to rain about 1500 and they shut down the operation. It rained hard for about two hours, and then on-and-off after dark. The hole filled up with water. My survival gear was pretty much useless; I had a package of fruit loops, about eight ounces of water, Band-Aids, and fishing gear. As soon as the last aircraft left I heard people talking, tailgates slamming, and engines revving up – the trucks were on the road again."

The situation continued for two more days, with the FACs arriving early and the weather forcing the SAR forces to back off. Smith stayed in his hole and avoided discovery, but by now he was pretty much convinced he would be taken prisoner. Still, the SAR effort continued; he later learned that on his fourth day on the ground Marine units successfully rescued another downed airman nearby: Air Force Lt.Col. Iceal Hambleton, the famous *Bat-21*:

"After the run of bad weather, I was beginning to think that I wasn't going to make it. The SAR group was making one hell of an effort, along with a large number of carrier-based aircraft, but I knew they couldn't keep it up. *Nail 46* came over about 0830 however, and said, 'It looks good, I think we can do a good tune on you today.' Little did I know that the SAR group had met the night before and determined that they would have to develop some sort of alternative – perhaps like Bat 21 – if they couldn't get me out that day. I had no way of knowing that then-Captain Bill Harris, the skipper of the *Coral Sea*, had called Seventh Air Force and asked why it was taking so long to get 'our boy out.' When the Air Force said they didn't have the assets to suppress the heavy ground fire enough to get a *Jolly* in and out safely, Captain Harris launched 78 aircraft from *Coral Sea* to help out. Jim Harding later told me that they would not have been able to get me out without that *Coral Sea* firepower.

"I had gotten to know Captain Harris on board ship and had developed a tremendous amount of respect for him as a leader and commander. He was always composed regardless of what was going on around him. I'm writing this story today because he spared no effort and his aircrews risked their lives to get me out."

For the next four to five hours the crews plastered the local North Vietnamese while maintaining a running commentary with Major Smith. About 1700 he got the call he had been waiting for: the *Jolly Green* was inbound, about five minutes out. By this time there were three or four *Nail FACs* directing at least four *Sandys*, as well as 10 to 15 other aircraft involved in flak suppression:

"I stood up for the first time in about 10 hours," says Smith, "Took out my flare gun and signal flares and – with the radio to my ear – listened to what seemed like controlled chaos. Overall, 25 to 30 fast moving aircraft were operating in a confined airspace for over an hour dropping all kinds of bombs, rockets, smoke, and cluster munitions. One man – *Sandy 01* – was orchestrating the whole show. Not one life or aircraft was lost, and they didn't hit the survivor. What a great tribute to the skill of the aircrews and the skill, guts, leadership, and determination of Major Jim Harding. I heard him tell *Sandy 02*, 'Go get *Jolly 32* and bring him in.' The rescue effort had never gotten this far before, and I had to keep telling myself: stay calm, don't lose it now, think about what you have to do to help the situation. I know the guys on the *Jolly* (piloted by Captain Ben Orrell and First Lieutenant Jim Casey) were sweating bullets just as I was.

"*Sandy 02* fired smoke rockets in front of *Jolly 32* to mark the way to my position. The helicopter began taking ground fire immediately upon starting the run-in. The rear ramp was down, and Sergeant Bill Brinson manned the mini-gun there; Airman First Class Bill Liles and Airman First Class Kenneth Cakebread were the door gunners. Brinson was hit in the knee early on the run-in but kept shooting, commenting to the rest of the crew, 'I'm alright, they just got me in the knee, but there's some holes in the helo.'"

Jolly 32 took 11 hits, including at least one through the windscreen, and *Sandy 01* had to guide the big helo in to Smith's location. Initially the Air Force crew could not see him in the gully, so Smith popped smoke, but the HH-53's downwash pushed the smoke all over the place. They then told the downed airman to pop the night end:

"Immediately, I turned my flare around and pulled the tab – just like a highway flare – and it ignited, showering me with sparks. I had been told not to chase the helo, to let it come to me, but when Orrell said – for the third time – that he still couldn't see me I decided it was time to move. I went up the hill and out into an open area pocked with craters and littered with fallen trees. Smoke hung in the air. I saw what appeared to be some sort of cloth tied around a tree. It looked like a trail marker; maybe that NVA did see me that first evening and they had decided to use me as bait for a deadly trap. At the top of the ridge I saw *Jolly 32* so low the rotor blades were cutting off tops of trees and slinging them in every direction. The helo was 50 to 60 yards distant and moving farther away. The door gunner/winch operator, Liles, was looking away from me. I ran toward the helo hollering on the radio, '*Right here, right here, behind you, behind you*!' There was so much noise on the radio I don't know how he heard me, even though I was screaming at the top of my lungs, but he turned and looked right at me and said 'I got him, I got him.' Orrell told him to lower the hoist."

The crews had been told repeatedly to let the rescue penetrator hit the ground first in order to dissipate any static electricity; needless to say, at the moment this was the least of Smith's concerns. He grabbed the cable with one hand while the penetrator was still five feet off the ground and snapped the link to his torso harness, later recalling he probably set the hook-up record. Liles took up the slack, and within short order the Marine was in the HH-53's cabin ... at which point Lile yelled, "Get the hell out of the way," and resumed firing his mini-gun:

"The interior was filled with smoke, empty shell casings flying all over, and three gunners firing in every direction. Light streamed in through bullet holes in the deck and overhead. Almost simultaneously Liles told Orrell, 'He's in the door, let's get the hell out of here.'"

The rescue helo departed low and fast and took fire all the way back to Thailand. It recovered at Nakon Phanom Royal Thai Air Force Base about 90 minutes later to a huge reception on the flight line. Smith remembers them as a:

"Very proud group of men and women who had worked day and night and risked their lives to rescue me, and succeeded. Their pride was exceeded only by the gratitude of one very humble Marine aviator. Someone who looked like Charlton Heston walked up to me. 'Hi,' he said. 'I'm *Sandy 01*.' It was Jim Harding. We threw our arms around each other. To this day I have no idea what we said. When he arrived overhead that first day I had pictured him as some old guy who got stuck in *Spads* instead of fighters, a perception quickly dispelled when I heard his engine quit, followed by his explanation to his wingman that he didn't know how much gas he had in his centerline tank, so he just let it run dry and then selected the other tank when the engine quit. When I heard that, I knew I was in good hands."

After two days in Thailand Maj. Smith returned to *Coral Sea* for another emotional homecoming. He then went to NAS Cubi Point for five days of recovery; while there he made the trip to Clark AFB, met Lt.Col. Hambleton, and compared notes about their respective rescues. Lt. Ketchie was never seen again and was eventually declared killed in action; notably, a few days later Jim Harding himself was shot down during another rescue attempt, but fortunately an Army helicopter was able to pull him off the ground.

Smith later got another major indication of the effort put into his rescue. On the fourth day the Air Force truly was low on assets due to other missions, and top cover for the rescue attempt had to shift to the Navy. Skipper Harris in Coral Sea realized his air wing could not do it alone and called the skipper in *Constellation* for his support. At first, *Connie's* commanding officer said he didn't have the assets. Harris's response was succinct: if it was one of the *Connie's* pilots on the ground, he'd give everything he had. After a pause *Constellation's* skipper pledged his entire air wing in support of the operation.

The war went on. On 3 May, -224 lost its second aircraft when *Bengal 501* was brought down by unknown circumstances. The pilot, 1st.Lt. Joseph McDonald – a recent addition to the squadron – was declared missing in action, and the B/N, Capt. David B. Williams, was listed as KIA. The Marines pressed on and, now effectively commanded by CAG Sheets, started doing some good work. According to Charlie Carr, their operations continued right into May, in and around other tasking:

"In April 1972, CAG and I led a division of Marine All Weather Attack Squadron -224 A-6As from the carrier USS *Coral Sea* to the Bai Thuong airfield in North Vietnam. Bai Thuong was a North Vietnamese jet fighter base northwest of Than Hoa and about 30 miles southwest of downtown Hanoi. On that occasion, our mission was successful and we hit the runway and support facilities and damaged a Soviet-built transport that was turning at the end of the runway. Again during May '72, CAG and I were leading an entire *Air Wing 15* strike against Nam Dinh ... (we) had already manned up, and the entire strike group started the engines. About halfway through the aircraft start sequence and system preflight we were told to shut down immediately and report back to AI (Air Intelligence). Something was up and it was hot. All the aircrews shut down and returned to AI. A very excited intelligence officer briefed the strike group that there were 28 MiGs on the deck at Bai Thuong and we were cleared to 'go get 'em.'"

Loaded with 16 *Rockeyes* each, the four *Bengal* A-6s headed out, along with A-7Es of VAs -22 and -94 and F-4Bs of VFs -51 and -111:

"After completion of the launch and rendezvous of the strike group," Charlie continues, "the two separate elements pushed to ensure timing over the target. This would ensure that aircraft making their runs on the target would not risk a mid-air collision with other aircraft in the strike group. The plan was for a five-ship A-6 *Intruder* division to go low (approximately 100-feet AGL) on a direct line from the ship to Bai Thuong airfield. The medium altitude divisions, led by two divisions of A-7s and F-4s, was to proceed on a course showing well south of the target area, then buttonhooking northeast out of the sun to compliment the A-6 ordnance on the airfield, and maybe shoot down some MiGs. The F-4s were configured for air-to-air combat with *Sidewinders* and *Sparrow* missiles.

"After the A-6 push from the rendezvous to the 'feet dry' ingress point, it was fairly simple to keep track of the nav and timing, ensuring deconfliction with the A-7 division in the target area. Inbound, about ten miles from the target, I saw two 'dots' at 12 O'clock, which quickly turned into NVA MiG aircraft. Moments later the F-4s, our TarCAP, called the multiple 'bogies' at 12 O'clock. By then, I was calling MiGs all over the goddamn place on 'Strike Common.' Evidently not believing our feint force, the MiGs had scrambled out of Bai Thuong, and by the time the A-6s hit the pop and aim points the MiGs were wheeling at about 1,000-feet over the runway. I think they realized that they might be in a trap, so they generally hit burner and headed northwest.

"Our A-6 division came in underneath the MiGs. We continued our bombing runs while we watched the MiGs jettison their drop tanks on top of us. We were clearing the target for the A-7s when we saw MiGs going for our Dash 3. CAG, being an old fighter pilot, told me to hold on tight and watch for the MiG up the right side as he got us between the MiG and the other A-6 (Dash 3). Next thing I know there is a MiG at a little high at 4 O'clock rolling in on us as we slowed down to 360-knots. We were now heading for the coast, which was bout 50 miles away, dragging this eager MiG pilot at about 100-feet AGL. By this time in the air war it was common practice for *Intruders* to carry one or two *Sidewinder* missiles, and our plan was to drag the MiG until he had to 'bingo' somewhere, and we would in turn bag him. I had already switched the ordnance panel from air-to-ground to air-to-air.

"After two unsuccessful passes by the MiG on our A-6, an F-4 – evidently seeing our plight – turned in and began chasing the

> MiG. He came up on strike announcing, 'For the A-6 heading southeast being fired on by a MiG. Get out of the way.' He had a good angle and had a good 'Fox 2' (*Sidewinder* firing solution). The next time the MiG rolled he shot rockets at us and CAG gave him a big move, and when the MiG rolled out he was sucked (well behind the wing line) and high. I heard the F-4 call '*Fox 2*' and saw the AIM-9 guide up the MiG's tailpip, and the tail came off in a ball of burning jet fuel and MiG parts. Nguyen's airplane came apart. No 'chute was seen, and it was safe to assume he didn't make it."

The kill by Lt.Cmdr. Jerry B. "Devil" Houston and Lt. Kevin T. Moore marked the second MiG shootdown of the cruise for VF-51. Houston, the *Screaming Eagles'* ops officer and a former F-8 *Crusader* driver, would later comment on the shootdown:

> "They were going about 465 to 500-knots, and we were past 600, about 100-feet off the ground. At that speed the MiG-17 lacks sufficient power assist on the controls. I called for the A-6 to break, but he kept right on in front of that MiG as the North Vietnamese pilot blazed away with his cannon. I didn't know that the A-6 pilot was our CAG, who had earlier spotted the MiG on his other two airplanes. He had purposely slid in between them to draw the MiG off.
>
> "The action was relayed to 'Red Crown' for confirmation," Carr concludes. "Devil Houston and his RIO, who were driving the *Phantom*, got credit for the shootdown, but CAG and I did get a victory roll on recovery (up the port side) and a piece of cake upon recovery. It was great fun."

During this hectic time the reinvigorated *Bengals* even mastered their B-Model *Intruders*, logging 95 *Iron Hand* missions over the next line period. They fired a total of 47 AGM-78s for an estimated 35% success rate against *Fan Song* radars; it was a very high rate of success for anti-radiation missiles and indicated the positive turn in the fortunes of -224. The squadron noted this in its history, stating:

> "... any variety of tasks were handled proficiently and effectively by what was now an experienced and professional combat squadron, VMA(AW)-224."

While the *Bengals* were keeping the North Vietnamese at bay and tussling with MiGs, other carriers responded to *Seventh Fleet's* "Hey Rube!" call. On 3 April *Kitty Hawk* with CVW-11 and VA-52 embarked joined *Hancock* and *Coral Sea*. Six days later *Constellation* returned with CVW-9 and VA-165. On 30 April 1972 *Midway* – with CVW-5 and VA-115 onboard – checked in at Yankee Station. *Midway* had deployed from Alameda seven weeks earlier than originally scheduled; her arrival gave CTF-77 *five* aircraft carriers to work with, four of which had *Intruder* squadrons onboard.

At the same time, the Air Force and Marine Corps also got back in the game in a big way. On 6 April 1972 the lead F-4Js of VMFA-115 and VMFA-232 arrived at Da Nang from Iwakuni. A few days later VMFA-212's F-4Bs arrived from MCAS Kaneohe Bay. Their arrival and the assignment of MAG-15 – which subsequently gained three A-4E squadrons – put Marine aviation firmly back into the war in Southeast Asia.

Second response: *Pocket Money*

In Paris on 2 May 1972 Henry Kissinger confronted Le Duc Tho concerning the invasion. Once again he was greeted with threats, insults, and attempts at intimidation. Again, the North Vietnamese representative made references to unrest in the United States and indicated his government felt the U.S. government no longer had the backing of its own people. North Vietnam was going to reunite the country under their rule, and there was nothing the U.S. could do about it.

Kissinger reported his conversations to President Nixon, adding the North Vietnamese were not interested in negotiations. The President decided enough was enough and ordered two additional operations that hopefully would get the attention of the PDRV's leaders. The first, to take place on 9 May, would mine the ports of North Vietnam, starting with Haiphong. The following day U.S. aircraft would start regularly hitting targets from Haiphong and Hanoi north to the Chinese border.

President Nixon's decision met with strong resistance from within his own cabinet. He was scheduled to visit with Leonid Brezhnev in approximately two weeks, and there were concerns the President's actions would result in cancellation of the talks. Harkening to the Johnson/McNamara era, there were other expressions of concern over the possibility of damage to Soviet shipping in Haiphong, or worse yet, the inadvertent injury or death of Soviet advisors. To top everything else, it was an election year, and both President and party had been roundly battered by the press and protesters over the conduct of the war. The President, in a decision later characterized by Dr. Kissinger as "... one of the finest hours of Nixon's presidency," ordered the mining to proceed. If all went as planned North Vietnam would shortly be starving and running out of armament, thus blunting its invasion of the south.

Early on the morning of 9 April 1972 multiple aircraft from CVW-11 in *Kitty Hawk* – including several VA-52 A-6s – launched, formed up, and proceeded into North Vietnam. Finding their primary target at Nam Dinh socked in they diverted to Thanh and Phu Qui. The bombs started falling on the backup targets at 0840 and 0845 locally. At that time a separate strike, launched from *Coral Sea* and consisting of three VMA(AW)-224 A-6As, six A-7Es from VAs -22 and -94, and a single *Black Raven* EKA-3B from VAQ-135 Det 3, departed marshal and also headed towards North Vietnam. Overhead, eight *Screaming Eagle* and *Sundowner* F-4Bs – fully loaded for bear with AIM-7s and AIM-9s – provided top cover. Nineteen minutes later, at 0859 local, the *Intruders* and *Corsairs* started depositing a total of 36 mines in Haiphong Harbor.

It was 9:00 p.m. in Washington DC. In a televised address President Nixon notified the nation and the world of the mining operation and advised foreign shipping in Haiphong they had 72 hours before the mines became active. According to Lt.j.g. Craig Weaver, a member of VF-51 onboard *Coral Sea*, upon hearing this announcement:

> "The ordnancemen looked at each other with a rather odd expression and finally told us that they had set the delays for 48 hours. No shipping tried to leave, so it was no problem. But for three days we had ordnance officers checking closely with intelligence on the status of shipping."

Be as it may, the mission – led by CAG Roger Sheets in a *Bengal* A-6A with regular B/N Charlie Carr by his side – was a complete success. Each aircraft carried four Mk.52-2 mines; the *Intruders* laid their 12 in the inner channel, while Cmdr. Leonard A. Giuliani – *Raven One* of VA-94 – coordinated the *SLUF's* drop of 24 in the outer channel. Carr says:

> "We utilized the A-6s in the inner harbor – inside Isle de Cac Ba – while the A-7s mined the outer harbor. Mining is overblown! All it is is putting ordnance on a target. Planning is a little different because of mine separation, etc. I had flown one MINEX while inport Subic, CAG had never flown one. We actually changed some one or two strings to conform to strike make-up and tactics."

During the proceedings the guided missile cruiser *Chicago* (CG-11) added to the excitement by using a *Talos* SAM to bag a MiG coming out of Phuc Yen. Even though he was a tad busy at the time Charlie recalls, "The shot was memorable."

The following day other CTF-77 attack squadrons started sowing additional Mk.52-2 and Mk.36 *Destructor* mines in other harbors, coastal waters, and estuaries, including Thanh Hoa, Dong Hoi, Vinh, Hon Gai, Quang Khe, and Cam Pha. A few ships left Haiphong on the 12th, but 32 others chose to remain in place and were bottled up for the duration. Ships that were inbound to Haiphong diverted to other ports:

> "The good news was the fact that ... ships were stuck in the harbor," Carr adds, "But the bad news was that any time we coasted out of the Haiphong area we had to count the ships. As you can imagine, the Viets didn't like this!"

Neither did a lot of other groups, commentators, and politicians back in the states; the uproar over this "escalation" of the war was loud and would continue through the events of the following days. Many speculated the Soviets would react by canceling President Nixon's planned visit to Moscow, but the Soviet government remained quiet about the mining. Despite the experts, the President's planned meetings with Party First Secretary Leonid Brezhnev started as scheduled on 22 May when Nixon flew to Moscow; in the process he became the first American President to visit that city. During the course of their week of meetings the two leaders discussed a wide range of issues and agreed to pursue a policy of détente. On 26 May 1972 the Nixon-Brezhnev sessions culminated with the signing of the Strategic Arms Limitation Treaty, a first between the two world powers.

As for the combat commanders in the Pacific, after years of recommendation and downright pleading they had finally achieved one of their goals. Transhipment of eastern block equipment, food, and war materials stopped except for those that could be transported through China by rail. Now it was time to proceed with phase two.

Operation *Linebacker*

On 10 May 1972, *Freedom Train* gave way to Operation *Linebacker*, a campaign of maximum effort strikes against targets in North Vietnam above the 20th parallel. Following *Pocket Money*, *Linebacker* was the second part of the plan to force North Vietnamese troops out of the south and bring their government back for meaningful negotiations. President Nixon had finally had quite enough, and now the North Vietnamese were going to pay for the reckoning.

CTF-77 had four carriers with *Intruder* squadrons on line. This time around the targeting restrictions were minimal; the vast majority of the industrial, military, and transportation sites that the combat commanders had long wanted were now free and open for attack. While the fighters were handling hordes of MiGs – on the first day of the operation the fighter crews blasted an incredible eight North Vietnamese fighters out of the sky – the attack squadrons went in to hammer their targets. Hell had truly broken loose over North Vietnam.

For VA-165 in *Constellation* the return to unrestricted combat meant a trip to POL facilities at Haiphong as part of an Alpha strike. Operating with the A-7Es of VAs –146 and -147, the *Boomers* flew through a storm of SA-2s and proceeded to thoroughly wreck the vital storage facility. The TarCAP of VF-92 and VF-96 *Phantoms* kept the enemy off their backs, and all of the attack aircraft returned safely to the carrier. The level of combat remained intense through this, *Connie's* sixth and last line period, including an attack later the same day on the rail yards at Hai Duong that saw the target obliterated. Unfortunately, the *Falcons* and *Silver Kings* lost one F-4J each on the raid to SAMs and AAA, but the TarCAP kept the MiGs off the *Intruders* and *Corsairs*. CAG-9, Cmdr. Lowell F. "Gus" Eggert – who led the attack in a VA-146 *Corsair* – would receive the Navy Cross for the mission. Subsequent strikes saw CVW-9 planes hit targets in the port of Hon Gay.

The carrier and Air Wing 9 concluded operations on 13 June 1972 and departed on 24 June 1972. The homecoming for VA-165 at Whidbey came on Saturday, 1 July, eight months to the day from their departure. Other than the one *Intruder* shot down on 30 December 1971, the *Boomers* had lost no aircraft in 169 days in combat; however, the squadron, its wing, and *Connie* would return one year later for one last turn in the crucible.

The *Bengals* of VMA(AW)-224 had a rougher time of it, but they too took the fight to the enemy with a vengeance, starting with an Alpha strike on Haiphong's railroad yard. Notably, on 10 May – the same day the *Phantom* versus MiG furballs were flying over the PDRV – CAG Sheets and Charlie Carr found themselves with a ringside seat for an event that would become Navy legend.

Flying *Bengal 500*, they were escorting squadron mate *Bengal 502* out of the target area and back to *Coral Sea*; *502* had taken hits resulting in a 2-foot hole in its horizontal stab and was marginally controllable. As they headed out for the Gulf the crews observed a Navy *Phantom* also outbound; it turned out to be *Showtime 100*, a VF-96 F-4J flown by Lt. Randy "Yank" Cunningham and Lt.j.g. William "Irish" Driscoll. Cunningham and Driscoll had just knocked down their third, fourth, and fifth MiGs, making them the first American aces of the Vietnam War.

As the two *Intruder* crews watched an SA-2 passed overhead and tracked *Showtime 100* from astern. Unable to contact the F-4 crew, they observed the SAM's detonation, which crippled the *Phantom*. Cunningham was able to wrestle the aircraft back over the Gulf before its tail exploded, at which point he and Driscoll ejected. Stuck with their own emergency, the crew of *Bengal 502* continued back to the boat with *500* accompanying. Fortunately, the three HC-7 HH-3As alerted to the drama were able to move in and successfully retrieve the two new aces. Cunningham and Driscoll were hauled out of the Gulf by *Big Mothers 62* and *65* respectively, and eventually returned by Marine CH-46s to a hero's welcome in *Constellation*. As for the stricken *Bengal 502*, it made it back to *Coral Sea* and successfully recovered. It wasn't quite "just another day at the office."

USS *Coral Sea* also did six periods on the line during its combat deployment, totaling 148 days. Among CVW-15, it was the attack squadrons that bore the brunt of the SAMs, with *Corsairs* constituting fully half of the wing's combat losses. Two of the aircraft shot down were - 224 *Intruders*. On 29 May 1972 Phil Schuyler and Marine Capt. Louis J. "Lou" Ferracane were forced to vacate their aircraft after getting shot up over the Uong Bi rail repair facility.

Two weeks prior a covey of A-7s had attempted to bomb the target but pulled off in the face of heavy antiaircraft fire. CAG Sheets, however, decided four A-6s could make it in, suppress the defenses, and allow an Alpha strike to take out the rail facility. Things got off to a bad start for Schuyler and Ferracane when they flew over an old French fort on an island off Haiphong and took a 23mm round through their wing that started a small fire. They pressed on, located the target, dumped their bombs, and immediately took four more 23mm rounds. Two hit the left wing, and a third impacted on the armor plating over one engine, while the fourth came into the cockpit, shattering the B/N's console and grazing his face. After determining they were still flying, the two turned towards their next target. After unloading their load of *Rockeyes* on the second AAA site Phil and Lou pulled up to turn for home; CAG Sheets called on them to reduce power, they did, and a big cloud of black smoke billowed from underneath their stricken *Intruder*.

Sheets and his B/N, Bill Angus, flew underneath to check out the battle damage and directed the crew of *Vulture Three* to jettison its MERs and drop tanks, which they did, almost hitting the CAG. He then barked, "*Vulture Three, Eject! Eject! Eject!*" much to Schuyler and Ferracane's surprise. After a brief debate they went for the handles and punched out. CAG later explained that he saw daylight under the aircraft from the trailing edge of the flap to within about four inches of the wing's leading edge; the fire had all but cut through the wing, and he was convinced the A-6 would go out of control at any minute. After about 20 minutes in Haiphong Harbor a friendly Navy helo moved in and retrieved both men. Both Schuyler and Ferracane returned to duty without much in the way of fuss, although Phil was med down for a week due to an apparent sprained back. He in fact flew seven more combat missions before *Coral Sea* departed Yankee Station on 11 June 1972.

Regrettably, that same day – the last of the carrier's fifth line period – Capts. Roger E. Wilson and Bill Angus were also hit by AAA, apparently while dueling the battery that got them. The target was Nam Dinh, about 60 miles south of Hanoi, and once again CAG Sheets led the way. To this day – nearly 30 years later – Angus can't remember much of the flight after the launch, only recalling reaching for the handle and coming to in the back of a North Vietnamese Army truck:

> "Either very late that Sunday night or shortly after midnight my memory returns as I am being forced into an interrogation room," he later wrote. "The only articles of clothing that I have on are my white boxer shorts and one of my favorite golf shirts. My left arm is a bloody mess, with a thin bamboo strip as a brace and wrapped loosely with gauze. Both arms had been pulled backwards, and they had been bound together at my elbows with thin wire that had cut into my flesh. Needless to say, I was still in shock and very disoriented."
>
> "I remember being forced to sit on a stool under a single dangling light bulb in front of a simple desk and the interrogation commencing. Thankfully, my amnesia helped me with the Code of Conduct during the multitude of quizzes that I went through during the first couple of days. In turn, my unwillingness to respond due to memory loss also created some painful moments inflicted by the guards. This turned out to be fairly standard new guy treatment and did not even remotely approximate the brutality that was inflicted upon the *FOGs* – endearingly, the F'ing Old Guys – that were shot down before the 1968 bombing halt.
>
> "As F'ing New Guys – those of us shot down from late 1971 to the end of the war – at the most we were exposed to an infrequent attention getting beating, as opposed to the prolonged character testing tortures that the *FOGs* underwent. It would be difficult to do justice either verbally or in writing to the tremendous courage and heroism exhibited by the *FOGs* – men that I respect tremendously."

Bill remained in solitary confinement for about two weeks before joining the other personnel of the Fourth Allied POW Wing at the Heartbreak Hotel section of the Hanoi Hilton. His first contact with a fellow American aviator came in the form of Maj. Bill Talley, an F-105G *Wild Weasel* pilot, and he later made the acquaintance of next door neighbor Lt.Cmdr. Al Nichols, a VA-56 A-7 driver shot down on 19 May.

"I wasn't alone after all!" Angus commented:

> "The relief that you experience when other Americans know of your existence is immeasurable.
>
> "My cracked left elbow and wrist healed on their own, and the boils that sprung forth finally went away. Unlike many of my roommates I didn't require the archaic and, for the most part, non-existent medical attention grudgingly offered infrequently by our North Vietnamese handlers. Our universe had been boiled down to a 75-foot by 25-foot prison cell. I can remember more than one very serious squadron meeting discussing the merits of using a washcloth and the proper way to dry our clothes. A discussion of this nature is where you could really differentiate the added value of Air Force Academy training. We had every type of personality in our room, and to our credit we were very respectful of each other. Thank God we didn't have to endure what the *FOGs* went through! I can, in all sincerity, say that I don't know how they did it."

On a more positive note, VMA(AW)-224 contributed to a double-MiG shootdown by CVW-15 *Phantoms* on 11 June; the circumstances were highly unusual, to put it mildly. VF-51 skipper Cmdr. Foster S. "Tooter" Teague, with RIO Lt. Ralph Howell and wingmen Lt. Winston Copeland/Lt.j.g. Don Bouchoux, were vectored towards Nam Dinh to cover an Alpha strike. Their controller onboard cruiser *Long Beach* (CGN-9) turned the two *Screaming Eagles* towards four MiGs. It quickly became apparent – and naturally, at the worst possible time – that *Long Beach* could not hear the two fighters. *Bengal Two* Lt.Col. Ralph Brubaker, flying a tanker hop in one of the squadron's KA-6Ds, keyed in and started relaying the calls from the cruiser to the fighters, even though he couldn't hear Teague and Copeland. In the resulting high-speed dustup both *Phantom* crews bagged a MiG-17, the fourth and fifth for CVW-15 this cruise. Regrettably, the -224 XO did not receive a formal "assist," but he had definitely earned one.

As for Navy exchange pilot Phil Schuyler, his war was definitely over. While in Cubi Point the flight surgeons ran him through several X-rays for his nagging sprained back ... and discovered he had a broken neck! The medics put him in a horse collar for a couple of months, but the bones never fully healed, and the proud warrior never flew combat aircraft again. He did receive the Silver Star for his extraordinary heroism during his mission of 29 May and, after retirement, became one of the leading fighter tactics instructors at NAS Oceana.

Bengals depart, *Knightriders* Arrive

By the end of the month other carriers arrived in the Tonkin Gulf to take up the load, and on 11 July *Coral Sea* departed CTF-77 and set sail for NAS Alameda. She returned to ConUS on 17 November, having made a nine-month cruise. The *Bengals* returned to Cherry Point shortly afterwards. For the men of VMA(AW)-224 – both Marine and Navy – the deployment had started on a dark note, but in spite of the flailing and four losses they had gained ultimate success. Moreover, they'd made a place for other Marine *Intruder* squadrons in carriers. As Charlie Carr later put it:

> "To make a long story short, at the end of the cruise, -224 ended up as the best squadron in the air wing. We'd put the big ball and anchor on the door of Ready Room Five. At the end of the deployment 'Big Bill' Harris, *Coral Sea's* skipper, came down and told us 'as long as *Coral Sea* sails the seas, that will remain on Ready Five.'"

Over on *Kitty Hawk*, *Carrier Air Wing 11* was still responding to a few schedule changes when the balloon went up. The carrier had departed NAS North Island on 17 February 1972 for its sixth Vietnam deployment and chopped to CTF-77 on 1 March. Once again, the *Knightrider* JO team of Larry Yarham and "Ferd" House were making the cruise, and again had "enjoyed" an interesting workup prior to the deployment. House recalls they were operating off Southern California during the wing weapons training exercise (WEPTRAEX) when the event took place:

> "Strip charts overflowing from my nav bags, kneeboard code word cards, 'Red Flag' special weapons handling procedures booklets, etc., etc. We emerged from the catwalk and found our trusty airplane. We were loaded with not one, but two B-61s ... sleek, chromed, shining beauties! It's the A-6 with the armed Marines all around it (I wonder if they had training rounds in those M16s). Preflight, startup, and nuclear weapons checklists all done by the book. Launch went right on time as per the air plan. Our mission was a two-plane for part of the way, a simulated drop on a Yuma target, and eventual recovery at NORIS (NAS North Island). Our wingman was an A-7. We rendezvoused overhead at the pre-briefed point and headed for the SoCal coast. We played the game as much as possible, with the low-level entry, TACAN to receive, IFF to

standby, and zip lip. The late afternoon weather was the usual haze with scattered clouds at coast-in.

"To this day, I don't know how we got lost," House comments. "I could blame it on a bad system, bad alignment, the ship's PIM, or whatever. We're DR nav'ing this route, trying to stay VFR, dodging a few clouds, with the A-7 flying off our right wing. You just kinda get this feeling that you're not where you think you are. Going from chart to ground and ground to chart ... I pointed to that big river heading in our direction out the pilot's window ... 'Is that the Rio Grande over there?'

"About that time my helmet hits the canopy and we're in a left turn, pulling 4 G's headed north. As the river goes under the wing one's mind kinda thinks how the headline in the *National Enquirer* would read: NAVY A-6 CREW NUKES MEXICO ... film at 11:00. Needless to say, we 'detached' the A-7 at that point, sooner than planned. I figured my career was history. The TACAN went to transmit/receive, and we 'found' ourselves (the IFF came back on when we were back well inside the good old U.S. of A). We landed at NORIS like nothing happened."

The remainder of workups were normal, and ship, wing, and squadron went to war. However, Fred says his buddy Yarham continues to remind him of this particular flight to this day.

For CVW-11 the air wing lineup was the same as the previous summer, with the exception of the duty *Vigi* and *Whale* squadrons, which were now RVAH-7 and VAQ-135 Det 1. As for the *Knightriders* of VA-52, the combat cruise marked their third with *Intruders* and sixth overall.

Kitty Hawk had completed its second line period on 25 March and departed the area for upkeep and liberty, but the onset of the North Vietnamese spring offensive resulted in its recall. The third line period started on 3 April and saw VA-52 and the other squadrons attacking NVA forces and lines of communication. A little over a month later, the *Knightriders* were in the thick of *Linebacker*, flying their initial raids against the main railroad and highway bridges leading out of Haiphong. Ferd recalls things were busy enough at the end of April, even before the mining of Haiphong and subsequent bombing campaign Up North:

"One night in the marshal stack ... with three carriers operating in the Tonkin Gulf at the same time, airspace and ship maneuvering water was limited. It seemed that every other time we would turn back in towards the ship (away from the beach), we would get a 'singer low' on the RHAW gear. Back on deck we plotted out that the marshall stack extended into a known SAM envelope! Needless to say, we both stayed awake in marshal that night!"

Four days before *Linebacker* kicked off, Yarham and House pulled a tanker hop that also ended up raising the pucker factor a bit. If anything, it served as a reminder that the situation in the Gulf was definitely not "as before":

"We were ... on a midnight-to-noon schedule, the seventh cycle of an eight-cycle *Kitty Hawk* flight plan ... in a KA-6 tanker enroute to the *Hawk's* BarCAP, our third hop of the day. Having two radios in the cockpit was such a novel concept to us A-6A types that it made the tanker hop somewhat entertaining. You can 'dial-in' all sorts of people, the other squadron's base freq, other carriers in the area, all sorts of people. Well, I just happened to hit on a freq that had the typical fighter-pilot jargon goings-on, 'I got a Judy left five, I'm going high ... you break right ... blah, blah.' I quickly cycled through the freq in search of something less annoying and obnoxious.

"About the same time, on the other radio, we get instructions for a 'steer' away from the BarCAP's usual location and vectored toward feet-dry direction. We were told to switch to a specific freq and did so, where the first words I hear were – you guessed it – '*Tanker Posit*.' Well, I'm tired. My body hasn't adjusted to the midnight to noon thing yet. I just turned, looked at Larry, and just shook my head. I was so tired of these fighter types, I refused to answer. Larry finally gave in and gave them a TACAN cut from a station they weren't receiving, yet I saw them and pointed at our 11 O'Clock to Larry (You could see them from many miles away because those 'tactical fighter-bombers' didn't have smokeless burner cans yet). Shortly thereafter, we hear, 'One's got contact, 10 right at 8, steer 125, judy ... two's got 'em, blah, blah, blah, blah,' the same old typical stuff. However, they seemed a bit more verbose than normal ... a bit more excited ... kinda like on adrenaline. It seemed like it went on forever, like it was never going to stop.

"When there was a bit of a lapse in their conversation with the lead's RIO, I keyed my mic and said, 'Gosh, do you guys get MiGs this way?' The response was, 'YUP, you bet!' Well, it turned out that freq that I had found was these guys, and they were actually mixing it up with two MiGs! They actually shot down two MiGs and really needed the gas!

"My only concern now," Ferd concludes, "Was they'd better hurry and recover, before their heads swelled so much they wouldn't be able to get their helmets off!"

The two victorious fighter crews were VF-114's Lt. Robert G. "Bob" Hughes/Lt.j.g. Adolph J. "Joe" Cruz in *Linfield 206* and Lt.Cmdr. Kenneth W. "Viper" Pettigrew/Lt.j.g. Michael J. "Wizard" McCabe in *Linfield 201*. The *Aardvarks'* spectacular two MiG-21 kills in one mission would prove to be the only ones for CVW-11 during the cruise.

Once the fourth carrier – *Midway* with CVW-5 – arrived and *Linebacker* kicked in it got really busy in the Gulf. With Alpha strikes and 2-v-many MiG sweeps during the day and single-ship ops all night, a person could get confused, as related by another *Knightrider*, Paul "Raoul" Bloch:

"The *STARM* patrol could be exciting in other ways. One day Lt.Cmdr. Mike Cockrell and Lt.j.g. Duncan Lewis were flying the figure eight pattern at the normal speed of about 300-knots, talking to their Air Force controller on the ground in Thailand. Suddenly, instead of a SAM indication from the east they received an 'air intercept' (AI) warning strobe from the west. They thought it might be a spurious signal, but asked their controller if he had anything out there. Receiving a negative reply, they continued the mission.

"The strobe continued to get longer, which on that piece of gear meant the threat was getting closer. Beginning to get concerned, Mike increased the speed somewhat and started moving the aircraft around a little bit. Suddenly, they received indications of a missile launch, and Mike broke the aircraft hard left. After the break, two air-to-air missiles went flying by the spot where the A-6 had been. Definitely moving the aircraft around now, they started yelling at their controller to find out what was going on. From him they learned that a different controller in the Air Defense Section had launched a section of F-4 CAP against what he believed was a North Vietnamese MiG (no doubt, a MiG flying a 300-kt figure-eight pattern at 20,000 feet over Laos). They now admitted the screw-up, and the F-4s were directed to return to base. Mike and Duncan also decided to call it a day.

"Bloch recalls the confused Air Force *Phantom* pilot's name was Murphy, adding, He received a great deal of flack back at his squadron for two reasons. First, he misidentified and shot at a 'MiG' that had a big ugly nose and refueling probe unlike any other aircraft in the world. Secondly, he missed. Needless to say, A-6 crews who subsequently flew *STARM* patrols made no assumptions about the locations of 'friendlies' in any direction. Somehow, this incident had added new meaning to the slogan, 'Sleep well America, your Air Force is alert!'"

Linebacker continued, and on 19 August the *Knightriders* suffered their only loss for the deployment when Lts. Roderick B. Lester and Harry S. Mossman went out in their A-6A and did not return. They were on a low-level armed reconnaissance mission in the area near Cam Pha, off Route 183, when the ship received a radio transmission, "Let's get the hell out of here." Another aircraft in the vicinity later reported seeing a flash under the thunderstorms in the area of the A-6's flight path. Their last known position was over the Gulf of Tonkin, and a later search mission found an oil slick and not much else. Both men were listed as MIA.

Kitty Hawk and CVW-11 ended up spending a total of 192 days on the line during the violent spring and summer of 1972, finally departing the Tonkin Gulf on 4 November. As was the experience in other air wings during this period, it was the two *Corsair* squadrons (VAs -192 and VA-195) that took the brunt of the losses, with the *Golden Dragons* losing four A-7Es in combat and two more in operational accidents. The *Dambusters* dumped one each respectively. As for the *Knightriders*, their return to NAS Whidbey Island on 27 November 1972 was a joyous one and marked the conclusion of their third and final trip to Vietnam in A-6s. The war wasn't quite over yet, but VA-52 had done its part.

However, the fourth *Intruder* squadron to participate in the initial phases of Linebacker – VA-115 in *Midway* with CVW-5 – would remain in Southeast Asia for some time to come. Now led by Cmdr. E Inman "Hoagy" Carmichael, the *Arabs* departed ConUS on 10 April 1972 and checked into Yankee Station for their first line period on 30 April. As with their sister squadrons, the circumstances this time around were quite different. John McMahon was one of the pilots making a second trip to SEA, and comments:

> "My first cruise with -115 we spent most of our time operating over the South. The second cruise took us into *Linebacker* and things *really* got ugly.
>
> "I particularly remember one strike – it was an Alpha strike, bunch of aircraft – everyone rendezvoused okay, and we were proceeding inbound when lead's aircraft went down. One of the -115 JGs assumed the lead, and *then* we got a target change relayed by the E-2. We turned south and got all stretched out to the point where the lead aircraft was at the IP while the tail aircraft were still making their turn. Just as we got there an Alpha Strike from *Hancock* came off the target with 25 to 30 aircraft. The whole thing just disintegrated. There were SAMs all over the place, aircraft all over the place, people trying to keep out of each other's way. At one point I looked up and saw a few Mk.82s coming *down* through what was left of our formation. Needless to say, quite a few people had to make a visit to the Admiral afterwards. 'Where'd the bombs land?' 'Hell if *I* know, Admiral.'"

For this deployment the *Arabs* operated a mix of A-6As and KA-6Ds. Unfortunately, their first loss was one of the tankers; on 2 May 1972 Lts. Rick Bendel and Jim Houser had to step out following an inflight electrical fire after launch, but fortunately both were recovered. John McMahon was in the tower and saw the whole thing; he recalls the *Intruder* launched off the cat, stayed level, the drop tanks came off, followed by smoke from the cockpit and the crew ejected.

The *Arabs* suffered a particularly personal tragedy in August, one that was later immortalized in a popular novel and movie. On a night mission over North Vietnam pilot Mike McCormack and B/N Ray Donnelly were trucking along when a random bullet penetrated the right side of the cockpit, hitting Donnelly in the neck. McCormack declared an emergency and flew the 100 miles back to *Midway*, all the while trying to apply first aid to his stricken bombardier. After landing the pilot remained in the cockpit and pushed away rescue personnel who were trying to attend to his mortally wounded partner. Several years later former VA-196 A-6 pilot, lawyer, and new author Steven Coonts would use a fictionalized account of McCormack's attempts to open his book *Flight of the Intruder*.

The next loss of the cruise did not occur until 24 October, and was again of an operational nature, but the end result was even more tragic. According to the accident report the crew of Lt. Bruce Kallsen and Lt.j.g. Mike Bixel made what appeared to be a normal approach and touchdown on *Midway*. However, once they hit the starboard wheel came off and the aircraft swerved to the right, clearing the arresting gear. Their *Intruder* ended up impacting the aircraft on the bow of the carrier; Bixel, the B/N, was seen to eject, but he went over the side and was lost at sea. Kallsen survived the mishap with minor injuries.

Four on the Line

With the initiation of *Linebacker* on 10 May 1972 the Navy had five carrier air wings available in the Tonkin Gulf: CVW-5 in *Midway*; CVW-9 in *Constellation*; CVW-11 in *Kitty Hawk*; CVW-15 in *Coral Sea*; and CVW-21 in *Hancock*. All but Air Wing-21, a *Skyhawk* and *Crusader* operation, had *Intruders* assigned. By the end of the month their squadrons flew a total of 3949 sorties over North Vietnam – an increase of almost 2700 over April's total – and 3290 strikes in South Vietnam. Working with the Marines and Air Force, CTF-77 had proven highly effective in blunting the NVA invasion, although there was still much to do.

Several of the carriers had operated through the six month mark of their cruises and were pushing seven and even eight months of deployment. This led to a problem for AirPac in that there were not enough carriers available for relief. *Oriskany* (CVW-19) and *Enterprise* (CVW-14) were next on the dance card, but they were still in the middle of their pre-deployment turnarounds and would not be able to depart ConUS for several months. As a result, AirLant stepped in, hurriedly dispatching two of its carriers and air wings to Southeast Asia. First to arrive was *Saratoga* with CVW-3 on 8 May 1972. Their arrival brought the total number of carriers on station to six, marking the greatest concentration of carrier air power since the war began, and probably the largest gathering since World War II.

For *Sara* – which had departed Mayport on 11 April 1972 and transited via the Cape of Good Hope – it was her first appearance in the Tonkin Gulf. The *Sunday Punchers* of VA-75, however, had been there twice before, including the original *Intruder* deployment five long, hard years ago. CAG-3 was *Intuder* vet Cdr. Deke Bordone. Initial operations took place from Dixie Station on 17 May and saw wing aircraft hammering the North Vietnamese invaders in the vicinity of An Loc, Kontum, and Quang Tri province. The *Punchers* went right to work, as did the other veteran squadrons in the wing: *Phantom* phylers VF-31 and 103; A-7A operators VAs -37 and -105; RVAH-1 (making their fourth appearance in SEA); and VAW-123. Air wing operations subsequently concentrated on targets in North Vietnam – such as POL facilities and highway bridges near Hanoi and Haiphong – with other activities occasionally interspersed.

One in the "other" category came on the night of 6 August and centered on the rescue of a VA-105 pilot. An SA-2 bagged the *SLUF* driver, Lt. James R. Lloyd, who went down 20 miles inland. While Lloyd attempted to evade his pursuers *Punchers* CO Cmdr. Charles M. Earnest and B/N Lt.Cmdr. Grady L. Jackson set up an orbit and assumed control of the rescue effort. Eventually the cavalry arrived; Earnest and Jackson guided in the HH-3A from HC-7 that, after several attempts, was able to pull Lloyd out. The rescue marked the deepest penetration of North Vietnam by the *Big Mothers* since 1967; Earnest and Jackson subsequently received the Silver Star for their direction of the successful retrieval.

The news was bad for VA-75 on the night of 6 September when an *Intruder* flown by Lt.Cmdr. Don F. Lindland and Lt. Roger J. Lerseth was shot out of the sky. Roger recalls their plane was the only one in the squadron that had not been fitted with the new backup hydraulic system, one of the combat-driven improvements added to the A-6:

> "I thought about that as we were spinning down into North Vietnam ... well, *damn!*," he comments. "We thought we'd been hit by AAA, but it turned out to be two SAMs. We learned that when they brought in the SAM people – I was their trophy – and they were all smiles and nodding."

VMA(AW)-224's Bill Angus was one of the Hanoi Hilton residents who greeted Lerseth after he checked in. Angus writes that at the time he was still trying to come to terms with his imprisonment while dealing with an inoperative memory:

> "Towards the end of the summer our group was joined by Roger Lerseth, a Navy A-6 B/N. Finally, I would find out how I'd gotten myself in this mess; my imagination was rampant. Unfortunately all Roger could share with me was the fact that something unusual had happened, even more cause for my paranoia. I was fully prepared to get back to the States some day and be told that I had flipped out and ejected from a totally good airplane. Nobody had seen or had any word on my very good friend and pilot, Roger Wilson."

Still, things started looking up for the POWs. Angus lists two positive developments: the removal of the internal barricades separating the rooms, and the release of three officers in October. The latter enabled his family to learn that he was still alive:

> "Since day one of my captivity, I was never permitted to write or receive mail," Bill recalls. "In fact, until my name showed up on the release manifest in January 1973 the NVA had never acknowledged my existence as a Prisoner of War. Mark Gartley left Hanoi with my name, and my family finally found out that I was at least alive four months later. This fact would prove to be monumental on my release date. Things were looking up. A consuming dream of mine was to get back home and buy my mom a home in the Scottsdale area with all the money I wasn't spending. Plus, we were raking in five bucks a day in per diem for substandard food and quarters!"

While Lerseth and Lindland "settled in" to their new surroundings, VA-75 continued to take the war to the enemy. Promptly at dusk on the night of 10 October several *Sunday Puncher* A-6As blasted the North Vietnamese airfield at Bai Thong. Bombardier/Navigators Lt.Cmdr. John A. Pieno and Lt.j.g. John R. Fuller received the Silver Star for their planning and execution of the mission.

Operations continued; as one carrier rotated into the theater another departed for the United States, and the next new carrier to report for duty with CTF-77 was another AirLant boat, *America* with CVW-8. She had been here twice before in 1968 and 1970; her A-6 squadron, VA-35, was making its fourth and last combat appearance in Vietnam with a mix of A-6As, Cs, and KA-6Ds. However, when *America* checked into the Gulf this time there was something decidedly different on the roof; instead of the familiar *Whales* there were four aircraft that resembled stretched *Intruders*. These were EA-6B *Prowlers* – aka the *Intruder Stationwagon* or *Queer Intruder* – making their first fleet deployment. Assigned to the *Scorpions* of VAQ-132, they gave CVW-8 and CTF-77 a flexible, and substantially enhanced electronic warfare capability.

With the heightened level of combat, the cruise proved to be a tough one for CVW-8. Within three months the wing lost eight aircraft, including five A-7Cs from VAs -82 and -86, one F-4J from VMFA-333, and an A-6A. The *Intruder* went down during a strike near Hanoi on the night of 16 September, killing *Black Panther One* Cmdr. Verne G. Donnelly and B/N Lt.Cmdr. Kenneth R. Buell. The last contact with the crew occurred at about 0150; seven minutes later another aircraft observed an explosion on the ground south of Haiphong, right about where the skipper's plane was supposed to be. Executive Officer Milton D. Beach formally assumed command on the 26th, and the *Panthers* pressed on.

The next two carriers to check in were the veteran *Oriskany* on 29 June 1972 and *Enterprise*, which came around for its sixth Vietnam combat cruise on 19 September. The big nuclear carrier had made a seven-month turnaround since its last deployment, and while CVW-14 was still the assigned air wing, it too now operated *Prowlers* vice *Whales*. The EA-6B squadron was VAQ-131, the former VAH-4. Now known as the *Lancers*, they took VAQ-130 Det 4's place in the wing. Also making the trip – their fifth – was the *Main Battery* of VA-196. Led by Cmdr. Howard I. Young, the squadron deployed with a mix of A-6As, Bs, and KA-6Ds.

Unfortunately, during pre-deployment workups off San Diego they had lost one of their B-models in an accident; the crash occurred on 30 July, approximately six weeks before *Enterprise* departed NAS Alameda. According to pilot Lt.Cmdr. Dick Toft the aircraft in question – BuNo 151559, one of the original ten A-6Bs – had spent some time on the hangar deck, was leaking all over the place, and *Big E's* skipper wanted it off his ship *now*. Toft launched with B/N Lt.j.g. John David Austin, the aircraft immediately pitched up, and they ejected. From what they were able to determine post-accident the *Intruder's* air navigation computer in the birdcage came loose on the cat stroke, locking the aircraft's controls and jamming the stick in the full aft position. The crew successfully got out at 700-feet, but Toft was seriously injured and went med down for six months. As a result he missed the cruise, but was later able to rejoin the *Milestones*.

Due to the increased level of combat in Southeast Asia, *Enterprise* and her primary escort, *Bainbridge* (DLGN-25), proceeded direct to Subic Bay, bypassing Hawaii. After a 12-day transit from California they arrived on 24 September and relieved *Hancock* and her escorts. *Enterprise* and CVW-14 started their first line period on *Yankee Station* on 3 October 1972 and eventually saw 183 days in combat.

With their arrival they joined a strong lineup consisting of *America* (CVW-8/VA-35), *Kitty Hawk* (CVW-11/VA-52), *Midway* (CVW-5/VA-115), *Oriskany* (CVW-19), and *Saratoga* (CVW-3/VA-75). The requirements of *Linebacker* had forced several of these carrier/air wing teams to execute a "drop everything and go" drill, but at least the Navymen were now used to it. However, the same could not exactly be said for their counterparts in the U.S. Air Force. In May 1970 Air Force assets in Southeast Asia included six tactical fighter wings, two special operations wings, and one dual fighter/reconnaissance wing. By the time the war kicked up again in the spring of 1972 the Blue Suiters were down to one fighter wing in South Vietnam and two fighter wings, the dual-purpose wing and a single special operations wing in Thailand.

The scene repeated elsewhere, as the other services started running aircraft back into combat. The Air Force's contribution to *Linebacker* reinforcement took the form of Operation *Constant Guard*, which moved 12 new F-4, F-105, and EB-66 squadrons into theater, as well as 168 KC-135A tankers and other support aircraft.

One other major component of the Air Force to get thrown in in a big way was *Strategic Air Command*. SAC's B-52 *Stratofortresses* (aka *BUFFs*) had long been involved in the Vietnam conflict, although to date their participation had been limited to operations over the low-threat South, known as *Arc Light*. The missions were often referred to as "monkey bombing" by tac air types of all services, but they were greatly appreciated by those on the ground, particularly during the siege of Khe Sanh.

Freedom Train and *Linebacker* changed the equation. Under *Bullet Shot* SAC quickly reinforced the dwindling number of B-52s in Guam and Thailand. On 17 April 1972 B-52s appeared over Hanoi and

Haiphong for the first time, much to the consternation of the North Vietnamese. By the end of the year the Air Force would regularly send streams of B-52Ds and Gs to both cities, as well as other targets Up North, with the *BUFF* crews from Andersen AFB in Guam making a butt-numbing 12-hour round trip. While their initial tactics – directed by SAC planners back in Omaha, and apparently based on the aerial campaign against Germany some 25 years before – would leave their smaller counterparts shaking their heads, the tenacity of the SAC crews would win the admiration of all, particularly when they started taking heavy losses. One group would particularly come to love the B-52 strikes: they were the POWs who – for once – could believe the war might finally be coming to an end:

> "Experiencing the Christmas B-52 bombings of Hanoi was incredible," says Bill Angus. "On the first night of the bombings and for the first five minutes several of us thought the A-6s were striking, but it was quickly apparent that this was something much bigger. From the barred windows in Room Five we could look due south, which was the direction from which the *BUFFs* were coming. Surface to Air Missiles were being fired in salvos by the hundreds, and the AAA made the sky look like a fireworks display. The only problem was that we were at ground level zero, and many of us had our doubts about the *BUFF's* accuracy.
>
> "The B-52 crews were great. Not only did they suffer far fewer losses of aircraft than anticipated, but they also hastened the end of the war. For once the guards in our camp were scared to death. Within minutes of the air raid sirens the NVA were six feet deep in covered bunkers, no grabbing their carbines and AK-47s and taking pot shots at the Air Pirates! President Nixon was even more of a hero to us after this show of force. Being part of and then watching planes get shot at from the ground created conflicting feelings for me. While I would far rather be in a jet streaking back to the ship or an air base someplace, having been shot down I felt lucky to be on the ground and alive. All I had to do was stay healthy and I'd get home someday."

Marines in the Rose Garden

During the time VMA(AW)-224 was embarked in *Coral Sea* the Marines had hurriedly sent other units back to Southeast Asia for duty from the beach. *Marine Aircraft Group 12* moved to Bien Hoa with two A-4E squadrons on 17 May 1972 for combat in the south, while VMAs -211 and -311 started operations on the 19th. For the *Tomcats* of -311 it was their fourth tour in Southeast Asia, while the *Wake Island Avengers* were back for the third time.

The Marines needed yet another base from which to support operations over the North. Thailand became the obvious candidate, but all of the Air Force bases there were operating at max capacity. So the Marines turned to a 10,000-ft Air Force emergency field located 14 miles northeast of Khon Kaen and 300 miles from Da Nang. It was a place only the Marines could love; it had no power, little water, and needed serious work to support a Marine aircraft group. The Navy's Seabees were called in, completed the runway, and on 29 April 1972 the first KC-130F of VMGR-152 Det D arrived with the first of what would become 3200 men on the field.

The "bare base" soon sprouted 310 "strong-back huts" (glorified tents), some administrative and maintenance structures – including corrugated aluminum "Wonder-Arch" hangars for jet aircraft – and storage for 360,000 gallons of jet fuel. While the official name of the new location was Nam Phong Royal Thai Air Force Base, it would forever be known to the occupants as the "Rose Garden" after a popular Lynn Anderson song of the day whose tag line was, "I never promised you a Rose Garden." MAG-15, with F-4-equipped VMFA-115, arrived on 16 June and flew combat sorties the following day. By the 20th the *Nighthawks* of VMA(AW)-533 had flown in with 12 A-6As to join four VMGR-152 KC-130F tankers and four CH-46Ds for local SAR duty. Another Phantom unit, VMFA-232, arrived soon after to add to MAG-15's punch. The *Nighthawks* went right to work, flying strikes throughout both Vietnams and Laos to stem the tide of the North's invasion. Aircraft frequently bombed targets in South Vietnam's old I Corps area, recovered at Da Nang for a load of bombs and fuel, then returned to Nam Phong via the combat route.

MAG-15 moved into Nam Phong AB, Thailand, in June 1972 with a single *Intruder* squadron: VMA(AW)-533, and a pair of *Phantom II* units (VMFAs-115 and -232). A *Nighthawk* A-6A drops a load of Mk.82s in company with *Red Devil* F-4Js. Leading the flight and providing the LORAN targeting cue through the clouds is an F-4D from the 433rd Tactical Fighter Squadron, 8th Tactical Fighter Wing. (U.S. Marines)

On 7 July 1972 the squadron suffered its first loss when Capt. Leonard Robertson and Capt. Al Kroboth went down. They were over their target in South Vietnam and getting ready to drop when heavy and accurate ground fire tagged their *Intruder*. Robertson did not make it, but Bombardier Kroboth somehow managed to eject, but ended up breaking his back, his neck, and his left scapula; when he came to a Viet Cong held him at gunpoint. Kroboth eventually wound up in the Hanoi Hilton, although the long walk was a dangerous one, above and beyond his severe injuries; as had happened with other shot down American airmen, all the way north villagers tried to kill him. He survived the ordeal – barely – and was released a year later.

Combat continued for all participants into the fall, while the services attempted to forge a somewhat stable situation in the south. As an example, on 20 October 1972 aircraft from CVW-3 in *Saratoga* responded to calls for assistance in the vicinities of Pleiku and My Thach. At the latter location *Sunday Puncher* A-6As and VA-37 and -105 A-7As pounded the North Vietnamese *48th Regiment, 320th Division*, which had cornered a smaller ARVN force. Following a six-hour aerial assault the NVA troops abandoned the field.

On 23 October 1972, the United States announced the conclusion of Operation *Linebacker*. Remarkably, the grand total for tactical sorties flown by carrier aircraft since the initiation of the campaign in May came in at 23,652. Once again there was a bombing pause, with all strikes suspended north of the 20th parallel. Through this action the Nixon Administration hoped to convince the North Vietnamese government of America's good will and strong desire to resume bargaining in Paris.

However, combat operations continued in South Vietnam and against lines of communications targets in Laos. CTF-77 returned to a pattern of alternating carriers on Yankee Station while everyone waited to see what diplomacy would achieve. *Hancock* and her battered *Carrier Air Wing 21* left the theater on 25 September, having lost 12 aircraft in combat and operational circumstances. *Hanna's* departure for home left *Kitty Hawk*, *Midway*, *Saratoga*, *Enterprise*, *Oriskany*, and *America* rotating through line periods or performing upkeep and liberty calls elsewhere in the vicinity.

As with previous bombing halts aircraft losses continued, although an analyst could probably claim they were occurring at a reduced rate. While the *Intruder* squadrons did not lose any aircraft in combat through the end of November, three A-6s did go down in operational accidents from three different carriers. The first was the VA-115 landing accident of 24 October onboard *Midway*. One week later VA-196 lost a tanker at NAS Cubi Point; the KA-6D stalled on takeoff and fell into the water, killing the crew of Lt. Clay Hiel and Lt. Scott Thomas.

The third loss was a particularly tragic one that hit the *Sunday Punchers*. On 28 November the squadron lost its second aircraft of the cruise when the pilot's Vertical Display Indicator (VDI) came loose on a night catapult shot off *Saratoga* and jammed the stick full aft. The B/N, Lt.Cmdr. Grady Jackson, ejected and was recovered, but the pilot, VA-75 skipper Cmdr. Charlie Earnest, was trapped in the aircraft and lost. The crew was the same one that had earlier been awarded the Silver Star for their night SAR effort on 6 August. With Earnest's loss Cmdr. William H. Greene, Jr., assumed command of the squadron.[1]

That same day another carrier joined the task force. *Ranger* with CVW-2 embarked had departed NAS Alameda on 16 November 1972 for its *seventh* combat deployment; VA-145 – led by Cmdr. Rupe Owens – was onboard for its third *Intruder* cruise to Vietnam. *Ranger* reported in with CTF-77 on 28 November and took her place in the Tonkin Gulf rotation on 9 December. Finally – and also on the 28th – *Kitty Hawk* tied up at NAS North Island. The carrier and *Air Wing 11* had been absent from Southern California for an incredible 10 1/2 months, during which time they had participated in both *Freedom Train* and *Linebacker*. Her CAG, Cmdr. Jim McKenzie, would later comment on the deployment, the carrier's sixth and last of the Vietnam War:

> "I became CAG shortly after *Linebacker I* started; *Kitty Hawk* was pulled out of REFTRA early and sent back over to the Tonkin Gulf, and I went over and relieved Hunt Hardisty the day before we mined Haiphong Harbor. That was the single most effective military operation I've ever seen. The ships that were in the harbor when the mines went down were still there when we left the line for the last time. During the cruise we'd do a day or two of cyclic ops – twelve on and twelve off – punctuated by a day with three daylight Alpha strikes (28 to 30 aircraft) for what we hoped was a worthwhile target. By the end of our line period we were able to fly around Hanoi with impunity. The only missiles left were in a ring around Hanoi, and there were damn few of those. It was really weird to fly around Hanoi and not hear anything in the headset, no RHAW, no tones, nothing."

CAG McKenzie adds that the changes wrought in the theater by the resumed bombing – and concurrent positive results – helped produce the carrier and wing's excellent morale:

> "Up until my time and for maybe two or three deployments afterwards -11 was the only wing to ever operate on *Kitty Hawk*. It was a good arrangement, because we knew each other, worked well together, and it really paid off. The plane captains were helping the crews push the bombs around, the green shirts helped keep the aircraft clean. Morale was wonderful; the wing really worked together. We had line periods back then that lasted longer than wars do now, and a number of people were concerned about that. Some of the flags expressed concern about our morale, fatigue levels, that kind of thing. Our approach was this: if you're going to do this stuff, *do it*. The morale remained high throughout the cruise; we got the best of everything. Anything I needed I got, be it money, parts, or getting the best people for the squadrons. The challenge was *right there*, right over the western horizon, and the results were immediate.
>
> "We had lieutenant commanders in the wing who had never made a peacetime cruise. Still, it was the finest deployment I ever made. We had quite a few Air Force exchange officers in the wing, and after going into combat with us they all said the same thing: 'Dear Air Force, *I want to stay with the Navy*.' When we left the line for the last time I told the captain the war was over. This could probably be the definition of insanity, but our last war cruise was *fun*, and *Kitty Hawk* was a good ship."

Heading into the end of the year the diplomats continued their "serious and frank discussions" in Paris, while the crews on Yankee Station continued to fly strikes into South Vietnam and Laos.

Referendum: The 1972 Elections

The President's announcement of the end of *Linebacker* came two weeks before the 1972 elections. It marked yet another twist in an already contentious political year. The 1968 contest between Nixon and Humphrey was judged by many to have been a referendum on the Johnson Administration's handling of the war. Indeed, part of Humphrey's difficulty during the campaign was his inability to distance himself from the Johnson policies. At one point, the vice president announced:

> "As President, I would stop the bombing of the North as an acceptable risk for peace, because I believe it could lead to success in the negotiations, and thereby shorten the war."

[1] Jackson, a former EA-1F NAO, would subsequently return to the EW world and command a *Prowler* squadron, VAQ-134 and the base at Diego Garcia. Promoted to Rear Admiral, he would later take charge of ComMatVaqWingPac.

Humphrey had lost, but the Johnson bombing halt – announced just days before the 1968 election – remained in effect for 3 1/2 years. The long sought Paris Peace talks had also taken place, but accomplished little if anything in the face of continued North Vietnamese intransigence. Now there was another campaign, and another national decision on the conduct of the war.

Several Democratic contenders, including Hubert Humphrey, Ed Muskie, and Ted Kennedy fell by the wayside during 1972's early primaries. Finally, at their convention in Miami Beach the Democrats selected South Dakota Senator George McGovern as their standard bearer. A member of the senate since 1962, McGovern had served in the Fifteenth Air Force during World War II as a B-24 pilot in the Mediterranean Theater of Operations. He built a rather quiet reputation for himself in public service, as well as early opposition to U.S. involvement in Southeast Asia. Now his views on the war were the party's views, and they centered on one item: end the war now and "bring the boys home."

Unfortunately for his party, McGovern's campaign was a disaster. A large part of the problem may have been in public perception, according to Democratic Congressman James O'Hara of Michigan, who at one point stated:

> "The American people made an association between McGovern and gay liberation, and welfare rights and pot-smoking and black militants, and women's' lib, and wise-ass college kids."

Vice Presidential candidate Senator Tom Eagleton of Missouri had to step aside after it was revealed he had been treated for mental illness; after a period, former Peace Corps director R. Sargent Shriver became his replacement. Even more damaging were some of McGovern's comments about how he would settle the Vietnam question, including a nagging controversy over the subject of amnesty for draft evaders.

For President Nixon's part, he could point to his rapprochement with both China and the Soviet Union. In addition, while the war still continued he *had* brought most of the troops home and turned over most of the effort to the South Vietnamese (albeit propped-up by massive American air power). The American electorate decided to stay the course, and on Tuesday, 7 November 1972, elected Nixon to a second term. The results were in all senses of the word a landslide. The President took 60.7 percent of the popular vote, 521 electoral votes, and 49 states, with a nearly 18 million vote plurality. The McGovern/Shriver ticket managed to win only the state of Massachusetts and the District of Columbia with 17 electoral votes (perhaps not coincidentally, in 1974 the Nixon Department of Defense closed the two major U.S. Air Force bases in Massachusetts, Westover AFB and Otis AFB, as well as the Boston Navy Yard at Charlestown, Mass.).

The President had his mandate and could now return his attention to the resolution of the war in Vietnam. However – and perhaps ominously for the victorious president – the Democrats retained both houses of the Congress.

Linebacker II

The Paris peace talks quietly resumed in the summer of 1972, but the North Vietnamese negotiators quickly returned to a pattern of obstinacy and demand. Items the two sides agreed upon at earlier sessions were suddenly reopened for debate, while PDRV delegate Madame Nguyen Thi Binh publicly reiterated her government's stance that the United States must unconditionally withdraw. Dr. Kissinger had no better luck in private sessions with Le Duc Tho. The subject of the release of the POWs was another stumbling block. North Vietnam still maintained they were war criminals, not prisoners, and would not be released until all American forces were removed from the region and support withdrawn from the government of South Vietnam.

Adding to the festivities, RVN President Nguyen Van Thieu submitted 69 major changes to the draft treaty. The South Vietnamese government's biggest complaint was the proposed extension of recognition to the National Liberation Front as a legitimate political entity in post-war Vietnam. Dr. Kissinger had his hands full trying to convince

VA-165 was among the *Intruder* squadrons involved in *Linebacker II* during the late stages of the Vietnam War. NG506 gets the run up signal from the catapult officer on *Constellation* during the 1972 deployment. CVW-9's performance during this cruise would lead to the ship/Air Wing team receiving a Presidential Unit Citation (PUC). (U.S. Navy)

the South Vietnamese there would be no separate peace between the United States and the PDRV, while at the same time attempting to forge an equitable document with the North.

On 26 October 1972 Kissinger announced, "Peace is at hand." However, his statement proved to be premature. U.S. analysts gathered evidence that once again the North was furiously attempting to rebuild and rearm itself. North Vietnam was also gambling that President Nixon would not dare order a resumption of bombing in the face of domestic unrest and anti-war protests. To that end, on 13 December 1972 Le Duc Tho and his aides walked out of the peace talks. The world stepped back and watched to see what would happen next.

They got their answer on 18 December with the commencement of *Linebacker II*. The transfer of American troops out of South Vietnam was immediately halted, and U.S. aircraft again went north of the 20th parallel in large numbers. The level of operations was intensified over *Linebacker I*, with concentrated air strikes against SAM and AAA sites, enemy army barracks, POL storage areas, Haiphong naval and shipyard areas, and railroad and truck terminals. In addition, the Navy replenished the minefields in Haiphong harbor and at other locations. The response from the anti-war movement and American media was immediate and loud, with several pundits damning the President for attempting to bomb the North into submission, one coining the phrase "war by tantrum." President Nixon, however, had decided that enough was enough, and the Navy, Marines, and Air Force were ordered to deliver the message. The six carriers on Yankee Station – five with *Intruders* in their air wings – handled the notification process with a vengeance, with the squadrons of CTF-77 flying 505 sorties over the following 11 days.

On the night of 20-21 December two *Intruders* went down while prosecuting targets. Operating from *Saratoga,* VA-75's Lt.Cmdrs. Robert S. Graustein and Barton S. Wade were coming off the target when their A-6A exploded amongst intense AAA fire. Their last contact with the strike was the call of "bombs away;" and both men were listed as missing in action. That same evening VA-196 also lost a bird during operations from *Enterprise*. The downed *Main Battery* A-6A was flown by squadron XO Cmdr. Gordon Nakagawa, with Lt. Ken Higdon as bombardier/navigator. Notably, their wingman for the strike on a small shipyard south of Haiphong was Lt. Steve Coonts, with B/N Bill Wagner. In Nakagawa's words:

> "Something was really screwed up; we had a weapons release malfunction and couldn't even do an emergency jettison. On our first run we used a manual range line with a small bridge as the IP; as a backup, I had the gunsight set 140-mils down. We came in, got the hack, and could see the target in the soft moonlight. It would've been a perfect hit, but there was no weapons release. I turned left to 240, leveled out, and Ken and I agreed to try to dump the bombs on Kien An Airfield, south of town. We were already within ten miles by that point with less than a minute to release. Ken said, 'Come right five degrees ... ready ... ready ...' BOOM! We took a 23mm hit – probably a *Gun Dish* – the right engine was hit, and flames punched through the airframe like a blowtorch. Sixteen 500-pound bombs, and we couldn't punch them off.
>
> "We turned back towards the coast-out point and climbed to 3500-feet, basically flying VFR with a BIG candle going. As we approached the coastline things started to fail; generators, the radars, etc. The overcast became denser – we couldn't see anything at all – and I knew that the bases were at 800 to 1000 feet, so we descended below the cloud cover and kept heading for the coast with the intent of ejecting over the Tonkin Gulf. Our big concern was the bombs on the right side; we didn't know how long the thermally protected Mk.82s would stay safe, what with the flames.
>
> "Coming down to 1000-feet I couldn't bring the aircraft's nose up. However, the rate of descent was slowly decreasing; I hit Ken on the leg and showed him the gauge. The plane leveled off, and we continued toward the coast at 240-knots, flying with only the trim tabs, as the stick was inoperative. About two to three miles from the coast the aircraft rolled to the left; as it went through 45 degrees all electrical power dumped. Ken punched out at the 90 degree mark, and I followed him at about 120 degrees. We landed about 150 yards from each other, Ken on a dike between some rice paddies, and me in the middle of a paddy."

Nakagawa pulled himself up out of the muck, called Higdon, and he said he had hurt his knee. After they rejoined it quickly became apparent that the B/N would not be able to move too far due to a major laceration on his right wrist, a souvenir of the ejection. Nakagawa used one of his leg restraints as a temporary tourniquet while they discussed their options:

> "There were still no people in the area," the XO continues. "The plane was burning about one to two miles away, a huge fire. About 20 minutes later an SA-2 in the vicinity of the wreckage cooked off. Apparently our *Intruder* came down near a SAM site, which made us feel pretty good."

The two downed crewmen determined that Higdon's injury made his evasion unthinkable. He decided to submit to capture in the hope of getting medical attention while Nakagawa attempted to escape to the coast:

> "I told Ken I'd take my raft and try to make the coastline. My biggest concern was all the mines we'd been dropping off the coast and in the rivers. I hoped the raft would provide enough cushioning in case a nearby mine exploded. However, I didn't get very far. When you evade through a rice paddy you leave a very muddy trail behind, and they tracked me down. There were thousands of people formed in lines; they had the B/N and knew the pilot had to be nearby. Eventually three militiamen found me. I heard them come up, so I carefully raised both hands and tried to stand up without making any threatening moves."

The capture ended their night. Nakagawa was marched to the next village, where he reunited with Ken Higdon, and they subsequently transferred to Hanoi. Cmdr. Jackson E. "Jack" Cartwright, a veteran attack pilot who had flown in the initial *Linebacker* strikes of 10 May 1972 with VA-165, subsequently reported to VA-196 to assume the duties of executive officer.

On 24 December 1972 the President directed a 36-hour Christmas stand down. The bombing of Hanoi and Haiphong and their suburbs resumed on the 26th with the single largest raid of the war, involving all Navy, Marine, and Air Force aircraft in the theater, including the junior service's F-111As from the *474th Tactical Fighter Wing* at Takhli RTAFB. B-52s staging from Thailand and Guam flew an incredible 116 sorties and – using modified tactics learned from the tacair types – overwhelmed the North Vietnamese defenses. They and the other strike aircraft were supported by electronic jamming from Navy *Prowlers* and *Skywarriors*, Marine EA-6As, and Air Force EB-66s. The North Vietnamese quickly ran out of SAMs, and by the end of the Christmas bombing were firing their few remaining SA-2s ballistically.

The violent aerial campaign ended on the evening of 29-30 December. There was one *Intruder* loss during the 11-day operation, suffered by VMA(AW)-533 on the night of the 27th in Route Package I. The *Hawks'* A-6A was on a road recce mission with 12 Mk.20 *Rockeyes* when it disappeared, apparently the victim of AAA. There were no communications and no wreckage, and both crewmen disappeared from the face of the earth.

When *Linebacker II* kicked off, the North Vietnamese responded with an intense public relations campaign that accused the United States of criminal activities and the other usual verbiage, including the bombing of innocents and wounding of POWs. By the 29th, however, the government of the People's Democratic Republic of Vietnam reportedly pleaded with the Nixon Administration to resume the peace talks. North Vietnam finally lay prostrate, with its Air Force grounded or destroyed – the only Navy MiG kill was scored by VF-142 on the 28th, while B-52 tail gunners claimed two MiGs – its industry and infrastructure lay in ruins, and its SA-2 stocks depleted.

After clearing up a few items with their North Vietnamese counterparts the U.S. agreed to meet, and on 6 January 1973 the talks resumed. On 23 January – and following some pressure on the South Vietnamese government by Kissinger deputy Alexander Haig – the parties in Paris finally signed the "Agreement on Ending the War and Restoring the Peace in Vietnam."

The End of a Long, Hard War

The peace agreement specified a ceasefire – to take place on 27 January – the release of the POWs, and U.S. assistance in clearing the mines from North Vietnamese waters. It also allowed North Vietnamese forces already in South Vietnam to remain in place with the understanding they would not be reinforced. According to the Nixon Administration the United States had finally achieved "Peace with Honor."

The United States continued offensive air operations over the north through the 15th, with Navy and Air Force crews authorized to pursue attacking MiGs and otherwise defend themselves. Reconnaissance and monitoring activities over North Vietnam did not conclude until the 23rd, the day of the signing. Operations over South Vietnam continued through the ceasefire on the 27th.

Operating from *Midway*, VA-115 lost their last plane of the war on 9 January 1973. An A-6A flown by Lt. Mike McCormick and Lt.j.g. R.A. "Al" Clark went after a radar target in North Vietnam; they were heard calling "feet wet," and shortly afterwards radioed they had cleared the target area and needed another aircraft. That was their last transmission; SAR forces never heard any beepers, nor saw any wreckage. McCormack and Clark were declared MIA and eventually declared killed in action. They were the last *Intruder* crewmen to die in combat in the Vietnam War; in 1979 NAS Whidbey Island would name its BOQ complex for McCormack.

On 20 January 1973, CVW-2 in *Ranger* provided an interesting denouement to the war when VA-145 led the first massive Navy laser bombing attacks against targets in North Vietnam. The *Swordsmen* had deployed with two of the Air Force developed Philco-Ford *Pave Knife* podded laser designators. So equipped and carrying modified Mk.83 and Mk.84 bombs, the squadron led VA-25 and -113 A-7Es against several bridges in the North. Strike Lead was *Swordsman One*, Cmdr. Rupe Owens. He writes:

> "*Pave Knife* was an Air Force system adapted for A-6 use. We launched A-6s equipped with *Pave Knife*, A-6 and A-7 bombers, *Corsairs* in the *Iron Hand* role, and F-4 flak suppressors, TarCAP and BarCAP, and destroyed 14 railroad and highway bridges in a three hour period. Using the designators, the A-6s illuminated targets for the other attack aircraft. Moving from target to target, the coordinated strike effectively closed all lines of communication in a route sector in a single mission."

With their mission magnificently complete, all of the aircraft returned to *Connie*. It was the last Navy bombing sortie against North Vietnam.

On 23 January – the day of the signing of the Paris Peace Accords and announcement of the impending ceasefire – the Navy established *Task Force 78* at Naval Station Charleston, SC. Under *Operation Endsweep*, task force personnel conducted the initial studies of Haiphong Harbor the first week of February. Full-scale mine-sweeping operations commenced 27 February and concluded at the end of July, with a break in March due to questions over Hanoi's release of the POWs.

During this period Yankee Station moved to a position off the northern coast of South Vietnam. Combat ops continued against Laos and Cambodia with sorties over South Vietnam through the 27th. On 24 January 1973 the *Black Panthers* of VA-35 sustained their second and last combat loss of the deployment when Lt. C.M. Graf and Lt. S.H. Hatfield were hit by AAA over the South. They stepped out and were rescued, in the process becoming the 160th and 161st successful recoveries by Navy SAR helicopters for the Vietnam War. It was also the last loss of an A-6 *Intruder* for the war.

At the end of January CTF-77 started releasing its carriers. The first to depart was *Saratoga*, which left the line on 8 January and departed the theater on the 16th. The *Sunday Punchers* concluded their third and last Vietnam War cruise with their return to Oceana on 13 February. Next to depart was VA-115 in *Midway*, which outchopped on 23 February; for their outstanding performance during the rough year – which included an early deployment – *Midway* and her assigned units received the Presidential Unit Citation, the third *Intruder* unit to receive this award during Vietnam. The *Arabs* returned to NAS Whidbey Island on 3 March 1973, having completed two combat deployments with *Intruders*.

Next to go were *America* and *Oriskany*, departing on 4 and 20 March respectively. For the *Panthers* of VA-35, the homecoming at Oceana on 24 March 1973 marked the conclusion of four combat deployments with *Intruders* with three different carriers and air wings. The *Main Battery* followed them back to the states, departing CTF-77 in *Enterprise* on 3 June. Notably, on 23 January *Enterprise's* CVW-14 sustained the final combat loss of the Vietnam War by a naval aircraft. A VF-143 *Pukin' Dog* F-4J was downed by AAA in the South; the RIO, Lt.Cmdr. Philip. A. Kientzler, was rescued, but pilot, former *Blue Angel* Cmdr. Harley H. Hall, was listed as MIA. His status was changed to killed in action on 29 February 1980; he was the last Naval Aviator killed in Vietnam.

VA-145 was the final *Intruder* squadron to head home following combat in Southeast Asia. Following a total of 120 days on Yankee Station spread over six line periods, *Ranger* departed CTF-77 on 14 June 1973. Total losses for CVW-2 during the deployment came in at one VA-113 A-7E, the operational loss of two VF-21 F-4Js in a midair collision on 29 January, and one VAQ-130 EKA-3B. Cmdr. Fred J. Metz – who relieved the legendary Rupe Owens on 29 February – brought the *Swordsmen* triumphantly home to Whidbey Island on 22 June 1973. It had been a memorable deployment for all concerned, including John Juan, who had just completed his fourth combat cruise, his second in *Intruders*:

> "I was there when Rupe made his famous LGB mission. Another time – in Laos – they told us over and over again, if you see flak, make one pass and keep going. So I did…and I looked back and there was Rupe, making repeated runs. He was going to get that guy. One mission against this island near Haiphong we were supposed to go and lay mines. All of these A-7s would come in from one direction, while us, the single A-6, came from the other direction. We missed the IP, had to go all the way around. Night VFR, overcast below the tops of the peaks, managed to come back around, and everyone was waiting for us. Not one shot, but we got back to the ship … naaaah, we don't want to do that kind of thing again.
>
> "I remember a tanker hop and seeing the B-52s getting shot down like big flares falling out of the sky," Juan concludes. "We used to load up with CBUs to go after the SAM sites. You carry enough of those things, you're bound to hit something. In all the missions I was on I never saw a SAM shot at us. It was a good thing: they were running out of SAMs."

By no means did *Task Force 77* disband. Operation *Endsweep* continued, as did operations in Laos and Cambodia, and in the interim *Constellation* and *Coral Sea* arrived in theater. As for the Marines, the two *Skyhawk* squadrons assigned to *Marine Aircraft Group 12* at Bien Hoa rotated home in January 1973, but MAG-15 remained at the "Rose Garden" through September. The last *Nighthawk* A-6A of VMA(AW)-533 finally returned to MCAS Iwakuni on 12 September 1973. By that date, the attention of the nation was focused on other areas.

Operation Homecoming

On 12 February 1973 Air Force C-141A number 66-0177 – assigned to the *63d Military Airlift Wing* at Norton AFB, CA – landed at Hanoi's Gia Lam Airport. Lined up on the tarmac were a group of 40 Americans dressed in gray windbreakers and dark slacks, the first of 591 released by the peace-loving government of the People's Democratic Republic of Vietnam. Greeting them were a group of U.S. military personnel led by Army Brig.Gen. Stan McClellan and several observers from neutral nations. The procedures were wholly formal and appropriate and – after some last minute dickering with the captors – the first group of POWs loaded onto the *Starlifter*, appropriately nicknamed "The Hanoi Taxi." Before they released their prisoners North Vietnamese personnel offered asylum to any American who wished to remain behind. None responded.

According to Lt.Cmdr. Robert B. Doremus, a VF-21 RIO who was shot down on 24 August 1965 while operating off *Midway*, he and his fellow prisoners showed no emotion to their captors. They maintained perfect military decorum until the plane lifted off and they heard the wheels go in the well. *Then* the cheering and backslapping started. "That cheer will be with me forever," he recalls.

The flight, which took place on Lincoln's birthday, was the first of 118 under Operation *Homecoming*. Upon their departure from Hanoi the men flew to Clark Air Base in the Philippines, where they found cheering crowds and the classic red carpet treatment. After the welcoming ceremony they went to the hospital for medical care and evaluation, followed by debriefing. As soon as it was practicable the men returned to the United States for reunions with their families and assignment to the military medical centers nearest their homes.

Through 29 March, in both the Philippines and the states, the crowds came out no matter what the weather or time of day. Many of the long-term prisoners had never seen a C-141 before, let alone miniskirts and other symbols of early 1970s American culture. It took a lengthy period of adjustment, but none cared. Marine Larry Friese, who had ejected from his crippled *Hawks* A-6A in February 1968, said:

> "There was a stinking cell that morning, and I ended the day with a fistful of $20 bills in one hand and a nice cold Michelob in the other."

Among the 145 naval aviation personnel who came home were VA-196 and VA-85 skippers Leo Profilet and Ken Coskey – prisoners since August 1967 and September 1968, respectively – and VA-196 XO Cmdr. Gordon Nakagawa, downed in December 1972 during *Linebacker II*. Profilet went on to become professor of Naval Science at the University of New Mexico in Albuquerque, and retired as a captain following a tour as Chief of Staff, *Commander Third U.S. Fleet* in Hawaii. Coskey also resumed his career, made captain, and following retirement became executive director of the Naval Historical Foundation. As for Nakagawa, he eventually returned to the *Main Battery* for his command tour – he did, in fact, return to Whidbey before the rest of the squadron – and following retirement became a professor at the Naval Postgraduate School in Monterey.

Mike Christian of VA-85, shot down on the squadron's second combat tour on 24 April 1967, was another to come home; he was released with his pilot, Irv Williams. While in captivity Mike became famous for fabricating U.S. flags for his fellow prisoners despite constant and repeated rounds of torture. According to POW and Medal of Honor recipient Lt.Col. Leo Thorsness – who served in captivity with Mike and would later write and speak about him – when President Carter later declared amnesty for the Vietnam draft dodgers, Christian expressed his anger and disgust by throwing all of his medals over the White House fence. Following his retirement Christian and a few other *Intruder* veterans opened the famous White Heron in Virginia Beach; the nightspot quickly became a favorite with attack crews from nearby NAS Oceana. Tragically, having survived six years of incarceration and torture, Mike would later die in an apartment fire.

For FOGs and FNGs alike, the last three months passed like some sort of combination of a nightmare and dream, according to VMA(AW)-224's Bill Angus. In some ways, their final release came very quickly, while in others – particularly for the old hands – it did not come fast enough:

> "Within a couple of weeks after the Christmas bombings ceased a mass move occurred within the Hanoi Hilton that was viewed properly as segregating the POWs into shoot-down order. It was not by coincidence that those of us that had never written a letter home were assembled one day and told to quickly write our seven-line missive.
>
> "The next day we assembled in military formation for the first time as the NVA read to us that the peace treaty had been signed and that hostilities had ceased. The Paris Peace Accords provided for a fourth of the Wing to be repatriated every fifteen days, with the injured/sick and women first, and then on a first in first home basis. Two months to go. Within days we were moved as a group again to a camp called the Zoo on the southwest outskirts of Hanoi, where we were reunited with other recent shootdowns. No Roger Wilson, nor was there ever any word of him. I still had no clue what had happened to us on 11 June."

During this period the Red Cross packages started appearing for the first time, as the North Vietnamese made an effort to upgrade their image as humanitarian captors. Finally, Freedom Day – 28 March 1973 in Angus' case – arrived. The North Vietnamese gathered the day's group of POWs, strip searched them to ensure they were not carrying any contraband, gave them new clothes, and put them on a bus for Gia Lam Airport. The images of the surrounding city of Hanoi were startling:

> "What a run-down and decrepit city. For the first time I was being transported during the day and without being blindfolded and bound. Hanoi had been a living hell, but at least I knew I would be gone within a couple of hours, while the NVA would remain. We crossed the Paul Doumer Bridge and viewed areas where the B-52s had obliterated everything. For whatever reason – and within minutes of the airport – we disembarked from the buses and had a 45-minute tea party. One final opportunity to experience the humane and lenient treatment of our hosts. A few minutes later our bus circled a hangar, and in front of us was the most beautiful airplane in the world, an Air Force C-141.
>
> "We assembled in formation by our date of capture, and when our names were read, we marched forward, saluted smartly, and passed into American hands. Our escorts then led us into our 'Flight to Freedom.' A fabulous day reached its zenith as we lifted off of the Gia Lam runway, and about twenty minutes later we exited the North Vietnamese airspace en-route to Clark Air Force Base in the Philippines. As our C-141 neared Clark, my excitement turned to anxiety, as I knew I was coming face to face with the circumstances of 11 June. I was still hoping against hope that Roger Wilson would be there to meet me. On our approach into Clark I started to hyperventilate, and once we were on the ground an Air Force nurse escorted me from the plane, through the receiving line, and onto another bus.

"We were then whisked to the Clark hospital, where the party began in earnest for everybody but me. The minute that I reached my hospital room I was approached by a Catholic priest who informed me that my mom had died from a stroke on November 12, 1972. What an emotional roller coaster I was on – from a high that I had never experienced before, to the saddest day of my life within a two-hour period. My first hot shower in almost ten months and I couldn't stop crying. My sister and brother-in-law (retired Army) had been both brilliant and kind. They intentionally spared me – in the many letters that I never received – word of my mom's death, a letter that would have been gleefully delivered by the NVA. Hanoi was miserable enough, even for my short stay, without being the recipient of devastating family news.

"I still had one bit of unfinished business to address: what had happened over Nam Dinh nine and a half months before. Thanks to a number of available sources, including Jerry Owen, who was my escort back to the States, the following occurred that day. As the strike group was turning up on the flight deck of the *Coral Sea* CAG and Charlie Carr's plane went down, and they passed the Alpha Strike lead to Roger and me. We led the strike group to the target, and as the lead aircraft we were the first to roll in. Just after releasing our ordinance AAA blew our left wing off at the wing root. The A-6 proceeded to snap roll itself into a small lake in a matter of seconds. Both Roger and I managed to eject; however, Roger's chute never deployed, and he died on impact.

"In May 1973 I spent an evening with Roger's mother, sister, and brother at their home in Virginia Beach. It was important for the Wilson family to hear directly from me regarding my memory loss and the fact that I could not answer any of their questions regarding the events of the day or Roger's subsequent status. The kindness and compassion of the Wilson family made a difficult and awkward evening bearable. Roger's remains were returned to the Norfolk area in the fall of 1988, bringing closure to a brave family."

End Game

On 29 March 1973 the few remaining U.S. combat forces in South Vietnam departed, and *U.S. Military Assistance Command Vietnam* disestablished. On that date formal U.S. involvement in the defense of the Republic of Vietnam, dating to 1961, ended.

Total U.S. military losses for the war included 47,369 combat-related deaths, 10,799 deaths from other circumstances, and 153,303 other casualties. Naval aviation lost 719 fixed-wing and 283 rotary-wing aircraft, 377 aviators killed in action, and approximately 100 missing in action. Between VA-75's combat introduction of the A-6A in June 1965 and the final departure of VMA(AW)-533 in September 1973, A-6 squadrons made 33 deployments to Vietnam in nine carriers. A total of 69 Intruders were lost in the conflict, yet the *Intruder* had made its mark and delivered:

> "The A-6 was too young for the Cuban Missile Crisis, but it was just right for Vietnam," comments John Diselrod. "In fact, when it first went into combat the system still had problems and the aircrews didn't have enough confidence in it to use it to its fullest. It took some real convincing by Capt. Bill Houser of the *Constellation* and Cmdr. Bill Small of VA-65 to have the Navy keep it in production. I think it proved itself then, and was still doing so thirty years later."

As Rupe Owens would later put it:

> "... the 'convincing' came in the actual combat performance of the A-6, the men who flew and maintained it, and those who supported it."

Still, for most everyone who flew and maintained the aircraft, the war left a bad taste. In the words of former *Black Falcon* CO Jerry Patterson:

> "Vietnam was a very incongruous war. We'd go out on a one and a half hour flight, have our 15 minutes of adrenaline time, 45 minutes of boredom, then back to the pattern. We'd go back to clean linen, eat on clean plates – those other poor guys were flopping around in the mud. A very strange war."

Iron Hand

When people talk of the Navy's anti-radar mission, the A-6 rarely comes up in the discussion, yet it was heavily involved in the role throughout its service. *Shrike*-carrying A-4s and A-7s typically come to mind, but the AGM-45 was also used extensively by the A-6A, B, and E models. The B model was the only Navy aircraft designed to carry the AGM-78 Standard ARM, which was the best anti-radiation missile of its generation. Also carried by the Air Force's legendary F-105G and F-4G *Wild Weasels*, the "Starm" had the reach and flexibility the *Shrike* did not. While the AGM-45 had a very narrow frequency range that couldn't be changed once airborne, the Standard ARM, which used an airframe based on the *Tartar* Naval SAM, featured a much wider flight envelope and a lot larger warhead. With the limited number of A-6Bs available, they never matched the number of Iron Hand sorties the *Skyhawks* and *Corsairs* flew, but the *Intruder* did remain the most capable defense-suppression aircraft the Navy owned through the Vietnam years.

With the demise of the A-6B, the Navy developed the AWG-21 weapons system that allowed the AGM-78 to be carried on a limited number of E-models. Through the late 1970s and early 1980s AirPac assigned its AWG-21 birds to VA-115 out of Japan, while AirLant rotated two or three airframes among the deployed squadrons.

The end of the Standard ARM came in 1986 when cracks in the propellant sections caused the entire stock to be grounded. By then, AirPac had centered their remaining airframes in Whidbey at VA-145, as CVW-2 went to an "all Grumman" configuration, where no other ARM shooters were available. The conversion of stable mate VAQ-131 to HARM-capable EA-6Bs made it unnecessary for the *Swordsmen* to continue with the "Starm" mission.

The SWIP A-6Es of the late 1980s added HARM to the *Intruder's* bag of tricks, and some were expended by the two SWIP units during *Desert Storm*, although in most cases the A-6s "hauled iron" and left the lethal anti-radar role to *Prowlers*, *Hornets*, and *Corsairs*.

The Iron Hand mission continues to be a tough one, not only for aircrew dealing with SAMs, but also for friendly forces who are occasionally mistaken as the enemy. Anti-radiation missiles are "equal opportunity" weapons and do not discriminate between friend and foe once launched. It is up to crew training and discretion to keep "blue on blue" (friendly on friendly) shots down. During Desert Storm there were as many as seven such incidents, including two near misses on U.S. ships.

In Vietnam, the *Leahy*-class frigate *Worden* (DLG-18, later re-designated a cruiser as CG-18) was hit on 15 April 1972 by two *Shrikes* fired by an Air Force F-105G *Wild Weasel*. The missiles impacted the superstructure, killed one man, and wounded everyone else on the bridge. Although still able to make way and never in any danger of sinking, the ship was unable to use any fire control equipment until shipyard repairs were made.

The A-6E SWIP was the ultimate expression of *Intruder Iron Hand* capability. Two squadrons took the type, along with the AGM-88 HARM, into combat over Iraq. Here a crew from VA-75 inspects their aircraft on the deck of the *Kennedy* prior to the first *Desert Storm* launch on 15 January 1991. (U.S. Navy, LCDR "Heyjoe" Parsons)

Although the AGM-45 *Shrike* was considered largely obsolescent by the early 1980s, it was still being used after introduction of the HARM. The missile would finally be retired after Desert Storm. *Blue Blaster* AB503 readies from launch off *John F. Kennedy* with a practice *Shrike* on its inboard station in spring of 1984. (Rick Burgess)

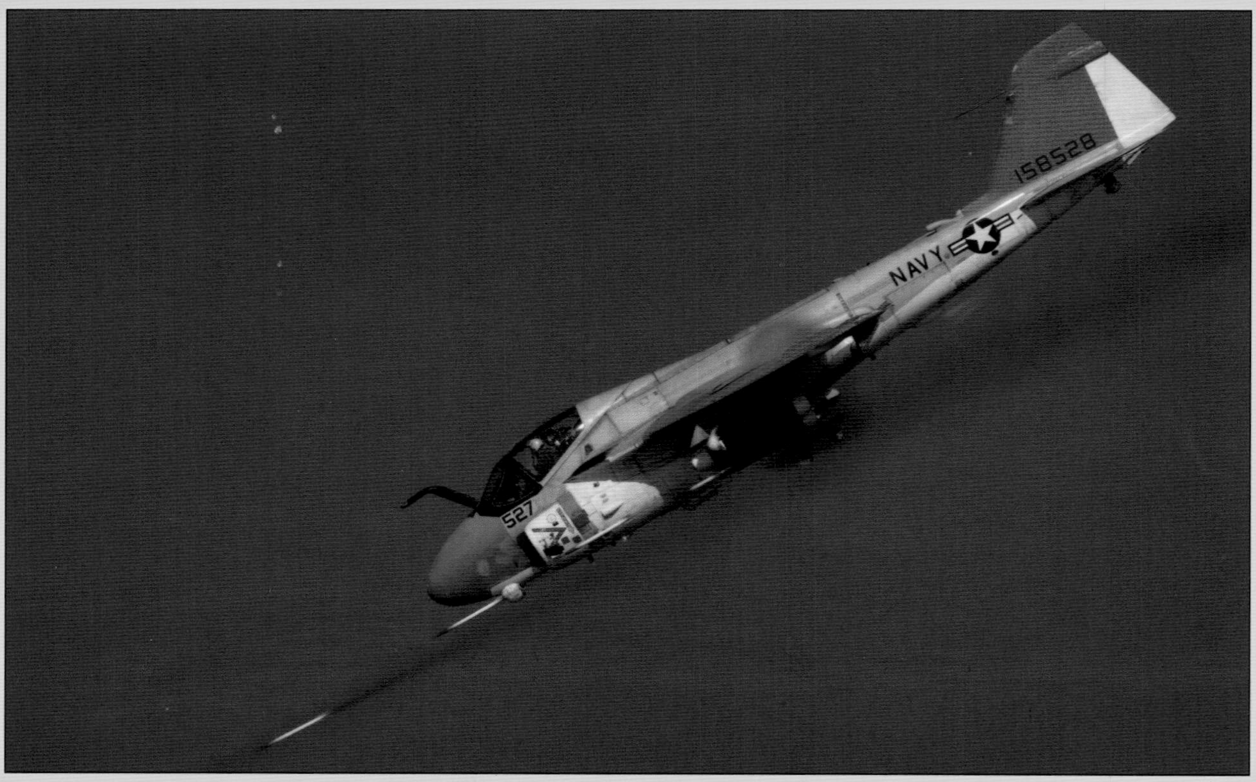

Throughout the *Intruder's* service the fleet squadrons were backed up by an extensive support network that helped develop tactics, evaluate weapons, and test new airframe configurations. These units included China Lake's VX-5, the Naval Weapons Evaluation Facility (NWEF) at Albuquerque, and the Naval Air Test Center (NATC) at Patuxent River, Maryland. Here a test-configured NATC A-6E TRAM fires a pod of 2.75-inch rockets. (U.S. Navy)

CHAPTER SIX

Violent Peace, 1973-1981

As could be expected, the war had left a bitter taste for all military personnel, and its aftereffects would last for years. Things were expected to slow down now that the war was over, although it did not quite work out that way.

Intruder pilot John Juan commanded VA-145 towards the end of the decade and recalls the period as being:

> "... pretty low key. It looked like the military was going to get smaller, so we had to start looking at improving our flexibility. Fortunately I had no major personnel problems beyond some minor drug stuff, but not everyone was so lucky. At least we didn't have any gigantic problems like those we had in the late 1960s and early 1970s."

The problems took many forms, ranging from an exodus of personnel to drug and racial difficulties and worn out equipment. The military draft ended in January 1973, and the nation's armed services switched to the "all volunteer force," although the general weariness with Vietnam and revulsion with things military made the recruitment of qualified, educated people problematic. Sporadic racial incidents – such as the heavily publicized "sit-down strike" in *Constellation* on 3 November 1972 – did not help. The *Connie* fiasco in particular had proven particularly galling to many in the Navy; it ended skipper JD Ward's exemplary career, but the worst punishment accorded to the participants was a general discharge.

To top it off, the Navy was physically worn out. For eight years the sea service had maintained a constant series of multiple battle group deployments to the Tonkin Gulf, all the while keeping an eye on China and the Soviet Union and reacting to other trouble spots worldwide. Its ships – many dating to World War II – were old and tired, and a large number were decommissioned following the war, reducing the fleet's flexibility. Famous names such as *Bon Homme Richard* and *Ticonderoga* left the fleet, as did aircraft types, such as the A-4 *Skyhawk* and F-8 *Crusader*, which were retired. Squadrons were lost as well, but none in the Medium Attack community.

Fortunately, at first glance there appeared little chance of renewed combat, but that too changed following a surprise war in the Mideast, a disastrous oil embargo, and an outbreak of international terrorism. Hammered by political and economic storms, the United States seemed ready to turn inward; however, events overseas did not allow a return to isolationism, and – as is usually the case – the nation's military was forced to respond throughout this period of "violent peace." The question was this: what sort of military did the post-Vietnam United States posses, and how would it perform?

Green Lizards Arrive

The Navy maintained a strong carrier presence in the South China Sea and Gulf of Tonkin through the conclusion of Operation *Endsweep* on 27 July 1973. On 15 August aerial combat operations in Cambodia concluded in response to a Congressional directive of 30 June. Between the two events CTF-77 was able to start sending carriers home.

The first to depart was *Oriskany*, with CVW-19 embarked. She checked off the line on 5 March 1973 and returned to NAS Alameda on 30 March, having completed seven combat cruises to Vietnam. *Enterprise* was next, departing the region with CVW-14 and VA-196 on 3 June; *the Main Battery's* return to NAS Whidbey Island on 12 June marked the conclusion of five A-6 combat deployments in three different carriers. VA-145 was next to depart, leaving the Tonkin Gulf with CVW-2 in *Ranger* on 14 June 1973. The *Swordsmen* had made three *Intruder* cruises to Vietnam from two decks. Three carriers took their places: *Constellation* with CVW-9 and VA-165 which checked in with CTF-77 on 31 January 1973; *Coral Sea* with CVW-15 and VA-95, arriving on 31 March; and *Hancock*, which arrived with CVW-21 on 1 June.

For the *Boomers* of -165 the return marked their sixth deployment to Southeast Asia with *Intruders*, tying them with VA-196. However, for the first time since their initial appearance with the type in 1967 they were not going into a shooting war. As for the *Green Lizards* of VA-95 – AirPac's sixth and newest A-6 squadron – this was their first time around, period.

The squadron was the third in Naval Aviation to bear the designation VA-95 and, as with their Oceana counterparts of VA-34, bore no literal ties to their predecessors. The first VA-95 had established as VT-20 on 15 October 1943, flew combat in the southwest Pacific with TBMs from *Enterprise* (CV-6) and *Lexington* (CV-16), and disestablished as VA-95 on 30 November 1949 at NAAS Charlestowne, RI. The second VA-95 established under CVG-9 on 26 March 1952 at NAS Alameda and adopted the "Green Lizards" nickname in 1963. The squadron made two Vietnam cruises – one each with A-1H/Js and A-4Bs – and two additional Med deployments with *Scooters* before disestablishing at Alameda on 1 April 1970. The "new" *Green Lizards* stood up at NAS Whidbey Island under Cmdr. George E. Wales on 1 April 1972, received their first A-6A Intruders the following June, and immediately prepared for their deployment with CVW-15. *Coral Sea* departed the pier at Alameda on 9 March 1973 for its seventh visit to Southeast Asia.

Unlike her six previous tours, the proud old carrier faced minimal combat this time and concentrated instead on supporting *Task Force 78* and the clearing of mines from Haiphong Harbor. Bombardier/navigator Terry Anderson recalls:

Established in 1972, the Whidbey-based *Green Lizards* made three WestPacs in *Coral Sea* before participating in two deployments in the late-1970s/early-1980s to the Med with *America*. NH505, an early A-6E, was shot at NAS Chase Field, TX, on 15 December 1979. The aircraft bears the "zap" markings of CVW-11 members VF-213, VA-192, and VS-33. (Rick Morgan)

"On the way over to Vietnam the war ended; all we did was ride the ship for seven to eight months. We were in the Gulf of Tonkin and did surface surveillance, and our bingo field was Cubi Point, 600 miles away – no joke! John Todd and I decided to test the profile once, and it turned out to be pretty accurate. Three of us qualified as fleet OOD underway. We were told we'd get no slack with the flight schedule, and we had a lot of nights with only two hours sleep."

So went the entire cruise, which passed uneventfully with no combat or operational losses, and on 30 October 1973 *Coral Sea* turned for home. According to pilot Steve Richmond – who was now a *Green Lizard* – the most exciting component of the deployment came in Pearl Harbor on the return leg:

"We were told we would have five hours of liberty because we were on our way back to the states, and that's all the time we needed to load some provisions onboard," Richmond writes. "Of course, all the JOs headed straight for the O'Club. As we were leaving, an announcement on the 1MC said to be back aboard at 1330 instead of 1500 since the ship would be leaving early. I thought it was just a scare tactic to get us aboard early. At the O'Club we tried to drink all their cold beer before time ran out. I fell in love with a waitress at the club about 1300 (this was between marriages for me). We went for a ride in her GTO while all the smart JOs went back to the ship. A little while later I noticed the cranes were lifting the brows from the ship. Oh, oh! We went screaming down to the dock. I jumped out and ran to the sponson where the quarterdeck was located and yelled, 'Throw me a line!'

"Meanwhile, 4,000 guys in whites are manning the rails and watching to see what this Lt. would do next. I made a running jump and caught this 1/2-in white nylon line (the ship was only a few feet from the pier). As soon as I caught the line, the four or five sailors on the other end started to pull me up. Unfortunately, I kept whacking my shoulder against the rounded bottom of the sponson. So it went ... '*Heave*' ... whack my shoulder. Swing back out where they could see me ... '*Heave*' ... whack my shoulder again. Finally, I got up and over the railing ... trying to cover my nametag with one hand and saluting with the other. My hat was long gone. It's a good thing my CO was up on the bridge and the skipper of the ship thought it was funny, or else I would have been in 'hack' again. My XO, Van Westfall, didn't think it was funny at all, and chewed me out as only he could. He said, 'That's strike *two*. One more and you're out of the squadron.

"Years later, when Van was XO of USS *Ranger* and I reported as flight deck officer, I asked him, 'New ball game, new strikes?' He said, 'You're still working on two strikes ... *same* ball game.'"

The men of VA-95 successfully concluded their initial A-6 deployment at NAS Whidbey Island on 8 November, returning about a month after VA-165. The *Boomers* had also experienced a quiet cruise, with 11 fairly short line periods and no losses for the squadron; CVW-9 suffered only one operational accident in January, when VA-147 dumped an A-7E. VA-95 and -165 had experienced the new norm: fewer appearances in the South China Sea, more liberty ports, and more operations elsewhere in the Pacific theater.

There were the occasional forays into something new and different, such as VA-52's late 1973 deployment with CVW-11 in *Kitty Hawk*. The carrier was the first in the Pacific Fleet to redesignate to CV – the event occurred on 29 April 1973 – and deployed in November on the first PacFleet CV deployment. The *Knightriders* did their part for the new mission, sortieing with a mixed bag of A-6As and KA-6Ds, as well as a new piece of antisubmarine warfare equipment, the Multi-Channel Jezebel Relay pod. The pod allowed A-6s to forward low-frequency signals from AN/SSQ-23, -28, and -38 passive sonobouys to the S-2s and the carrier's Anti-Submarine Classification Analysis Center.

As with *Saratoga's* experience in the Med in 1971, things got a little crowded on *Kitty Hawk's* flight deck. Besides the *Knightriders* CVW-11 had its standard attack carrier loadout of two F-4J squadrons (VFs -114 and -213), two light attack with A-7Es (VAs -192 and -195), one *Prowler* (VAQ-136), one E-2A *Hummer* squadron (VAW-114), and one *Vigi* outfit (RVAH-7), plus the sporadic appearance of VQ-1 Det 63's EA-3B. Joining this group were *two* S-2G operators: the *Sawbucks* of VS-37 and *Red Griffins* of VS-38, along with *Black Knight* SH-3Ds of HS-4. According to career *Stoof* driver Cmdr. Doug Seigfried, it made for some trying times:

"The trouble on the initial CV cruises was the mix of aircraft was too much to handle. One original plan put three S-2 squadrons and one H-3 squadron in each wing, and we had the assets, but it was just too much. They did initially try putting three S-2 squadrons on *Kitty Hawk*, and it was an absolute mess. On the first cruise in *Kitty Hawk* CAG-11 was Ham Byng. On that cruise they unloaded the S-2s at Cubi while they went CVA'ing. On the second cruise they were going to do it again, but *Com7Flt* called and said, 'No, now remember what the CV designation stands for' so they unloaded one of the A-7 squadrons at Cubi. It was a terrible cruise; money was tight, and we typically did about 1.5 weeks out of port followed by three weeks in. About the time everyone got back in step and started acting like a CV/CVW team, we'd go back into port.

"The ASW ops with the A-6s were sort of a jury-rigged setup,"

Seigfried adds:

> "Poor old VA-52 would launch with sonobouys, and our skipper (Brice Marshal) would go up to CAG and say, 'Gee CAG, this is great, we're launching with bombs ... and what's that the A-6s are carrying?' CAG's remarks can't be repeated. They tried, but they either didn't know where they'd put the sonobouys or they didn't know their numbers. We'd have two No.9s out there, didn't know which one was which, so we'd have to check through 16 frequencies to determine which sonobouy was which. Then we'd have to determine where the submarine was and which sonobouy was tracking it."

Yet the crews – both *Intruder* and *Tracker* – plugged away at the mission, setting the groundwork for conversion of the other Pacific Fleet carriers to multi-purpose CV standard. Doug's proud of the role the *Stoofs* and *SeaKings* played, but still shakes his head, adding:

> "They just didn't have enough room for everyone. Our JOs never even had permanent room assignments."

While the Navy was shaking out its carrier air wing makeup and mission assignments, the Marines also made adjustments. On 12 September 1973 the last VMA(AW)-533 *Intruders* left the "Rose Garden" at Nam Phong, Thailand, and headed back to MCAS Iwakuni. The departure of -533 from Southeast Asia consolidated the Fleet Marine Force's six operational *Intruder* squadrons at three locations: -533 at Iwakuni; VMA(AW)s -121, -224, and -332 at Cherry Point; and VMA(AW)s -225 and -242 at MCAS El Toro, CA. The assignment of the *Bats* and *Vikings* to MAG-13 at El Toro – in June 1970 and April 1971, respectively – gave the *Third Marine Aircraft Wing* its first full-time A-6 squadrons since the introduction of the type to the Marine Corps. However, due to shortages both squadrons were slow in remanning and returning to a full operational status. In the end the Marines decided to shut down VMA(AW)-225; the *Vikings* slipped into cadre status on 1 January 1972 and formally deactivated on 23 June 1972.

Fortunately for Navy and Marine *Intruder* squadrons alike, the subsequent months passed in relative quiet. They had earned the rest after nearly eight years of combat, during which time they had proven the worth and value of the A-6A many times over. Now it was time to bring in the next variant.

Problems with the *Intruder* ...

Despite its eventual success in Vietnam, the A-6A and its variants (A-6B and A-6C) had proven sorely lacking in several areas. The early aircraft had several design deficiencies that were only discovered in the hot, humid, and violent context of Vietnam combat. Among these were a horrendous radar cross section; in Daryl Kerr's words, "We stuck out – we were always targetable," although admittedly at the time efforts towards reducing combat aircraft radar cross-signatures (RCS) were still a good decade off. However, the *Intruder* also had a "huge" electronics signature due to mission requirements; to perform the job the radars had to radiate, which gave the North Vietnamese a good fix on the A-6 and usually other aircraft in the vicinity.

Other problems with the early *Intruders* included design flaws in the powerplant and its accessories. The J52-P-8A's burner cans did not efficiently burn fuel, and at anything over 90 percent power the A-6 left a smoke trail, similar to that which afflicted the *Phantoms*. According to Kerr the Navy tried several fuel additives, but nothing succeeded in reducing the problem until the later development of the J52-P-408. In addition, there were problems with the aircraft's V-band bleed air clamps that drove the air conditioning, constant-speed drives, and other ancillary equipment. They had a nasty tendency to fail, which vented hot (over 600° F) engine air into the surrounding airframe with predictable results. Through mid-1973 the medium attack community lost at least five aircraft due to sudden uncontrollable inflight fires.

The system problems were particularly notable, and the squadrons regularly stressed them in their post-cruise reports. According to Kerr the plane was originally designed as a level bomber, and in most combat missions the A-6 delivered from a level attitude. With a tight system the plane could consistently score hits within 100 feet of the target. However, as he recalls, once you trapped aboard, the stuff started coming apart:

> "The system consisted of tubes and first-generation transistors, and not many of those. I flew all 85 of my combat cruise

1975 – Vietnam is over, and the Navy has returned to a more-or-less peaceful routine. This *Swordsman* A-6A was the 400th *Intruder* built, and is seen cruising the skies on 1 October 1975 between deployments with CVW-2 in *Ranger*. (U.S. Navy)

missions with Al Koehler, and I can't recall a single mission where everything worked. We had a great maintenance department; those kids worked their hearts out.

"First to go was the track radar, usually followed by the INS. We'd come off the cat, lose all the 'neat stuff,' and end up having to do a radar drop just like the A-3s, with estimated slant ranges. Our accuracy was no better than the A-3s as a result. The computer must have failed every other flight. DIANE, the so-called digital system, *wasn't* – it was all analog. It wasn't unusual to go in over the target and have to do a visual drop. Fortunately, the A-6 crews were talented and worked like crazy. The B/Ns took pride in their hits, but it was darn hard to get those hits. The guys who transferred from the A-3s did well in the degraded mode, as it was a step up from the *Skywarrior's* system. On our 1970 combat cruise seven of the new B/Ns were ex-enlisted A-3 B/Ns – a bunch of LDOs – who were Ensigns in their late 20s and 30s. Those guys could really bomb well. They were the best bombers in the squadron because they were used to bombing straight radar."

Several others echo Kerr's comments, including former VA-165 and VA-42 CO Dick Zick and pioneer VA-42 and VA-85 B/N Ted Been. Zick says:

"The A-6A wasn't ready to go when it first came out. They almost killed themselves trying to do their jobs over there. It was pretty nightmarish, some of the things they had to do to keep the electronics working.

"Systems such as the inertial navigation were primitive back then and required the ship to steam a constant heading for almost 20 minutes for a valid alignment," comments Been. "Ships' captains were not fond of this idea. It was also not very fond of radical G forces, either. The track radar was very unreliable and often took the system into 'East Jesus' when you turned it 'on.' Then there was the computer, with its Forward Pedestal drum memory; remember, this was 1950s technology. The computer hated cat shots, and if it did survive the takeoff it was certain not to survive the trap. The computer had a MTBF (Mean Time Between Failure) of about 12 minutes. It's really a shame, because the guys on the ground put their hearts and souls into trying to keep those systems up."

The Marines had to work similar miracles with their aircraft, and they had the advantage of operating from fixed bases. Former Marine systems maintainer Bob "Gorilla" Guerra remembers some of the things he and his cohorts had to do with VMA(AW)-242's birds at Da Nang in 1967:

"VMA(AW)-242 deployed to Da Nang in 1966 as the first USMC A-6 squadron in-country. Leading the deployment was a forward cadre sent on ahead to set up the Electronic Systems Maintenance vans right behind the -242 hangar, and where all of the avionics test and repair took place on black boxes pulled at the flight line.

"It was October, the weather was less than optimal (like that's a revelation), and we soon discovered the Inertial Navigation System (INS) Platform Test Station may be in trouble. The Platform is the heart of the INS, which houses the gyros and accelerometers used to measure angular rate and acceleration, as well as gimbal mechanisms to keep the internal workings stable during any aircraft maneuver. Anyway, to properly calibrate the Platform the Test Stand must know its position (Lat/Long) precisely, and it must be aligned to True North. Now there's the rub! The plot thickens.

"Da Nang, RVN, is just above the equator at about 16° N latitude," he continues, "Which put the North Star somewhere around nose high. So what, you say! Well, without a good 'Polaris Shoot' there is no reference to True North, and hence no calibration of Inertial Platforms. For several nights we went out with trusty theodolite and attempted to find the North Star, but to no avail. At ground level it was constantly shrouded in haze/clouds just above the horizon. The only Time Tables we had were for Polaris, and we had no other celestial charts, so we couldn't use another star, dust off the trig tables, and get a bearing that way either. So, what to do?

"As luck would have it we had the smartest avionics-trained Marine available in the person of Gunny Tom Verhovcheck from H&MS, who suggested that we take the Platform Test Stand (humungous 3-axis steel monster), manually move it around to approximately north, and then use the gyrocompassing capabilities of every Inertial Platform to 'find north' for us. And, since the stand was not perfectly level – which was another criteria for alignment and calibration – we used the accelerometers within the platform in two axes to allow a manual leveling of the stand each time a Platform was tested. Tedious is too kind a description, but Tom conceived of the idea, I believe, while we were still at Cherry Point prior to deployment, and it worked. The small army of contractor reps who deployed with us included Litton Guidance & Control Systems guys who accepted the new procedures (like they had a choice), and onward we went. The new procedure unfortunately took twice as long to calibrate each platform, but at least we had a solution. The only way this scheme wouldn't work was if all you had was a truly broken platform and no other to help align the Test Stand. That scenario never occurred, as we were, unfortunately, rich in platforms needing to be re-calibrated all the time.

"We no sooner got the Inertial Platform Test Station up and running when the aircrews TransPac'd the birds in and work began in earnest. The inertial system on the A-6A was very touchy and somewhat unreliable, and platforms were always in for repair. I can only remember a few times when I actually had to replace a gyro, accelerometer, or other component. It was mostly a task of getting the calibration to stay put for more than a few missions and not drift off. Gyro replacement was not authorized at the Intermediate Maintenance level, but we did a few anyway, rather than lose the platforms to the depot and risk never getting them back."

Guerra says the field expedient method worked fairly smoothly until mid-1967, when the Inertial Platform Test Station's transformer blew out. He and Gunny Verhovcheck responded by loading eight platforms in a C-130 and paying a friendly visit to NAS Cubi Point:

"The Navy was less than thrilled, and we had to work around their schedules. They had their own problems. We could only work on/calibrate one platform a day in the early morning, and not all days were available to us. Being creative and resourceful, we discovered that the golf course and Olongapo City needed modifications and proceeded along those lines during 'off hours.' It was back to Da Nang after a few weeks, and the rest of the tour was electronically uneventful."

"The fact remains," Bob concludes:

"That all parts of the Marine Corps have worked around problems, no matter what type or wherever found. 'Creative' acquisitions and the use of resources to their fullest has always been our trademark. There were many stories like this where necessity and innovation created solutions which went outside the book and SOP to get the mission done. It was an interesting, complex time."

Finally, the A-model had basic design problems affecting the airframe. Daryl Kerr rates the A-6A's wings – which were a "skin-stressed" design – as the biggest single design flaw. As designed, the wing had no main spar, and the wing surface took all of the flight stresses; unfortunately, it also had a lot of holes for the slats, flaps, hard points, and speed brakes. According to Kerr, the wings as delivered were certified for +6 Gs:

> "We overstressed the aircraft regularly from the day it was delivered. We were told not to put rolling Gs (where heavy G-loading is placed on the airframe while not wings level, thereby putting severe and uneven strain on the wings) on the airframe, but we had to due to the mission and the deliveries we were making. After the introduction of the Alpha strike mission in 1968-69 – we were launching with 28 Mk.82s – we couldn't help overstressing the wings on pullout. We agreed to try not to overstress the plane in combat, but it was difficult."

For its first several years in the fleet each and every *Intruder* had to have its wings checked after each flight for stress cracks or other structural problems. According to then COMFAIRWHIDBEY A-6 Maintenance Officer, Jim Vannice, this was a major part of his job, in and around Administrative Material Inspections (AdMats) and other close looks at wing aircraft conditions. He comments that the A-6 and its squadrons were coming into their own, but the problems with cracked wing spars caused all sorts of difficulties, including at least one instance where it directly contributed to the loss of an aircraft and crew. On 19 August 1969 a wing failure claimed the life of VA-145 XO Cmdr. Dick Walls and his B/N, Lt. B. Conchrun. They died at Boardman Range in northern Oregon when the wing of 152633 separated during their pull off target; Daryl Kerr, also in the pattern, witnessed the accident.

There were others. On 6 February 1970 VA-196 pilot Lt.Cmdr. "Pee Wee" Reese and Lt.j.g. Don Fraser ejected over Laos. The Navy officially recorded the loss of BuNo 152937 as due to AAA, but following his recovery Reese reported their was no ground fire, the wing "just came off." Another disaster at Whidbey Island on 12 June 1970 marred a squadron change of command ceremony. Cmdr. John Wunsch was relieving Cmdr. Lou Dittmar as the skipper of VA-196; that same day VA-145 lost an aircraft at Boardman when the wing of BuNo 156998 separated in the pattern, killing B/N Lt. "Stone" Van Stone. Dittmar recalls:

> "We had all of the Grumman people there for the change of command. They immediately got involved in the accident investigation. I went to NavAir as the A-6 class desk officer and also got involved. We eventually determined the problem was with the holes in the wings, and worked to get them fixed."

Capt. Kerr feels the same structural problem may have directly contributed to the numerous "they flew out and just didn't come back" losses suffered during Vietnam. This design flaw would continue to wreak havoc with the aircraft until the late 1980s, when the SWIP wings started coming on line. In the interim, the Navy and Grumman worked up a replacement that addressed the majority of the A-6A's shortcomings; it received the designation A-6E.

... and the Solution: the A-6E

The Navy first decided to proceed with an upgraded follow-on to the A-6A in 1968. Grumman's intention with the improved *Intruder* was to correct the deficiencies identified in the original design, while replacing 1950s-era systems with more modern and reliable counterparts. A-6A BuNo 155678 was selected for modification, test, and evaluation, and made its first flight on 27 February 1970 with Grumman test crew Joe Burke and Jim Johnson at the controls.

From nose to tail, the E-model's changes were major, reflecting both combat experience and ten years' worth of technological advances. A single multi-mode search radar, the Norden AN/APQ-148, replaced the A-6A's troublesome AN/APQ-92 search and APQ-112 track radars. The new radar was capable of simultaneous operation in the ground mapping, terrain clearance, and track-while-scan modes, using a separate phase interferometer to provide target elevation and terrain clearance data. A second generation, truly digital attack/navigation system replaced the analog DIANE, with transistors replacing vacuum tubes, yielding a true digital solid-state random-access processor. This was combined with a centralized solid-state armament control unit with simplified interfaces between the two major components and all subsystems. A Built-In-Test (BIT) system eliminated all line test requirements and enabled the B/N to do a complete avionics test before launch.

The E-Model also incorporated several airframe and powerplant enhancements, including the installation of two Pratt & Whitney 9,300-lbst J-52-P-8B engines. Grumman installed new constant speed drives while straightening out or otherwise simplifying much of the *Intruder's* internal plumbing for better efficiency and safety. Another modification included the addition of a self-contained ground blower/air-conditioning unit, and finally Grumman deleted the long dormant fuselage speed brakes, replacing them with blank steel plates.

Testing of the prototype A-6E confirmed both Grumman and the Navy's expectations: the new aircraft provided a quantum leap improvement in system reliability and bombing accuracy while reducing maintenance hours and ground support requirements. The Navy made a startup order for 129 new *Intruders*, and the first production aircraft, BuNo 158041, made its initial flight on 26 September 1971. Production examples subsequently arrived at VA-42 on 2 December 1971, with VA-85 receiving the first fleet E-models exactly one week later. To celebrate the occasion the *Black Falcons* held a ceremony to commemorate the arrival of their improved planes; during the festivities the wife of *Forrestal* CO Capt. R.F. "Dutch" Schoultz broke a bottle of champagne over the launch bar of an A-6E. The *Falcons* would also be the first to deploy with the A-6E, departing in *Forrestal* on 22 September 1972.

From this small start the A-6E spread throughout the medium attack community, with most squadrons receiving their first examples as they returned from cruise. Concurrently, the Navy and Grumman agreed on a program of rebuilding and upgrading selected A-6As to the A-6E standard. Eventually 200 As received the treatment.

Surprisingly, while many maintainers and junior crews greeted the new *Intruder* with open arms, the sentiment was not exactly universal. In fact, there were some in the community who looked askance at the improvements brought by the new and improved E-model. According to Capt. Dick Toft, a major source of pride with the early B/Ns was their ability to get their bombs on target despite multiple system failures; many of these A-6A crews practiced backup deliveries more than primary deliveries. However, when the A-6E hit the fleet some of the "old-timers" looked down their noses at the new guys, who flew with consistently good systems, never had to flail around in-close to the target, and never – but never – had to fly a manual range line delivery. He adds that he will always remember the first time he saw A-6s of any type, not suspecting he would ever become involved with the aircraft himself:

> "I was with *HATRON 2*, and we escorted several *Intruders* on a TransPac. We were looking around the cockpit at each other wondering, 'What's an 'INS'? What's a 'package?'"

George Strohsahl, who returned to VA-65 as executive officer in June 1974, had his own opinions of their new A-6Es based on his view from the left seat:

> "The big change? It worked," he comments. "When the A was fully up and working, I actually preferred the A. You had an independent sensor that gave you information for low altitude flight; you could back up the B/N by calling up the search radar. I felt the

pilot had a better low-altitude capability with the A. However, the A system was never working, so what good was it? The E was much better in staying up; thoroughly a better job.

"I saw something the community never talked about; we lost some of the extreme low altitude night ingress capability because you had to depend on a really good SRTC and a shit hot B/N to keep you out of the hills. I'm not sure a lot of people would argue with that, but it was like what we had with the A-4. Back on my first tour (in A-4s) I had responsibility for developing tactics for that very rudimentary system and got very proficient."

Notably, the A-6's AN/ASN-31 inertial navigation system remained one of the *Intruder's* weak points, as were the weak wings. The INS still had problems aligning properly, leading to delays on the flight deck, heightened tension, the works. Grumman did not completely solve the problem until the mid-1970s introduction of the A-6E CAINS (Carrier Airborne Inertial Navigation System) variant. CAINS employed the Litton Industries AN/ASN-92, which also went in the F-14 *Tomcat* and S-3 *Viking*. By use of an external plug into the carrier's navigation systems the *Intruder* B/N could perform a quick and highly accurate fix for the onboard INS, further improving system effectiveness. The only external evidence of the new installation was a large cooling scoop on the left rear upper fuselage; for years, the scoop served as the easiest means of telling the difference between a straight A-6E and a CAINS-equipped bird.

The A-6E CAINS also spread through the fleet, with each squadron receiving their upgraded aircraft in turn. According to Daryl Kerr, VA-52 received its aircraft in early 1976 during his tour as the *Knightrider's* commanding officer, and:

"One of our planes went down to Boardman Range and scored *12* consecutive bulls on systems drops. It was unheard of. That aircraft contributed to VA-52 winning four consecutive *Intruder* bombing derbies at Whidbey Island."

With the A-6E, the *Intruder* and the medium attack community had apparently achieved maturity. However, something even better was coming down the pike.

The Development of TRAM

During this period of technical largess a group of *Intruder* operators and combat veterans serving in Washington D.C. on the OpNav and Naval Air Systems Command staffs continued to gin up ways of improving the mighty *Intruder*. They had flown the tough missions, fought both the enemy and systems problems, and knew exactly what they wanted in an improved all-weather attack aircraft. Capt. Charlie Hunter, who served as one of the early A-6E Program Coordinators, recalls:

"As the program coordinator I was in charge of about half of everything," he recalls. "They'd already sold the programs for the new radar, the new computer, and the new weapons release system. We completed that, and they started cranking them out. I was also fortunate enough then to work with Lyle Bull on getting the TRAM program going, so I finished one aspect of the A-6E and worked to get another one started before I departed for my next tour."

Now a lieutenant commander, Bull arrived at OP-50C (the aircraft requirements directorate of OpNav) in April 1970 and relieved Cmdr. Mike Hall as the TRIM Program Manager. Despite its mediocre performance in Vietnam, the concept of the TRIM system – employed with the A-6C – still had promise, and Bull decided to concentrate on incorporating some sort of improved system into the A-6E. The resulting project became TRAM, for Target Recognition Attack Multisensor.

"I'd seen some infrared imagery, and knew that adding that capability to the A-6 would drastically boost its combat capability," he comments:

"I also knew there was space in the nose of the *Intruder* where the track radar used to be, but the constraint was weight, which we determined to be about 400-pounds." Lyle also decided the system had to be internal; as he puts it, "Pods and sailors are not compatible."

Bull's major briefing points were that the TRAM system would save lives and deliver drastically improved bombing accuracy, on the order of *ten foot* miss distances. He took his presentation to the decision makers in the acquisition and test and evaluation sides of the house, and subsequently received $15 million for the project. However, the money came with a warning delivered by OP-50, Rear Adm. Donald C. "Red Dog" Davis; Davis advised Bull the funding would come from other people's R&D programs, and he stood to make a lot of enemies, particularly at NAVAIR. Bull should proceed with the understanding that several people would be, "after your ass, so watch yourself."

The TRAM proposal called for a ground stabilized infrared system with laser designator and full lower-hemisphere coverage. Hughes

Grumman introduced the A-6E TRAM in 1974. Prototype 155673 is shown at Peconic Field. (Grumman History Center via Tom Kaminski)

and Texas Instruments stepped forward and said they could do it, while Grumman agreed to coordinate the work and integrate the resulting design into the A-6. TI had developed the original TRIM pod and used the same basic concept for TRAM; the company's proposal featured a faceted window turret with trainable infrared sensor employing Mercury Cadmium Telluride, a state of the art material that allowed for growth in sensor capability. Conversely, Hughes' proposal employed a Mercury Doped Germanium detector with a fixed window mounted on a gimbal. Tests revealed the Texas Instruments' design gave the best infrared imagery. On the down side, it suffered from picture quality problems, including an occasional loss of IR imagery, and was limited to only two fields of view: wide or narrow. Conversely, Hughes' proposal had continuous zoom capability with an infinite range of variations. According to Lyle Bull, their design was so capable, "It delivered 11 more shades of gray than what the human eye could detect." Hughes got the nod from NAVAIRSYSCOM and development continued.

The next step was to develop and test the radome-mounted TRAM pod. Bull recalls:

> "The early tests were performed with a fully retractable pod. There was concern that the pod would cause major aerodynamic problems, possibly to the extent of blanking out the engine intakes. Grumman flew test flights to check out the overall effect on the airframe, with the idea being that if the engines flamed out following pod extension they'd have time to do an air start. As it turned out, there were no problems."

When integrated with TRAM the Norden AN/APQ-148 radar became the APQ-156. To complete the package the Navy added the AN/ASN-92 Inertial Navigation System and a Laser Quadrant Receiver; the .5 milliradian beam laser – also developed by Hughes – had an effective range of 12.5 miles and provided superimposed targeting information to the B/N's display. According to one veteran bombardier, using the system was remarkably simple:

> "Turn on the ground-stabilized laser, designate, and drop. It made working with the FACs a *lot* easier."

The prototype A-6E/TRAM, BuNo 155673, flew at Calverton on 22 March 1974. The full system integration tests continued through 1975 and gave every indication TRAM was a winner, so the Navy proceeded with acquisition. The first production TRAM bird (BuNo 160995) made its initial flight on 29 November 1978; VA-42 at Oceana took possession of its first aircraft, BuNo 155710, on 1 December. CAINS made it to the fleet well before the turrets did, and the last medium attack squadron did not convert to the all-up TRAM aircraft until the early 1980s. By that time a Direction and Ranging Set (DRS) and improved Airborne Moving Target Indicator (AMTI) were added to the package. These served to make an already impressive attack system even more capable, culminating in *Desert Storm*, where A-6s made 85 percent of all laser designations.

As for Lyle Bull, he remained with the project for three years as Attack Weapon Systems Program Coordinator, later working on the A-7 podded TRAM package and the A-4 Angle-Rate Bombing System (ARBS). He received the Meritorious Service Medal for his efforts with the A-6E TRAM project, while the project itself received the Weapon System Award for Outstanding Systems Achievement from the Order of Daedalians.

Matwing and "Vacuum-Pac"

The A-6E brought a substantial improvement in *Intruder* capabilities and was warmly received by the squadrons and their assigned air wings. The TRAM system promised even more, but both upgrades took some time to achieve full operational service.

By early 1973 the medium attack community had reached a point of stability. At NAS Oceana six *Intruder* squadrons (VAs -34, -35, -65, -75, -85, and -176) reported to *Medium Attack Wing One*, while at NAS Whidbey Island *Commander, Fleet Air Whidbey* exercised administrative control over another six squadrons (VAs -52, -95, -115, -145, -165, and -196). The Fleet Marine Force maintained five all-weather attack squadrons at MCAS Iwakuni, MCAS Cherry Point, and MCAS El Toro. The RAGs (VAs -42, -128, and VMAT(AW)-202) continued turning out qualified aircrews and maintainers as before at their respective locations.

The Navy was still fine-tuning its A-6 organization as the year opened, and on 5 January 1973, in ceremonies at NAS Oceana, former VA-35 CO Capt. Herm Turk relieved Cmdr. Dick Zick as COMMATWING ONE. Zick had assumed command of the RAG on 31 August 1972 following his surprise tour as Air Boss in *Independence*; concerning his tour as both FRS and wing commander he comments:

> "It was a small 'bonus command' that worked out pretty well, no big problems. In fact, it was probably better, because you had overall control of the situation. There weren't all these little things that I had to ask the wing commander about because *I* was the wing commander. We handled everything in house. If I needed to go any further AirLant was right there, unlike Whidbey, where AirPac was at the other corner of the country.
>
> "It was kind of tough, in that I was from the west coast and wasn't familiar with the targets. Other than that it was a typical RAG, with stuff going on all over the place, and we just tried to keep it in the middle. Both the wing and squadrons had lost some aircraft due to accidents. The CO of VA-35 lived right across the street from us at Oceana; that was the hardest part, making all of those calls on the families with Nancy. We came in, started at the bottom, rebuilt the place, and away we went."

Following his relief at VA-42 Cmdr. Zick transferred to NavAirLant, where he worked for Rear Adm. Paul Peck as:

> "Number two supply officer. He wanted to prove that aviators could do staff work like supply. He was wrong, but it was an interesting tour."

During this period MATWING's west-coast counterpart, COMFAIRWHIDBEY, also underwent a transition. Dating to 15 February 1954, the wing had traditionally controlled the heavy attack community on the Rock and started gaining *Intruder* squadrons in November 1966, with the arrival of VA-196 from Oceana. On 1 July 1972 *Tactical Electronic Warfare Wing 13* – also housed at Whidbey – disestablished, resulting in the assignment of multiple *Whale* and *Prowler* squadrons to COMFAIR. On 30 June 1973 COMFAIRWHIDBEY was redesignated *Commander, Medium Attack, Tactical Electronic Warfare Wing, US Pacific Fleet*, bringing the organization's designation in line with the other functional wings. Abbreviated as COMMATVAQWINGPAC – and quickly dubbed "Vacuum-Pac" by the locals – the organization settled in under the command of Rear Adm. John. M. Tierney. Within short order the wing was administering to the needs of seven *Intruder* squadrons, six *Prowler* squadrons – including the RAG, VAQ-129 – and two remaining EKA-3B squadrons at NAS Alameda, VAQs -130 and -135.

However, within a few months one A-6 squadron would depart Whidbey for what could be termed an "extended" deployment. In late September 1973 the VA-115 *Arabs* left Oak Harbor for assignment with *Carrier Air Wing 5* in Japan, a result of an accord between the U.S. and Japanese governments allowing the forward basing of an aircraft carrier. *Midway* was selected, and on 30 June 1973 she departed NAS

The Navy's "Foreign Legion" was CVW-5, based out of Japan with *Midway* (CVA-41) from 1973. A-6A NF501 is on approach to Atsugi, Japan, in May 1975. (Toshiki Kudo)

Alameda and transferred to Naval Station Yokosuka. The squadrons assigned to CVW-5 followed, checking into NAF Atsugi as their new land base; VA-115 arrived at Atsugi on 5 October, joining VFs -151 and -161, VAs -56 and -93, VAW-115, an HC-1 detachment with SH-3Gs, and a VFP-63 det with RF-8Gs.

The arrival of a complete carrier and carrier air wing – with escorts – in Japan gave the post-Vietnam Navy a major boost in regional capability. Carrier historian Cmdr. Pete Clayton rated *Midway's* assignment to Yokosuka as "invaluable," enabling a reduction in:

> "... the deployment cycles of her sister Pacific Fleet carriers, allowing them to each receive extended in-depth overhauls and greatly increasing their homeport time for routine maintenance and training."

With VA-115 embarked, *Midway* would serve for another 16 years as the Pacific Fleet's "fireman," responding to calls when needed. However, the next call for carrier aviation came on the other side of the world in the Middle East, and once again, attack personnel found plenty of trade.

Action in the Med

On 6 October 1973 Syria and Egypt attacked Israel, kicking off the Yom Kippur War. The U.S. responded by sending *Independence* – with CVW-7 and VA-65 embarked – *Franklin D. Roosevelt* with CVW-6/VA-176, and *Guadalcanal* (LPH-7) to take up station in the vicinity of Crete. *John F. Kennedy*, with CVW-3 and VA-34 onboard, was ordered out of the Norwegian Sea and proceeded to a point off Gibraltar at a high rate of speed.

Initially the war was touch and go for the Israelis, with major fighting on the Golan Heights and tank warfare in the Sinai Peninsula. In the face of the *Hey'l H'Avir's* severe aircraft losses – modern Soviet SAMS and heavy AAA were knocking down Israeli *Skyhawks* and *Phantoms* in large numbers – and under the threat of possible Soviet involvement, the United States initiated resupply operations while pursuing a diplomatic solution. Within days of the outbreak of hostilities, Navy pilots started shuttling A-4s to Israel as combat replacements. America's allies around the Mediterranean denied landing and refueling rights, so the *Skyhawks* staged through the Azores and then received a series of midair refuelings the length of the Med from VA-34 *Intruders* operating near Gibraltar; VA-176 aircraft flying in the vicinity of Sicily, and VA-65 tankers off of Crete. The Navy aircraft also topped off fighters escorting Secretary of State Henry Kissinger's aircraft as he headed into Israel. Thus resupplied by the United States, the Israelis counterattacked and eventually took back the Golan Heights and surrounded an entire Egyptian army in the Sinai. All parties declared a ceasefire on 24 October, and United Nations peacekeeping forces entered the region, prelude to the signing of a peace agreement on 17 January 1974.

Sixth Fleet resumed normal operations, if briefly, but within a few months two other nations in the region neared the brink of war. The object of contention this time was the island of Cyprus, long a source of friction between Greece and Turkey. The President of the island nation, Archbishop Makarios III, had stockpiled weapons for two years due to concerns over a possible right-wing coup. Responding to requests by the Turkish and Greek governments, Makarios surrendered the weapons to United Nations peacekeepers in early 1974. However, the Greek government announced this was not enough, and demanded the formation of a truly unified Cypriot government that, conveniently, would be dominated by Greek Cypriots.

In July 1974 Makarios demanded Greece remove 650 Army officers who were under assignment to the Cyprus National Guard, stating they were there to overthrow his government. As it turned out he was correct; the government in Athens authorized a coup, removed Makarios, and installed Nicos Sampson as the new president of Cyprus. On 19 July a sniper killed the U.S. Ambassador to the embattled island, Rodger P. Davies. The following day the Turks landed 6000 troops on the island – including tanks and under the cover of air and naval forces – and open factional warfare ensued.

The situation quickly deteriorated to the extent the Greek government threatened to mobilize against Turkey, its erstwhile NATO ally. In Cyprus the new American ambassador quickly recommended an evacuation of U.S. and foreign nationals; in response Sixth Fleet ordered in *Forrestal* and *Independence*. Led by Cmdr. Don Boecker, the *Black Falcons* operated from *Forrestal* in providing cover for the evacuation and subsequent monitoring operations. *Indy* joined them in early August; VA-65 was onboard, led by Cmdr. Charles D. Hawkins, Jr., and XO Cmdr. George Strohsahl. The latter says that what with troubles breaking out all over the Med the *Tigers'* 1974 cruise proved to be rather confused, if not entertaining:

> "Before we pulled out *Indy* was being worked up to go to a new homeport in Greece, so VA-65 was on tap to become an over-

seas squadron. That was terminated before the detailed plans got underway; it was unfortunate, because I was really looking forward to it. What stands out from the cruise was this: a tradition had been established of lots of dependents going over to meet the ship in Greece. The funny thing was, going into the Med the Greeks and Turks went at it again over Cyprus, and we never saw Greece during that deployment. A lot of the *wives* did. We chased ourselves all over the Med, doing contingency planning for strikes, playing cat and mouse with the Soviet forces, and we did a lot of 'flanchoring' – flying at anchor. Actually it was a pretty good cruise."

Starting in Geneva on 30 July, the concerned powers met with British Foreign Minister James Callaghan and attempted to work out a ceasefire and withdrawal. All the while the Turkish and Turkish Cypriot forces rolled back the Greeks, eventually forcing them into the southern two-thirds of the island. Negotiations continued, and toward the end of the year combat operations finally subsided, with the Turks firmly in control of the northern third of the island, running roughly from Morphou Bay through Nicosia to Famagusta.

However, tensions in the region continued well into 1975, during which time *Saratoga* relieved the "FID" in January. Getting *Sara* to the Med had taken a bit of doing; at one point during its 1973-1974 overhaul at Norfolk Naval Shipyard the ship had caught fire. The carrier's crew, shipyard workers, Portsmouth fireman, and others managed to extinguish the flames after 11 hours, and fortunately no one was injured, but the damage was extensive. Patched back together, *Saratoga* deployed on 27 September 1974. *Forrestal* had her own problems; mid-cruise one of her four shafts warped, requiring the ship to pull a section of the shaft and replace it while still deployed.

On 13 February Turkey announced the formation of the independent Turkish Federated State of Cyprus; Congress responded by suspending all military aid to Turkey, which in turn responded by removing support functions from American bases within Turkey. Much political back and forth, negotiation, and charge/countercharge ensued. Sixth Fleet's carrier air wings kept a wary eye on not just Cyprus, but also the Middle East, where political and religious factions kept everyone on the edge; unfortunately, these contentious times continued well into the next decade.

Back in the states operations continued at a relatively quiet pace, at least for the time being. VA-85 returned to NAS Oceana on 22 September 1975 having completed back-to-back summer trips to a very unstable Mediterranean, plus a brief show the flag visit to the West African coast on the second deployment. The Navy noted the squadron's effectiveness., awarding it the ComNavAirLant Battle "E" for the year. The *Falcons* would take the Battle E again in 1976, in the process logging the highest score ever recorded by an AirLant *Intruder* squadron in an Operational Readiness Evaluation.

At MCAS Cherry Point the Marine *Intruder* operators had cause to celebrate: on 28 March 1974 VMAT(AW)-202 received its first A-6Es and started transition training with the improved variant. The first FMF squadron, VMA(AW)-224, received its E-models the following June. Also during this period the *Green Knights* engaged in something different: tests with the Marquardt/Hughes Mk.4 20mm HIPEG gun pod. A few of Grumman's proposals for the *Intruder* had included installation of gun armament of one form or another, but of course no guns were ever fitted. However, over a one year period at Cherry Point VMA(AW)-121 tested the pod on their A-6s. Navy and Marine F-4s and A-4s had sporadically employed the pod during Vietnam. It was a twin-barrel 20mm cannon vice a rotating multi-barrel cannon, not un-

If any one place could be called the home of Marine *Intruders*, it was MCAS Cherry Point, NC. Three A-6s from VMA(AW)-332, VMA(AW)-224, and VMCJ-2 are seen here during the mid-1970s overhead the base. (U.S. Marines)

like the more familiar M61 *Vulcan*, and gained a reputation for being balky in flight and not particularly accurate.[1] Both the Navy and Marines declined further development of the weapon for their A-6s.

In a more successful evaluation for the Marines, VMA(AW)-533 qualified onboard *Constellation* in August 1974. It was the first Marine *Intruder* squadron to do so following the *Bengals'* deployment in *Coral Sea* two years previously, and apparently pulled off their deck training without a hitch. However, it would be another 10 years before a Marine A-6 squadron returned to sea as an integral component of a carrier air wing.

For the Navy, carrier deployments continued as a way of life. On 17 September 1974 *Enterprise* pulled out of NAS Alameda for her first post-Vietnam WestPac cruise. The departure was notable for both the Navy and *Carrier Air Wing 14*, as it marked the first operational deployment of the new F-14A *Tomcat*, operated by the VF-1 *Wolf Pack* and the *Bounty Hunters* of VF-2. The *Main Battery* of VA-196 also made its first post-war deployment in Enterprise, now led by Cmdr. Gordon Nakagawa, with Cmdr. Lyle Bull as executive officer. Following the precedent set by VA-165 some years previously, -196 also had a couple of Air Force exchange officers along for the ride; one was Maj. Bob Holloway, a former RF-4C backseater with the *432d Tactical Reconnaissance Wing* at Udorn RTAFB who later flew F-111As with the *366th Tactical Fighter Wing* at Mountain Home AFB, Idaho:

> "I'm pretty sure I was among the second group of Air Force exchange officers in the A-6 community," Holloway comments, "Following the trail blazed by Larry Beasely and Doyle Ballentine. I went through VA-128 while Bob Miles was CO and Daryl Kerr was XO, then went to VA-196 under Jack Cartwright and Gordon Nakagawa. Cartwright left, Nakagawa became CO, and Lyle Bull came in as XO. We didn't get along too well; Bull was big and intimidating, and there were a lot of – how should I put it – 'interservice questions.' Nakagawa still appeared to be suffering some effects from his imprisonment in North Vietnam. My impression was he took a lot of heat over there because of his oriental background, and they singled him out. We were also aware his parents had been imprisoned during World War II at Tule Lake, OR, or some place, but he wouldn't talk about it. I do remember that if you got on his bad side, his eyes would flash red and then he'd eat your lunch.
>
> "Otherwise, it was a good tour. In fact, it was interesting, because we ended up going into the Indian Ocean and stuck our nose into the (Persian) Gulf. The carrier's CO would not go into the Gulf proper, because it was too restrictive. At least, that was Navy doctrine. We also had an Air Force pilot in the squadron, Chuck Mosely. My regular pilot was Lt. Jim Huff. We were the last Air Force guys to fly the A-model *Intruder*, which was too bad. I flew in the E once and it was nice. It was an interesting time, and the A-6 was a great machine. I really enjoyed them."

Among the Navy B/Ns in the squadron was Dan Wright, who had previously made a combat cruise with the *Arabs*. He says he was supposed to head over to VA-165 following a tour as maintenance officer at VA-128, but Lyle Bull, "... got my orders changed, and I went to VA-196. He tended to do that." Wright had served under Bull at the RAG, and apparently made a good impression. He remembers there was a lot of speculation at the time about where *Enterprise* was going to end up, what with the ongoing conniptions between the two Vietnams and rumors of an Indian Ocean trip:

> "Our first cruise with Gordy (Nakagawa) there were all sorts of rumors flying around. We were going to go around the world and join with *Nimitz* to form the first two-nuclear carrier battle group, that sort of thing. We didn't go around the world, but we did go into the IO to Mombassa (Kenya). Then a big typhoon came through Mauritius and we did an emergency sortie to render aid. Actually, we primarily went to beat the Russians, who were planning on doing the same things. It was a rare thing operating in the IO at the time, because there were no diverts. We couldn't even go near Diego Garcia because it was secret. We did do some trolling missions off Vietnam to see if they'd come out and play. This was the first deployment for the F-14; the idea being to drag the North Vietnamese out and let the *Tomcats* handle them. They called us 'brazen provocateurs.' We got a big kick out of that, even got a great patch made up with 'Brazen Provocateurs' on it."

It fell to VA-165 – assigned with CVW-9 in *Constellation* – to become the first *Intruder* squadron in the Persian Gulf. In November 1974 the carrier became the first in 26 years to enter the restricted waterway while participating in Exercise *Midlink 74*. The operation included naval units from the United Kingdom, Iran, Pakistan, and Turkey, and was the largest of its type ever held in the region. Otherwise, the deployments of 1974 continued quietly for the *Boomers* and *Milestones*, with the IO operations serving as a precursor for the future. *Enterprise*, for example, spent 24 of her 32 days in the IO doing flight operations with the nearest divert fields over 700 miles away, but otherwise there was plenty of quality time in the various liberty ports away from the Indian Ocean. However, things picked up directly, and once again involved Vietnam and multiple carriers.

Last Stand In Southeast Asia

In late 1974 and early 1975 the political and military situation in Southeast Asia took a major downturn, as communist forces attempted to consolidate their control over both Cambodia and Vietnam. Once again, *Intruder* squadrons found themselves directly involved.

In the almost two years since the American pullout from South Vietnam, both the North Vietnamese and the Khmer Rouge had used the opportunity to rebuild and consolidate. By January 1975 the Khmers controlled fully 80 percent of the nation of Cambodia and were obviously poised for a final assault on the nation's capital, Phnom Penh. Their North Vietnamese counterparts had been equally busy, rebuilding their industrial base and constructing additional supply routes through Cambodia to South Vietnam. According to some observers, by 1975 the main component of the Ho Chi Minh trail had all the appearances of a four-lane highway, albeit paved with crushed gravel.

The governments of Cambodia and the Republic of Vietnam looked on the developments with great fear and trepidation and turned towards their American allies for assistance. What they received in response was thundering silence.

When the U.S. retired from South Vietnam in 1973, then-President Nixon had guaranteed his nation would respond appropriately if the North Vietnamese ever threatened again. However, he had been hamstrung by Watergate, and in August 1974 was out of office. In addition, on 7 November 1973 Congress overrode his veto of the war powers act, which curbed the President's ability to commit armed forces abroad without Congressional approval. President Ford's administration was inclined to help out once again in Southeast Asia, but was preoccupied with continuing damage control from Watergate and the resultant presidential pardon of Richard Nixon, problems in the Middle East, rising unemployment, and the effects of the 1973 Arab oil embargo. Between this focus with problems at home and continued Congressional and public antipathy with Southeast Asia, it was obvious the United States would not move to reinforce the region. Adding to the impact, Congress cut the FY74 appropriation of

[1] The USAF used a podded 20mm gun based on the M-61 on its F-4 *Phantoms*. The Air Force's SUU-16 and -23 pods are frequently confused with the Navy's Mk.4 HIPEG.

$1 billion for military aid to South Vietnam to $700 million for FY75 and gave every indication it would cut the aid even further. Both the North Vietnamese and the Khmer Rouge took note, and planned accordingly.

As far as the DPRV was concerned the war for reunification had never ended. The International Commission of Control and Supervision (ICCS) – made up of observers from Hungary, Indonesia, Iran, and Poland – had proven impotent in preventing armed clashes in the south. Further attempts or proposals for national referendums on the future of Vietnam had also failed due to obstinacy on both sides. As a direct result, by late 1973 North Vietnam figured it was time to reunite the country by force of arms. The initial attacks came in late December 1974 in Phouc Long Province, at the south end of the Ho Chi Minh trail complex, and barely 50 miles north of Saigon. RVN army and air force units fought back valiantly, but were hindered by a lack of armaments and equipment and mass desertions. Engaging heavy tanks, massed artillery up to and including 130mm weapons, and surface-to-air missiles, the South Vietnamese forces fell back. Phouc Binh fell on 27 December 1974, and on 6 January 1975 the Democratic People's Republic of Vietnam announced the "liberation" of Phouc Long Province.

President Ford and his advisors took note of the invasion and looked around for available assets. On the day Phouc Long fell Seventh Fleet had three carriers: *Enterprise* with CVW-14 and VA-196; *Midway* with CVW-5 and VA-115; and *Coral Sea* with CVW-15 and VA-95 embarked. *Enterprise* was halfway through her deployment, while *Coral Sea* had been in theater for about a month; for her part, *Midway* was scheduled to return to Yokosuka at the end of February. Also available was the 31st Marine Amphibious Unit embarked in several amphibs centered on *Okinawa* (LPH-3).

The Air Force's assets in the region were also somewhat limited, particularly in comparison to the large numbers of units assigned during the height of the Vietnam War. By January 1975 Seventh Air Force had operational control over the 56th Special Operations Wing at Nakhon Phanom RTAFB, and three Tactical Fighter Wings flying F-111As, A-7Ds, and F-4s.

The U.S. government ordered its forces to stand by and observe the proceedings while it attempted to formulate a plan. While protesting the North Vietnamese invasion of South Vietnam, President Ford went to Congress for the first of several requests for more military aid to the region. Public protests and resounding denials from the newly elected and heavily Democratic 94th Congress greeted his entreaties. Indeed, most in Congress felt the nation was tired of anything to do with Vietnam, and several members of both houses announced the U.S. had no humanitarian or national security reasons to continue involvement in Southeast Asia.

The situation was equally dismal in Cambodia, as the Khmer Rouge continued applying inexorable pressure to the government of Premier Lon Nol. U.S. bombing of the insurgents ended on 15 August 1973, followed by financial aid and logistics flights by contractor organizations; however, the U.S. government also sharply reduced funding to the Cambodian government, which created an even more traumatic impact on the poorly led and equipped Cambodian Army. Their leaders found themselves in the same boat as their South Vietnamese neighbors, ie, wondering what had happened to their allies the Americans. By the first of the year the Khmer Rouge owned 80 percent of Cambodia and cut off access to Phnom Penh via the Mekong River. It was never made public, but the United States had already effectively written off Cambodia. With that nation a wash and the Republic of Vietnam quickly going down, U.S. planning quickly shifted from reinforcement and support to evacuation.

Eagle Pull, *Frequent Wind*, and *Mayaguez*

On 8 March 1975, following a few weeks of respite, the final NVA offensive against South Vietnam kicked off. Within short order Ban Me Thuiot, Pleiku, and other cities and villages in the Central Highlands came under attack. Refugees started clogging all roads and trails heading south, and the Army of the Republic of Vietnam began to buckle, albeit with sporadic pockets of heroic resistance. The North Vietnamese quickly realized things were going much better than expected and reacted accordingly, pushing south along the coast. Soon Hue, Da Nang, Nha Trang, Xuan Loc, and Bien Hoa came under attack.

The U.S. had several evacuation plans in hand, ranging from an orderly and somewhat peaceful withdrawal from South Vietnam and Cambodia to a worst case evacuation under fire. As the situation in both nations continued to worsen the latter plan came to the forefront, and embassy staffs in Saigon and Phnom Penh made what preparations they could. Concurrently, the Navy ordered *Hancock* into the region, officially to relieve *Enterprise*, which was scheduled to depart at the end of April. However, upon "*Hanna's*" arrival with CVW-21 *Enterprise* was held, giving *Commander Task Force 77* four carriers to work with. Components of CTF-76 – the amphibious task force – steamed nearby and also waited.

Phnom Penh fell first; as with South Vietnam, President Ford had requested additional military and financial aid for the Cambodian government and was roundly rejected by Congress. On 1 April 1975 Premier Lon Nol fled the country. Over the following 10 days the four carrier air wings found little trade while monitoring the situation, but when it did come the evacuation of Phnom Penh – designated *Eagle Pull* – came quickly. U.S. Ambassador John Gunther Dean ordered the operation on 11 April following the closure of Pochentong Airport by Khmer Rouge gunners. A soccer field a few hundred yards from the embassy served as a makeshift evacuation center under the protection of Marines from the 31st Marine Amphibious Unit, with additional cover from carrier and Air Force aircraft.

Using CH-53s from HMH-463 in *Okinawa* the Marines removed a total of 276 people from Phnom Penh, including 82 Americans, 159 Cambodians with family or government ties to the U.S., and 35 from other nations. Ambassador Dean departed in the last flight at about 1015 on 14 April; Cambodian President Saukhm Koy went with him. Additional *Sea Stallions* from HMH-462 then pulled out the Marine defenders, who by this point were under Khmer Rouge mortar fire.

Phnom Penh fell to Pol Pot and the Khmer Rouge on the 17th. The victors immediately announced the formation of the nation of Kampuchea and ordered the city's residents out of town and into the country for reeducation. Over the following 12 months – designated "Revolutionary Year Zero" by the new regime – the Khmer Rouge slaughtered more than one million of their Cambodian countrymen, specifically targeting former military personnel, civil servants, educators, and anyone else considered "counter-revolutionary." The killing lasted for years.

One week after the capture of Phnom Penh it was Saigon's turn. Evacuation flights had been going on for some time from several locations, including Da Nang before it fell, but by mid-April the North Vietnamese Army fully surrounded the South Vietnamese capital. In preparation the U.S. ordered *Midway*, *Coral Sea*, *Hancock*, *Enterprise*, and *Okinawa* to a new position in the South China Sea and told the crews to get ready; *Enterprise* had been in Subic Bay only a few hours when she was ordered to haul out and head back to Vietnam at high speed. Characteristically, the North Vietnamese government protested the redeployment of the carriers as a violation of the 1973 Paris Peace Accords.

On 21 April 1975 Nguyen Van Thieu resigned as President of the Republic of Vietnam and was replaced by Vice President Tran Van Huong. Shortly afterwards Thieu left the country onboard a U.S. aircraft headed for Taiwan. The following day the evacuation kicked into overdrive, with personnel staging through Tan Son Nhut Airport, on the north side of Saigon. In order to protect U.S. and Vietnamese nationals during the process President Ford proposed sending in Army and Marine forces to secure the Saigon area. Congress treated his request as a call for a new "invasion" of Vietnam by the U.S. military and shouted it down. Junior congresswoman Bella Abzug (D-NY) compared Ford's request to "a second Gulf of Tonkin" resolution.

On 26 April the NVA started its final drive on Saigon, termed the "Ho Chi Minh Campaign." To this point the evacuation from the city's environs had been reasonably orderly, but now it turned into a flood, with tens of thousands of South Vietnamese attempting to flee on the U.S. airlift. The first rockets fell on Tan Son Nhut within 24 hours, destroying aircraft and structures and killing scores. Heavy artillery shelling followed the next day, as well as the incongruous sight of several captured ex-USAF/VNAF A-37s bombing the field. One rocket attack killed two U.S. Marine corporals, the last American military personnel to die in Vietnam.

The panic level increased as the U.S. shifted its main operation – now titled *Frequent Wind* – to the Defense Attaché Office compound. After clearing out the remaining personnel and Marines from Tan Son Nhut and the DAO complex, the final operations shifted to the U.S. embassy in downtown Saigon. Over the 28th and 29th some 1373 Americans evacuated from Saigon and went to units of CTFs -76 and -77, along with approximately 6000 Vietnamese. An additional 60,000 Vietnamese escaped in boats, made a bee-line to the U.S. fleet, and were also fished out of the sea.

On Tuesday, 29 April 1975, at 1952 local the last Marine guards locked the lower doors of the U.S. embassy, retreated to the roof of the long-time symbol of American presence in the Republic of Vietnam, and departed. Shortly before noon the next day NVA tanks rolled into Saigon and crashed through the gates of the presidential palace. President Gen. Dong Van Minh – who had replaced Tran Van Huong on the 28th – met with representatives from the victorious Democratic People's Republic of Vietnam and formally surrendered his nation. The Republic of Vietnam, formed in 1954, was dead.

Out in the fleet the situation was a madhouse, with hundreds of South Vietnamese helicopters – and the odd Cessna O-1A, one example of which successfully landed on *Midway* – flying out to land on or near any available Navy ship. To clear space, Navy crews pushed the helos over the side once the pilots and their families got out. Several aircraft made crash landings alongside the ships, with the people onboard jumping and swimming free. Fortunately, *Midway* had a relatively clear deck to handle the influx; the carrier had parked VA-115 and the other CVW-5 squadrons prior to heading to the region, and instead loaded Air Force CH-53s.

U.S. Air Force facilities in Thailand received a similar stream of planes flying in from the former RVN, as well as 30-plus vessels of the former South Vietnamese Navy. Conversely, the North Vietnamese captured hundreds of aircraft, at least five ships – including the former U.S. Navy minesweeper *Sentry* (MSF-299) and destroyer escort *Forster* (DER-334) – and innumerable patrol craft. The Navy managed to save some of the aircraft; on 2 May *Midway* unloaded 40 VNAF helicopters at U-Tapao RTAFB. In return the carrier picked up over 95 fixed-wing aircraft, including F-5As and A-37s, for transfer to Guam. The victorious North Vietnamese demanded their return; the Thai government respectfully declined and turned the aircraft over to the United States. The Vietnamese refugees went into relocation camps, and thousands eventually immigrated to the United States, although additional thousands eventually fled Vietnam by boat. As things got sorted out a bit CTF-77 finally released several ships for port calls to deliver their passengers or to make the long ride home.

Among the ships that headed back to ConUS was *Enterprise*, which onloaded one of the Marine *Sea Stallion* squadrons for the trip back to the states. For many of the older hands onboar – veterans of multiple combat deployments over the previous ten years – the collapse of South Vietnam had proven supremely frustrating.

Main Battery XO Lyle Bull, who would relieve Cdr. Gordon Nakagawa the following June, was one of those who flew support missions during the evacuations. He recalls:

> "It was one of the saddest days we ever had. To have such an investment in that war and those people, and then to basically forget about all the reasons we were there. I don't think the impact was as great for the younger guys. I think it was us guys who'd made multiple cruises there and had lost friends. The investment in life was great...and to pack up and go away and forget all the Vietnamese who were fighting the communists. We just deserted them."

Enterprise and CVW-14 returned home on 20 May 1975. By the time it pulled back to the pier one last final shooting conflict broke out in Southeast Asia, and this time the outcome was somewhat different.

On 12 May Kampuchean troops boarded and seized the U.S.-registered container ship *Mayaguez* in the Gulf of Thailand, roughly 60 nautical miles from the coast of Cambodia. The crew immediately called for help, and the United States started gathering forces again for a rescue mission. The next day an AC-130 assigned to the *16th Special Operations Squadron* located the vessel running in company with several gunboats. The *Spectre* gunship received indications the Kampucheans were running for port and attacked the gunboats; *3d Tactical Fighter Squadron* A-7Ds operating out of Thailand followed up with attacks on the 14th, which sank several of the smaller craft.

Analysts determined *Mayaguez's* crew had been put ashore on Koh Tang Island, and the mission became two-fold: rescue the civilian sailors and retrieve their ship. Hence, at 0700 local on the 15th 11 HH- and CH-53Cs from the Air Force's *40th Aerospace Rescue & Recovery Squadron* and *21st Special Operations Squadron* landed a combat force from the 2nd Battalion, 9th Marines on Koh Tang. Concurrently, additional Marines went onboard *Harold E. Holt* (FF-1074) to affect the recovery of the container ship.

The action on the island quickly became a major firefight, as Kampuchean defenders riddled the Marines and their Air Force helicopters. By the end of the day, eight of the *Super Jollys* were either shot down or seriously damaged. Navy and Air Force aircraft were precluded from flying close air support for the embattled Marines due to concerns over the captured sailors assumed to be on the island. There were no such restrictions against hitting mainland targets that might interfere with the operation, and carrier aircraft from *Coral Sea* went into Cambodia, along with USAF planes from Thailand. Several VA-95 A-6As and VA-22 and -94 A-7Es pounded Kampuchean air and naval facilities at Ream during an initial strike, then returned later to hit an oil depot at Kompong Son. The operation marked the combat debut for the *Green Lizards*, who were making their second deployment with *Intruders*.

> "It was the end of the cruise, and we were on our way to Perth to celebrate the Battle of the Coral Sea," Terry Anderson recalls. "Got up, saw the sun coming up on the *starboard* side of the ship, hey, wait a minute. At 0800 the captain got on the 1MC, told us of the *Mayaguez*, and here we were ready for Perth and Perth was ready for us."

The U.S. lifted restrictions against strikes on the island at 0945 local when a Thai fishing boat turned the 39 civilian crewmen over to *Henry B. Wilson* (DDG-7); apparently, the Kampucheans had released them the day before. U.S. forces were now free to concentrate on extracting the Marines from the island, and within short order waves of Navy and Air Force aircraft started hammering the kidnappers. The strikes included aircraft from *Coral Sea* working the pattern with F-111As from the 347th TFW, AC-130As, A-7Ds, and F-4D *Phantoms* from the 4th and *421st Tactical Fighter Squadrons*:

"Jerry Rogers – as I recall – was the only one to drop bombs," Anderson adds. "It was a two-plane, I flew a tanker. We could see Kho Tang Island and saw the destroyers pounding it. There were a lot of CH-53s on the ground. Afterwards we had the CH-53s on our deck with bullet holes; they really got shot up."

The first two attempts to recover the Marines failed. About 100 remained on the island pinned down by heavy fire, two helicopters lay on the beach wrecked, and every other helo was full of holes. On the third try (at 1745), an Air Force C-130 rolled a 15,000-pound "Daisy Cutter" bomb off its ramp, which upon detonating knocked down the defenders. Rescue aircraft swooped in, pulled the Marines off the island, and took them to *Coral Sea*. The final toll for the abortive rescue was 15 killed, three missing, and over 50 wounded; Lance Cpl. Joseph Hargrove, Pfc. Gary Hall, and Pvt. Danny Marshall were never seen again, although two decades later search teams found evidence that indicated at least two of them were taken to Cambodia and executed.

For the Ford Administration and the military the rescue was a success. The *Mayaguez* and its crew were back in the hands of their owner, while the military had proven it could still mount an operation. The reaction among press and political pundits back home, however, was more in keeping with the general attitudes of the time. Many attacked the operation, questioning the loss of equipment and personnel, while linking the *Mayaguez* rescue to the recent American "defeat" in Vietnam. One nationally known cartoonist used his daily strip to attack the Ford Administration, referring to the operation as "Frequent Manhood." Ignoring the controversy, naval forces in the region picked up, cleaned up, and attempted to return to a peacetime steaming routine. *Coral Sea* and her assigned units, including VA-95, later received the Meritorious Unit Commendation for their actions during the *Mayaguez* rescue. She returned to NAS Alameda on 2 July 1975 – via Perth, finally – and the *Green Lizards* finally concluded their second *Intruder* cruise.

The return to Whidbey also marked Terry Anderson's last deployment with the Navy. He did a subsequent logistics support tour at NavAirSysCom in Washington, D.C., and then went into the reserves, primarily for family reasons.

"Bob Knowles was my detailer," he comments:

"And he told me I was going to be the weapons officer of the carrier group on USS *Midway*. I would've had to pack up my family in Alameda and spend a year in Atsugi and decided against it.

"By the end of his fifth cruise I was getting assertive," says his wife, Betty. "He was ready to come home, I wanted a vacation. I enjoyed military life; one option was to stay in Virginia, but there would be more separations. I wasn't ready for that and didn't want to raise my kids in California."

Anderson ended up going to Boeing in Seattle, where he went into flight operations and manuals, writing particularly for the 737 program. He retired in 1994 after 16 years; ironically, his twin brother also retired from Boeing the same day.

With *Coral Sea* back at Alameda and the *Lizards* safely ensconced on the Rock, other carriers cycled out and in to maintain a U.S. presence in the region, although most everyone realized it would never quite be the same. Notably, one of the other departing carriers was *Hancock*, which finally turned for home and a well deserved retirement in October. Her 30 January 1976 decommissioning concluded a long line of proud service dating to 15 April 1944. *Hancock's* assigned air wing, CVW-21, also went by the boards on 21 December 1975, ending 20 years of service to the nation. Other ships, wings, and squadrons followed as the Navy contracted to fit the post-war world.

For most of the remainder of 1975 *Kitty Hawk*, with CVW-11 and VA-52 embarked, monitored the region. One way or another, after years of frustration, confusion, and violence, the war in Southeast Asia was finally over.

More Postwar "Routine"

The Navy of 1975 – and indeed all of the services that year – entered the postwar period facing declining budgets, a change of administra-

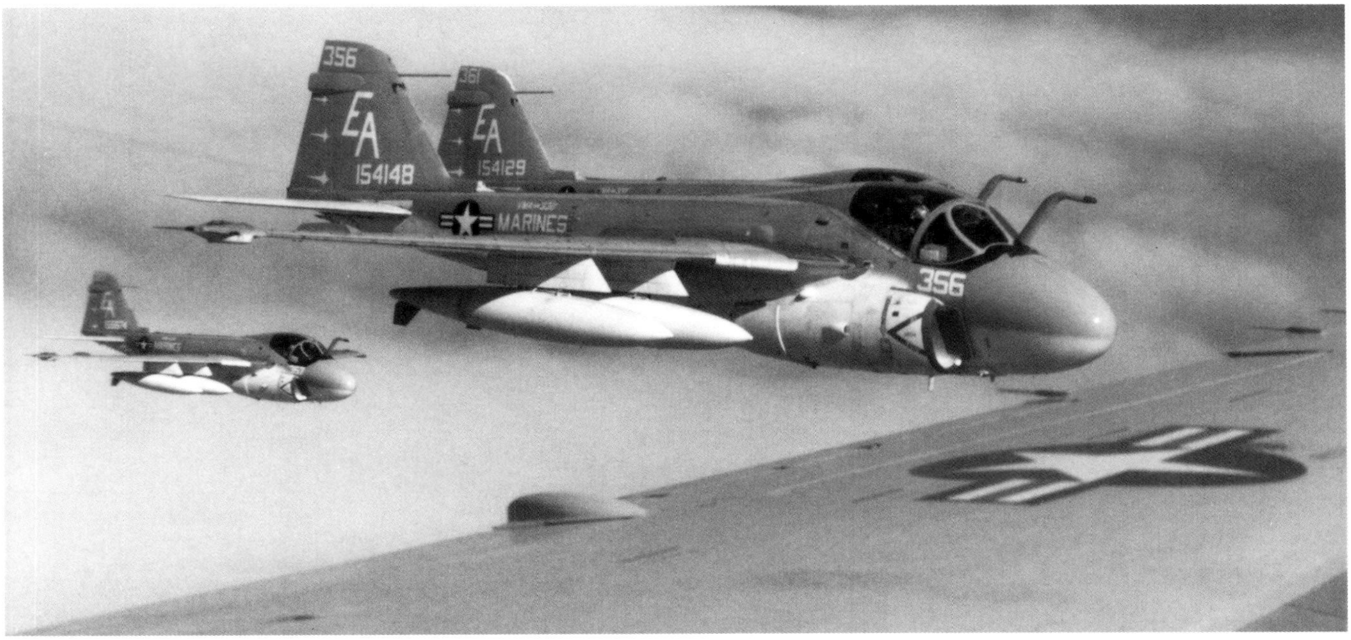

With the end of the Vietnam War the Marines began the Unit Deployment Program (UDP), where squadrons were rotated to MCAS Iwakuni, Japan, to serve with the 1st MAW. Aircraft from VMA(AW)-332 are seen here in transit from Japan back to CONUS in May 1980, while flying wing off a Navy C-9B that was acting as navigational pathfinder. (U.S. Navy, PH1 TC Mitchell)

tions, and problems getting good people to sign up and stay in. Combined with the wholesale retirement of tired and obsolescent equipment, the circumstances made the last half of the 1970s a particularly difficult time for all concerned. Where the nation's concentration had focused on Southeast Asia for the past decade, events elsewhere – such as the 1973 Arab-Israeli, or Yom Kippur, War and resultant energy crisis – now drew America's attention to the Middle East. There would be no sudden outbreak of peace, what with problems roiling there and in other locations around the world.

Still, ships deployed, personnel flew, operated, and maintained, and occasionally things "just happened." On 3 January 1975 another tired old warrior, *Franklin D. Roosevelt*, departed on what proved to be her next-to-last cruise. At 30 years of age she was not as extensively modified as sister ships *Midway* and *Coral Sea*, and all agreed *FDR* was in particularly poor shape. In other words, she'd been "rode hard and put away wet."

Making the last of its four deployments in *FDR* was *Carrier Air Wing 6*, which numbered the *Thunderbolts* of VA-176 among its squadrons. While not assigned to the wing or squadron, veteran A-6 pilot Lt. J.B. Dadson was one of those called on to prepare *FDR* for the cruise, its last with A-6s onboard:

> "I was at TacWingsLant when I got assigned TAD to the carrier group to work up the old *Roosevelt* for her cruise," he comments. "As a lieutenant I became the flag plans officer. They told me, 'We don't have the time to work this out, here's the books, *you* figure it out.' I got smart quick, but the learning curve was steep. We pulled it off; VA-176 and the wing had a good cruise."

Roosevelt showed the flag in the Mediterranean for the standard six months, returning to port in early July with no major incidents. With the end of cruise CVW-6 and the *Thunderbolts* transferred to *Independence*; they would sally forth again in April 1976, while *Roosevelt* (now under the command of Deke Bordone) would get called out of the barn one last time in October 1976 with a cobbled together *Carrier Air Wing 19*. The wing, normally assigned to NavAirPac, departed with two *Phantom* and three *Corsair* squadrons (VFs -51 and -111, VAs -153, -155, and -215), an E-1B det from RVAW-110, and VMA-231 with its AV-8As. The cruise, which was highly experimental, marked the first operational deployment of *Harriers* as part of a carrier air wing. Following the conclusion of the deployment, in April 1977 the carrier, the air wing, and the three *Corsair II* squadrons went to the scrappers.

Other carriers and other *Intruder* squadrons continued to carry the load in the Mediterranean, keeping an eye on the Middle East while the Israelis and their neighbors hammered out the aftermath of the 1973 war. While not on the combat line, they spent plenty of time at the various and assorted "great liberty ports," where young first-tour aviators and sailors received valuable exposure to local customs.

Life in both fleets was still anything but fun and games; there were the occasional losses, although a few bordered on comic. On 10 May 1975, while operating from *John F. Kennedy*, a VA-34 pilot ejected from his KA-6D without warning his B/N first. As the B/N – Lt. Ron R. Ayers – told his NFO class at Pensacola a couple of years later, the pilot locked the brakes on the cat stroke, heard the distinctive "pop! pop!" of the tires bursting, and didn't say anything, didn't do anything, he *just ejected*, "...and there I was, pilot in command of an A-6 with no warning." According to Ayers, he decided to give the tanker back to the taxpayers and also ejected, landing back on the flight deck of *Kennedy* and sustaining injuries in the process. However, he said the flight deck crew still had to restrain him from "greeting" his pilot once the helo dropped the aviator back on the deck, as "... all I wanted to do...was *kill* him..." Reportedly the two never flew together again.

More often than not the results could be tragic. During the 1975 Navy and Marine transition from A-6As to the improved A-6Es the Navy lost a total of seven *Intruders*. While a couple of the incidents were due to headwork errors, two were airframe-related. On 14 April 1975 VA-196 dumped a KA-6D off NAS Cubi Point when the starboard engine caught fire on final approach. Fortunately the crew of LT Jim Miller and Lt.Cmdr. Dan Wright were able to eject; unfortunately, they went out a tad late, and Dan got banged around a bit, as he recalls:

> "We knew we were going back to Vietnam to cover the evacuation, so we did work-ups at Cubi. We did seven approaches, landed and refueled, then went out again. The second time around I'm doing my standard thing in the cockpit, checking the switches and instruments. We rolled into the groove and had a power fluctuation. Now what the hell is going on? I looked out at the right wing, and there were flames all over the place. We're in the final throes of landing, so I said maybe we can land this thing. Right then the 'Get out! Get out!' calls started coming in. I looked at Jim Miller, he's rowing the stick around the cockpit and has no control. I think he ejected first; I hit on the first swing of the chute – those were the old GRU-5 seats, and we were almost out of the envelope."

"I landed right in the middle of a road," he continues:

> "Landed like a ton of crap. I was lying there and heard the fire engines and assumed I was on the access road around the end of Cubi's runway. I thought if I stayed there I'd get run over by a goddam fire engine, so I managed to release my chute, crawled to the side of the road, and stood up. I walked to the top of the hill and learned I was on the road down to Dungaree Beach. This kid in a pickup came driving by, picked me up, and said he'd take me to the ship. I said no, I think I'd better go to the dispensary. We then picked up Jim. The aircraft went in a building where the Marines kept their flight gear to support squadrons coming down from Okinawa for training. All of the flight gear was destroyed; the crash killed a couple of people and severely injured a couple of others. Completely shut them down for months.
>
> "I ended up with something like a 90 percent compression of my lower back," Wright concludes. "So when the squadron went off to Vietnam I was lying in the hospital for about a week on my back. I hit so hard both legs were black and blue to my knees, and I also broke a toe. I finally got up and around and would catch a cab and go down to the BOQ swimming pool. The ship came in, and they were going to AirEvac me back to the states, but they couldn't because of all the Vietnamese refugees getting shuttled around. I asked to go back on the ship, they said no, you can't handle ladders. I pointed out I'd be able to get from my stateroom to the ready room to the wardroom just fine. The guys from the squadron pitched in, got me back on the ship, I rode it back to Hawaii, and then took a flight back to the states. I could just see myself being stuck in the PI for months."

While both Miller and Wright survived, a pair of *Blue Blaster* aviators were not as fortunate on 25 April 1975 when the left wing of their A-6E (BuNo 152640, a rebuilt A-model on loan from VA-75) separated during a pullout near MCAS New River. The pilot rode the plane in, but his B/N, Lt.j.g. B. Kociemba, was able to eject and survived. Again, the A-6E was a major leap forward in avionics, maintainability, and combat capability, but there were still problems with the wings. Sadly, this sort of incident would occur again.

Six *Green Knights* form an arrowhead formation of early A-6Es between their 1981 and 1982 UDP trips to Iwakuni. (U.S. Marine Corps)

In the interim, transition to the upgraded variant continued with a mix of new construction and rebuilt aircraft. At NAS Oceana the *Green Pawns* were just about done, with every squadron flying A-6Es except for VA-176, which received its first aircraft in July 1975. Progress was a little slower on the west coast. VA-128 had completed upgrade training for VAs -52 and -165; VA-196 started replacing its A-6As in July 1975, followed by VA-95 (February 1976), VA-145 (September 1976), and finally VA-115; as they were forward-deployed, the *Arabs* did not get their first E-models until 1977. The actual trade-in of their A-6As and Bs took place while the squadron was in port with *Midway* at Naval Station Subic Bay, while formal transition training occurred at NAF Atsugi from May to August 1977. With the completion of their training all Navy *Intruder* squadrons flew the A-6E, and several started to get their upgraded CAINS birds.

Life in the RAGs continued to be lively on occasion. During one CQ period B/N John "Too Tall" Indorf was flying along with pilot Lt.Cmdr. Tracy "Trash" Smith when things quickly went to worms. He writes they were at flight level 280:

> "... 420 KTAS+, 40 nm south of SEA (Seattle) in solid IMC inertial direct to a SoCal CQ Det. Suddenly, there was a loud explosion and the cockpit changed from a relatively quiet, warm cozy place to an ear-splitting, frigid icebox. The entire plexiglass assembly of the B/N's canopy had just become another 'Things Falling Off Aircraft (TFOA)' statistic. When it departed the aircraft it took the pitot static boom on the vertical stabilizer with it.
>
> "Squawking 7700 (emergency), flying on AOA, and using pointy-talkee language in the cockpit, we turned back toward NUW and managed to find VMC at about 4,000 AGL on the radalt. Communicating with center using STBY and 7700, we opted to pass TCM (McChord AFB) and the other diverts and RTB. We regained comm when we dirtied up, but a power failure at the Rock precluded comm with the SDO on base radio. NUW tower told the SDO over the phone we were returning with an unidentified problem. The ODO, Al 'Big Bird' Lundy, met us in the line, gave us the obligatory 'WTFO' reserve salute look, didn't blink when I tossed my navbag down to him, but did a world class double-take when I climbed out of the jet and he noticed the canopy was still closed!"

Elsewhere, at MCAS Cherry Point VMAT(AW)-202 also worked to introduce the improved edition to the Fleet Marine Force squadrons. Assigned to Marine Combat Crew Readiness Training Group 20, the *Double Eagles* helped VMA(AW)-332 swap out for A-6Es in February 1975 and followed up with VMA(AW)-533 at the end of the year. To get their aircraft the *Night Hawks* had to depart MCAS Iwakuni; their return to Cherry Point marked their first U.S. landfall since March 1967. The *Bengals* of VMA(AW)-224, making a Unit Deployment Program (UDP) "cruise," replaced them at Iwakuni. These rotational deployments of Marine A-6 squadrons to 1st MAW in Japan, usually for six months at a time, would continue until March 1992.

Tragically for the community, 1975 also brought the single greatest aircraft disaster in all-weather attack history, and it occurred at Cherry Point. During takeoff for a routine training hop on 15 October a VMAT(AW)-202 TC-4C crashed, killing all nine onboard. The loss of life in the crash of *Tick* BuNo 155723 remained the greatest sustained by any squadron on a single day during the entire history of the *Intruder*.

Lizards See China

The community picked up and pressed on at various locations around the world. In WestPac, VA-95 had wrapped up its second cruise on 2 July. Participation in *Frequent Wind* notwithstanding, the *Green Lizards* had undergone a memorable deployment for other reasons, as related by Lts. Steve Richmond and Kris "Ack" Ackerbauer:

> "We launched off the *Coral Sea* as a target for a Taiwanese air defense exercise," writes Ackerbauer in *The Intruder's Lighter Side*. "Unfortunately, the Air Boss gave us the wrong launch position on the 5MC (flight deck loud speaker). We were one degree farther north (ie, 60 nm closer to Red China) than we thought. Our low alt ingress (with radar off in order to delay our detection), combined with a bit of salt spray on the windscreen caused us to be only about nine miles from land before we could see it."

"When we manned for the launch the ship's posit was announced over the 5MC by the Air Boss," adds Richmond:

"I saw Ack write it on his hand. We got an alignment and launched off on a DR heading for our first turn point. I made Ack turn off the radar, TACAN to receive, IFF off; we were really going to sneak up on those guys! We saw a coastline, but we weren't sure what it was. I eased up a little higher and got a TACAN azimuth-only lock-on pointing back behind our wing. Bad news! This coastline must not be Taiwan, but Red China. We turned and started to 'buster' for the *Coral Sea* just as we were intercepted by a couple of Nationalist Chinese fighters."

"Unfortunately, we were expecting to see Taiwan on our right and saw Red China off to our left," 'Ack' continues:

"Almost simultaneously we heard lots of garble on guard and 'Hot Dog' calls, none of whose meaning was briefed by CVIC, etc. We were intercepted by numerous F-5s who either thought we were lost or disoriented. One took trail position, while the others tried to get us to follow them. They gave us the international sign for 'you have been intercepted and we want you to land there,' pointing to Taipei. When they did that, Steve gave me some great advice: just look straight ahead, don't even look at them."

"They were calling us on guard, using our BuNo, telling us to land at Taipei," Richmond continues:

"Fat chance! Ack and I knew our careers were over, but we were going to tell our story on the boat. I told Ack not to answer on the radio and not to look at the guys joining on our wing. We just proceeded 'back sheep.' I was never so happy to see two F-4s join up on us and escort us back to the ship. That got rid of the ChiNats."

According to Ackerbauer, the two *Intruder* crewmen then proceeded back to the carrier at a high rate of speed via the 'scenic route'.

"As we disregarded their signals," he continues:

"We proceeded south, overflying the entire island of Taiwan at medium altitude. Once feet wet we descended and flew to the ship as low and fast as our trusty *Intruder* could go. Soon they turned around, as we ran them out of gas. We thought we'd be home free and nobody would be the wiser. Not so fast ... Immediately after we got on deck they wanted to see Richmond and Ackerbauer in the war room. It seems that the SDO had heard that we'd landed at Taipei. Intel had learned that the Red Chinese had launched every up MiG on the west coast to have a piece of us. Luckily, we had turned east just in time, and they were unable to catch us. We had obviously violated Red Chinese airspace and were close to being shot down.

"Our skipper, Cdr. Westfall stuck up for us and threw his command pin on the table with the words, 'You can take this pin, but you won't touch my guys.' We were called to the war room just about every night at midnight for a week while the State Department and DOD asked us inane and irrelevant questions about the incident. Fortunately that was our only punishment, because the skipper stuck up for his boys and we never read about it again ... except in the welcome home skit."

Richmond confirms the sequence:

"Fast forward to the Admiral's stateroom, where we were drilled on and off for seven nights in a row. The ChiNats thought we were trying to start WWIII. Our skipper, Van Westfall, backed us 100 percent. At one point he tossed his command button on the table and told the Admiral, 'Don't try to hang this crew. I'll put the navigator, the ship, everybody involved on report!' The Admiral basically answered, 'Not to worry ... nobody's going to hang.' Turns out he was right, but at the time it looked like the end of two illustrious careers. Take my word for it. You *never* want to be involved in an 'International Incident.'"

According to Skipper Westfall, the hardest part was trying to show the chief of staff – a fighter type – how to use a DR plotter while explaining navigation and a bad INS alignment. "Thank god for CAG (Hoagy Carmichael)," he adds. "We all survived."

Apparently the *Lizards* liked to play as hard as they flew, as witnessed by the following vignettes provided by Westfall. The first occurred the previous year while he was XO under Cmdr. W.D. "Zip" Zirbel:

"Liberty in San Diego was wonderful, except when fourteen *Lizards* in one rental car were stopped by a California state patrolman after a night on the town. As he stood there shaking his head, he said, 'Take two guys out of the four in the trunk and get the hell out of here.' Being good Naval officers, we complied and returned to Miramar."

Later, as *Lizard One*, Cmdr. Westfall pursued an aggressive program of debriefings and professional training designed to enhance squadron morale, leading to two other sea stories:

"Happy hours were great, except when a young JO – T. Toms, a qualified parachute instructor – says, 'I'll teach you and we'll all have fun.' Instruction followed and jump day arrived. Steve Richmond, a 'hot shit,' broke his leg trying to show up the skipper. John Schork drifted into the trees, not once but twice, as the group is yelling '*Cross your legs*.'

"Happy hours are great for innovative ideas, problem solving, career-enhancing talks. 'How about a motorcycle group in the 4th of July parade, wearing SH flight suits, and T. Toms can parachute into the parade?' 'Skipper, you can lead the group ... just ride straight and don't fall down.' 'Hot Shoe Porter' organized and practiced the *Lizards*. Parade day, all went well with circles, weaves, and figure 8s; T. Toms landed safely in the bank parking lot off Highway 20/Pioneer Way. Debrief was conducted at the Oak Harbor Tavern (where else!?). A *Lizard* came in ... we won a trophy as the unique group in the parade. Command guidance: 'Go get it and bring it back. We're in the middle of a debrief.' Those were the days."

Hot Spots Around the World

In mid-1975 events in the Middle East once again grabbed the world's attention. The focus this time was the nation of Lebanon, which was undergoing an extremely violent civil war. The presence of U.S. nationals and other interests in the country mandated a response, including a possible evacuation operation. As had happened in 1958, a large component of America's response centered on carrier air power.

Wedged in between Syria, Israel, and Jordan, Lebanon possessed a long history of religious and political problems. In the late 1950s the country found itself caught between the forces of Lebanese nationalism and a growing pan-Arabism movement centered on Egyptian President Gamal Abdul Nasser. The movement had already resulted in the merger of Egypt and Syria as the United Arab Republic, and many groups in Lebanon pushed inclusion in the UAR by whatever means necessary. A period of protests, political violence, and multiple assassinations ensued. On 14 July 1958 Pan-Arab extremists assassinated the entire Iraqi monarchy, including King Faisal, his uncle Crown Prince Abdullah, and Premier Nur as-Said. The President of Lebanon, Camille Shamun, received strong indications he and his administration were next, so he appealed to France, Great Britain,

The *Tigers* of VA-65 were famous for their bright orange tail markings. They were also the first *Intruder* unit to deploy on the new nuke carrier *Dwight D. Eisenhower* (CVN-69), in January 1979. Lead aircraft AG512 carries a load of inert Mk.83 1000-pound bombs. (U.S. Navy)

and the United States for help. President Dwight D. Eisenhower responded by landing the Marines on 15 July. They departed after the unrest abated, and in time the United Arab Republic itself dissolved.

Now, 18 years later problems started anew, and again a lot of bloodletting was involved. This time around the country was self-destructing along religious and ethnic lines, with the Christian Maronite-dominated Phalangist movement fighting multiple Muslim-affiliated groups. The Palestinian Liberation Organization and its military front, the PLF, weighed in on the side of the Muslims in early 1976 and the fighting worsened. In May 1976 the Syrian Army intervened; notably, it came in on the side of the Christians, as the Syrian government had major concerns over the PLO taking control of its neighbor. Multiple militia groups from both sides rampaged through the country committing atrocities and massacres. In short order both the Lebanese government and its Army collapsed, and the nation splintered. Within a year, an estimated 60,000 Lebanese of all backgrounds and faiths were killed.

On 27 October 1975 *John F. Kennedy* – with CVW-1 and VA-34 embarked – was ordered to take station off Lebanon. There it stood by with an amphibious ready group centered on *Inchon* (LPH-12) and prepared for possible evacuation operations. *Independence* – with CVW-7 and VA-65 – subsequently joined, as did JFK's relief, *Saratoga*, with CVW-3 and VA-75. What with the situation in Lebanon, *Sara's* – and VA-75's – attention remained focused on operations in that area throughout the cruise; these included one at-sea period of 42 days straight off the war-torn nation. However, in April 1976 *Sara* diverted east to cover the island of Crete in case Americans needed evacuation from *that* island, which was also suffering from internal problems.

On the plus side, *Saratoga* was able to visit Split, Yugoslavia, during the cruise, reportedly marking the first American carrier port call at that location since *Forrestal* in 1960. To top off the deployment *Sara* celebrated the nation's bicentennial by pulling liberty in the fabled principality of Monaco. Afterwards she returned to the states, tying up at Mayport on 28 July 1976.

Back in the Med the situation in Lebanon continued to deteriorate, and on 13 June 1976 terrorists assassinated the American ambassador to that splintered nation. It fell to the newly arrived *America* battle group, operating with an amphibious force centered on *Coronado* (LPD-11), to perform the evacuation. *Intruders* from VA-176 joined other CVW-6 aircraft in covering the operation, which took place on 3 July. By the end of the day a total of 160 Americans and 148 personnel from other powers left Beirut. The Arab League finally enforced a truce in Lebanon, but the violence of ethnic/religious conflict continued to fester in the region; Israel got dragged into the conflict in 1978 and again in 1982, leading to the establishment of a multi-national peacekeeping force in the early 1980s. The remainder of the decade remained tough for all of the participants, including the carrier aviators, who were still struggling with declining budgets and elderly equipment.

According to *Tiger* CO George Strohsahl, VA-65's October 1975 to May 1976 Med deployment in *Independence* was "not as good" as the previous year's tour:

> "First of all, we were deployed without the normal pump-up of people you get. I don't know if higher authority was trying to show how bad it could get, but I had a couple of shops I couldn't operate two shifts with. Both cruises ended up with large numbers of admin discharges, druggies, and others just trying to get out of the Navy. Lastly, they were hot-decking the KA-6Ds. We got some turnover KA-6s to get us up to our usual four, and two of them weren't usable. The problem was corrosion; one was punched through the floorboard on the B/N's side. We ended up parking both at the rework site at Brindisi (Italy). Since we had the tanker commitment we ended up putting buddy stores on all our Es, flying double-cycle, three-tank contingency tankers. That really curtailed the amount of tactical training we could do. We got a lot of hours, but not a lot of traps.
>
> "That had been preceded by one of the many incidents of A-6 wing cracks. We spent the workups with most of the aircraft down for wing cracks, with depot field teams trying to come up with a remedy. Most of the squadron didn't get their strike training quals. The planes were good bombers, still fairly new A-6Es. At Fallon we out-bombed the A-7s using radar offsets. We couldn't dive bomb due to the cracks in the wings, but we had a tighter CEP than the A-7s had in visual. CAG didn't give us the trophy because our tactic was 'not an approved tactic.'
>
> "We were undermanned, under-equipped, and I didn't enjoy it. It was a peacetime cruise – a repeat of the previous cruise – but we did a lot of exercises and a lot of time in port. There was a lot of pain due to shortfalls. Those were tough years, and then the Carter years just killed us."

After turning VA-65 over to Cmdr. Don Hahn on 14 September 1976, Strohsahl reported to Medium Attack Wing 1 as Chief of Staff:

> "I continued to fly, observed various exercises, and kept current with the squadrons. My memory of that time was a lot of people busting their hump to keep the community going."

Despite these problems, the Sixth Fleet carriers still had to keep an eye on the region, all the while monitoring an expanding Soviet Mediterranean Squadron and racing from one troublesome location to another. They were not alone in their efforts; their Seventh Fleet counterparts also found themselves charging around from time to time on the other side of the world.

One pattern that became standard for the AirPac air wings and squadrons was deployment to the Indian Ocean. In July 1976 *Ranger*, with CVW-2 and VA-145 onboard, was ordered to depart the South China Sea and make all due haste to the east coast of Africa. On the 3rd Israeli commandos landed at Entebbe, Uganda, and freed over 100 hostages held by pro-Palestinian guerrillas; they killed all of the terrorists and did a good job of shooting up the airport and Ugandan military guards during the proceedings. In response, Ugandan president-for-life Idi Amin Dada threatened both the Americans within Uganda and the nation of Kenya, which had provided assistance to the Israelis. Tensions here also continued into the following year, with *Enterprise* (CVW-14/VA-196) eventually relieving *Ranger*.

On 21 August 1976 another incident took place that also required the presence of an aircraft carrier. This time the duty deck was *Midway*, which was ordered to take station following the murder of two U.S. Army officers by North Korean troops in the DMZ. The U.S.' reaction drew world attention to the Sea of Japan, and for a while it looked like there might be another shooting war on the Korean Peninsula. However, the Ford Administration chose not to take military action beyond a short-term reinforcement and the threat subsided.

For the crews deployed around the world at the tip of the spear, responding to threats and crises, it undoubtedly seemed like civilization was going off the deep end in quick order. However they kept at it, training, preparing, and performing their assigned missions with a strong sense of pride, whatever the circumstances. As VA-196 B/N John Indorf recalls:

> "(We were) TransPac'ing from Barbers Point to Cubi, without tankers, what a great deal for a nugget aircrew! The stopover at Midway Island was too short and the RON at Wake Island was a blast. However, the most memorable part of the trip for us came several days after arriving at Cubi. Someone from Guam mailed us a copy of the evening paper published the day we departed NAS Agana. On the front page there was a picture of our departure flyby, with our C-9 in the lead of a V formation of three A-6s tucked in tight on each wing. The caption under the picture read: 'Secret Bomber Escorted By 6 Fighters Enroute to Korea.' The day before we arrived at Agana, some border guards butchered a couple of GIs in the infamous tree chopping incident along the Korean DMZ. It was the only time in our careers we didn't mind being associated with fighter pukes."

The situation worldwide did not become stable any time soon. In fact, it got worse – much worse – in short order.

Showing the Flag

As the nation entered the midpoint of the Carter Presidency it continued to keep a close watch on several locations around the world, including Lebanon, Uganda and Kenya, the Korean Peninsula, and elsewhere. There were also rumblings of communist insurrection in Central America, but the major problems in that part of the western hemisphere were still a few years off. Despite the national "malaise" – as described by President Carter – and ongoing funding, personnel, energy, and equipment problems, the Navy was managing to hold its own, but it was not easy.

The *Intruder* community started 1978 rather shakily, with the loss of a VA-75 KA-6D (BuNo 151791) on 18 January; the circumstances came under the "oops!" category. A crew from VRF-31 was ferrying the aircraft from Oceana to Naval Air Rework Facility Alameda for depot-level maintenance and stopped for one night enroute at Buckley Air National Guard Base outside of Denver. They took off the following morning without topping off the tanks and ran out of gas two miles short of Alameda's runway. Both men ejected safely – one onlooker swore they went over the Bay Bridge, while the abandoned *Intruder* went *under* – and were quickly pulled from the water by the Coast Guard, while gawkers observed the proceedings from Yerba Buena Island. Afterwards the *San Francisco Examiner* noted the ferry crew had managed 2,518 miles of their 2,520-mile journey. The following day salvage crews pulled the KA-6D from 35-feet of water in San Francisco Bay and delivered it to NARF Alameda, which struck the aircraft from further service.

Unexpected swim calls notwithstanding, the *Intruder* community continued to polish its skills while deploying to the far flung corners of the world. In the Mediterranean, problems continued in Lebanon, culminating in an Israeli invasion in March 1978. The Israelis eventually pulled out in favor of a 6000-man United Nations peacekeeping force, but the region continued to bear considerable watching. On the plus side, during the year Israeli Prime Minister Menachem Begin and Egyptian President Anwar Sadat began peace negotiations, leading to the latter's visit to Tel Aviv. Their talks culminated in the signing of a peace accord at Camp David, MD, on 26 March 1979.

During this period the *Green Lizards* of Whidbey made a "guest appearance" with Sixth Fleet, embarking with CVW-11 in *America*. Led by Cmdr. Ken G. Craig, VA-95 departed for the first of two consecutive Med cruises on 13 March 1979. Early in the deployment Cmdr. Dick Toft relieved Craig as commanding officer; he says one of those things subsequently happened which made him proud of his JOs:

> "Lt. Bob Brown launched one day with his B/N, and while they were climbing over the ship both flaperons popped up. The obvious initial action was to make sure the flaperons were turned off. The crew checked and they were, but for whatever reason the flaperons had extended and were staying up. Brown had to use 100 percent power to maintain controlled flight, and obviously the fuel rate was astronomical."

The skipper – who was airborne at the time – communicated with his maintenance people and the Grumman tech rep, and they suggested Brown turn off all of his electrical systems, but that didn't solve the problem. The *Lizards* then sent up a KA-6D for an attempt at aerial refueling, but Brown – with no stick input and only rudder control – was unable to maintain adequate control for a hookup. With the situation going beyond tense Toft recovered and stayed in his A-6 in order to maintain communications with his squadronmates and the others. Finally, Brown called and said he was down to 1600-pounds, and it was now or never. According to Toft, *America* did not have time to rig a barricade:

> "Fortunately he flew a beautiful pass and trapped, saving the A-6. Post-flight checks revealed a short in the emergency battery bus, which energized the flaperon circuit even when the power was turned off. It was a brilliant feat of airmanship in keeping his cool under stress. He made my tour as skipper."

Brown received the Air Medal for his efforts.

VA-95 made it through the remainder of workups and cruise without further incidents, and returned home to Whidbey Island in Septem-

Marine *Intruder* squadrons transitioned to the A-6E through the mid- to late-1970s. Colorful markings were still the norm through the period, as displayed by *Bat* DT12 at NAS Chase Field, TX, on 20 February 1980, and *Bengal* WK06 NAS Key West, FL, in November of the same year. (Both: Rick Morgan)

ber 1979. By that date Pacific Fleet had a full plate; fully five years after the end of the Vietnam War the deployments still tended to be long, and trips to the Indian Ocean had become the norm. There was plenty going on in the region to keep the Navy's attention; Libyan-supported Muslim separatists were stirring up trouble in the Philippines, Vietnam had invaded Kampuchea, and thousands of Vietnamese were attempting to escape their disrupted country by small boats. Adding to the festivities, communist insurgencies broke out in Central America *and* the Soviet Union invaded Afghanistan at the end of 1979. As a result, carrier battle groups made like fire trucks, racing from one ocean to another to keep an eye on things.

However, the biggest international disaster effecting U.S. diplomacy occurred in a wholly unexpected region of the world. It resulted in a lengthy full-time carrier presence in the western Indian Ocean, and ultimately contributed to the departure of the Carter Administration.

Iran

One region of the world that had shown a modicum of stability over the previous years was the portion of the Persian Gulf and North Arabian Sea dominated by Iran. Led by Mohammed Rezah Shah Pahlavi, the nation laid claim to the proud and warlike tradition of Persian leaders, such as Cyrus the Great, and as such had worked towards becoming the dominant nation in their corner of Southwest Asia.

After his rise to power in 1941 the Shah proved to be a staunch ally of the United States, signing his nation's first defense agreement with the U.S. in 1959. Over time he built a substantial military using aircraft, such as the F-4 *Phantom II*, F-5 *Freedom Fighter*, and F-14 *Tomcat* (notably, the *Tom's* only export use), and U.S. armor and weapons. Concurrently, Iran's Navy dominated the region's waterways with a modern mix of surface combatants armed with effective anti-air and anti-ship weapons, including *Harpoons*. In the mid-1970s the Shah ordered six guided missile destroyers based on the U.S. Navy's *Spruance*-class that would have been the envy of almost every other navy in the world.

The United States supported the Shah and groomed his nation as the strategic bulwark in the region, providing a balance to Iraq and other countries. The majority of his military personnel trained in the United States, taking their flight instruction alongside their American counterparts at locations like Pensacola, while senior officers took advanced studies at the major command and staff schools. Thus trained and using primarily American equipment, the Iranian forces regularly took part in multi-national defense exercises in the region, such as the *Midlink* series. At home the Pahlavi government attempted to bring Iran into the 20th century through education and economic and social change. However, his regime did not tolerate any form of political opposition and banned all parties except one, while using violence and intimidation through the Savak, Iran's secret police. When a wave of Muslim fundamentalism swept through the country in 1978 the Shah's government was ill put to handle the challenge. The repercussions lasted a long time and directly impacted the United States.

The actions of the Ayatollah Ruhollah Khomeini, exiled by the Shah since 1963, served as the flashpoint. For over a decade he exported messages and support to his followers in Iran with the intent of overthrowing the Pahlevi Dynasty and forming an Islamic government. At his behest Shiite fundamentalists started rioting in 1978 and were quickly joined by striking workers. On 8 September the Iranian government declared martial law in 12 cities; it followed with the establishment of a military government on 6 November. Prime Minister Shahpur Baktiar was subsequently appointed to run the government in the absence of the ailing Shah. The unrest quickly escalated into more strikes and open warfare on the streets of Iran's cities. In response, on 27 December 1978 the United States directed *Constellation* and her escorts to move to a position near Singapore. While not cleared into the Indian Ocean, the carrier – with CVW-9 and VA-165 embarked – was ordered to stand by in the South China Sea and be prepared for possible operations in the vicinity of Iran. Nothing came of this initial reaction by the U.S. government, and *Connie* cleared the vicinity of Singapore on 28 January.

By that date the Iranian government had all but collapsed. The Shah, suffering from terminal cancer, left the country with his family on 16 January and took refuge overseas. The United States concurrently ordered all non-essential personnel to leave Iran by 30 January. The following day Khomeini left Paris for a triumphant return to Tehran, some 15 years following his banishment by the government. On 11 February Khomeini's forces routed the Shah's Imperial Guard, and the last vestiges of the old government collapsed. In their stead the Ayatollah and his supporters established the *Jomhoori-e-Islami-e-Iran*, or the Islamic Republic of Iran. The radicals who had fought the Iranian military became the Ayatollah's new defenders of the faith, assuming power in their own right as the Revolutionary Guard; while not necessarily trained in the military arts, they were reliable where the new Islamic republic was concerned. A blood bath ensued, as the Guard hunted down, tortured, and executed members of the Savak while purging – and occasionally executing – experienced military personnel.

The U.S. responded by the traditional means, running carriers into the Indian Ocean and North Arabian Sea while quietly getting people out of the country. What ensued, however, was another CV "Chinese fire drill," as Navy units rushed from one location to the other, covered gaps, and responded to a series of crises above and beyond Iran.

On 21 February 1979 *Ranger* deployed from North Island; onboard were CVW-2 and VA-145, with the latter under Cmdr. John Juan. While all pre-cruise planning had been for a "standard WestPac," the situation instead dictated that they head for the Indian Ocean. Unfortunately, a couple of weeks later *Ranger* ran into a tanker in the Straits of Malacca and quickly found itself turning for Subic Bay and repairs.

Cmdr. Juan was another one of those pilots with an "interesting background." Passed over once while assigned to VA-128, he subsequently served as XO of the *Golden Intruders* under Rupe Owens. Miraculously, he made commander on the second look, and when time came for him to screen for command he received VR-1, a transport unit

The early 1980s witnessed a wholesale change in the appearance of *Intruders*, as the bright Vietnam-era paint began to be replaced by subdued, low visibility markings applied to the standard gull gray and white paint scheme. VA-75 AC502 and VA-42 AD504 were both shot at NAS Key West in 1981. The formal two-tone dull gray "Tactical Paint Scheme" (TPS) was being introduced into the fleet at the same period. (Rick Morgan)

at NAS Norfolk. He rated it as a good tour, particularly for a former *Spad* and *Intruder* pilot:

> "They weren't used to having any fleet guys around," Juan recalls. "We had these young guys, fresh out of TRACOM and enroute to the airlines, and I said, 'Let's go to happy hour.' What? A CO who goes to happy hour? I got there and said we're going to do things like in the fleet. It was a great tour, we got to go to great places. We took people to all sorts of places, made extended tours to the Med, pathfinder flights across the Atlantic via Bermuda and Lajes, then spent two to three weeks over there. We always came back with a load of booze. However, VR-1 was going away – the mission was going to the reserves – and the detailer called and said, 'Have I got a deal for you! How about Whidbey Island?' I called the guy who had our house and said don't vacate the lease, we're on our way back."

John ended up going to VA-145, where Cmdr. Rick Hauck was preparing to relieve Cmdr. Vince Huth. However, NASA selected Hauck for the astronaut program, and Juan went to the squadron in his stead:

> "Dave Williams had already checked in as the next XO, but I was senior to him. So I went in, took an unbelievably short course at VA-128, and reported to -145 as the CO. Obviously, under the circumstances I had to rely on the people under me. For example, I had five or six guys who had been admiral's aides, so you can see the quality of people we had. A bunch made CO themselves. We also had a couple of Air Force officers in my squadron; they were great, it was a really good program."

The group included Juan's regular B/N, Jack J. "Black Jack" Samar, who later commanded the reborn VA-155. One first tour JO named Randy "Darth" Dearth later commanded VA-95, served as the last skipper of VA-128, and eventually led *Carrier Air Wing 2*. Juan's XO, Dave Williams, later served as commanding officer of NAS Whidbey Island:

> "The cruise reminded me of the early *Spad* cruises, ie, no war," Juan continues. "We were supposed to go to the IO, but *Ranger* hit a tanker. I remember thinking, 'If I can feel a bump and I'm on a carrier, we must've hit something big.' We got the immediate smell of fuel. So no IO; we went to Subic for immediate repairs, then went into drydock in Yokosuka. It was the first time I'd ever seen a carrier up on blocks, quite impressive."

While *Ranger* diverted to Yokosuka, ongoing disturbances between the two Yemens broke into open warfare. Hence, on 7 March AirPac ordered *Connie* out of the Philippines to monitor both Iran and the situation between North and South Yemen. She remained on station performing "special operations (non combat)" until 16 April, when the *Midway* arrived. The *Boomers'* return to the Rock on 17 May 1979 put an end to an eight month cruise in *Constellation*, with a fair chunk of the deployment spent in the IO. Meanwhile, *Kitty Hawk* prepared to pull out for her turn in the rotation; the big carrier departed Southern California on 30 May 1979 with *Carrier Air Wing 15* replacing *Carrier Air Wing 11* on her deck. VA-52 went along for its sixth consecutive cruise on *Kitty Hawk*, having transferred or swapped over from CVW-11 to CVW-15.

With this lineup the Navy attempted to maintain one carrier in the Indian Ocean and North Arabian Sea through the end of the year, but events kept happening elsewhere. The crews of CVW-15 spent a better part of their summer in the South China Sea assisting the Vietnamese boat people, who numbered in the thousands. No matter how many were spotted at sea by aircrews and subsequently pulled from the water by the units of the battle group, more seemed to appear every day, usually in extremis. All fled the "workers' paradise" of the reunited Vietnam, some four years after the final collapse of the Republic of Vietnam. The repaired *Ranger* also performed contingency operations through the end of her deployment, but did not make it into the Indian Ocean this time around. For the *Swordsmen* it made for an interesting if disruptive cruise, what with the collision at sea. However, John Juan recalls his squadron was tight and worked well together, whatever the circumstances. Interestingly enough, he liked the late 1970s Navy.

"The late 1970s was still a good time to be in the Navy," he comments:

> "Still not overpowered by too many regulations and too many programs. The ones we had … well, I really, really hate touchy-feely stuff, but there were some benefits in some of the programs. It gave us a better snapshot of where we were. The best part was we always had Oak Harbor to come back to, where you could fly and do things. Not like back east."

Ranger headed home in early September 1979, and Juan turned things over to his XO, giving Dave Williams an early start on command:

> "On my way home I'd already decided in my mind I was going to retire," Juan comments. "So I told Dave, 'Here's a career enhancing job. I'm going home, you take the squadron back.' I took a TransPac back to Whidbey."

VA-145 returned to the Rock on 22 September 1979 and held its change of command six days later on the 28th. From Whidbey Cmdr. Juan transferred to *Enterprise* as Air Boss, albeit under unique circumstances:

> "The detailer called and said 'What would you like?' 'Anything close to Whidbey Island?' 'You're probably not going to like this: USS *Enterprise* in Bremerton.' I said, 'Yeah! Yeah! Yeah!" I'd been on *Saratoga* in the Philadelphia Navy Shipyard and figured any other yard had to be better.
>
> "I spent a year there as Air Officer for a CVN in the yard; I commuted like a bunch of people. We concentrated on training and stuff like that. Otherwise, I spent the majority of my time as habitability officer, building bunks and heads."

John Juan retired from the Navy in 1981 and settled in Oak Harbor. Of all the aircraft he flew, in a career that took him from A-1s to A-6As to C-9As to A-6Es, he rated the E-model as his favorite, adding:

> "I flew *Spads* with a plotting board, looking at the waves and the winds. I had one sump warning light the entire time I flew them; the 'proper' response was to unscrew the light, because you were already 'single-engine.' My second time in VA-145 we had the A-6E, and the system was up more than it was down. It was a big step up from the A-6A. The only time I had problems in the A-6 was on a mini-Alpha strike at Fallon and we got a fire warning light. There were no other indications, so I shut the engine down while carrying a full load of bombs – no problem – and we made it to the target and back. When we came back I learned we were this close to having it burn through. We managed, though."

Having assumed the duties of *Swordsman One*, Cmdr. Dave Williams prepared his squadron for another deployment. He too rated his command tour highly, commenting:

> "If I could point at one tour that I enjoyed more and one that meant more, it was this tour. We had an outstanding airplane, outstanding people – both officer and enlisted – and we proved we could do things that not even we thought we could do."

His *Swordsmen* and *Ranger* finally made it to the IO by late 1980; by the time of their arrival, however, the situation with Iran had changed once again.

The circumstances for the deployed Seventh Fleet units first took a major downturn in October 1979 when anti-U.S. demonstrations started breaking out in Iran. In partial response *Midway* pulled out of Yokosuka and went back to the Indian Ocean. Early the following month the ailing Shah of Iran went to America for treatment of his cancer, and the level of protests escalated substantially. The bottom fell out on 4 November 1979. Following a series of vitriolic speeches by the Ayatollah attacking the former Shah and his American hosts several hundred "students" stormed and captured the U.S. embassy in Teheran. A total of 66 Americans – from the Ambassador on down to the Marine guards – were taken hostage; the Iranians subsequently released 13, but held the remaining 53 as bargaining chips. According to Khomeini and his spokesmen, they would release the hostages following a formal U.S. apology for its long time support of the Shah, the surrender of his "stolen" wealth, and his return to Iran to stand trial for his "crimes."

The U.S.'s immediate response was – once again – to send in the carriers. The first to respond was *Midway*, which immediately proceeded to the North Arabian Sea. *Kitty Hawk* followed her on 3 December; she put two carrier task forces in the region for the first time since World War II.

One of the pilots assigned to VA-115 was nugget Larry Munns; he comments:

> "We were in the Indian Ocean already. *Midway* was famous for doing the 'October to December IO' cruise before the IO was famous. We always missed Thanksgiving, but were always home for Christmas. I was supposed to detach and be back in the states in time for Christmas, but I stayed, the first of several deployments with no end. *Kitty Hawk* didn't relieve us, they joined us on station. We were wondering when we were going to do something.
>
> "It was a little bit different. I'd been on the Midway '77 IO cruise when we were friends with Iran. We had flown low levels in Iran, had met with them and used their facilities, even put people ashore at Bandar Abbas. Now, two years later we're going back into these places, possibly to bomb them. There wasn't a lot of frustration yet; we were ready to do something and had the capability. VA-115 was unique at that time. We had a couple of *Pave Knife* (laser pods) planes, and also AWG-21 (A-6E Standard ARM birds). We had four aircraft and the east coast had four aircraft. Those weapons were in our plans."

On 13 November *Coral Sea* sortied from NAS Alameda with CVW-14 and VA-196 onboard, also heading for the Indian Ocean.

There they found long hours at sea and no diverts while the State Department worked to secure the release of the hostages. In the face of Iranian intransigence against the "great Satan" it quickly became apparent deployments to the IO would probably not end any time soon.

On top of the general frustration over the seizure of their fellow Americans, the cruise took a particularly nasty turn for VA-52 right before the end of the year. During a launch from *Kitty Hawk* on 29 December the KA-6D flown by squadron CO Cmdr. Walter D. Williams and B/N Lt.Cmdr. Bruce Miller suffered a soft cat shot and went in the water. Sadly, neither man survived; even more tragic was the fact Cmdr. Williams had relieved Cmdr. James R. McGuire only seven weeks earlier. XO Pete Rice, who had previously made his department head tour with -52, fleeted up to command.

Finally, just when everyone thought it could not get any worse, *Coral Sea* diverted to the Sea of Japan to monitor a deteriorating situation in the Republic of Korea. On 26 October Kim Jae Kyu – the head of the KCIA – and five followers assassinated South Korean president Gen. Park Chung-Hee and his bodyguards. During the ensuing confusion and disorder ROK Chief of Military Intelligence Gen. Chun Doo Hwan placed the country under martial law. For all concerned the events in Korea marked a truly bleak two-year period, but the nadir undoubtedly arrived on 4 November with the seizure of the hostages.

1980: "Shuttle Diplomacy" and *Desert One*

Through the end of 1980 the Navy continued to cycle carriers through the Indian Ocean, regularly operating two in the region, with one CV to the south and the other placed at a northerly location close to Iran. Ships assigned to the latter position – which became known as "Gonzo Station" – quickly became familiar with extended periods at sea between port calls.

As the year opened, it was already apparent the AirPac carriers could not maintain a full-time presence in the IO without leaving gaps somewhere else. Therefore, in January 1980 AirLant's *Nimitz* – with CVW-8 and the *Black Panthers* of VA-35 embarked – left the Med and made the long voyage around the Cape of Good Hope. Its departure left *Forrestal* – with CVW-17 and VA-85 – as the only U.S. carrier operating in the Med. After *Nimitz* arrived on 22 January *Kitty Hawk* turned for home, having completed 64 days straight in the North Arabian Sea. Two weeks later, on 5 February *Coral Sea* relieved *Midway* as the other duty carrier in the Indian Ocean.

President Carter ended his attempts at diplomacy with the Iranians in early April 1980, and on 11 April he authorized his advisors to proceed with a military rescue attempt. Originally developed by National Security Advisor Zbignew Brzezinksi as "Rice Bowl," the operation – which kicked off from *Nimitz*, *Coral Sea*, and other locations on 24 April – received the more formal title of *Eagle Claw*. Marine pilots flying eight Navy RH-53Ds launched into Iran from the deck of *Nimitz* and joined with Air Force C-130s at the refueling point in country. Meanwhile, both carriers armed, fueled, and prepared their aircraft – marked with distinctive black-orange-black wing stripes for identification purposes[2] – for possible action against Iranian targets. If the rescue forces had to fight their way out of Iran, the Navy would be more than ready to oblige, and there was little doubt that *Intruders* would have been attacking targets well into the country.

Unfortunately, the rescue attempt degenerated into a major fiasco. Several helicopters suffered mechanical problems enroute, reducing their numbers; at *Desert One*, the refueling and rally point inside Iran, mission commander Col. Charlie Beckwith – a legendary Army Special Forces officer and creator of Delta Force, the Army's new counter-terrorism unit – engaged in a spirited discussion with his superiors in Washington DC over whether the operation should continue. Many of the participants felt they could continue with the five RH-53s that were still flyable; others felt they could not press on safely into Tehran. After a lengthy debate President Carter and Secretary of Defense Harold Brown decided to go with Beckwith's decision to abort and ordered the mission scrubbed. During the refueling and egress, however, one of the *Sea Stallions* collided with a C-130, resulting in an explosion and fire that killed eight U.S. servicemen and injured another five. The survivors destroyed the remaining helos in place and departed Iran in the other C-130s.

The disaster created a major international relations black eye for the Carter administration; post-mission analysis questioned both the attempt to run the op from Washington DC – which recalled actions of the Johnson Administration during Vietnam – and the apparent "everyone gets a piece of the action" approach to the joint operations, ie., Navy helicopters, Marine pilots, Air Force transports, and Army troops, and overall control. A Pentagon panel led by retired Adm. James L. Holloway III later identified 23 areas "that trouble us professionally –

---

[2] Both *Nimitz* and *Coral Sea* had aircraft that were so marked. The stripes were applied to help tell US aircraft from Iranian- who also operated *Phantoms* and *Tomcats*.

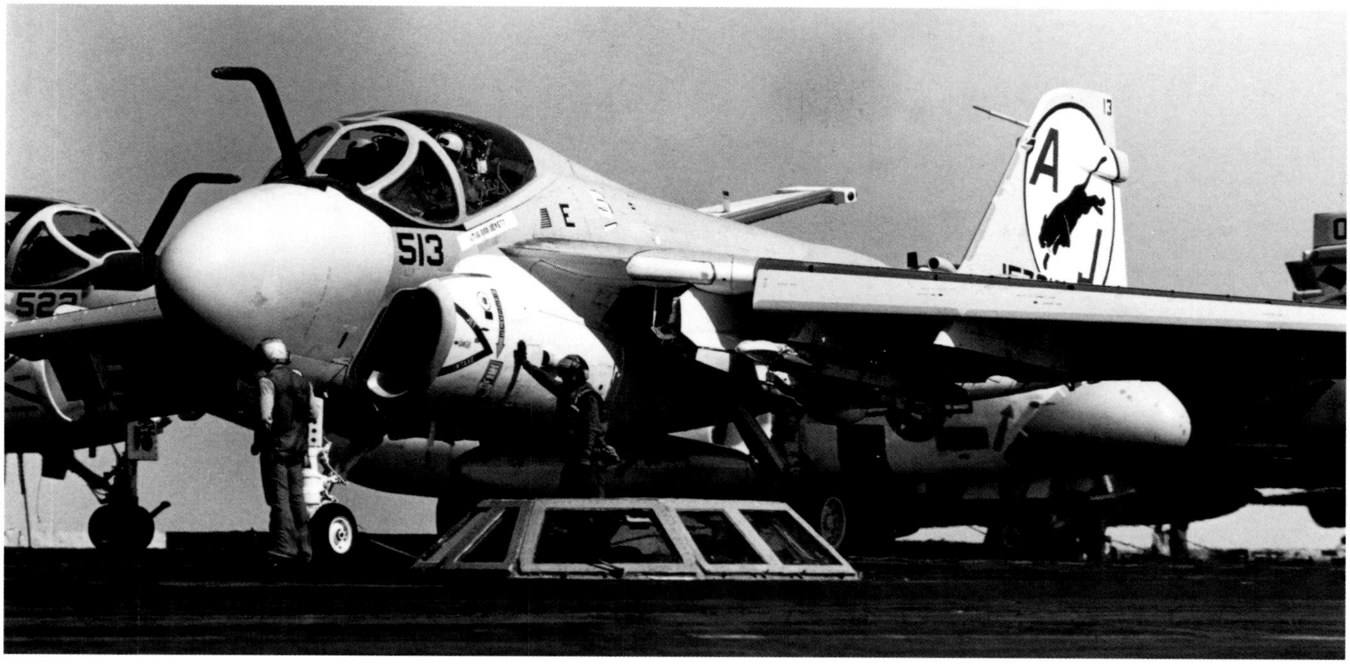

Wearing the *Black Panthers'* classy markings, a VA-35 A-6E taxis up to *Nimitz's* cat one during the ship's 1980 deployment. The *Panthers* spent almost 20 years with CVW-8 while operating from three different carriers. (U.S. Navy, PH2 James Brown)

Areas in which there appeared to be weaknesses." On the plus side, the investigative panel announced that the "helicopter crews demonstrated a strong dedication toward mission accomplishment by their reluctance to abort under unusually difficult conditions." President Carter later assumed full responsibility for the failed rescue attempt, Beckwith blamed the Marine helicopter drivers – commenting they, among others, lacked "the guts to try" – and the Congress eventually mandated the establishment of a joint special operations command to prevent further disasters.

On the carriers, the crews heard the events and post-mission finger pointing unfold with anger and disgust, mirroring the national response. Post-mission photo and television coverage of Iranian Mullahs kicking around American servicemen's bodies at the accident site did not help the national image of its government and military. With the hostages still firmly in Iranian control – and violent anti-American demonstrations a daily staple on the evening news – the Carter Administration licked its wounds, while carriers from both fleets made the long trip to the Indian Ocean, assuming the duty from one another after extended periods at sea with no liberty but the occasional single can of beer for each crewman.

Coral Sea remained in the northern Indian Ocean for an additional week following the desert disaster. On 30 April *Constellation* – again with CVW-9 and VA-165 – relieved the *Coral-Maru* after 89 days on station. However, while enroute back to Alameda *Coral Sea* diverted to the Sea of Japan, where she joined *Midway*. The massacre of several hundred people in the city of Kwangju had demonstrated once again the need for carrier air power being in several places at one time. Once the crisis subsided – problems flared in Korea again the following year – *Coral Sea* was finally dispatched home. The carrier completed her seven month cruise on 11 June 1980.

As for *Connie*, she "enjoyed" an auspicious stay in the IO and NAS, which included 110 days underway time. While engaged in an underway replenishment in mid-May she collided with the merchant vessel *Banglar Joy*. Somehow the civilian managed to get among the Navy vessels and whacked the carrier underneath the angle on the port side. Fortunately, no one was hurt, and the damage was minimal. At the end of July *Constellation* detached from Gonzo and started back for relief by *Midway*. Regrettably, this time *Midway* went bump in the night, colliding with a merchant; she turned back to Subic, unloaded damaged aircraft, and started repairs. *Connie* pulled into Singapore for liberty while the Navy figured out what it was going to do, then went right back to the western IO for one last tour. She finally concluded her 7 1/2 month deployment at NAS North Island on 15 October 1980.

The rotation of AirLant units continued while their AirPac counterparts bounced off the odd rogue merchant ship. On 8 May 1980 *Dwight D. Eisenhower* – making her second deployment since commissioning – arrived and relieved *Nimitz*. *Nimitz* started back for Norfolk after 115 continuous days at sea, but *Ike* subsequently broke the record. Her IO deployment with CVW-7 and VA-65, which until 22 December incorporated only one brief port visit for liberty purposes, was a staggering 246 days underway.

The rotation continued, but all efforts towards Iran were limited to diplomatic activities, and they seemed to achieve nothing but more vitriol. Back in the states most folks were preoccupied with the upcoming presidential election, which saw California Governor Ronald Reagan challenge President Carter. Governor Reagan ran on a platform of military preparedness and bringing honor and pride back to the American people, among other items. Above and beyond his background as an actor, detractors claimed the man was dangerous and would probably involve America in a war with the Soviet Union. It made for a spirited campaign.

Slightly on the plus side, the *Intruder's* systems continued to demonstrate capabilities in areas not even dreamed of by the designers. Between March and May 1980 VA-128 used its TC-4Cs to monitor hot spots on a peak in the Washington Cascades named Mount St. Helens. Working with the U.S. Geological Survey, *Golden Intruders* crews used the *Tick's* FLIR sensor to monitor St. Helens' impending eruption, keeping track of heat patterns and determining the probable areas of greatest danger. Their work enabled the government scientists to provide warning to surrounding communities.

VA-128 skipper Cmdr. Daryl Kerr was flying with four scientists from the University of Washington on 18 May when the mountain blew up; he recalls it was:

> "Something else. After a while I recommended we return to Whidbey Island, as the volume of smoke and ash in the air was causing some safety of flight concerns."

Two eruptions followed St. Helens' initial explosion, which killed 60 and caused an economic loss estimated at nearly $3 billion. The devastation was massive, but -128's efforts helped mitigate the effects.

On the other side of the world carriers came and carriers went. On 17 August 1980 *Midway* relieved *Constellation* in the IO, and was in turn relieved by *Ranger* on 6 November, having spent 157 days on the line over two deployments. *Ranger's* arrival with CVW-2 marked its first trip to the theater in four years, but the *Swordsmen* would come to know the region well.

On Tuesday, 4 November 1980, Ronald Wilson Reagan defeated President James Earl Carter by over eight million votes, taking 489 electoral votes to Carter's 49. One month later, on 8 December, *Independence* – with CVW-6 and VA-176 onboard – relieved *Eisenhower* in the IO.

Resolution in Iran, Problems Elsewhere

On Inauguration Day, 20 January 1981, Ronald Reagan took the oath of office as the 40th President of the United States of America. That evening the President announced Iran had released all of the hostages – including the 12 active duty Navy and Marine Corps personnel – after 444 days of captivity. Following medical and other evaluations in Germany the hostages returned to the United States to resounding cheers and celebration. As part of the agreement between the two nations to free the hostages, the U.S. had agreed to free up $8 billion in Iranian assets seized when the hostages were taken; Iran needed the money desperately, as it was now involved in a shooting war with Iraq.

At least the LantFlt carriers did not necessarily have to make the long trip around the Cape of Good Hope anymore. On 4 May 1981 *America* transited the Suez Canal, making it the first aircraft carrier through "the Ditch" since *Intrepid* performed the feat during the Six Day War of June 1967. Notably, *Carrier Air Wing 11*, with VA-95, was making the cruise, the second of two for the west coast air wing with east coast carrier. Later in the month *Independence* – with CVW-6 and VA-176 embarked – repeated the trip, albeit heading northbound this time. The Israelis had attacked Syrian SAM sites in Lebanon, and the U.S. wanted two carriers in the Mediterranean again. The contretemps between Syria and Israel – with the various Lebanese factions fighting in the middle – continued well into the year.

In and around charging from one crisis point to another, the Navy was about to undergo major changes. John F. Lehman, the new Secretary of the Navy, was pushing for more ships, more aircraft, and enough personnel to enable a "600-ship Navy." Indeed, the Reagan Administration's budget proposals supported expansion for all of the services; for many observers the dark days of the post-Vietnam/Carter defense department appeared to be ending.

One component of Lehman's plans for his services – he'd added "Secretary of the Marine Corps" to his title – involved taking the fight to the Soviet Union when and if necessary. Hence, over the following years carriers and their air wings found themselves operating more often in both the North Atlantic and North Pacific, close to Soviet bastions. The Soviets reacted as could be expected, while pundits predicted disaster at sea. The medium attack crews, however, dressed warmly, knuckled down, and pursued training centered on showing the flag and taking offensive action, not reaction.

Cmdr. Dan Wright, who assumed command of VA-85 in August 1980 from Ron Zlatoper, was later assigned to *Carrier Group Eight* as Air Operations Officer and got to see the change in the Navy's direction from both the command and operational staff levels. Reflecting on this period, he says:

> "In retrospect, we always gave the enemy more credit than what they were actually capable of doing. During the War at Sea (WAS) period it was amazing we didn't kill more people than we did. We'd do EMCON recoveries, go out 50 miles from the carrier with the tanker, and plan on a rendezvous. The time comes, no aircraft, 'well, we're obviously not where *we're* supposed to be, or they're not where *they're* supposed to be. Now, where's the ship?' We'd feel our way back in ... 'Hey! There's a bunch of planes, this must be the place!'"

There were occasional hiccups concurrent with the rapid growth of the Navy and Marine Corps, but notably the A-6 community suffered only two crashes during 1981. It marked the lowest yearly total – encompassing both peace and war – since the community's establishment 18 years earlier. Given adequate funding, equipment, and training, the medium attack responded well. Still, on 2 January 1981 a VMAT(AW)-202 A-6E crashed at the Dare County Range in northeast North Carolina; while making a run the *Intruder* rolled left and then plunged into the ground while performing low altitude deliveries. Sadly, the crew of replacement pilot 1st Lt. Chris Taylor and instructor B/N Capt. Doug Doran was killed.

Four months later, onboard *Nimitz*, VA-35 personnel found themselves responding to a much greater aviation accident, one that did not directly involve their *Intruders*. *Nimitz* had completed a short, two-month Northern Atlantic deployment in October 1980 as part of the ongoing process of tweaking the Soviet Bear. On 26 May 1981, during workups for a planned Med cruise, disaster struck: an EA-6B *Prowler* from VMAQ-2 at Cherry Point crashed into parked aircraft on the carrier's deck while trying to land. The flight deck erupted in flames and explosions, and three of the crew – Capts. Steve White, Myers Armstrong, and Larry Cragin – died, while others on deck were severely injured. Working with other members of *Carrier Air Wing 8* and *Nimitz*, numerous *Black Panthers* responded, fighting fires, clearing away wreckage, and tending to the wounded. Fortunately, no VA-35 personnel were hurt in the crash and ensuing fire; the Navy later recognized 130 *Panthers* for their actions that night. The damaged *Nimitz* returned to Norfolk for repairs, and eventually deployed on 3 August.

By that date the duty carrier in the Mediterranean was *Forrestal*, which had deployed from Naval Station Mayport on 2 March. From May through June *FID* operated from a point off Lebanon, keeping an eye on the ongoing confrontation between Israel and Lebanon. After leaving that post in early July the carrier moved west, towards Sicily. There, on 6 July 1981, the *Intruder* community suffered its second and final loss of the year; the recipients were *Black Falcon* CO Cmdr. Dan A. Wright and his pilot, Lt.Cmdr. Gary L. Stubbs.

As background, Wright points out that he was a rarity for the period: a west coast B/N commanding an east coast squadron:

> "I'd left -196 and gone to FICPAC (Fleet Intelligence Center Pacific) at Pearl for 18 months, and then screened for command. I was the only guy west of the Mississippi to screen, and I wanted to go back to Whidbey. They told me, 'No, we don't have any seats for you there.' Funny ... they took Pat Hauert from the Test & Eval force in Norfolk and gave him VA-95, and they took me from Hawaii and put me in Oceana. It was interesting being put on the east coast. With Doug Griffith – who was going to VA-176 – we were the bastard stepchildren of the west coast. I went to VA-85, we went to the Med, and we really didn't follow our schedule. Things kept getting switched around. It changed three or four times, but it was a good cruise otherwise."

His squadron would sustain an accident during his second *Forrestal* cruise, in July 1981. While operating near Sicily their aircraft suffered a dual-engine flameout. Attempts to restart the engines failed and, at the appropriate moment, the skipper and his driver stepped out.

"We'd just been in port," Wright comments:

> "And were the third launch. The plane we were flying in had been on the first launch. They put it on the bow and started refueling it, but didn't complete the refueling. They then moved it back to the rear, parked it next to a KA-6D, unplugged the tanker, and started refueling us. As it turned out, we took on something like 35 gallons of sea water-contaminated fuel. We were out there bombing the spar, and hey, we've got a problem. We requested a straight-in, as we had a rough-running engine. We were placed in high holding, then low holding, then Gary said, 'Hey, we have another problem here. The engines won't come up to speed.' He tried it on one engine and – bang, bang – we lost the engine. 'Skipper, we have another problem; I can't get the right engine past 80 percent either.'
>
> "We cleared off the racks and started getting more help than we could've possibly used. A couple of aircraft took station on us, including an S-3. Then the call came from the ship: 'go to Sigonella. Fellas, we have a 16,000-ft mountain betwen us and Sigonella, and there's no way we're going to clear it.' We kept getting lower and slower, but stayed with it until we hit 150 kts, at which point Gary lost control of the aircraft and ejected. I screwed my back up again."

Fortunately, the damage to Dan's back was not too extensive, and he was able to return to flight status within short order. He notes that during the subsequent search for answers concerning his *Intruder's* loss, a few interesting facts turned up:

> "The maintenance chief came up to me and said, 'Hey skipper, look at this.' We had an aircraft with a ruptured fuel cell. We looked inside and found salt crystals all over the place. It turned out several other engines in the squadron had been having problems with surging. So the question became, where did the fuel come from? Apparently it all came from the same fuel cell on *Forrestal*, the one we always used to fill the tankers. Then the report came in and said 'crew error,' ie, the engines must've FOD'd on takeoff, and it was our fault because we didn't refuse the launch. I was livid and fired a response back; the carrier's skipper had to calm me down.
>
> "End of story: the fuels chief went to check out *Forrestal* – it was one of the older carriers with the fuel lines mounted outside the hull – and found the JP-5 line on the exterior of the ship was riddled with holes. When the carrier rolled a certain way while underway seawater would enter the line. We called in Pratt & Whitney, and they confirmed salt water contamination would do the damage we'd experienced."

Apparently the damage to the JP-5 feed line was not detected during *Forrestal's* SLEP (Service Life Extension Program) at the Philadelphia Naval Shipyard. According to Wright, such problems were regrettably common, adding:

> "They had some real disasters during that whole period of SLEP. USS *Saratoga* had similar problems when it came out of the yard."

Still, *Forrestal* was deployed and capable of continuing flight operations, which was a good thing. When *Nimitz* arrived Sixth Fleet decided to hold an exercise that required the presence of both carriers and their assigned battle groups. The activity was a Freedom of Navigation (FON) exercise in the Gulf of Sidra, north of Libya, and involved air operations and missile ShootEx's.

The op followed an extended period of *Burning Sand* missions, ie, electronic reconnaissance by land-based Navy and Air Force aircraft of Libyan emitters and communications. These missions had already served to rile the leader of the Socialist People's Libyan Arab Republic, Col. Muammar al-Qaddafi; the American aircraft also regularly drew the attention of the Libyan Air Force, the *Al Quwwat Al Jawwiya Al Libya*. As a result, the reconnaissance aircraft required escorts, which the carriers typically provided.

CVW-17 in *Forrestal* had drawn the duty on its previous cruise, according to Dan Wright:

> "On the second cruise on *Forrestal* we'd done a couple of *Burning Sands* missions escorting EP-3s out of Rota and Air Force RC-135s out of Greece while they worked off Libya. We had done a couple the first cruise and were supposed to do a *bunch* the second cruise. We had F-4s aboard (VF-74 and VMFA-115), and I told CAG I needed a fifth tanker, and I wouldn't give up a bomber for it. The first time out we pumped something like 30,000-pounds of fuel. We kept one tanker in Sigonella when we weren't flying *Burning Sands*, but the rest of the time we needed that fifth tanker."

Col. Qaddafi took great exception to the reconnaissance flights, but was particularly adamant about any planned surface exercise in the Gulf of Sidra. He claimed the Gulf – which stretched roughly from Zâwiyat-al-Baydâ in the east to Tripoli in the west – as Libyan territorial waters. The U.S. Navy intended to show otherwise. Qaddafi, suspected of links to terrorist organizations, threatened the United States with a grave response if it came into his nation's waters.

He lost, or at least two of his pilots did. On 19 August Libyan Su-22s engaged two VF-41 *Tomcats* flying CAP off *Nimitz*. The *Black Ace* crews responded appropriately and knocked the two Sukhois down, marking the Navy's first air-to-air victory since the VF-161 shootdown of a North Vietnamese MiG-21 12 January 1973. Wright recalls what ensued for the U.S. Navy units was:

> "... absolute chaos, with ships rushing in. See, one of the things that happened at that time – on the east coast – they'd take any flag officer and put him in charge of these multi-unit battle forces. They weren't always well versed on carrier aviation. I had *Standard ARM* shooters and told him we could go out against the patrol boats, keep an eye on them, determine what (radar systems) they were using, and take out their radars if necessary. No, they kept putting us over a *Foxtrot* submarine. Some of the staff never understood that. The other thing – and it's interesting from the A-6 perspective – is we knew they might do something. What should be our response? The typical thing involved the whole air wing. I briefed Adm. Easterling and told him if we tried that we'd lose a lot of aircraft. 'Okay, what would you do?' Five A-6s in low, at night, pop over the hill and blast the target. Perhaps two EA-6Bs as stand-off jammers, and a couple of CAP stations to de-louse."

However, the Libyans backed down under a cloud of propaganda and threatening pronouncements, and the exercise eventually concluded. The reaction in the United States was mixed, with liberals expressing outrage and fear over the Navy's "provocative" act, while making similar comments about the President and his approach to international affairs. Conversely, there was great celebration in the fighter community. More quietly, many analysts and military personnel took note of one major point: the participants in the Gulf of Sidra FON exercise – aerial, surface, or whoever – operated under a firm ROE that allowed self defense without first calling back to the fleet commander and Washington for permission. For those that remembered Vietnam the change was startling, and served to boost confidence and morale.

Having participated in operations at both ends of the Med – including a consecutive 54 days off Lebanon – *Forrestal* departed in the fall of 1981 and returned to the U.S. on 15 September via a NATO exercise in the Northern Atlantic. It had been an interesting cruise for all concerned *Black Falcons*, including the skipper, who had survived his second *Intruder* ejection. On 10 December 1981 XO Cmdr. John I. "Jack" Dow fleeted up to the command slot, and Dan Wright departed for other challenges, including the CARGRU 8 tour. As with many former COs, he rated squadron command as the highlight of his career:

"It was a good squadron with great people. I used to say, you can give me all the parts and all the aircraft, but it doesn't matter unless I have good people. The big thing then is to let the people do their job. We had a new guy – Jerry McWithey – a fresh-caught JO. I made him nuke officer, and he ran with it, a perfect example. We did other things like putting puzzles in the family grams for the kids back home. We also held regular captain's call in the station theater with the squadron MCPO, all the wives, the E's, the chiefs, and the officers. The people I had were just superb."

Wright adds one other event really stood out from his tour as *Black Falcon 1* that helped him feel his tour was worthwhile; if nothing else, it demonstrated a good relationship between the skipper and the troops.

"We had one plane that would not release any ordnance," he says:

"We'd tried everything, and it just wouldn't work. Naturally, it always checked good on deck, no problem, then we'd fly and it wouldn't drop. So I told the maintenance guys to get it in the hangar and set it up like it was in flight and try again. I went down there, and it was on jacks and dropping just fine, and my guy leaned out of the cockpit and said, 'See, skipper? We *told* you there wasn't a problem.' I yelled back, 'How often do you fly with the cockpit open?' They closed the canopy, and it wouldn't drop the ordnance; it turned out to be some sort of short in the canopy circuit. Amazing."

The times were changing for the fleet, and would continue to improve into the foreseeable future. However, the positive times would also come with other challenges – and combat – in several corners of the world. The fighter pukes could justifiably puff out their chests, but the truth was future endeavors would focus on attack aviation, preferably of the medium/all-weather variety. John Indorf put it thusly, concerning the last phases of VA-52's 1981 deployment in *Coral Sea*:

"A couple of us ship's company A-6 bubbas were manning the rail on the starboard side of *Coral Sea* as she made her first approach to the Subic side pier following the change of command, where one of our *Intruder* heroes, Capt. Dick Dunleavy, turned command over to Capt. Jerry Johnson, an A-7 driver. We commented how the bridge was overflowing with light attack weenies, ie., the skippers, strike ops, VA (light attack) COs and XOs, assistant navigator, and a couple of the shooters. On the pier to greet these former shipmates was now COMUSNAVPHIL, Rear. Adm. Dick Dunleavy.

"It was apparent that the 'light attack' bridge team lost control of the vessel in the steady wind. The tugs couldn't keep us from clobbering the pier. With screaming go-go-dancers jumping off their platforms, the Subic band running for their lives, crusty Subic CPOs cussing the inept *Coral Sea* shiphandlers, and wood splinters raining down on the Admiral (who was shaking his head in disgust and ducking for the safety of his staff car) the A-6 bubbas had the biggest laugh of the deployment.

"This proved, once again, that when you have a tough job to do, don't send light attack ... send the *INTRUDERS*!"

Intruder in the Weeds

The Grumman A-6 *Intruder* was built to carry a lot of bombs swiftly down low. The design's side-by-side seating developed a close-knit bond between its crew that created a powerful weapon when properly used. Its big, thick wings gave it a stable ride at low altitudes, and the type's crews trained for the mission both day and night. Low altitude provides surprise, removes a lot of surface to air missiles from the picture, and allows the use of terrain masking to hide strike aircraft from radar. If at high altitude the *Intruder* was vulnerable to enemy aircraft, few air-to-air missiles of its day would work at 200-feet, and even fewer enemy fighter pilots were willing to attempt a gun run when collision with the ground was a distinct possibility.

Low altitude flying is an intense, physically taxing environment. But flying low at high speeds has its own hazards. Low-level flight provides intense training, and things other than the ground could pose danger: power lines, light aircraft, and birds could all ruin your day. As they say, the record for lowest flight can only be tied – never exceeded. During the three years between May 1984 and May 1987, thirteen *Intruders* were lost while conducting tactical flying in the low altitude environment. At Whidbey low-level losses included two aircraft with both crew within a week while flying night terrain clearance missions on the IR-344; their impact points were within a few miles of each other. In late 1991 a VA-155 aircraft crashed into the Columbia River while executing a hard turn on the VR-1351 route. The pilot, a highly regarded *Desert Storm* veteran, died with his B/N, who was a recent RAG grad. A VA-95 aircraft had a mid-air collision with a Grumman Ag-Cat crop duster in 1993 near Colfax, WA, while on the IR-340. Both aircraft were destroyed, the Navy crew ejecting, while the civilian pilot was critically injured when his bi-plane hit the ground.

Many mishaps were ruled "crew error," with disaster resulting from either bad judgment or a moment's inattention. Impact with the ground has a probability of kill (Pk) of almost 100%, which is far higher than enemy defenses. Nonetheless, there were few training missions more highly sought after by aircrew than low levels. They are that challenging, and fun, to fly. An aircraft running at 420kts covers seven miles a minute. Navigation is crucial – a minor error in heading will place the aircraft miles off course within minutes at that speed. A mile off track could mean all the difference in the world when it came to finding a target or avoiding defenses. In peacetime it could mean a flight violation. It could be life or death in combat.

The United States is covered with Federal Aviation Agency (FAA) approved Military Training Routes (MTRs), which are the officially sanctioned low level routes that aircraft can fly on. As far as tactical jets are concerned, there are two flavors. The IRs are Instrument Routes, which require an IFR clearance. These can be flown either in good weather or bad, in day or night, assuming the crew is qualified and the aircraft has the proper equipment in working order. While frequently flown in day/visual conditions, the IR routes were also used by A-6 crews for night terrain clearance training missions.

VR s (Visual Routes) are flown in good weather, when Visual Flight Rules (VFR) can be maintained. Most of the routes allow airspeeds up to 480kts or higher. Altitudes are normally allowed down to 200 to 300 feet AGL, although some portions will have higher minimum legal altitudes to avoid specific areas, such as towns or wildlife areas. Not that *Intruders* couldn't go lower, as in some military training areas operations down to 50 feet or less have been recorded. In combat, experienced crews were expected to fly at their "comfort level," which was as low as they felt confident while still retaining the ability to find their target and hit it.

On the east coast, Oceana and Cherry Point crews typically used a variety of routes through the Piedmont and Appalachians, as well as the coastal plains. North Carolina's Great Dismal Swamp's low, marshy ground swallowed several *Intruders* over the years, almost always with the loss of their aircrew.

The wing at Whidbey Island "owned," or scheduled, a variety of routes in Washington, Oregon, and Nevada. Almost all involved 20 to 30 minutes of low altitude work, and most terminated in a bombing range, either at Boardman or in Fallon, NV. Terrain varied from the peaks and valleys of the Cascade Mountains to high desert plateaus. A "classic" Whidbey low-level flight would include an IR or VR to the Boardman range, where practice bombs would be dropped, and a return via the VR-1355 route,

A *Thunderbolt Intruder* carries out a low level over the desert with a load of inert 500-pound bombs during the early-1980s. (U.S. Navy)

which flew up the spine of the Cascade Mountains through some of the most spectacular scenery in the country. It was the kind of flight that Whidbey aviators fought for.

Evolution of the low altitude threat during the 1980s largely changed the tactics and reliance on low altitude penetrations. In Vietnam the threat of SA-2s and MiGs at higher altitudes made the low route the preferred method of attack. Most experienced *Intruder* crews in Vietnam preferred the lone night low level attack to mass gaggle "Alpha Strikes" where they were mixed in with other aircraft in daylight.

The incredible proliferation of anti-aircraft guns and man-portable, infra red (IR)-guided SAMs, as well as the ability of defense suppression aircraft to keep radar SAMs at bay have made the use of low altitude ingress not nearly as safe as it once was. In *Desert Storm*, three *Intruders* were shot down or damaged in the first few nights conducting low altitude ingresses (a fate shared with coalition *Tornadoes* and B-52s). By the end of the war's first week practically all operations were carried out at medium to high altitudes, 10,000 ft and higher.

None the less, the requirement to "train low" still exists for the *Intruder's* successors, partially because the quality of training cannot be matched by high altitude flight, and it also keeps a very important tactic in a combat pilot's "bag of tricks," as they face a new generation of SAM systems that may challenge our use of the high altitudes in the future combat environment.

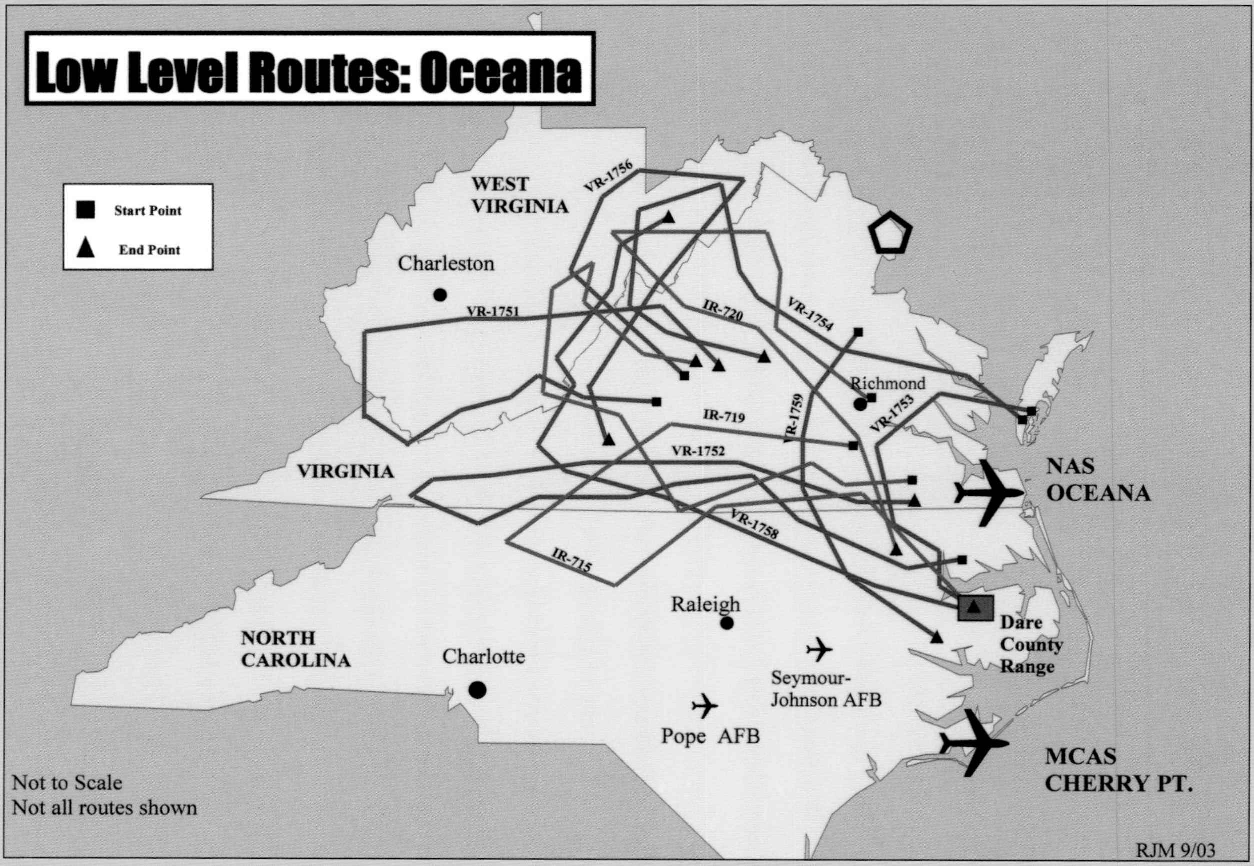

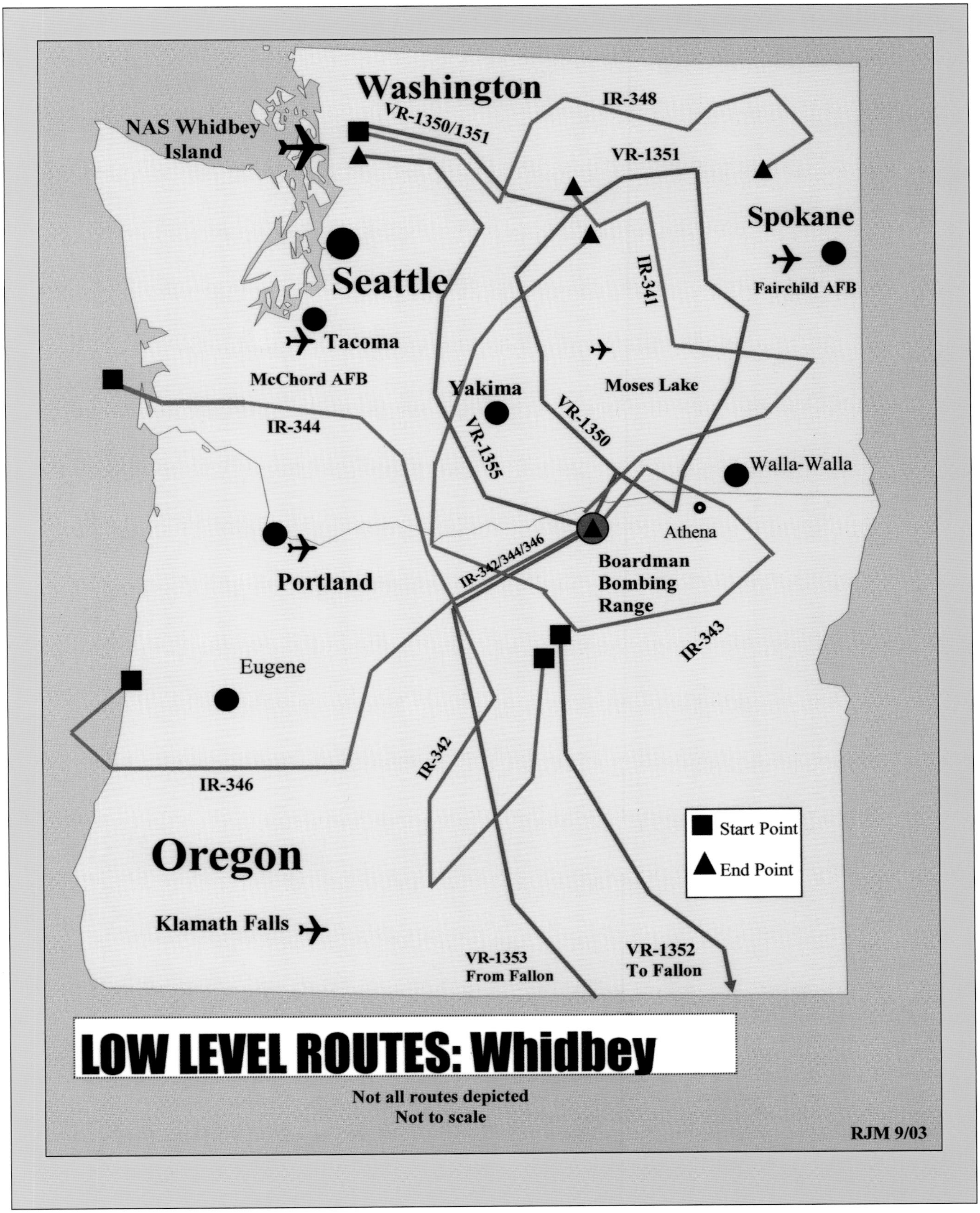
Washington
NAS Whidbey Island
VR-1350/1351
IR-348
VR-1351
Spokane
Fairchild AFB
Seattle
IR-341
Tacoma
McChord AFB
Yakima
Moses Lake
VR-1350
IR-344
VR-1355
Walla-Walla
Athena
IR-342/344/346
Boardman Bombing Range
Portland
IR-343
Eugene
IR-342
IR-346
Start Point
End Point
Oregon
Klamath Falls
VR-1353 From Fallon
VR-1352 To Fallon
LOW LEVEL ROUTES: Whidbey
Not all routes depicted
Not to scale
RJM 9/03

CHAPTER SEVEN

Small Wars, 1982-1985

On 20 January 1982, the United States marked the completion of the Reagan Administration's second full year at the helm. For the American military it had been a good 24 months; aircraft and ships were on order, pay and benefits were improving, and good people once again considered service life for a career. Buoyed by the enthusiastic and upbeat President, an activist Secretary of the Navy, and dramatically increasing defense budgets, U.S. serviceman were finally beginning to turn their back on the dark days of the mid-to-late 1970s. However, while fortunes and morale improved, the world of the early 1980s was still a very dangerous place. The Marines helped hammer home the point in February when the service lost two *Intruders* in short order.

On the fourth MAG-12 CO Col. Ken Bateman and VMA(AW)-533 skipper Lt.Col. Charlie Carr had to eject out of their *Hawks* A-6E after takeoff at Iwakuni. According to Carr, the hop started normally enough and then quickly deteriorated:

"If you've never flown out of Iwakuni, one of the standard departures out of there is the Yama 2. You take off to the south, make a left turn almost right away to 5,000-ft, then make a right turn to the intercept heading. It was a dog shit day, and we were leading a couple of sections down to the bombing range. One of the pilots showed up not feeling too well, so I called the group commander, Ken Bateman, and invited him along. We all briefed; he had to get his poopy suit on, so I told others in the flight, 'Take off and we'll catch up with you.'

Now, this aircraft had had a total pitot static failure on a previous flight, but supposedly it had been repaired," Charlie continues. "We took off and just about right away went into the clag on the Yama 2 departure, making the turns here and there. I started cranking up the radar, doing the usual housekeeping stuff. All of a sudden Ken said, 'Which way is up?' 'Ohhhhh ... man!' I looked over there, and the VGI was showing left wing down, while the VDI – which I could barely see – was showing the opposite. We'd been in the clag making turns for, oh, about a minute or two. Right away I thought it was a frozen VDI. I slammed it into MAG-VGI, said follow the VDI, he started to pick up the left wing, and the wing appeared to stabilize. By now I'm really scanning the instruments – the VSI showed something like 1000 to 1500-feet up – then all of a sudden it started to decay a bit. Ken pulled the nose up a bit, then the aircraft went into something that felt like a pre-stall shudder. Immediately the VSI went from 1500-feet up to straight down – it pegged the meter – and I looked at the VGI and I see all kinds of black, so I yelled '*eject, eject, eject.*' Out we went."

The two men punched out into the clouds and fortunately got clean chutes, although they were knocked around a bit. Col. Carr says he realized something was wrong when he looked down at his right hand and noticed one finger hanging by a flap of skin:

"We went out at about 1,000-ft, at maximum speed downwards and right on the edge of the envelope. I saw Ken's chute coming down, then I looked at my finger and said, 'Oh shit.' It was my right nose-picking finger, the index finger. I'd ejected before and got trapped under the canopy, and this was not going to happen again. I released the right koch fitting with my left hand because I couldn't use the right hand, then inflated my LPA. As soon as I figured I was close to the water I released my chute, and fortunately fell only about ten feet. I got caught in some shroud lines, but nothing serious. I started looking for Ken. It was misty, and the visibility was not good, so I got my radio out and started maydaying. Now, a guy by the name of Joe English was the ops officer at Iwakuni. Radar called him and said, 'We just lost one of your friends off the scope,' but they'd managed to pinpoint the location. It just so happened the 'Angel' (rescue helo) was preparing for a test hop, so they were sent out for us.

"I could hear them coming in, but man, there are a lot of *big* boats out there, and I know they can't see me, so I started swimming for this small island best I could. All of a sudden here comes the helicopter, breaking out of the scud, and Ken's behind me about two to three hundred yards and he lets a pencil flare go. I showed the helo my finger, they picked up Ken, then came back and put the swimmer in to rescue me. When we got back to Iwakuni my ops O came in, and I said, 'Get the XO back from Japan,' because I knew I was going to be in the hospital for a while. Actually, they told me they could send me anywhere I wanted, such as Tripler – the big U.S. Army hospital on Oahu – or some other place. I looked at them and said, 'Tell you what: why don't you cut off the finger, then use a cigarette lighter or something like it to cauterize the stump.' They just kind of looked at each other, then looked at me ... 'Uh, okay colonel ... ' They ended up sending me to a hospital in Japan where they attempted to reattach the finger. That night I was back in the hospital at Iwakuni.

"The next morning, Ken and I were in adjacent beds, and he looked at me and said, 'Next time you have a hole in your schedule, *don't* call me."

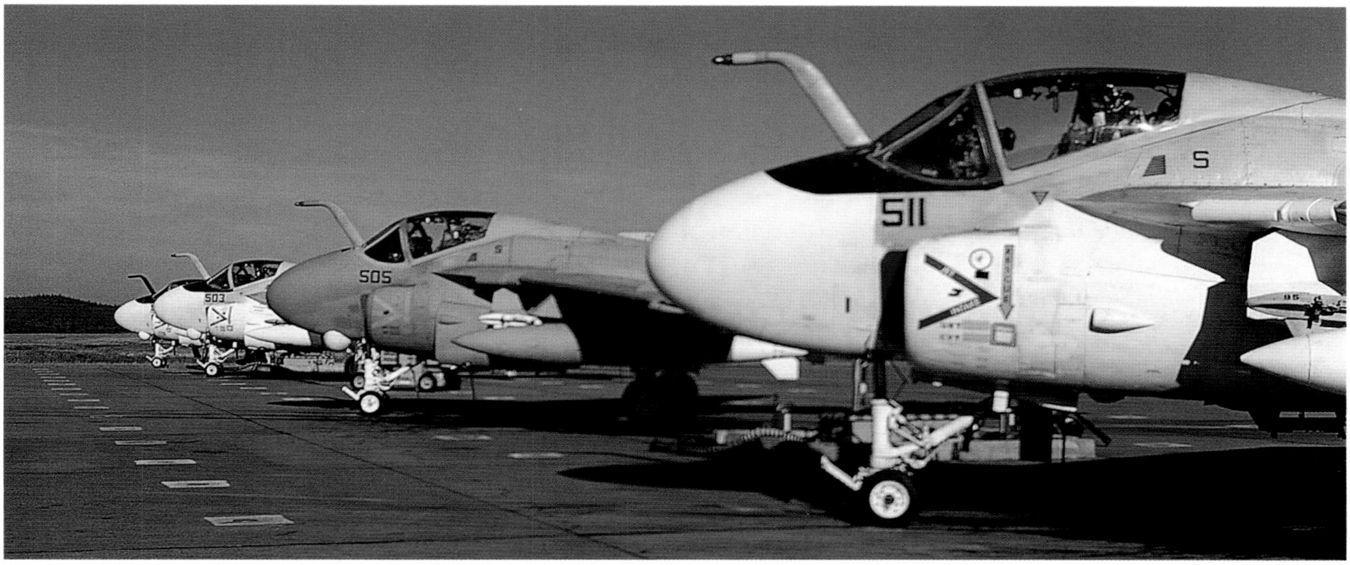

Swordsmen Intruders on the ramp at Whidbey, September 1983. (Rick Morgan)

Fortunately, neither man was more seriously hurt, above and beyond Charlie's lost finger. Eight days later, however, the outcome was different for Capt J.E. Bull and 1st Lt. J.J. Spegle. Their VMAT(AW)-202 A-6E crashed for unknown reasons while making a low altitude training flight about forty north of Jacksonville, NC. Neither Marine survived.

The two accidents were the only *Intruder* losses for the entire year, which otherwise proved to be rather hectic in the usual corners of the globe. The war between Iran and Iraq notwithstanding, the AirPac carriers and squadrons were "enjoying" a relatively quiet period, although extended underway periods in the Indian Ocean were still the norm. Their counterparts in the Med, however, were forced to step up the pace due to ongoing problems in Lebanon.

After the Israeli invasion of that tattered country in mid-1982 the U.S. dispatched a series of battle groups to keep an eye on things and assist with yet another evacuation from Beirut. *Dwight D. Eisenhower* went in first with CVW-7 and VA-65 onboard; *John F. Kennedy* (CVW-3/VA-75), *Forrestal* (CVW-17/VA-85), and *America* (CVW-1/VA-34) followed the big nuclear carrier to Bagel Station. The *Tigers*, led by Cmdr. Dickey P. Davis, and the *Sunday Punchers* under Cmdr. Earl Wolfgang supported the actual evacuation, which took place on 24 June.

August 1982 saw the establishment of a tenuous peace in Lebanon and the introduction into Beirut of several hundred peacekeepers, including 800 Marines. VA-85 and the other squadrons of CVW-17 flew protective cover for the Marines as they moved ashore and set up operations; concurrently, Israeli Army units left the environs of Beirut and moved to the southern portion of the country. Working with troops from other nations – primarily French – the Americans set up patrol areas and worked to keep the warring parties separate. However, the situation continued to deteriorate and took some startling turns along the way. In August the Lebanese parliament elected Christian militia leader Bashir Gemayel as president. The following month, barely four days from his inauguration, he was assassinated by a car bomb. In his stead Gemayel's brother Amin assumed leadership of the country and worked to consolidate Christian control of both the government and the Lebanese Army. Israel tacitly assisted his government, while Syria attempted to provide a counterbalance by aiding the Palestinian Liberation Organization and the other assorted Muslim factions.

The circumstances in country remained fluid, and on 22 September *America* – with Cmdr. Ben Liner's *Blue Blasters* onboard – was ordered to Lebanon from the Norwegian Sea. Along with the other components of *Carrier Air Wing 1*, VA-34 started flying surveillance and support missions as soon as it checked into Bagel Station; *America's* arrival bolstered the efforts of CVW-17 in *Forrestal*, which had been the only carrier in the vicinity since the departure of *Eisenhower* and *Kennedy* in early July. Ominously, aircrews from the two ships started mentioning sporadic ground fire while flying their missions; apparently, not all Lebanese were happy to see the Americans in and over their country. The Marines of the Battalion Landing Team in Beirut also started running into problems, ranging from noisy protests to rock throwing. Throw in the weapon of choice in and around Beirut – the car or truck bomb – and the long-term outlook was anything but promising.

Forrestal eventually departed the eastern Med and returned to Naval Station Mayport on 16 November 1982. For Jack Dow's *Black Falcons* the return to Oceana marked the completion of their twelfth *Intruder* cruise and eighth consecutive trip with the *FID* and CVW-17. For their next go around they would transfer to CVW-3 and *John F. Kennedy*; as for *Forrestal,* she would check into Philadelphia Naval Shipyard for SLEP. The transfer left *Carrier Air Wing 17* without either an assigned A-6 squadron or deck to operate from, but the Navy had already worked out a solution.

In December 1982 VMA(AW)-533 joined the wing, making it the second Marine *Intruder* operator to come under assignment to a Navy CVW. The *Night Hawks* worked up at Cherry Point with regular field qualifications at MCALF Bogue Field, and then challenged the "ramp monster," completing CQs in *Lexington* (AVT-16) in the Gulf of Mexico. Towards the end of 1983 -533 was assigned several KA-6Ds and mastered the air wing tanking mission; a few months later the *Hawks* and CVW-17 deployed in *Saratoga*.

Of Sandanistas and World Cruises

While the Reagan administration was tussling with the problems in the MidEast, its attention was also drawn to two locations much closer to the continental United States: Grenada and Nicaragua. In the Caribbean and Central America, Communism – as always happily supported by the Soviet Union and Cuba – was a growth industry, and a lot of people in Washington D.C. watched with growing concern.

Both countries had recently adopted communist governments, although, in typical fashion, not exactly in a manner that helped either population. Nicaragua was now run by the Sandanistas and their leader, Danial Ortega. In Grenada, the New Jewel Movement, under Maurice

Only two Marine units flew tankers, both to support carrier deployments. The *Nighthawks* were assigned several KA-6Ds for its 1984 deployment with CVW-17 and *Saratoga*. AA513 taxis at Pensacola in May 1983. (Rick Morgan)

Bishop, had overthrown the previous administration and thrown in its lot with Castro's Cuba. Both countries thereby received the attention of the American government. The details of how and why they had chosen the communist route were irrelevant to the men in Naval Aviation, although most aviators did realize that any possible military reaction would probably involve carriers and offer yet another chance to steam off exotic locales and plan for possible combat.

In July 1983 *Ranger* – with CVW-9 and VA-165 embarked – departed San Diego for Nicaraguan waters. After spending almost a month tooling around *Ranger* was relieved by *Coral Sea* – with CVW-14 and VA-196 – and headed west for a standard WestPac/IO cruise. *Coral Sea* then did its turn monitoring the Sandanistas until it too departed for the Western Pacific and Indian Ocean ... and then kept going, on an around-the-world cruise. The Navy had plans for the old carrier, including a limited overhaul – there was no SLEP for *Coral Sea* – and a new air wing. According to the skipper, Cmdr. Ken Pyle, it turned out to be quite a deployment, one of the most memorable he had ever made. He also rates the CarGru/CV-43/CVW-14 grouping as one of the tightest, most cohesive teams he has ever worked with.

"Working with all four fleets was pretty amazing," Pyle comments:

> "We had a very strenuous workup cycle, saw all kinds of weather all around the world, and then a great exercise against the *Carl Vinson* in the Arabian Sea. That old oil burner *Coral Sea* found the *Vinson* first, although they followed our recon back and hit us first. We landed the first blow in the fight, but I'm not sure we won the 'war.' Deception is always the biggest part of any war at sea exercise. This was one of the most well-planned, well-executed air wing/ship/staff teams I'd ever been associated with. Our captain, (Capt. Jeremy) 'Bear' Taylor, was intimately involved in making it happen. He was a rabid A-7 guy, but he wanted to win more than anything, and he always relied on the *Intruders* to 'get out and touch them.' I ended up working for him later in the Pentagon; he was a great leader and he had a great staff.
>
> "We went down to Nicaragua. That was different. There was no Order Of Battle that you had to worry about, so we spent most of our time just trying to figure out what we were there to do – other than rattle the sabers – and try not to do something stupid."

Once they departed Central America there wasn't a lot to see or do while enroute to the Persian Gulf other than fly. The carrier managed one liberty call in Singapore, but from there until the Med it was max underway and min sightseeing:

> "Between Singapore and Naples there wasn't anything," Pyle comments. "I think we got to go to Israel, but there was something going on in the western Med and we had to get rolling right away. We went to Naples for a couple of days to kind of get the ship spruced up, then spent the Fourth of July weekend in Cannes. We were something like the first carrier in nine years to pull in there for the holiday. My wife loves to tell a story. She met us over there – as did most of the wives – and it was the South of France in July, hard to keep the family away. It was a wonderful deal. I kept talking about how this wonderful Mayor DePuis had built up for the visit over the weeks leading up to our visit. He'd put the red carpet out and had all sorts of special ceremonies set up for us. The second day we had the reception for the staff and squadron COs and their wives. Someone announced, 'And now the Mayor of Cannes ... *Madame* DePuis. My wife elbowed me and said, 'See? It took a woman to put this on."

From the south of France *Coral Sea* finally made the final turn for home, exiting through the Straits of Gibraltar. Ken Pyle brought the worldly *Main Battery* back to Whidbey in mid-September ... following a 48-hour delay enroute for participation in the annual Tailhook convention in Las Vegas:

> "We won the McCluskey Award that year," says the skipper. "We flew the squadron off – everything except the four tankers – flew to Corpus Christi, went to China Lake, and then put about 20 squadron members into cars for the drive to Las Vegas. We were there to accept the award.
>
> "That was the kind of cooperation we got from 'Bear' Taylor and the air wing commander, Roy Cash. We had a lot of help getting that sort of thing done. It was just a great team. We had the old F-4s and the old A-6 and were the low-tech guys, but we worked really hard. That was probably the highlight."

To take *Coral Sea's* place in the Pacific Fleet the Navy dispatched *Carl Vinson* (CVN-70), also via the long way around. Commissioned on 13 March 1982, *Vinson* was the Navy's newest carrier and making its first deployment. She was also the first aircraft carrier named for a U.S. representative, in this case a Georgian who first entered the House in 1914 and did not retire until 1962. To be sure Congressman Vinson – who was still alive when the ship was christened on 18 January 1974, yet another first – was a staunch supporter of a strong Navy throughout his career in politics, but the naming of one of the nation's largest, most powerful, and most prestigious combatants for a living and largely obscure politician caused many a naval historian and traditionalist to grumble.

Whatever the circumstances of its name, the Navy was glad to have CVN-70, and following a series of workups, which included operations with CVWR-20, dispatched her on 1 October 1982. Onboard for the maiden cruise were AirPac's *Carrier Air Wing 15* and VA-52, late of *Kitty Hawk*. The *Knightrider's* deployment would mark their first time around the world, and also the first of six consecutive cruises in *Vinson*.

The lengthy deployment would prove to be a relatively quiet one, with only one major contingency operation towards the end. On 1 September 1983 Soviet fighters shot down a Korean Airlines Boeing 747 (flight No. 007) after it strayed over the heavily defended and highly secretive Kamchatka Peninsula. The shootdown killed 240 passengers and 29 crew, and led to serious diplomatic problems between the Republic of Korea and the Soviet Union. The incident also contributed to further poor relations between the Soviets – who claimed the ill fated airliner was actually on an intelligence gathering mission for the Americans – and the United States. With tensions in the region already high, on 9 October North Korean commandos attempted to assassinate South Korean president Chun Doo Hwan during a state visit to Rangoon, Burma. The bomb killed four members of Hwan's cabinet and 14 other South Koreans, but the president survived. The Republic of Korea quickly tracked down and caught the perpetrators, while the United States dispatched *Vinson* to the Sea of Japan. On 12 October *Vinson* departed the region and resumed her trek to the United States, tying up at NAS Alameda on 29 October. The first cruise for the Navy's newest carrier lasted eight months.

Enter the Warhorses

John Lehman's Navy was still growing by leaps and bounds; as one Naval Officer later put it, "It was as if the money would never run out." Among other things, the Secretary's plans specified 600 combatants, with 15 carrier battle groups and four "surface action groups" centered on the reactivated *Iowa*-class battleships. To equip the carriers Naval Aviation needed more aircraft, weapons, personnel, and equipment, and "Prince John" intended to deliver.

One tactic was to build a series of new *Nimitz*-class carriers as a multi-ship buy, thereby saving money in the long run. Therefore, President Reagan's FY83 budget authorized two new CVNs: *Theodore Roosevelt* (CVN-71) and *Abraham Lincoln* (CVN-72), with long-lead funding for what would become *George Washington* (CVN-73). It marked the first time since World War II that the Navy ordered two carriers in a single fiscal year; total savings for the multi-ship order were estimated at over $1 billion. Notably, CVN-71 had been previously approved by Congress but vetoed by President Carter.

As of 1983 the Navy had fourteen carriers and thirteen carrier air wings in the regular service. *Coral Sea*, having moved to the Atlantic Fleet at the end of its 1982 deployment, needed an air wing. Carrier Air Wing 13 was established in March 1984 to join the ship, and was organized to what would be called a "Coral Sea Wing" configuration, with four new strike fighter squadrons[1] and a single, new A-6 unit.

On 7 October 1983 VA-55 stood up in ceremonies at NAS Oceana as the first new *Intruder* squadron since VA-95's establishment in April 1972. Adopting the name and rampant seahorse emblem of their predecessor, the *Warhorses* set to work under Cmdr. Stan W. Bryant.

The previous VA-55 was an old line attack outfit, dating to their establishment at NAAF Pungo, VA, on 15 February 1943 as *Torpedo Squadron Five*. Nicknamed the *Torpcats* and initially equipped with TBF-1 and TBF-1C *Avengers*, the squadron went into combat with CVG-5 in *Yorktown* (CV-10) in the summer of 1943. In March 1945 both squadron and air wing were onboard *Franklin* (CV-13) when "Big Ben" was taken out of the war by Japanese bombs.

The *Warhorses* of VA-55 were established in October 1983 as the first new *Intruder* unit at Oceana in 13 years. Joining the new CVW-13, the unit deployed in *Coral Sea* (CV-43). (Bob Lawson)

Postwar, the squadron operated from several stations in the western United States and Hawaii while flying a mix of TBM variants before redesignating as VA-6A on 15 November 1946. Still flying "Turkeys," it made one WestPac cruise in *Shangri-La* in 1947 and redesignated to VA-55 on 16 August 1948 at NAS San Diego.

Over time the squadron continued their association with attack aviation, eventually adopting the *Warhorse* name and cycling through several models of the AD *Skyraider*. In 1959, following two years in FJ-4Bs, VA-55 shifted to the A4D-2; it remained in *Skyhawks* for another 16 years, flying four different versions. The squadron made its first of nine Vietnam deployments with CVW-5 in *Ticonderoga* in 1964 and wrapped up with CVW-21 in *Hancock* on the famous "last Scooter" cruise of 1975. The *Warhorses* disestablished with stablemates VA-212 and VA-164 on 1 December 1975.

Bureaucratically, this edition of VA-55 could claim no connection with their notable predecessor – at least, officially. Whatever the circumstances or rules, the new *Warhorses* gladly took possession of the proud old name and famous emblem long identified with attack aviation. They were welcomed warmly by their fellow A-6 squadrons at Oceana.

However, any celebration of the recent growth in the A-6 community was muted barely two weeks later. On 23 October 1983 a truck bomb destroyed the building housing the Marine peacekeeping forces in Beirut, Lebanon, killing 241 Marines. Two days later, the United States invaded Grenada.

Pile on in Grenada

25 October 1983 the world awoke to find that the U.S. military had swept down on the obscure Caribbean island nation of Grenada. While the United States was stunned by the turn of events in the Middle East, the invasion news roughly 48 hours later absolutely flabbergasted many. Critics of both President Reagan and his military quickly savaged the decision to invade, referring to it as a knee-jerk reaction to the disaster in Beirut. In truth, the operation in the Caribbean had been a long time coming.

The Reagan administration had been watching events on Grenada with growing concern for some time. The country received military aid from the Soviet Union and Cuba since 1979, and now they were building an airfield that U.S. analysts determined could easily handle Soviet

[1] Only two new Navy Strike Fighter units would make the first cruise- VFA-131 and -132. The other two intended squadrons, oddly numbered out of sequence VFA-136 and 137, were delayed for personnel until 1985 and two Marine units, VMFA-323 and -314, took their place.

VA-176 participated in its first combat as an *Intruder* squadron in Grenada in 1983. AE507 is shown at Fallon during WepTraEx carrying a live Mk.84 2000-pound bomb in May 1980. (Michael Grove)

long-range aviation assets, including *Bears*. The country's increasingly Marxist government, led by Prime Minister Maurice Bishop's "New Jewel Movement," replaced the country's constitution with "People's Laws," shut down the opposition press, and started filling up the prisons. In addition, more foreign advisors and military equipment started appearing on the island. On 19 October Bishop was murdered by radical elements within his own government.

Grenada's neighbors responded by meeting in Barbados on the 20th to discuss the turn of events. The participants included representatives from Antigua and Barbados, Dominica, St. Kitts-Nevis, St. Lucia, Montserrat, St. Vincent, the Bahamas, Barbados, Belize, Guyana, Jamaica, and Trinidad and Tobago. In and around arguing over the need for military action they expelled Grenada from the Caribbean Community (CARIBCOM), imposed economic sanctions, and turned to the United States for help. A formal request for assistance went to President Reagan from the Organization of East Caribbean States (OECS) under the signature of Prime Minister Eugenie Charles of Dominica. Following discussions with the joint chiefs, President Reagan ordered the commencement of Operation *Urgent Fury*, to include participation by the OECS nations, but notably without prior consultation with Great Britain, which still maintained a Governor General on Grenada. On 24 October President Reagan notified Prime Minister Margaret Thatcher by cable that the United States was preparing to invade the strife torn island; by the time the message arrived at 10 Downing Street the first American special operations units were attempting to move onto the island.

Independence, with CVW-6 and VA-176 assigned, had departed Norfolk on 18 October enroute to a standard Med cruise. Once underway the carrier diverted southward and came under assignment to the joint forces preparing for the invasion of Grenada. For the *Thunderbolts* and other members of *Air Wing 6*, what had started out as just another deployment now turned into the very real prospect of combat, the first encountered by VA-176 since its 1966 Vietnam cruise with CVW-10 while flying A-1 *Skyraiders*.

Operation *Urgent Fury*

Urgent Fury kicked off on 25 October with the invasion of Grenada by over 1900 troops, predominantly American, but with the participation of OECS' military units, including personnel from Barbados, Dominica, St. Lucia, St. Vincent, and Jamaica. Notably, the latter nation had earlier undergone its own dalliance with the Cuban version of communism from 1972 to 1980.

The official purpose of the invasion was three-fold: to settle the political strife and loss of individual liberties afflicting Grenada; to help reestablish democracy; and to rescue the approximately 1000 Americans – primarily medical students – stranded on the island. On the 24th the Grenadan government – aware that something was up – had appealed to Great Britain to warn off the United States, but it was too late. Prime Minister Thatcher did respond angrily to the U.S.'s *fait accomplit*, but otherwise the British government sat out the proceedings. As for the Cubans already on the island, Fidel Castro ordered them to defend themselves, but otherwise make no effort to interfere with U.S. rescue operations. The murder of Bishop and his followers had horrified Castro, and he made it quite plain to Coard and Hudson that Cuba would not reinforce the small nation.

Led by SEALs and Army Special Forces and Ranger units, the allies quickly pumped in Marine and Army combat forces, including the *82d Airborne Division's* ready brigade. Aircraft from *Independence's* CVW-6 and the Air Force's *1st Special Operations Wing* provided the air cover, while multiple patrol squadrons and embarked LAMPS helo units kept an eye on the approaches to the island. Marine helicopters from composite HMM-261, embarked in the *Guam* (LPH-9) Amphibious Ready Group, flew combat support and assault missions alongside Army *Blackhawks* from the *82d Airborne* and *160th Aviation Regiment*. To ensure non-interference by Cuba, the Air Force based several F-15s from the *33d Tactical Fighter Wing* and E-3 AWACS aircraft from the *552d Airborne Warning and Control Wing* at NAS Roosevelt Roads, which flew regular patrols over the proceedings. The threat was predicted to be light to medium AAA (up to 23mm), as well as possible hand-held SAMs; potentially lethal, but not exactly what had been faced in Vietnam.

Overall command for the operation fell to COMSECONDFLT Vice Adm. Joe Metcalf, embarked in *Guam* as *Commander Joint Task Force 120*. His staff included an Army major general named H. Norman Schwartzkopf, who was pulled from assignment as division commander at Fort Stewart, GA, for duties as Army advisor. Rear Adm. Richard Berg commanded *Task Group 20.5* in *Independence*. From the start several aspects of the invasion went poorly. Early on the morning of the 24th several SEALs drowned in heavy seas off Point Salines. Subsequent heliborne assaults by Marine and Army units ran into a buzz saw of ground fire that downed several aircraft and severely damaged others. Communications between the various forces and their commanders proved to be a major problem, resulting in not enough support in some areas and shifting targets and landing zones.[2]

[2] According to Rupe Owens, who was CINCLANT Operations at that time, states that, about two days out of Norfolk, *Independence* was instructed to "steam 180 at 12 kts and await further orders". Indy's SATCOM gear was down at the time, which lead to serious communications problems throughout the event.

Led by skipper Cmdr. Mike Currie, the *Thunderbolts* of VA-176 initially flew a mix of close air support and reconnaissance missions, but the focus quickly turned more to CAS when it became apparent there was going to be localized resistance. The Army Ranger's assault on Point Salines airfield – one of the original bones of contention – ran into trouble early, and VA-176 was one of the squadrons that responded, dumping ordnance on Cuban units defending the site; apparently, the engineers were fully combat qualified and well armed. The *Thunderbolts* later turned their attention to operations in the vicinity of St. Georges, where Marine ground units ran into problems. Throughout the week long operation A-6s and other attack assets continued to provide fast, accurate, and plentiful CAS as needed.

The *Thunderbolts'* flight deck coordinator during the surprise combat deployment was AEC Denny Franklin, a native of Guadalupe, Iowa, a former college football player, and Navy SAR swimmer. He had been to Vietnam with VA-65 during its 1968-1969 cruise in *Kitty Hawk* and had later served with VA-34 before reporting to VA-176. He'd also been recently been picked up for LDO and was making his last *Intruder* deployment before moving up to the officer ranks:

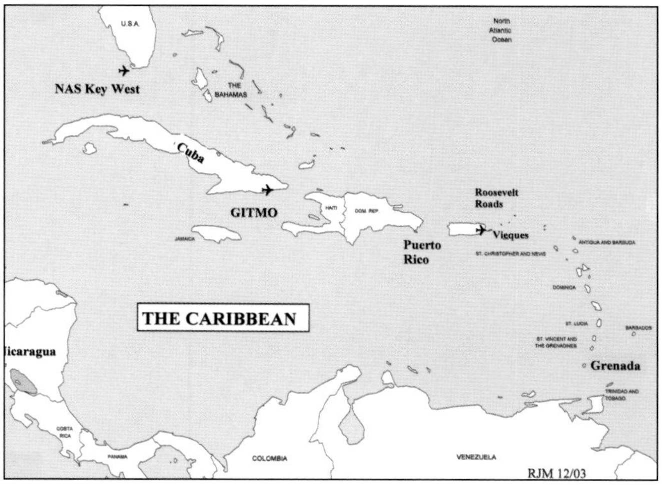

> "One of the big things I remember out of that tour was we'd pulled out for the Med," says Franklin. "The next morning, I walked up on the flight deck and the sun was in the wrong place. We had the planes torn apart, because usually we had 10 days prior to inchop. I went to the maintenance chief, said we had a problem, and we both went right to the skipper's room and beat on the door, waking him up.
>
> "'Where are we going? The planes are torn up.'
>
> "'We're going to war.'
>
> "That was it, no speeches," he continues. "We had our planes so torn down – I mean, engines out and everything – and we had about 72 hours to build them back up. We'd done so much intensive training before the deployment that we hadn't had time to do the maintenance. That was normal for cruise; you get the final preparations done on the way over.
>
> "There were combat vets in the squadron, but we had two guys turn in their wings something like 10 minutes before the first strike launched," he continues. "These guys just flat came out and said they were scared to death. Everyone respected them, but they were gone on the next COD. Everyone then manned up, and we never lost an airplane."

Despite the problems with communications and confusion – all to be expected in the new era of "jointness" and "everyone plays a part" – Cuban and Grenadan resistance eventually ended. Point Salines and Pearl Airfields came under allied control by the end of the first day, but the recovery of the students and foreign nationals took another 48 hours. By the 27th most of the New Jewel Movement's leadership was hurriedly looking to either evacuate or blend in with the population. The students and other Americans were rounded up and returned to the states, where commentators waited to get the true story of the invasion; much to the media's surprise, the first few students off the planes actually stopped and kissed the ground as they debarked, rendering many of the President's critics speechless.

On 2 November the United States declared the island and *Task Force 120* formally disbanded. Once relieved by Army personnel the Marines of the 22nd MAU re-embarked on their amphibs and resumed their trip to the Med, where they would relieve the survivors of the Beirut bombing. The U.S. repatriated the Cubans to their home island, having suffered a total of 24 personnel killed during the operation. The Grenadans lost 67 military and civilian personnel, while U.S. losses were set at 18, with another 150 wounded. All of the representatives of the Soviet Union and its aligned nation were also sent home. In the aftermath Bernard Coard, Hudson Austin, and a score of other New Jewel Movement, PRAF, and PRM leaders were jailed and tried, with several sentenced to death by hanging.

For the U.S. military, despite the problems and confusion the Grenadan operation was hailed as a major victory and proof that the United States once again was "standing tall," to use President Reagan's words. For many critics in and out of the military, however, the operation pointed out major problems with the concept of jointness, ranging from communications to lines of command to poor planing and intelligence. At least four U.S. aircraft, all helicopters, had been shot down or otherwise destroyed during the operation. (several being Army UH-60 *Blackhawks* that collided while under fire), and there had been 18 American service personnel killed in action (most of which were in the lost helicopters). Be as it may, the services ended up handing out nearly *35,000* awards and decorations to the participants in *Urgent Fury*, many of which went to staff members who never left the United States. Adding a surreal touch, after all was said and done Admiral Metcalf and several of his aides were busted for bringing captured AK-47s back into the U.S. Metcalf survived the *faux pas* and assumed the duties of Deputy Chief of Naval Operations-Surface Warfare.

As for *Independence* and the crews of CVW-6, they shifted sights and departed the Caribbean for their originally scheduled Mediterranean deployment. While it may have seemed the *Valions* and *Golden Warriors* of VAs -15 and -87 had taken most of the positive press for the air wing during *Urgent Fury*, the *Thunderbolts* knew they had done more than their share, and had done it well. However, things were destined to get interesting again for the squadron. Once on station in the eastern Med *Indy* joined *John F. Kennedy* for what would become another round of combat. The nation was now going to turn back to Lebanon and try to achieve some sort of revenge, or at least a resolution; what followed was an out and out disaster, both politically and for Naval Aviation.

Disaster over Lebanon

The Marines went ashore in Beirut on 25 August 1982 as part of the international peacekeeping force. Surprisingly, one of the strongest proponents of the introduction of foreign troops into Lebanon had been PLO Chairman Yassir Arafat, who pushed the idea during negotiations over Israel and Syria's withdrawal from Beirut. President Reagan had agreed, with the understanding the troops would pull out within 30 days. The French and Italians, who provided military personnel for the mission, also agreed.

On 14 September, shortly after the Marines departed, a car bomb killed Lebanese president-elect Bashir Gemayel. Israel responded by invading West Beirut, which brought the Syrians back into the country from the north, and shortly afterwards Lebanese Christian militias massacred the inhabitants of two Muslim communities in a section of Beirut. The United States, France, and Italy quickly put their troops back into Beirut and attempted to separate the warring factions, but this time there was a big difference: their tour of duty in the shattered nation was officially open-ended.

For the Christian government of Amin Gemayel the return of the U.S. military offered a good opportunity to retrain and rearm the Lebanese armed forces on the American model. The United States accepted the request, and soon Marines trained and patrolled with their Lebanese counterparts. For the Muslims, however, it made the Americans an object of hatred, because it appeared the U.S. was taking sides or, worse yet, supporting the extermination of the followers of Islam. In the spring of 1983 Lebanese youth started throwing stones and other objects at Marine patrols, much to the surprise of the Americans. A grenade attack on a patrol on 16 March wounded five U.S. servicemen. On the afternoon of 18 April a truck bomb destroyed the U.S. embassy in Beirut, killing 54. Despite these events – and the obvious portent of things to come – the Marines continued to labor under a very strict ROE, which meant their weapons were usually unloaded.

In September 1983, following the signing of a peace agreement between the Gemayel government and Israel, the latter nation again pulled back from the Shouf Mountains to Southern Lebanon. Militia groups from both sides flowed into the vacuum, and the level of violence and bloodshed escalated, including open combat with Lebanese Army units. On 19 September a U.S. official in Beirut – Special Envoy Robert McFarlane, himself a former Marine – ordered a fire support mission to bail out a Lebanese Army patrol that was under attack. Over the strong objections of local Marine commanders *Virginia* (CGN-38), *Arthur Radford* (DD-968), *John Rodgers* (DD-983), and *Bowen* (FF-1079) opened fire with their 5-inch guns. Post-mission analysis later determined the Lebanese units were never in as much trouble as they had indicated, but by that time the damage had been done. This direct American action against the Muslims in support of the Lebanese served to further generate anger against the United States. Unfortunately for the Marines in Beirut, *they* were the United States of America in the eyes of many, and they paid the price.

On 23 October 1983 a suicide bomber with visions of Paradise drove a truck carrying six tons of dynamite through the main gate of the Marine headquarters in Beirut. He managed to plant his rolling bomb squarely in the front door of the structure before it detonated; the resulting explosion obliterated the building, killed 241 Marines, and destroyed U.S. foreign policy in Lebanon. A similar attack elsewhere in Beirut killed 58 French soldiers.

In a later address President Reagan vowed to track down and bring to punishment those who were responsible for the bombing, but to this point U.S. efforts against international terrorism had brought minimal results. However, everyone agreed *something* had to be done, while the Marines picked up and moved out of Lebanon. As a result, once again the nation's leaders asked, "where are the carriers?" Two were on station in the Med: *Dwight D. Eisenhower* with CVW-7 and VA-65 onboard; and *John F. Kennedy* with CVW-3. Notably, the latter carrier was operating with the now famous "All Grumman Air Wing" made up of two squadrons of *Intruders* (VAs -75 and -85) and no light attack assets. Former *Buckeye* CO Dan Wright was Air Operations Officer for *Commander, Carrier Group 8*, and recalls the deployment of two A-6 squadrons in one air wing was "interesting":

> "I think it was John Mazach's air wing," he comments. "He was an A-7 guy. His Ops O was also an A-7 guy, but he became seriously ill and the admiral suggested we put in an A-6 guy as a replacement. Rob Weber got selected."

Weber, the first XO and second CO of the *Warhorses* of VA-55, subsequently reported aboard:

> "One of the reasons for putting the two squadrons into one air wing was that we recognized there was going to be a transition period from the A-7 to the F-18 with all of the squadrons transitioning. There was going to be a deficiency in the number of aircraft and the number of squadrons, so we put in two squadrons of A-6s to keep the numbers up. A couple of things stand out: for one, we put this together, and a lot of people predicted it would be an absolute disaster, 'All of those big airplanes on one flight deck,' 'there will be more accidents,' that sort of thing. It just didn't happen; the flight deck crews had fewer different types of aircraft to worry about, and everything worked out fine. The other thing was working out the tanker assignments. One squadron, VA-85, was given 14 A-6Es, while the tankers went to the other squadron (VA-75). The tankers were old and worn out, and as a result the other squadron had a lower (sortie) completion rate. For that one cruise VA-85 had a completion rate in the high 90s."

Wright adds:

> "We did find out that UnReps could be a problem. They were big aircraft, and we had to run them together, as there were too many tails hanging over the side of the flight deck. The other possible drawback was that we had to plan missions very carefully, because we didn't have quite the same numbers in the air wing. We worked around it."

While there were plenty of Vietnam vets on the staffs, in the wings, and in the squadrons, the majority of the attack personnel – with the exception of the recently qualified *Thunderbolt* crews – had never logged green ink (combat) missions before. One of these junior officers was Pete Frano, making his first cruise with the *Punchers* in *Kennedy*.

"I was commissioned in 1981," he comments:

> "Wrapped up at Pensacola in July 1982, graduated from VA-42 in July 1983, and went on to the last phase of training with VA-75 on 1 August 1983. A-6s were the most appealing aircraft because I was in the front seat. Originally I thought I was going to P-3s, as I thought it'd be the easiest for my wife to handle. However, once I started to fly jets it was fun, so I decided to go that way. Also, there were two instructors at VT-86 who were A-6 guys, and they were very good; they made a big impression on me. I don't know if anyone has ever done a study on that, but *who* was there definitely skewed who went to what communities. These were the right guys to steer me in the right direction, because I wanted to be just like them.
>
> "On that cruise in September 1983 our first port visit was Rio de Janeiro; we were the first carrier to visit there in three years. We did the crossing the line ceremony going down, and that was a real eye-opener. There were not a lot of shellbacks on board, maybe a quarter. They started on us in shifts down in the hangar bay ... let's just say my knees were sore, by butt was sore, and I smelled terrible. We had a change of command in Rio (Cmdr. James Glover for Cmdr. Earl Wolfgang, on 13 October), and a few staff types had a 'private' crossing-the-line ceremony on the trip back north. We continued the cruise, went to the Med, and then the Marine barracks got bombed in Beirut. It was also a real eye-opener; holy shit, we might be going into combat."

While the government attempted to come up with options for appropriate responses to the Marine barracks bombing, the two air wings regularly ran reconnaissance flights over the beach. Occasionally they drew a few pot shots from the ground, which didn't set well with the president and his advisors. As a result they told the Navy to start working up an air strike, and Rear Adm. Jerry O. "Sluf" Tuttle, *ComCarGruEight*, was given the lead.

According to Dan Wright, working for Tuttle was:

> "... extremely satisfying ... a gratifying period. I'd heard some bad things about him, but if we had to go to war he's the one I'd want to go to war with. We were always doing something, developing new concepts, new comm systems, that sort of thing. We developed the FIST system using HF to transmit photo imagery to the ships. We did it through the Monterey weather system. You've heard of JOTS? Officially it stood for the Joint Operations Tactical System, or something like that, but it actually stood for the 'Jerry O. Tuttle System.' We also did a lot of work with anti-air warfare; he helped prove the *Ticonderoga* and Aegis system when it arrived."

Using his professional assets – which besides Dan Wright included former *Blue Blaster* skipper and current *Kennedy* CO Capt. Gary Wheatley and *Eisenhower's* CAG-7, former VA-65 skipper Capt. Joe Prueher – Adm. Tuttle came up with an initial plan. CVW-3's *Black Falcons* under Cmdr. Kirby E. "Skip" Hughes II had the most *Intruders* available and would be go in at night against a terrorist camp in Lebanon. The only other participants would be the *Rooks* of VAQ-137 providing electronic warfare support, plus *Tomcats* from VFs -11 and -31 to handle any Syrian aircraft that might be foolhardy enough to pursue *Intruders* coming off the target. The admiral submitted the plan and the staff waited, and then waited some more while the government concentrated on the Grenada invasion and other activities. In the meantime, the *Buckeyes* kept a number of A-6Es bombed up and ready to go, just in case the word came down to hit the terrorists. Eventually, *Independence* arrived with VA-176 and relieved *Eisenhower*, which turned for Norfolk and home. *Indy's* crews were brought up to date on the planning process, but in the absence of any further guidance from Washington Cmdr. John Mazach's CVW-3 in *John F. Kennedy* still had the lead on any operations.

On 3 December 1983 an F-14 TARPS bird operating from *Kennedy* reported it had been fired on by both AAA *and* shoulder-fired surface-to-air missiles. In response the National Command Authority – with strong input from the State Department – ordered a retaliatory strike for 4 December with some changes to the existing plan. The raid was upgraded to Alpha Strike status – in other words, multiple aircraft from both decks – scheduled for broad daylight, and the target list included Syrian antiaircraft positions in the vicinity of Hammana. The word went down the line to prepare for an 1100 time over target.

According to Pete Frano, *Kennedy* was nearby enjoying a liberty port when the word came to them:

> "We'd just pulled into Haifa when we got the word. We pulled out the following Friday, and that Sunday did the strike. I was on the CO's strike team. We broke the squadron up into four groups, led by the CO, XO, and two senior department heads. I was lucky enough to be on the CO's team: Skipper Jim Glover with B/N Lt.Cmdr. John Tindall, the XO, Lt.Cmdr. Rich 'PBR' Pabst with B/N Ian 'Jabo' Jablonski in the other plane, and Lt. Bob Solik and I."

A few hours before the planned launch Washington called and rescheduled the strike for 0700, chopping a full four hours of the two air wings' prep time; to this day, no one has ever fully explained why the target time was moved up. Now everyone had to scramble. The two carriers readied a total of 28 aircraft with a mix of ordnance, including five and seven A-6Es from VAs -75 and -85 in *Kennedy* and five A-6Es from VA-176 in *Independence*, plus an additional two KA-6Ds from the *Sunday Punchers*. CAG Mazach led the crowd from *Kennedy* in an *Intruder*, while his CAG-6 counterpart, Cmdr. Ed "Hunyak" Andrews, led *Indy's* contingent in a VA-15 *Valion* A-7E. All three *Intruder* COs led from the front, including Cmdr. Hughes and pilot Lt.Cmdr. Mark McNally in *Buckeye One*, Jim Glover in *Flying Ace One* with "Tiddles" Tindall, and Cmdr. Mike Currie in the van of the *Thunderbolts*.

With an 0700 TOT the strike crew found themselves flying right into the sunrise and – adding to the moment – at an altitude of 20,000-feet, also specified by higher authority. They rolled in on the alerted Druse Muslim targets to find a hail of gun and missile fire, not just from SA-7s, but also including Syrian SA-6s and SA-9s. They pressed on; the five VA-176 A-6s, for example, dropped 25 Mk.82s, 23 Mk.83s, and 30 Mk.20 *Rockeyes* while expending 30 rounds of RR129 chaff and 57 rounds of Mk.46 flares. The strike recorded hits, but two aircraft did not make it back. Frano comments:

> "We were 'Tail End Charlie' over the target. As we were heading in we heard the calls over strike frequency: CAG Hunyak said he was hit and was ejecting. We knew we were going in. CAG Mazach led the charge, flying an A-6 from VA-85. Mazach was a classic – my first CAG – and he smoked these big long cigars. He flew all the time, and was exactly what I thought a CAG should be."

"Hunyak" Andrews' *Corsair* was shot up, but he managed to eject before his plane quit flying and was subsequently recovered. Not as fortunate were VA-85's Lt. Mark Lange and Lt. Robert Goodman, who were hit while prosecuting an SA-9 installation. Only Goodman survived to be taken prisoner; apparently, the Syrians let Lange bleed to death while they concentrated on manhandling the B/N, who had just come down in their midst. After much negotiation, debate, and demands, Rev. Jesse Jackson went to Damascus and gained Goodman's release on 3 January 1984; notably, the Syrian government uniformly ignored the Reagan Administration's demands for the young bomb/nav's release.[3]

It was a bleak showing for all concerned and weighed heavily on the participants through the end of their deployments, and for several years to come. Most agreed the original strike plans had been good, but they had been abandoned for political reasons. For the senior crews it must have been sadly reminiscent of the Vietnam experience, and not what they'd come to expect from the Reagan government.

Professionally, there was the strong feeling that we had to do business differently. The Lebanon strikes sealed a decision that had been a long time coming – that there needed to be a better way to train our aviators and prepare them for combat. On 15 Sept 1984 the service established the Naval Strike Warfare Center (NSWC) at NAS Fallon Nevada. Better known as "Strike U" (or, as dubbed by naval writer Barrett Tillman, "Clobber College"), the command grew quickly from a small group of zealots occupying a beat up old building in the desert to a multi-million dollar training facility that brought together all of the warfare disciplines. Strike U took over annual Air Wing training and instituted a standardized syllabus that replaced the CVW-planned WepTraEx that had been the norm since Vietnam. For some old hands it was a shock – no more barely planned mid-day low levels and dogfighting in the Austin MOA followed by a night in Reno. Most events were now heavily, and professionally, planned, with formal debriefs afterwards. The two-week deployment to Fallon became busier and a whole lot more productive in most eyes, and the entire Air Wing benefited from it. Other courses were offered specifically for Strike Leads, CAG, and Battle Staffs, all to educate them on the new tactics and technology that were becoming available to fight with.

If Strike U was becoming the Navy's graduate-level education for its aviation warriors, each of the type wings were also establishing un-

---

[3] Goodman's release led to a community story, perhaps apocryphal, where when told by a Syrian guard that "Jackson is coming to get you", he supposedly replied, "Oh yeah – which one? Reggie or Michael?"

The Dual A-6 Squadron Air Wings

In the mid-1980s Medium Attack was a growth industry. John Lehman, a Reserve A-6 B/N, was the Secretary of the Navy, the Reagan defense budgets promised new growth in Naval Air that would include two new Air Wings, and the A-6F promised the production line would remain open for some time to come. The Navy's standard Air Wing configuration at this time was five large squadrons (two Fighter, two Light Attack/Strike Fighter, one Medium Attack), and what was perceived as the support outfits, made up of VAW, VS, VAQ, and HS squadrons. It was during this period that plans were made to add an additional A-6 squadron to every air wing. To do this would require the establishment of at least twelve new medium attack squadrons, six at each base.

So why grow to two A-6 units per wing? Sources vary as to why the Navy wanted two *Intruder* units in their air wings, but they include:

1. Provide more strike aircraft at sea.
2. Provide more command opportunities: Doubling the units also doubled the number of career critical commanding officer slots available for A-6 officers.
3. The Soviet threat: According to one former Congressional staff member, the Navy had studies that said that over 20 A-6s were required to put enough iron on a Soviet *Kirov*-class Battle Cruiser to sink it.
4. Leverage: Other reports have stated that the Navy wanted to show McDonnell-Douglas that it could do just fine without large numbers of FA-18s, so the "All Grumman Wing" was established, with only two A-6 units making up the entire attack component.

The first two new units would equip two new air wings, with the remainder being added to the existing CVWs. Oceana's VA-55 was first, for Air Wing-13, and Whidbey soon followed, with VA-185 in the new CVW-10. VA-155 was next, at Whidbey, and VA-36 in Virginia, with VA-12 scheduled to follow. In Whidbey the functional wing, under RADM Fred Metz, made formal plans to resurrect the designations of five former attack squadrons as A-6 units (VA-93, 152, 153, 164, 212, and 215, all numbers used by former AirPac A-1, A-4, or A-7 squadrons). But then the roof fell in. The A-6F was cancelled, and the dual air wing plan ended at five of the 14 CVWs in the Navy at the time.

There were three distinct configurations for the dual *Intruder* wings, informally given the names of the first carriers they were assigned to: "Coral Sea"; "Kennedy"; and "Roosevelt." Notably, the three remaining dual A-6 wings in the Navy would fight together in the Persian Gulf during *Desert Storm*. It would be their swansong though, as they did not last long after the war. *Ranger*, with CVW-2 and VA-145/155 onboard, would make the last dual A-6 cruise in 1993.

KENNEDY AIR WING

Squadron configuration:-2 F-14, 2 A-6, no VFA/VAL. 1 each, VAW, VAQ, VS, HS

J.F.KENNEDY	CVW-3	9/83-4/84	MED/Lebanon	VA-75, VA-85
J.F KENNEDY	CVW-3	8/86-3/87	MED	VA-75, VMA-(AW)-533
J.F.KENNEDY	CVW-3	8/88-2/89	MED	VA-75, VMA-(AW)-533
RANGER	CVW-2	10/11 86	WPAC Surge	VA-145, VMA-(AW)-121
RANGER	CVW-2	7-12/87	WPAC/IO	VA-145, VMA-(AW)-121
RANGER	CVW-2	2-8/89	WPAC/IO	VA-145, VMA-(AW)-121
RANGER	CVW-2	12/90-6/91	D.STORM	VA-145, VA-155
RANGER	CVW-2	1992-93	WPAC/IO	VA-145, VA-155

Notes: Frequently called the "all Grumman Air Wing," which was technically a misnomer, as Lockheed (S-3) and Sikorsky (H-3) aircraft were also assigned. *JFK* made three deployments with this configuration and participated in the December 1983 strikes into Lebanon. The Marine *Intruder* unit was replaced by two squadrons of A-7s in 1989. *Ranger* and CVW-2 spent seven years with a dual A-6 configuration and participated in *Desert Storm*. FA-18s joined the wing in 1993 when it moved to *Constellation*.

Two separate *Intruder* squadrons in one Air Wing are exemplified by this shot of VA-75 and VMA (AW)-533 aircraft on *Kennedy* in 1986. (Rick Burgess)

***CORAL SEA* AIR WING**

Squadron configuration:-3 FA-18, 2 A-6. 1 each, VAW, VAQ, HS. No VS.

CORAL SEA	CVW-13	9/87-3/88	Med	VA-55, VA-65
CORAL SEA	CVW-13	5-9/89	Med	VA-55, VA-65

Coral Sea/CVW-13 transitioned from a four FA-18 and single A-6 configuration to 3 and 2 for its 1987 deployment. It kept that format for its final cruise, in 1989. The ship was decommissioned and VA-55 disestablished afterwards, while VA-65 moved to CVW-8.

MIDWAY	CVW-5	1987-91	WPAC/D.Storm	VA-115, VA-185

Forward deployed to Japan throughout the period covered. CVW-5 replaced F-4s and A-7s with three FA-18 units, while VA-185 moved from Whidbey Island to join 115. As with *Coral Sea*, the ship's size dictated smaller units, typically nine *Intruders* per squadron. VA-115 absorbed 185 in 1991 as the wing reverted to a 1 VA, 2 VF, 2 VFA configuration.

***ROOSEVELT* "Notional" AIR WING**

Squadron configuration: 2 F-14, 2 FA-18, 2 A-6, 1 VAW, HS, VS, VAQ each.

THEODORE ROOSEVELT	CVW-8	12/88-6/89	Med	VA-35, 36
THEODORE ROOSEVELT	CVW-8	12/90-5/91	Desert Storm	VA-65, 36

ONLY ONE – the navy's biggest air wing of its era. CVW-8 was the blueprint for the planned Lehman period air wings, and reportedly what the remainder of the wings were supposed to become, before budget realities forced otherwise. The plan is frequently referred to as the "notional" air wing in Navy documents, and as such was the largest in the service at that time. The wing performed well during *Desert Storm*, but was modified in 1990 for the "Special MAGTF" experiment, where a Marine helo unit and FA-18 unit were assigned to the ship. VA-65 was disestablished to make room, and an F-14 squadron (VF-41) was beached. The experiment lasted only one cruise, provided mixed results, and CVW-8 converted to a 2 VF, 3 VFA configuration.

dergraduate courses for their specific type aircraft. At Oceana Medium Attack Weapons School U.S. Atlantic Fleet (MAWSLANT) was established, while a West Coast counterpart, MAWSPAC, was formed at Whidbey. These schools carried out *Intruder*-specific tactics and weapons training that, in time, would be standardized with Fallon to provide a new crop of warriors to the fleet.

1984: *Night Hawks* Go to Sea

Following the debacle in Lebanon, 1984 passed relatively without incident. Throughout the year AirLant and AirPac continued to cycle their carriers through the Med and WestPac, with both commands alternating ships in the Indian Ocean. Besides the ongoing violence in Lebanon and in the Red Sea region between Iran and Iraq, there was plenty of other activity in the world to keep everyone's attention focused.

Four years into its war with Iraq, Iran had reached the point of relying on human wave attacks in the land battle; all that achieved was the piling up of thousands upon thousands of young Iranian bodies. The air war between the two nations was equally dismal for the Islamic Republic, with its once impressive assortment of U.S. aircraft and weaponry becoming increasingly unserviceable due to lack of parts. One area where Iran held the upper hand was through its Navy, which could still put an assortment of combatants – and a large number of high-speed boats operated by the Revolutionary Guard – into the Red Sea. When they didn't run across their Iraqi counterparts they harassed commercial traffic, particularly ships they felt were engaged in trade with the enemy. Eventually this grew into specific threats against tanker traffic, and it did not matter whose ship it was. For the time being the Americans were content to keep an eye on the region and let the two

The formation of the Naval Strike and Air Warfare Center (NSAWC) at Fallon in 1984 marked a major change in the way Carrier Air Wings were trained for deployment. "Strike U" had its own fleet of aircraft, which included *Intruders, Corsairs*, and *Hornets*. This May 1987 shot, taken over their home base, was taken by the incomparable aviation photographer Bob Lawson.

countries bash each other's brains in, but there was obvious concern about a possible disruption in the world's oil traffic.

As for the two carriers that participated in the Lebanon raid, they resumed normal steaming, completed their deployments, and headed for the barn. There were no further combat sorties over Beirut or other targets in Lebanon; *New Jersey* shelled several targets over 8-9 February, but that mission proved something of an afterthought *vis a vis* U.S. policy. The last Marines and most of the other personnel finally pulled out of the thoroughly devastated nation over 17 to 26 February, with the final U.S. personnel departing by 31 July. *Independence* was the first to return to Norfolk, arriving at NOB's carrier piers on 11 April 1984 following a swing through the North Atlantic. It had been quite a cruise for VA-176 and the other squadrons of CVW-6, including two rounds of combat. However, proving once again their was no rest for the weary, the *Indy/Air Wing 6* team would deploy again in just five months, and this time they would add the Indian Ocean to their places to visit. In the interim, on 26 July *Thunderbolts* XO Cmdr. James E. Hurston relieved. Mike Currie as VA-176's skipper in ceremonies at Oceana.

"Big John" was next to come home, returning to Tidewater on 2 May 1984 after its two *Intruder* squadrons flew off to Virginia Beach. For Pete Frano it had been a:

> "... Great first cruise, pretty exciting. After Lebanon we started doing twenty-four on and twenty-four off with *Saratoga*. We flew for about a month – through Christmas – with no port visit until Naples in February. It turned out to be an eight-month cruise with only 15 days in port. They needed to keep us handy, what with the terrorist threat."

Kennedy's "All Grumman Air Wing" had also undergone an auspicious deployment that – the outcome in Lebanon notwithstanding – continued to appeal to naval planners. More "dual A-6 air wings" were in the offing.

With *Indy* and *Kennedy* safely home it fell to *Sara* and *America* to take up the spear. *Saratoga* was first out, departing in February for points east with CVW-17 embarked. Providing a proper medium/all-weather attack balance to the SLUFs of VAs -81 and -83 was VMA(AW)-533, operating nine A-6Es and five tankers. Notably, VMAQ-2 Det X-Ray also made the trip, providing the air wing's organic electronic warfare capability in place of the usual Navy VAQ. *America* followed, departing ConUS on 24 April 1984 with CVW-1 and VA-34 onboard. Five months into a deployment which took them from the Caribbean to the Med and on into the Indian Ocean, the *Blue Blasters* gathered for an underway change of command. During the 7 August ceremony Cmdr. James B. Dadson – the same JB who had achieved fame as the youngest attack pilot during the Vietnam War – relieved Cmdr. Garth Van Sickle. Richard G. "Richey" Coleman joined the squadron as JB's executive officer for the cruise, which ran until November.

The *Tigers* of VA-65 also took a turn in the Mediterranean during the year, following a brief North Atlantic cruise and change of command. On 3 May 1984, five days before the squadron left Oceana for the northern climes, Cmdr. William J. "Fox" Fallon relieved Cmdr. Bob Houser as *Tiger One*. The squadron returned to Virginia Beach on 20 June, and fifteen weeks later – on 10 October 1984 – pulled out in *Dwight D. Eisenhower* heading eastbound. That same month *Sara* returned to Mayport, having completed an eight-month Med deployment.

For the Marines of VMA(AW)-533 it had been a good – and fortunately quiet – cruise with no incidents. Despite their prior lack of boat time with A-6s, the squadron managed 5121 hours and 1240 total traps with a boarding rate of 94 percent. As only the second Marine *Intruder* squadron to deploy as part of a carrier air wing – 12 years after the famous VMA(AW)-224 1972 cruise in *Coral Sea* – the *Night Hawks* had performed very well. Once home in Cherry Point the squadron found out that, instead of preparing for a UDP pump to Iwakuni, they were going to go back to the boat – this time joining VA-75 in CVW-3, where they replaced VA-85, which was headed back to CVW-17.

Whidbey Fun and Games

On the West Coast the COMMATVAQWINGPAC *Intruders* maintained a similar schedule of operations out of Whidbey and Yokosuka, with WestPacs, rotations through the Indian Ocean, and several squadron changes of command. On 29 February 1984 *Ranger* came home to North Island following its slow pass by Nicaragua and a lengthy period in the IO. The Navy tacked on the latter assignment after the Iranians started making noise about sinking tankers or otherwise blocking the Red Sea and/or the Straits of Hormuz.

By the time the *Boomers* of VA-165 got back to the Rock they had been out of country for seven-months-plus. Still, they could claim the honor of being the first A-6 squadron to spend part of its deployment monitoring the festivities in Central America, and to cap it off they took

VA-115, with CVW-5 in *Midway*, made numerous deployments through the 1980s from its Atsugi, Japan, home. NF502 is about to launch from the cat 1 with sixteen Mk.82 bombs. (U.S. Navy, via Steve Barnes)

Intruder barricade events were rare, but always exciting. Approaches into the net called for flawless coordination between the aviators, deck crew, and Landing Signal Officers. Showing how it's done, a *Blue Blaster* with an unsafe nose gear takes the *America's* barricade in summer 1984 while in the Indian Ocean. A perfect approach, engagement into the net, a call for tilly, and the Crash & Salvage gang takes over. As for 158045, she was not much worse for wear and would return to the skies, finally being retired to AMARC in July 1995. (Rick Burgess)

the Battle E and a Meritorious Unit Commendation. Fortunately for them and the other squadrons in CVW-9, they would not deploy again for well over a year.

VA-145 relieved VA-165 as the duty *Intruder* squadron in WestPac and the IO, once again outbound with *Carrier Air Wing 2* in *Kitty Hawk*. The *Swordsmen's* previous deployment – under Cmdr. Chris G. Overton – had run from April to October 1982 and saw a tour in the IO. Now they were going back for another look, this time with new skipper Cmdr. Arthur N. "Bud" Langston, known to many in the medium attack community as "Buffalo," or "Buff." Skipper Langston was something of a legend in the *Intruder* community, considered by many to be larger than life, and occasionally known for doing interesting things with the aircraft ... *and* his B/Ns. During a prior tour with VA-52 in the mid-1970s he'd taken a "stash Ensign" up one day for a fam hop in one of the *Knightriders'* tankers. Years later the future B/N – who, like all stashes, was spending time in a fleet squadron while waiting for his school date at Pensacola – recalled a most unusual introduction to Naval Aviation courtesy of the Buffalo:

> "The other two Ensigns and I took turns getting hops in the KA-6D. Actually, one other guy and I fought for flight time; the third guy wasn't really interested in the *Intruder* and wanted to fly E-2s. Anyway, we both flew with several pilots and quickly determined who our favorites were, but generally they were all good. They treated us well, like regular members of the squadron, made sure we had Sierra Hotel flight jackets and name tags, and ensured we were properly trained, both professionally and socially ... usually at the O' Club.
>
> "The day of my first flight with the squadron our schedules officer, Bob Gibson, put me in a plane with Langston. I didn't know much about him, but hell, flight time was flight time, even in the right seat of a KA. I don't remember much of the flight; I think we just took off and tooled around Puget Sound and the Cascades for about an hour, then started back for Whidbey. *Then* it got interesting.
>
> "Some of the pilots had us try communications, while others let us just sit in the right seat and take it all in," he continues. "On this hop Langston did all the work. We lined up for final to runway 31 at NUW, dirtied up – flats, slats, gear, etc – and then he called out a checklist item: 'Ensure controls are free to the left,' or something to that effect. Wham! He slams the stick over and we quickly roll inverted, with everything normally hanging down now hanging up. I believe I thought something to the effect of, 'Hey! *This* is neat!' After we rolled around he called out, 'Check controls to the right' and around we went again. I remember looking *up* out of the canopy at the cars passing under us on Highway 20. After we landed and got out of the plane I was really hyped – my first flight in the *Intruder*, the best plane in the known universe, and all that. I felt like a true *Knightrider* and fairly bounced into the hangar. There was our skipper, Daryl Kerr, standing with his hands on his hips, looking out toward the ramp. With a big smile I walked up and waited for him to congratulate me on my first hop in the *Intruder*, but without shifting his gaze he just quietly said, 'Go wait inside.' I thought it was kind of curious, but didn't say anything. In retrospect, I did get some strange looks from the other officers and some of the enlisted men after I got back in the squadron spaces.
>
> "It was some time later I learned Skipper Kerr felt the need to have a private session with Lt. Langston; safety of flight issue or something like that. All I know is I didn't fly with Langston again, and I don't think the other two stashes did either."

"Buffalo" survived his "go forth and sin no more" session with the skipper and went on to bigger and better things, assuming command of VA-145 on 7 December 1984. Over the following years he occasionally added to the legend as he moved up the career path, eventually making flag rank.

While there are no reports of other Whidbey squadrons flying tankers in the pattern upside down over Washington State Route 20, it is known they continued deploying during this semi-quiet part of the 1980s. On 30 May 1984 *Enterprise* pulled out from NAS Alameda and landed the squadrons of CVW-11 aboard, including the *Green Lizards* of VA-95. A few months into the cruise they too held an underway change of command, with Cmdr. John S. McMahon relieving Cmdr. Raymond T. Wockjik on 1 August. Two months prior to coming home *Enterprise* was joined by *Carl Vinson*, making its first full deployment as a unit of the Pacific Fleet. She departed Alameda on 18 October 1984 with CVW-11 and VA-52 embarked, the latter under the leadership of Cmdr. Donald L. "Sully" Sullivan. Both carriers made full WestPac/IO tours, with stops in the usual places and a fair amount of time in the North Arabian Sea.

As for Yokosuka's *Midway*, she did what she did best, deploying twice during the year for a total of seven months. On the first go-around – from December to May – *Midway* maintained a carrier presence in the IO for five months. The second cruise, which ran only from mid-October to mid-December, was a two-month WestPac. Obviously, having a carrier in Japan full time was continuing to prove beneficial for the Navy.

Finally, in December 1984 AirPac gained its first All Grumman Air Wing with the assistance of *III Marine Air Wing*. The *Green Knights* of VMA(AW)-121 chopped to CVW-2, joining VA-145 as the second A-6 squadron in the wing, while replacing the *Blue Diamonds* and *Argonauts* of VAs -146 and -147. Under assignment to *Ranger*, the wing would participate in an experimental "surge ops" program, where *Ranger* would be kept in a high state of readiness for emergency use in the Pacific theater. In effect, CVW-2 and the carrier served as the West Coast's counterparts to CVW-5 and *Midway*, not necessarily spending a lot of time on deployment, but ready for any eventuality.

While there were still the occasional seven-month deployments, most of the carriers and air wings were enjoying longer periods between cruises. The semi-relaxed schedule was another aspect of the "Reagan Revolution" that was directly benefiting the Navy, and it showed in retention statistics, parts availability, and in other areas. In general, the entire nation was feeling good about its day-to-day existence and the future. However, among the electorate there were still concerns over the President's generally inept handling of international terrorism, while some continued to express fear over the ongoing defense buildup. Naturally, many critics blamed President Reagan for the ongoing Cold War and other tensions in the world and not the Soviet Union.

Still, the Reagan revolution and defense buildup would continue for at least a few more years. In the interim, the President's staff once again tackled the question of terrorism, and once again the Navy took the lead. The *Intruder* would continue to play a major part in foreign relations, and would even see some additional growth, but during 1985 it would also gain a new stable mate in the deployed air wings.

CHAPTER EIGHT

Terrorism and Retribution, 1985-1988

As 1985 opened the carrier Navy still found itself gainfully employed around the world, keeping an eye on the usual suspects at the usual trouble spots. As always, it appeared to many observers that just as soon as one region quieted down another fired up, but then, that's what carrier battle groups are for. During the year, terrorism continued to occupy a large part of the nation's attention and time. In Lebanon kidnapping and hostage taking were the daily norm, as the factions worked to apply leverage against Israel and its western allies. Within short order airline hijackings to the Middle East also became commonplace; unfortunately, most of these activities involved Americans, and that meant a response of some sort from the government.

In the meantime, the carrier Navy added a new weapon to its flight decks, the FA-18 *Hornet*. The *Hornet's* arrival had an obvious and immediate impact on the light attack warriors of the A-7 community, who happily got in line for transition to a *real no-kidding fighter* (albeit with a co-equal attack mission, or at least that was the party line). Equally impacted, however, were the medium attack crews, and most were none too happy about the circumstances.

The *Hornet* Cometh

On 21 February 1985 *Constellation* departed San Diego with CVW-14 and VA-196 embarked. The deployment marked the air wing's first since making the 1983 around-the-world cruise in *Coral Sea*, and the *Main Battery's* first in *Connie* since Vietnam. Also onboard were the Navy's first two fleet FA-18 *Hornet* squadrons: the *Fist of the Fleet* of VFA-25 and *Stingers* of VFA-113. A good time was expected by all.

The -18 was an outgrowth of Northrop's YF-17A, which competed with General Dynamic's YF-16A in the U.S. Air Force's Lightweight Fighter program of the early 1970s. The GD project won out with the junior service, but the government – in the interest of developing a "high-low" mix of fighter assets – directed the Navy to select one of the designs as a potential replacement for the F-4 and A-7. In fact, Secretary of Defense James Schlesinger and the Senate Joint Committee on Appropriations all but told the Navy they would buy whichever plane the Air Force selected. General Dynamics went so far as to team up with Vought to produce a "navalized" version of the F-16.

Much to the surprise of many, naval aviation went with an upgraded variant of Northrop's P630/YF-17 design, developed by McDonnell-Douglas with Northrop. Tentatively designated the F-18, the new aircraft was initially programmed for two major variants: a dedicated fighter version, and one designed specifically as an attack platform. However, in short order the two missions were merged and the idea of separate variants discarded. Designated by the Navy "F/A" to indicate its dual fighter/attack capability (and in spite of a superceding DOD document directing it be called "FA," with no slash), the service selected the *Hornet* in January 1976 as the replacement for several Navy and all Marine *Phantom* squadrons, and all *Corsair* squadrons. The aircraft featured two GE F404-GE-400 afterburner-equipped turbofans, AN/APG-65 multifunction radar, an M61 rotary cannon, provision for two AIM-9 Sidewinders, two AIM-7 Sparrows, and five other stations for a total of 17,000-pounds of ordnance.

The first flight for the prototype FA-18A came at St. Louis' Lambert Field on 18 November 1978 and was followed in early 1979 by initial evaluation at the Naval Air Test Center at Pax River. On 13 November 1980, at NAS Lemoore, the Navy established VFA-125 as the initial joint USN-Marine *Hornet* RAG. Nicknamed the *Rough Raiders* – like their numerical predecessor, the old A-4 and A-7A/B RAG – the squadron received its first aircraft the following February. By late 1983 five squadrons had completed transition training: VMFAs -314, 323, and 531 – all ex-*Phantom* operators – and VAs -25 and -113.

Generally, the boys in light attack were ecstatic with the new arrival and eagerly pursued transition. Here was a fighter that could yank and bank with the best of them, shoot the enemy in the face, and then accurately put bombs on target, all with only one aviator onboard. What the plane supposedly lacked, though, was combat range, particularly when compared to the *Corsairs* it replaced. Equally important, the FA-

Lt.Cmdr. Fred Block of VA-196 signals "9000 pounds of fuel onboard" to the VAQ-139 *Prowler* on his wing while in low holding over *Constellation* on 11 May 1985. (Rick Morgan)

The arrival of the McDonnell-Douglas FA-18 *Hornet* "Strike-Fighter" to the fleet changed the entire complexion of Naval Aviation. VA-196 shared the decks of *Constellation* with VFA-25 and -113 in 1985 for the type's first deployment. VFA-25 FA-18A NK404 is seen over the ramp on 20 April 1985. (Rick Morgan)

18 would not carry a buddy store (at least initially). Therefore, the "good times" presented by the arrival of the FA-18s did not quite extend to Cmdr. Harry "Bud" Jupin's VA-196, which became the first *Intruder* squadron impacted by the new "strike fighters." Overnight the air wing went from having only two squadrons that required routine tanking to four, while their pool of available tankers went from both A-7s and A-6s to just *Intruders*.

The *Milestones* deployed with a 9&4 contingent of bombers and tankers, except that *all* bombers normally carried buddy stores for use as potential tankers. Three or four would usually be configured as "maxi tankers," with four 300- or even 400-gallon external fuel tanks, and the other always carried two drops and a D-704, leaving only two stations for ordnance. Among the tricks used to keep the crews in qual was the assignment of "mini-tankers" that would be configured with two drops, a buddy store, and a rack of Mk76 practice bombs. After launch they passed 4.0 to the "Fighter-Attack Guys" (aka "FAGs"), and then bombed the wake or smoke targets for the rest of the cycle. More than one *Milestone* crew briefed one mission event, only to be called back overhead to stream their buddy store's hose when the assigned tanker went down.

It was a bitter pill for the squadron, but they took up the new challenge and made it work. They turned in a very credible performance and helped write the book on how to keep their own training qualifications up, while also keeping the air wing fueled. LtCdr. Floyd "Pink" Cordell was a *Main Battery* B/N lieutenant who states the hardest part was "blue water" (no divert available) night operations, when they were required to keep two tankers airborne and a third on deck as a manned spare. Crews assigned to night tanking duties were normally only cruise-experienced aviators who had proven they could handle hawking a low-fuel *Hornet* or *Tomcat* around the pattern low at night, or be trusted to bring the tanker itself back aboard – without fanfare – as the last one airborne. While being designated for night tanker duty meant you'd arrived as a trusted carrier pilot, it also meant you would now pull more than your share of the disliked mission, which could routinely alternate between tedious and hair-raising.

As if being on the first *Hornet* cruise wasn't enough, a VA-196 *Intruder* flown by Lts. John "Eyeballs" Ivbuls and Robin "Jimi" Hendrix had a "minor" mid-air with a VFA-113 FA-18A one night over Thumrait target in Oman. Too many aircraft had been scheduled over the target at one time, and several were orbiting nearby waiting their turns. For whatever reason the *Milestone* and *Stinger* aircraft tried to occupy the same air space simultaneously and hit each other about 90 degrees off. Amazingly, the A-6 only lost its tail fin cap – about six inches worth – while the *Hornet* gave up a bomb rack and external fuel tank. Both aircraft landed successfully back on *Connie*.

Whenever they had the chance, VA-196 remained happy to demonstrate that the A-6 was still the first choice for long-range ordnance delivery in the air wing. *Intruders* fleet wide were fated to carry the tanker load until the S-3 was plumbed for a buddy store several years later; it could have been a major morale issue, but Bud Jupin's *Milestones* had a most successful cruise.

The Mideast Heats Up

While the *Hornet* was making its triumphant debut in the Pacific, the Navy continued to have concerns elsewhere. Again, the focus was on troubles in the Middle East, and FA-18s would not play a part there for some time yet.

On 14 March 1985 Lebanese terrorists kidnapped the CIA's Beirut station chief, William Buckley; they subsequently tortured Buckley, and the CIA was unable to do anything about it. On 14 June members of a terrorist group known as Amal hijacked TWA flight 847, a Boeing 727, while enroute from Athens to Rome. They diverted the Boeing to Beirut, and upon landing removed six Americans with Jewish-sounding names or government IDs; another 39 were kept onboard, including several U.S. military personnel. The hijackers then settled in for a lengthy stay and issued their demands, the primary one being the immediate release of several hundred Israeli-held Palestinian prisoners. In order to prove they were serious they chose a passenger and executed him. The victim was SW2(DV) Robert D. Stethem, a diver assigned to Underwater Construction Team One who was returning from an assignment in Nea Makri, Greece. His murderers dumped the body on the tarmac, and the Reagan Administration started issuing its own threats.

Nimitz, with CVW-8 and VA-35 embarked, arrived off the coast of Lebanon while the drama continued to unfold but played no role. The United States stuck to its official policy of not negotiating with terrorists, but on 2 July the hostages were finally released after Israel quietly freed some of its prisoners. Officially, the Israeli government made the decision on its own volition without any pressure or request

A *Milestone* A-6E is seen cruising off Oman during the squadron's 1985 Indian Ocean deployment. NK502's pilot, Lt. Steve Garcia, would lose his life on 8 August 1989 while practicing for an airshow at Whidbey with VA-128. (Rick Morgan)

from America. Petty Officer Stethem was buried at Arlington National Cemetery on 20 June and was subsequently awarded the Purple Heart and Bronze Star posthumously. Several years later an *Arleigh Burke*-class guided missile destroyer, DDG-63, would receive his name.

Despite the release of a few terrorists – or perhaps as a result – violence continued in the region. In September 1985 the U.S. embassy in Beirut was bombed again, and the following month terrorists led by Muhammad Abdul Abbas seized the cruise ship *Achille Lauro* in the Mediterranean. At one point during the hijacking they murdered American retiree Leon Klinghoffer, 69 and confined to a wheelchair; his body was then dumped overboard, still in its chair. The terrorists subsequently put the ship into an Egyptian port and disembarked under promise of safe passage from the Egyptian government. The U.S. quickly demanded their extradition, but the Egyptian government instead arranged for their transportation out of the country. While the State Department protested, President Reagan's advisors quickly decided to take the murderers by other means.

On the evening of 10 October 1985 EgyptAir flight 2483 launched with four of the terrorists onboard. U.S. intelligence assets had determined in advance the aircraft would head for Tunisia and knew the time of departure and flight routing. *Carrier Air Wing 17* in *Saratoga* was tasked with the intercept and sent several F-14A *Tomcats* from VFs -74 and -103 after the Boeing 737. Relying on VA-85 *Intruders* for tanking, as well as VAW-125 *Hawkeyes* for control, the *Toms* caught the airliner off Crete. After boxing in the airliner they directed it to Sigonella, where U.S. military personnel waited to take the terrorists into custody. A terse standoff followed on the ramp; Italian military personnel refused to turn the terrorists over to the Americans. At one point the U.S. prepared to send Special Forces units into Sigonella to forcibly take the terrorists, but that plan was canceled. The Italian government retained custody, and following a trial convicted the murderers – minus Abbas, who was not on the plane – and sentenced them to prison. One of the participants, Youssef Magied al-Molqi, was later released from the prison in Rome on a furlough and promptly disappeared. He was later captured in Spain.

Still, the event marked a major victory in the Reagan war against terrorism. Following the return of the TWA flight 847 hostages President Reagan had announced:

> "Let me make it plain to the assassins in Beirut and their accomplices that America will never make concessions to terrorists ... we're especially not going to tolerate these acts from outlaw states run by the strangest collection of misfits, Looney Tunes, and squalid criminals since the advent of the Third Reich."

Publicly the administration remained adamant, and the newfound American response to terrorism would remain quick and – if necessary – decisive. Quietly, however, the United States approached Iran for assistance in gaining the release of the growing number of hostages. The resulting negotiations would not become public knowledge for several years. In the interim, the interception and midair "apprehension" of EgyptAir 2483 brought great celebration among carrier aviators. While the fighter pukes basked in the glory of the event, the medium attack crews knew they had played a major part too, along with their VAW brethren.[1]

For many among the A-6 community it was the best of times. Even when they weren't flinging bombs or prosecuting the target at night and low-level, they knew they were still the tip of the spear and the best the Navy had to offer. For some, the adventure – to coin a phrase – was just beginning; for others, their long medium attack careers were finally coming to an end. One of the latter was the skipper of VA-34, J.B. Dadson. He had seen it all, from the left seat of the *Intruder*, from Vietnam through the dark years of the late 1970s, and on into the growth period of the early 1980s, and as he recalls it had been quite a ride:

> "I was on my second deployment with VA-176 in *Indy* and had orders in hand to MATWING, but they got canceled when I was out there. I ended up going to the Mine Warfare Inspection Group in Charleston as the air delivery guy, and ended up doing all of the inspections on the carriers and for the VP squadrons. In

[1] Most reports credit CVW-1's EA-6B squadron, VAQ-137, with providing communications jamming during this mission. In fact, according to both the squadron CO and XO, they were on deck throughout the event, the Egypt Air's crew apparently believing they were being jammed – and the media reporting what they thought was the obvious cause.

> the 11 months I was there I was on the road about 190 days! I finally screened for command, went to -34, and joined them in the IO."

During the *Blue Blasters'* 1985 cruise in *America* Dadson managed a couple of personal milestones in and around leading his men through several special activities, including flight operations from a Norwegian fjord in September. For most of the cruise his regular B/N was a young Lt.j.g. named Rick "Pokey" Keller, but on 1 August 1985 J.B. recorded his 1000th career trap with Rear Adm. Dick Dunleavy[2] in the right seat. Roughly a month later, on 30 August J.B. bagged his one-thousandth *Intruder* trap, this time with *America's* CO, Capt. Richard C. "Sweet Pea" Allen – dual qualified as A-6 pilot and B/N – manning the scope.

Following his turnover to "Richey" Coleman, JB managed a few more years of flying, in and around other duties:

> "I went back to *America* as operations officer," he recalls, "but we spent the better part of our time in the yard at Portsmouth. I then went to TACTRAGRULANT at Dam Neck as the first strike warfare officer they ever had. They didn't have too many aviators assigned, but I wrote the strike warfare syllabus and managed to get it interjected into their war games. Before I left I was the senior instructor. I flew my last A-6 hop in October 1992, 25 years to the day from my first flight, and then hung it up. I then made one last tour as NAVAIRLANT Chief of Staff for shore activities before retiring in the summer of 1993.
>
> "I think I was pretty fortunate when you come down to it," JB continues, "Because it (the A-6) was something. The longer you were with it, the more you thought of it. The crew concept was something; every time you flew with somebody it just got better and better. The only way to fly the mission is with two people working close. I had the opportunity to fly with some of the finest people, and worked with the most dedicated maintenance people. Those maintenance people had a special work ethic and bond; while other guys were out on liberty in four-section duty we were port and starboard, working weekends. It brought us together. Our ops on the *Coral Sea* proved it. We did the same things as the newer carriers."

Concerning his command tour as leader of the *Blue Blasters*, Dadson says that – as with many of his contemporaries – it was *the* career highlight:

> "I wouldn't trade our ready room full of JOs for anyone. I had to work hard to throttle them back. I worked with a great bunch, and it's been great to see them go on to bigger things."

Ironworks to the Med: *Attain Document*

Early in 1986, the Reagan Administration's push against international terrorism entered a new phase and concurrently identified a new target: Libya, and its maximum leader, Colonel Muammar Qadaffi.

For some time U.S. intelligence sources and the State Department had been working to trace the source of terrorist training and funding, and most of the trails led right back to Tripoli. Late in 1985 terrorists under Abu Nidal hijacked an EgyptAir plane, and 60 died during a subsequent rescue attempt by Egyptian Commandos. On 27 December 1985, in separate attacks on the airports in Rome and Vienna, 20 people were killed. The signs again indicated Libyan involvement, and as a result the U.S. Government publicly named Qadaffi as one of the number one exporters of terror. Concurrently, President Reagan signed National Security Directive 207, which put into motion options for a military strike against Libya and terrorist targets within the country. From this point it was just a matter of time before Libyan and U.S. forces clashed. However, there was no immediate, direct combat; instead, the American intent was to rattle Qadaffi's cage, and once again medium attack took the lead.

As the year opened most of the A-6 squadrons were actually "enjoying" their standard inter-cruise training periods. In AirPac VA-165 – now led by B/N Cmdr. Robert T. "Buff" Knowles – had just wrapped up its tenth *Intruder* cruise. The *Boomers* returned home to Whidbey on 21 December following five months in WestPac and the IO with CVW-9 in *Kitty Hawk*. Their spot in the lineup was taken by Cmdr. John McMahon's *Lizards*, which departed ConUS in *Enterprise* with CVW-11 on 15 January 1986.

In the Atlantic Fleet, VA-35 had also just completed a standard deployment to the Carib and Med, returning to Tidewater in *Nimitz* on 4 October 1985. Led by Cmdr. Ronald S. Pearson, in January the *Black Panthers* became the first *Intruder* squadron to report to a "Super CAG" through the reorganization of CVW-8. The Navy's "Super CAG" concept upgraded the air wing commander's slot to a captain, while adding a deputy air wing commander, or DCAG. The service's official intent was to "improve combat efficiency and capability of the air wing" by elevating the CAG to a level where he could report directly to the battle group commander. Several of the less reverent, however, pointed out the upgraded billet also conveniently increased the number of major sea command captain's billets for aviators. In any event, on 8 January 1986, in ceremonies at NAS Oceana, career fighter pilot Capt. Frederick L. "Bad Fred" Lewis assumed the conn of CVW-8 as the first of the Super CAGs. In time the other air wings followed, and several *Intruder* personnel fleeted up to both the Super CAG and DCAG slots.

As for keeping an eye on the Med, it fell to two squadrons: VA-85 in USS *Saratoga*, and VA-55 in *Coral Sea*. "Sara" had pulled out first, departing Mayport on 25 August 1985 with CVW-17 and Cmdr. Robert W. Day's *Black Falcons* onboard. *Coral Sea* followed them out of the barn on 1 October; Rob Weber's VA-55 provided the medium attack punch in the otherwise VFA strike component. The *Warhorses'* departure from Oceana marked their maiden voyage as an A-6 squadron; the deployment was also *Coral Sea's* first to the Med since 1957, and constituted the FA-18's operational debut in the Atlantic. Four *Hornet* squadrons were in CVW-13: brand new VFA-131 and -132, as well as two Marine units from El Toro, California (VMFA-314 and 323).

Little did anyone know things were about to pick up in a big way. A war of words was well underway between the U.S. and Libya; according to Qadaffi, the U.S. was an interloper in the Med, particularly when it came to operations in the Gulf of Sidra. He publicly stated that any American units or aircraft crossing 32 degrees 30 minutes north latitude would find themselves transiting a "line of death," and would face grievous consequences. The U.S. took the opposite stance and declared the contested area as international waters, and therefore open to all traffic. On top of all this, America was lining up opinions against Libya as a rogue nation while quietly trying to get its citizens out of the country.

Things were rapidly coming to a boil, but one of the two Sixth Fleet ready decks – *Coral Sea* – was lacking something that might come in handy in the event of open hostilities between the two nations: *Prowlers*, as the elderly ship's small flight deck supposedly did not allow space for them. The *Guardians* of CVW-13 were in fact the only air wing in the Navy without EA-6Bs assigned. Once it became apparent *Coral Sea* would probably get involved in strikes the fleet commander reportedly asked the air wing what it needed to conduct such ops. According to legend, EA-6Bs was number one the list; the JCS heard the cry, and on New Year's Eve 1985 made the call to Whidbey: "SEND *PROWLERS*." VAQ-135, home at Whidbey for New Years while in between workups with CVW-1, was ordered to the Med in response.

[2] New York native Dunleavy was former skipper of VA-176 and was the first Naval Flight Officer to command an aircraft carrier, *Coral Sea*, in 1979. He would rise to the rank of Vice Admiral and would end up as OP-05, the chief of naval aviation in the office of the CNO.

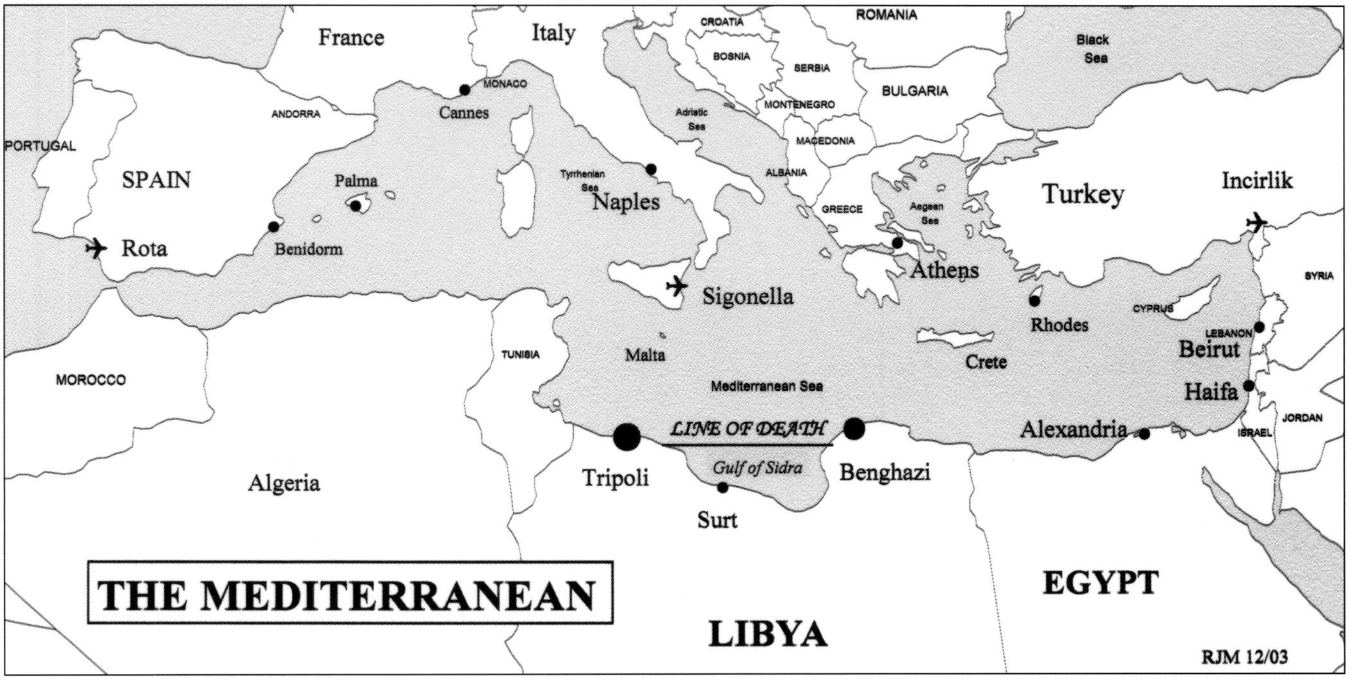

Within 72 hours the *Black Ravens*, reinforced with aircraft and personnel from four other VAQs, had five EA-6Bs ready for action in the Med.

On 26 January – and now with two fully equipped air wings of both *Intruders* and *Prowlers* available – Sixth Fleet initiated Operation *Attain Document*. *COMSIXFLT*, Vice Adm. Frank Kelso II, monitored the operation from the flagship *Coronado* (AGF-11), while Rear Adm. David Jeremiah – *Commander Cruiser Destroyer Group Eight* – ran the tactical end of the operation as CTF60 from *Saratoga*. Embarked in *Coral Sea* was Rear Adm. Jerry Breast, *COMCARGRU Two*.

That day TF60 crossed Qadaffi's "Line of Death" – the line that ran along the 32d parallel from Tripoli to Zawiyat al-Bayda – on what was officially termed a Freedom of Navigation operation. Qadaffi, wearing an absurd uniform that looked like a cross between a Good Humor ice cream man and a "Village People" stage costume, boarded the *La Combattante II* PGG *Waheed* at Misratah, 125-miles east of Tripoli, and set sail to Benghazi, proclaiming to several assembled TV cameras that he was heading into harm's way "... to prove this is our territorial waters." However, once his patrol craft rounded the point and left sight of the port the crew beached the boat and Qadaffi disembarked. Otherwise, during an operation which ran through 15 February, the only confrontation was provided by Navy and Libyan fighters keeping an eye on each other. The operation did prove the United States was prepared to operate in the Gulf of Sidra, and helped fuel rhetoric over the issue. Ultimately, it set the stage for subsequent operations.

As an aside, through this period and for some time afterwards the Navy maintained a plan to keep several *Intruders* bombed up and ready to go in case a specific terrorist target needed removal. According to Pete Frano, VA-75 – in the midst of a long inter-deployment "lull" – ended up contributing to the effort, which started in mid-1985.

"For our second 'deployment' with VA-75 we did something different by augmenting VA-65 in *Eisenhower*," he recalls:

> "We replaced some VA-34 people with eight crews for a total of 20 aircraft onboard. We had to have ten A-6Es ready at all times to hit a target. Ten A-6s would have gone over the target with ten Mk.83s each, over 100,000-pounds of bombs. J.B. had said at the end of his six months, 'That's it, I'm taking my boys home.' VA-75 was in the middle of a long turnaround with *Kennedy* in the yard, so off we went. At least we got a lot of flying time. That contingency strike stayed in the Med for years."

Intruders across the Line of Death: *Prairie Fire…*

Two months later, during a third round of *Attain Document* operations in the Gulf of Sidra, the Libyans got frisky. This time around the Navy had sent in a surface action group under *Commander, Destroyer Squadron 20* incorporating *Ticonderoga* (CG-47), *Caron* (DD-970), and other combatants. While the surface warriors tooled around south of 32 degrees latitude Navy aircraft patrolled above, challenging the various SAM sites and trying to get a rise out of the Libyans. By now *Task Force 60* included a third carrier; *America*, under Rear Adm. Henry "Hank" Mauz, *Commander Cruiser-Destroyer Group 12*. The squadrons with CVW-1 included "Richey" Coleman's VA-34 *Blue Blasters*. The Navy's total force for the exercise included 26 ships, 250 aircraft, and over 27,000 men ... and they were loaded for bear. For the *Blasters*, their arrival was perfectly timed for full participation in the upcoming event.

At 1453 local on 24 March 1986 an SA-5 *Gammon* (Soviet S-200) SAM site near Sirte hosed off a couple of rounds at some *Tomcats* that were bird dogging a MiG-25. Sixth Fleet responded with retaliatory strikes – well planned and well prepared-for retaliatory strikes – under the title of Operation *Prairie Fire*. While VA-81 *Corsairs* served as decoys, several VA-83 A-7Es fired *HARMs* with the jamming and targeting assistance of the *Rooks* of VAQ-137, and the SAM site dropped off line.

Libya didn't back down. At about 2100 Navy units detected the Libyan PGG *Waheed* (No. 526) sortieing from Misratah and heading in the general direction of the battle group. The Anti-Surface Warfare Coordinator (ASUWC) ordered in several *Intruders* for the intercept. The *Blasters* piled on first, firing *Harpoons*, marking the first-ever warshots for the AGM-84. They were followed eight minutes later by several *Black Falcon* A-6Es delivering *Rockeye*. The 350-ton *Otomat* SSM-equipped *Waheed* blew up and quickly went down; of its crew of 40 some 16 survived to sing the praises of their national leader. At about

2200 Navy crews dodged another SA-5 shot, which brought a quick visit from more SLUFs flinging *HARMs*. The site went down again and may have been damaged, but the following morning it came back on line. Two additional AGM-88 shots shut it down, permanently this time.

At about 2335 the night of the 24th VA-85 came *this close* to bagging another missile boat. The recipient of their professional attention, *Nanuchka II* PGG *Ean Mara* (416), just happened to be in the wrong place at the wrong time and collected several *Rockeye* hits for its trouble. Listing and smoking, the vessel was able to stagger back to Benghazi. According to *Jane's Fighting Ships*, the Soviets later towed it to Leningrad for repairs, and it returned to the Libyan Navy in 1991 under the name *Tarik ibn Ziyad*.

Not to be outdone, cruisers *Richmond K. Turner* (CG-20) and *Yorktown* (CG-48) both engaged surface targets with RGM-84 *Harpoons*. The Blackshoes claimed hits and kills, although no debris was found the next morning, and all Libyan combatants were reportedly accounted for. No Libyan vessels were ever confirmed as being hit by these attacks. The next morning (25 March) a mix of VA-55 and VA-85 *Intruders* applied a final lesson. Another *Nanuchka II*, the *Ean Zaquit* (419), stood out from Benghazi, and the *Warhorses* pounded it. Employing *Rockeye*, they scored a mission kill, and then the *Falcons* finished off the hapless boat with *Harpoon*. With the loss of two units of his Navy in barely 24 hours Colonel Qadaffi backed off, declared victory, and conducted a SAREX. Admiral Kelso declared a Great Naval Victory, and *Saratoga* headed for the barn – returning to Mayport on 16 April – while *America*, *Coral Sea*, and their escorts departed the area for port calls.[3]

For the carriers the ShootEx with the Libyans had constituted their first combat operations since 1973 when they were operating off Vietnam. Bob Day's *Black Falcons* had last seen action in 1983 over Beirut, but for both the *Blue Blasters* and *Warhorses* – 13 years apart in age – the combat experience was a new one. Despite the disparity in backgrounds and intervening years, Naval Aviation proudly noted its crews, aircraft, and personnel still had what it takes.

… and *El Dorado Canyon*

Despite coming out on the short end of repeated set-tos with the Americans – or perhaps, as a result – Qadaffi continued his support of terrorism. Several more incidents occurred through the spring, including an inflight bomb explosion onboard TWA Flight 840 over Greece. Four Americans died in that terrorist attack, but the renewal of terrorism culminated in the April 5th bombing of *La Belle Discotheque* in West Berlin at 0149. American soldier Sgt. Kenneth T. Ford and a Turkish woman died in the explosion. An additional 230 people, including 79 Americans, were wounded. U.S. and West German officials quickly

[3] There has been some confusion as to the actual identities of the Libyan corvettes hit during *Prairie Fire*. "Jane's Fighting Ships" states that the *Ean Ziquit* was sunk on the 24th while *Ean Mara* was heavily damaged on the 25th, towed home and subsequently rebuilt in the Soviet Union and renamed *Tariq Ibn Ziyad*. The Navy's official publication *Swift and Effective Retribution* (Naval History Office, 1996) states that the first ship was damaged while the other was sunk.

VA-55 was heavily involved in the action against Libya during 1986. Here an A-6E prepares for a shot from *Coral Sea*'s cat 1. (U.S. Navy, PH3 (AW) Wayne Edwards)

announced they had intercepted communications proving Libyan involvement in the bombings (although some German intelligence sources later stated they had no such indications). Within short order the Reagan Administration approved final preparations for another operation against Qadaffi, this time specifically targeting sites which "supported worldwide terrorism" and, in particular, the "mad dog of the Middle East."

In the Med *Task Force 60* – now commanded by Rear Adm. Mauz – once again moved into position and prepared for action. With *Saratoga's* departure the brunt of the action would be carried by *Coral Sea* and *America*, but this time around they had a little help from an unexpected quarter; Air Force F-111Fs of the *48th Tactical Fighter Wing*, operating from RAF Lakenheath in England, were going to participate.

The inclusion of the Air Force strike package remains controversial. The mission called for the simultaneous targeting of five individual sites, all affiliated with terrorist training and support, but not involving Regular Libyan Army positions, in order to increase the potential of an Army-led coup against Qadaffi. Officially, theater commanders determined five different locations was more than two squadrons of A-6s could handle; needless to say, the Navy crews thought otherwise, but those in charge requested the participation of the junior service to ensure proper coverage of the targets. Most Navy personnel remained unconvinced, and to this day maintain the inclusion of the F-111s was yet another example of "jointness" run amuck.

In any event, the blue suiters started planning ... and quickly ran into several roadblocks presented by America's erstwhile allies. All of the 48th TFW's original plans revolved around a direct flight from England to Libya via France. However, the U.S.' "allies" knew something was coming, and when the administration approached them about overflight rights and other cooperation they ran into a stone wall. The British, led by Prime Minister Margaret Thatcher, quickly agreed to the use of their island as a staging base for the "Aardvarks" and their support aircraft. The French, in turn, said they would not allow the strike package to cross their airspace enroute to Qadaffi; neither would the German, Italian, Greek, or Spanish governments, all of which were concerned about terrorist reprisals. The Air Force regrouped and came up with a plan that allowed their aircraft to fly the entire mission in international airspace once they departed England: south past France; Spain, and Portugal; through the Straits of Gibraltar; and on to Libya. This routing required 13 hours of flight time, about 6,400 miles, and around 10 inflight refuelings.

As the plan finally worked out VA-34 and VA-55 split targets in the vicinity of Benghazi, while the Air Force handled those around Tripoli. Seven *Blue Blaster* crews received a particularly difficult target – the al-Jamahiriyah Military Barracks, which housed Qadaffi's bodyguards in downtown Benghazi; concurrently eight VA-55 *Intruders* would bomb Benina Airfield. *Prowlers* and *Hornets* from both decks were tasked with SEAD support, with the "Ageless Warrior's" FA-18s handling CAP. In Tripoli nine F-111Fs would target the Bab al Aziziyah Barracks, which housed Libyan intelligence activities and one of Qadaffi's residences; three more would key on the Murat Sidi Bilala terrorist training facility, and the remaining three would hit Tripoli airfield (ironically, the latter was the Air Force's former Wheelus Airbase, which had served as a training and weaponry facility for USAFE units up until Qadaffi's accession to power). A brace of *Tomcats* from *America* were assigned the CAP duties for the Air Force component of the operation, while a mix of EF-111As, VMAQ-2 EA-6Bs off *America*, as well as HARM-shooting A-7Es, handled SEAD requirements. The planned time-on-target for all elements was 0200 local, 15 April 1986.

The "Aardvarks" of the 48th launched first, departing Lakenheath at 1936 hours on the 14th. A total of 24 F-111Fs plus five EF-111As from *the 42nd Electronic Combat Squadron* launched for the mission, along with 17 KC-10As and 13 KC-135s from RAF Mildenhall and RAF Fairford. After the first refueling six "Aardvarks" and a single "Spark 'Vark" turned and headed back to England, having fulfilled their duties as mission spares. Sharp-eyed English aircraft spotters noted the increased tempo and bomb-carrying "'Varks" and quickly put the word out that something was afoot, a security problem the carriers didn't have to face; Libya failed to notice, in any event.

The Navy's aircraft started launching at 0045 on 15 April. According to *Blue Blaster* B/N Lt. Dee Mewbourne there were a few immediate difficulties:

> "We had some problems with getting good targeting information from the ship. There wasn't much in the intell library on Libya or the targets in question. Still, we planned a high-speed attack, launched under EMCON, and performed a *very* interesting high-speed rendezvous. We went in as low and as covert as possible. The SINS wasn't working and the radars weren't lit off until we approached the coast.
>
> "We managed to fly past Benghazi; as we turned into the coast we noted the radar predictions didn't match what we were seeing. Aircraft one and three turned back north then east, and entered the target area as planned. The four others came in from the opposite direction and entered the target area as planned. The CO (Cmdr. Coleman, with Lt.Cmdr. Bill 'Frog Balls' Ballard) dropped the first bombs and No. 2 did a visual delivery. We did a backup delivery and got good hits on the target, as did everyone. The last three aircraft used Offset Aim Points and plastered the front gate."

VA-34's Lts. Joe Kuzmik and Bob Ayres were in dash six; Kuzmik has similar recollections of their squadron's raid on Benghazi:

> "I was a pretty junior B/N in VA-34 and didn't expect to fly the strike. We'd spent a lot of time working up plans for a variety of targets, and only found out that the F-111s would be involved about two days prior. At that time we were told to expect only four *Intruders* over our targets in Benghazi, which left me out. About a day prior we were told to send six, and my pilot, Bob Ayres, and I were laid on as dash-last. The target was the Revolutionary Guards barracks in downtown Benghazi, which was directed by higher authority, as was the bomb load. Four aircraft carried 16 Mk.82 *Snakeyes*, the last two eight Mk.83 1000-pounders with high drag mine fins (true *Snakeye* fins being unavailable for the Mk.83 at the time). Our plan was to go in very low in a bomber stream, a series of aircraft in a line. We would be the last over the target and well after the Skipper, which meant the air defenses would be fully alerted by the time we got there.
>
> "We covey launched from *America* with three bombers chasing a single KA-6D, executed a 1000-foot night over-water rendezvous, took gas, and setup for the push time. This sounds a lot easier than it really was. After the push we went in at 500-feet and were at 300-ft by coast-in. We weren't even feet dry when we could see the skipper's bombs going off, as well as some AAA and at least two SA-2s airborne. This was still five minutes before our TOT, so we knew things would be fully stirred up by the time we got there."

Kuzmik and Ayres were absolutely *right* about the reception, as Joe continues:

> "From feet dry we were 15 seconds to TOT, so we elevated to 500-ft AGL and I found the predicted radar points. The FLIR was down, which was no big deal, since I wasn't planning to use it in a fully lit-up city anyway. The bombs came off at the right time, and we made a 5G turn back to the water. I'd never seen that much ordnance explode at night before, and that – along with the flashes from our chaff squibs – initially made me think we were taking a lot of AAA, or maybe even on fire.
>
> "It was during the egress that we ran into trouble."

He adds:

> "We got indications of SAM activity and started jinking. We ballooned up to 1,200-ft and immediately were locked onto by an SA-3 site. They shot at least two missiles at us. I'm screaming at the pilot to get back down and move the aircraft, which he does, but not before one of the missiles passes right behind us, where it explodes. I've never seen anything move that fast in my life – from a dot on the canopy to a streak right by the aircraft: *VERY FAST*! Meanwhile, we were headed back to the ground in a big hurry, and we bottomed out at about 150-ft with a 5G pullout. We'd almost hit the ground trying to dodge the SAMS but had made it."

Rob Weber's *Warhorses* were scheduled to launch eight A-6Es from *Coral Sea* and strike Benghazi's Benina Airfield, while *Air Wing 13* EA-6Bs and FA-18s kept the Libyan air defense units' heads down. Two *Intruders* aborted after launch, in strict accordance with the ROE, which required "fully up" systems to reduce the chances of bombs hitting civilian targets. The remaining six A-6s dropped their loads of cluster weapons dead on target and were credited with the destruction of three MiG-23 *Floggers*, two Mi-8 helos, and one Dutch-built Fokker F-27 turboprop transport. They also inflicted damage on several other aircraft and hangar facilities at the field while trashing the runways. Reports on the urban area targets were more mixed, but afterwards VA-34 reported:

> "... (the squadron) struck Libyan terrorist barracks and aircraft storage facilities, inflicting over 70 casualties to the enemy and virtually eliminating Libyan MiG-23 spare parts inventory."

In a tremendous demonstration of professional airmanship the Air Force's strike package also hit its targets as scheduled following four inflight refuelings each – 15 hours and over 2,800 miles. Two of the F-111s could not push due to mechanical problems, leaving seven to bomb Al Aziziyah; three more failed to drop and aborted feet wet in accordance with the ROE. The other Air Force elements saw more success, with the best – and most spectacular – results gained at Tripoli's airfield. Dropping parachute-retarded Mk.82 *Snakeyes*, they managed to destroy three Il-76s, a Fiat G.222, and a Boeing 727, while damaging three other Ilyushin transports. At the Sidi Balal terrorist training facility, the Air Force crews also managed creditable hits using 2000-pound GBU-10 *Paveway 2* LGBs, among other ordnance. Notably, during the process someone also managed to dump one bomb near the French embassy, a fact that was highly appreciated by Navy and Air Force personnel alike.

All of the mission aircraft – both Navy and Air Force – recovered, with the exception of one F-111F. *Karma 52*, flown by Capts. Fernando Ribas-Dominicci and Weapon Systems Officer Paul Lorence, who did not survive the flight and was either hit or flew into the water avoiding fire. Still, 13 of the 48th TFW "Aardvarks" had made it over their targets in the Tripoli area out of 18 "go" aircraft launched from England with the support of four EF-111A *Ravens*, 29 tankers, and *America's* CVW-1. The Air Force portion of the 6000-mile raid took almost 15 hours to conduct and remains the longest strike in history ever conducted by tactical aircraft.

After the strikes Thatcher stood by her actions despite a hail of political attacks in Parliament and noisy protests in England. She told Parliament:

> "We had to consider the wider implications, including our relations with other countries, and we had to weigh the importance for our security of our alliance with the U.S. and the American role in the defense of Europe. Had Britain refused permission for the use of these aircraft, the American operation would still have taken place, but more lives would probably have been lost on the ground and in the air."

The response from the other NATO allies was more critical. Greek Prime Minster Andreas Papandreou referred to the raid as a "threat to peace," while Danish Prime Minister Poul Schlueter stated, "I do not believe a sole military solution is possible." German Chancellor Helmut Kohl was somewhat more understanding, commenting:

> "Force is not a very promising way of dealing with things. Most of the victims of this terrorism are American, and I understand the growing outrage among Americans."

There were several protests throughout Europe, including one riot in West Berlin that saw several U.S.-owned business and banks vandalized.

As for the United States, the Reagan administration announced the attacks were designed specifically to "destroy the nerve center of terrorism ... and demonstrate resolve." Somewhat in keeping with the precedent set by Grenada, both services offered up medals all around. However, the 48th TFW was awarded a Navy Meritorious Unit Commendation for its participation.

Still, the "Aardvark" – and "Spark 'Vark" – crews had completed the longest combat-related tactical fighter mission in history with highly satisfactory results. More notably, *El Dorado Canyon* was the first strike to have major press and public relations implications through the judicious post-combat use of target video. Some of the best shots were of Il-76s going up in smoke and flames, all of which was provided by the AN/AVQ-26 *Pave Tack* pods on the F-111s. The Navy took note of this lesson and would apply it five years later over Iraq and Kuwait.

Situation Normal

Following the raid the frequency of terrorist attacks on Americans overseas subsided dramatically. Qadaffi himself disappeared for 24 hours after the raid, leading some to believe he was either severely wounded or dead. Once he did turn up his spokesmen announced 130 Libyan civilians had died in the raids, including the leader's adopted infant daughter. There was no popular overthrow, but apparently *El Dorado Canyon* finally put the fear of death in Qadaffi, and he settled down, save for occasional bursts of vitriol. Apparently even the Soviets got fed up with his propensity for sticking his head in the beehive and the general ineptness of the Libyan military, and they quietly started pulling out. Even support from other Arab countries was restrained.

Still, Sixth Fleet continued to keep the big stick within striking distance of Libya and other hot spots. On 28 April *Enterprise* – with CVW-11 and VA-95 onboard – transited the Suez Canal northbound. Dispatched from the IO to relieve *Coral Sea*, she was the first nuclear carrier ever to go through the "ditch." The big CVN started her transit at 0300 and made it through in about 12 hours. Properly relieved, the tired – but still potent – *Coral Sea*/CVW-13 *Guardians* team finally pulled out and headed for home, having participated in two short, sharp actions. Early-deployment flailing over lack of organic *Prowler* support aside, the wing and its squadrons had done exceedingly well for themselves.

For the *Warhorses* of VA-55 it had been one hell of a cruise, and quite an initiation into fleet medium attack operations. Their return to Oceana on 19 May 1986 came with no small amount of celebration, but there was some other business to attend to. On 27 June XO Cmdr. Warren C. "Gator" Chewning said the traditional "I relieve you sir" during ceremonies at the Tidewater air station and replaced Rob Weber as *Warhorse One*, while ace A-6 LSO Cmdr. Ralph E. "Benny" Suggs moved into the squadron's XO slot.

Back in the Med, *America* and *Enterprise* manned the watch, with *Forrestal* and CVW-6 relieving the nuke in June. Onboard *FID* was VA-176, now led by Cmdr. Gary W. Stubbs and making its first ever

Forrestal deployment. They remained in the theater through the late fall. *America* and CVW-1 had seen enough excitement for one deployment, and the remainder of their 1986 cruise stayed fairly quiet. Of course, there are always exceptions, as recalled by then-JO Rick "Pokey" Keller:

> "During *El Dorado Canyon* I was a second cruise lieutenant, one of about ten or twelve who had come into the squadron in 1983. I had flown with the CO, then the XO, and by 1986 was flying with a Marine exchange pilot. Initially it was going to be a four-plane raid, and then they pumped it up to six, but they looked at the crews, and seeing as how I'd just changed pilots, I didn't get to go. We flew a SuCAP mission with *Harpoon* and *Rockeye*, about 20 miles down towards Benghazi, and then across to Tripoli, but no one was moving that day."

When the fun times did come for Keller, it was from a somewhat unexpected direction:

> "I had one bad pilot during my time with VA-34, and it was a Marine LSO from VMA(AW)-533. He was a typical new guy, nervous ... we hit the KC-135 the first time and plugged, no problem, but were not involved in the Libyan stuff. We missed shooting *Harpoon* by one cycle, dammit! We listened to the guys while we were in marshall, and it was frustrating.
>
> "Anyway, during this period he'd go med-down periodically, but we didn't think anything of it," Keller comments. "We recovered at 0400, and at 0845 got rousted out ... we were going in to hit a terrorist camp, fourth plane in a four-plane strike ... we're planning, he's glassy-eyed, catatonic. Forget this, we were assigned to the ready-five tanker. We're sitting in the cockpit, he's bent forward over the stick, has his arms over his head. I called him on the ICS – 'Hey, you okay?' – 'No, Pokey, I'm scared.' 'That's okay, it's normal to be scared.' 'No, I am *really* scared.'
>
> "I made my big mistake. Instead of calling CAG and saying we were down for gutlessness in the cockpit, I gave him my best John Wayne Speech: God, Country, they're relying on us, now let's get going and do the mission. I should've called CAG. We launched, and he was like a voice-actuated autopilot. We damn near had a mid-air ... had this A-7 hooked up, I was working on something when all of a sudden he pulls 6 Gs. I look up and see the underside of another A-7 *right in front of us*. The A-7 on the basket stayed hooked up, called the other guy, and talked him through it. He bottomed out at about 1,000-feet."

That was it for the hop, according to Keller:

> "Back to the ship, took a taxi one-wire. I kissed the deck and told the ops officer I wasn't going flying with *this* guy again. The guy kept saying he wanted to go back to the Marines. However, VMAQ-2 Det Y – onboard – they wanted a court-martial. They did provide the officers for the FNAEB (Field Naval Aviator Evaluation Board), and he ended up in intelligence. Afterwards I was supposed to switch to (XO Cmdr. Bernis H.) 'Butch' Bailey; oh good! I like that! Instead, I got crewed with Doug McCain, the senator's son."

The *Blasters* returned to Oceana on 10 September 1986, also having had an exceptional cruise, and shortly afterwards Keller detached for instructor duty at VA-42. Looking back, he feels his first fleet tour with VA-34 was a great one, occasional contretemps aside:

> "J.B. Dadson was a character," he recalls. "The JOs loved him, the troops loved him, and the squadron did well under him. We did even better under 'Breeze" Coleman."

The squadron did lose one aircraft, KA-6D 151783, immediately prior to the combat deployment. Keller remembers it went down off Roosevelt Roads:

> "The pilot was Jim Casey and the B/N Sam Young; they were in a tanker at Rosey, hot seating, and didn't make the launch. They took a 54K downwind shot on Cat 4 with a soft nose strut, and the wheel failed and FOD'd both engines. Everyone started yelling 'GET OUT! GET OUT' so they bailed. The B/N was a nugget, about 1 1/2 months in the squadron. This was before they had SEA-WARs (a salt activated parachute release system), and he got tangled in the chute. He pulled loose and later said he was thinking, 'Thank God I'm alive and I'm not going to get run over by the ship.' The raft was about 35-feet down when he pulled the CO_2. The raft comes up at a high rate of speed and hits him, and he thinks, 'OHMYGAWD, A SHARK!,' gets out his K-Bar knife, and stabs the raft four times as it comes flying out of the water."

Both Casey and Young returned to flight status the following day. As for the squadron, the mishap was the only real negative on an otherwise eventful cruise, at least until after the cruise:

> "We were MATWING's 1986 nominee for the Battle E," Keller concludes. "After we lost the plane during ORE, AIRLANT sent the nomination back, saying 'you have *got* to be kidding!'"

More Marines at Sea

At the time of VMAQ-2 det's assignment to CVW-1 with the *Blue Blasters*, the concept of putting Marine *Intruders* to sea was still something of a sporadic thing, almost a concept of wonder and disbelief. To be sure, between VMA(AW)-224's epochal combat cruise in *Coral Sea* during the late Vietnam War and -533's deployment with CVW-17 in *Saratoga* twelve years later, Marine A-6 crews had proven themselves more than capable around the boat. However, John Lehman wanted more *Intruders* at sea, and to that end was working up plans to establish more squadrons. In the interim, his staff types realized they would need to come up with someone to replace VA-85 in the "All Grumman Air Wing," CVW-3. Who to send? The *Night Hawks* of -533, barely two years after their previous deployment.

The assignment of VMA(AW)-533 to *John F. Kennedy*, where it joined VA-75, meant both Fleet Marine Force Atlantic (FMFLant) and FMFPac now had A-6 squadrons assigned to carrier air wings on a "semi-permanent" basis. What with El Toro's VMA(AW)-121 under assignment to CVW-2 – alongside the *Swordsmen* – seeing "globe, anchor, and eagle," A-6s at sea all of a sudden got to be a common occurrence. However, even with the *Green Knights* in extended – and seemingly endless – workups for WestPac surge ops and -533's fairly recent boat time, there were still a lot of new Marines to run through CarQuals. That fact undoubtedly contributed to the high number of gray hairs and worry lines on the faces of the RAG and air wing LSOs, as related through several vignettes in *The Intruder's Lighter Side* by Brian "Neubs" Neunaber, a *Golden Intruder* LSO:

> "The Marines got serious about CQ in the 'eighties. Just how serious we saw when we disqualed our first USMC Major. We were pleased HQ Marine Corps elected to extend this gent's training track to allow a second attempt, and he did great. Nevertheless, seeing these guys check into CQ Phase with 3,000 A-6 hours and only *six* traps in their logbooks caused us some concern.
>
> "One day at Coupeville a different Marine O-4 pilot and his Cat I USMC B/N entered a four-plane pattern for RWY 32. With a full pattern and the RWY 32 keyhole cutout in the trees I didn't see these guys until they were through the 45...and they were low and deep:
>
> "'811, Intruder Ball.'

During the 1980s the idea of a Marine squadron joining a Navy air wing for carrier duty was still considered by many to be something of an "unnatural act." Nonetheless, several Marine *Intruder* (as well as *Phantom* and *Hornet)* squadrons made successful deployments with their Naval brethren during this period. VMA (AW)-533 was lucky enough to make three deployments over a five-year period. This *Intruder* is seen trapping on *Kennedy.* (Rick Burgess)

> "'811, you're long in the groove, wave off! You're too low. Check your altitude. 'What's the problem?'
>
> "'Paddles, Paddles ... we have an airspeed problem.'
>
> "Now wings level, 1.5 miles in the groove at about 250 feet, I could see the resolution to the problem from the LSO shack ... '811, drop your flaps and slats and let me know how it works out.'"

Of course, the Navy crews also kept them jumping on occasion, as recalled by another former VA-128 LSO:

> "Paddles was on station for OLF Coupeville Runway 32. An aircraft reported '5 miles North of Runway 32' and paddles responded, 'Continue, you're number one, not in sight.' About a minute goes by and the aircraft reports, 'Initial.' Paddles replies, 'Continue, still not in sight.' Now paddles turns on the runway lights to full bright, in addition to the carrier box lights. Another minute goes by and the aircraft reports, 'Numbers.' Paddles, now wondering what's going on, asks, 'Not in sight ... check external lights on and say your TACAN cut from Whidbey.' The reply was 'the 190 for 20' (I'm not quite sure what TACAN he had selected, or whether he spoke the truth).
>
> "As the story unfolded later, I was told there weren't enough phones at base ops for all the noise complaints. All the complaints came from the Bremerton area (over 40 miles south of Whidbey), so they must have attempted an approach to Bremerton Muni. After realizing their major FUBAR, they got vectors to Coupeville. Sure would have liked to have been a fly on the wall for that debrief!"

Truth be known, Bremerton Municipal Airport – now Bremerton National – *did* serve the Navy during World War II as an outlying training field and support installation for Puget Sound Navy Yard. An honest mistake ... "Neubs" Neunaber has several similar stories concerning the Navy crews running through the program:

"I've conducted A-6 FCLP at Oceana, Fentress OLF, Whitehouse OLF, Choctaw OLF, NAS Sigonella, El Centro, Miramar, Whidbey, Coupeville OLF, Grant County (Wash.) Municipal, Ephrata (Wash.) Municipal ... and once on Camano Island (Camano Island, for the uninitiated, is the other half of Island County, Washington. Whidbey, in the Puget Sound, is the larger of the two islands, while Camano is nestled up against the coast between Mount Vernon and Everett/Marysville). The scene is set on a dark, but VFR night, when we were bouncing at Coupeville on Runway 14," Neunaber writes:

> "It had gotten kinda quiet, waiting for the next student to show, when I hear:
>
> "'805, Initial.'
>
> "'805, Paddles. You're Number One for Runway 14, the pattern is clear, winds are calm. Continue, I don't see you.'
>
> "Seconds later ...
>
> "'805, abeam, pilot is ... (he still denies it to this day) ... and then, '805, Intruder Ball, six point five.'
>
> "'805, are you sure you've got the ball? I don't see you.'
>
> "'805, Ball.'"
>
> "By now I thought, he must be 'in the middle,' then another few too many seconds...
>
> "'WAVE OFF 805, COME TEN MILES WEST. YOU'RE OVER CAMANO ISLAND!'
>
> "Noise complaints, anyone?"

FCLPs were one thing; getting the replacements out to the boat was another, as Neunaber continues:

> "In 20 CQ Dets at VA-128 I only ever had one student cause guys to vacate the LSO Platform via the Safety Net. 'Slam' Dunkle was waving (or is it waiving? I've always wondered), and I was backing up this gent's first night pass while on a SOCAL CQ Det. The pass went like this:
>
> <u>Settle,</u> Drift right start.
> I hit the waveoff lights and let out with three 'waveoff' calls.
> Pull nose up on low, right-to-left in-the-middle to in-close.

"As the aircraft was low and right, the pilot advanced the throttles to military and yanked back on the stick. But now he was so cocked up that MRT did not sound right ... I estimate that he went to 27-28 units AOA – and stayed there. After the three 'waveoff' calls, I went into 12 to 15 calls for 'POWER' – then not understanding why those J-52 motors didn't offer the glorious sound of freedom.

"Well, 12 LSOs and LSO-wannabees jumped into the net. Me and the 'Slammer' just stood there, as the jet continued left. We could only duck, as the aircraft's tailhook passed about five feet over the LSO Platform JBD. His next pass was also a waveoff. One LSO went into the net again ... and we sent this FRP and his IBN to the beach for the night. He qual'd that det ... with no magic LSO math."[4]

On 18 August 1986 *John F. Kennedy* moved away from the pier at NAVSTA Norfolk and commenced a "standard" Med deployment. The battle group left under Rear Adm. G.A. Sharp, *Commander Cruiser-Destroyer Group Two*; CAG-3 was noted *Intruder* pilot Cmdr. Robert E. "Bob" Houser. Upon their arrival they relieved *America* and CVW-1, and the *Blue Blasters* finally returned home to Oceana on 10 September, having completed one hell of a combat cruise. For skipper Rich Coleman the deployment also brought two notable milestones: his 700th trap onboard CV-66 – making him the first to log that number of traps on that deck – and over 4,000 hours of flight time. The following June Coleman completed an auspicious tour, turning VA-34 over to Cmdr. Butch Bailey, with Cmdr. Eugene K. Nielsen assuming the responsibilities of executive officer. Coleman moved up to the DCAG slot in CVW-3.

Back on *Big John*, the *Sunday Punchers* and *Night Hawks* settled into the cruise routine, participating in multiple exercises, such as NATO's *Display Determination* and *African Eagle* while taking in the various and sundry liberty ports around the Med. For the Marines in particular the deployment provided a unique opportunity to plan and fly with allied air forces and aircraft, including operations with Turkish F-104Gs, French F-8E(FN)s, and Royal Moroccan F-5Es. The Navy's eleventh and newest *Prowler* squadron, VAQ-140, also made the trip; equipped with ICAP-II (Improved Capability) birds, the *Patriots* went to sea with the AGM-88 *HARM* anti-radiation missile, a first for the community. However, in order to ensure an orderly *Iron Hand* transition within the air wing the Navy also sent along half a squadron's worth of interlopers of the Vought persuasion: several A-7Es assigned to "Attack Det 66."

The presence of the SLUFs in the "All Grumman Air Wing" was in itself an interesting sidebar of the deployment. With only four aircraft in CVW-3 capable of shooting the HARM (A-6s would not get the weapon until the SWIP, several years down the road), AirLant decided to place a six aircraft detachment of A-7Es onboard to supplement the Wing's lethal SEAD capability. The *Waldomen* of VA-66 were selected, even though the squadron was scheduled to soon disestablish. Thus, with only 96 hours notice 12 Os and 118 Es from the squadron gathered up a passel of *Corsairs* and went aboard *Kennedy* to back up the *Patriots*.

[4] Whidbey's bounce field at Coupeville is located roughly mid-way down the island in a mixed rural/residential area. To the south the island's "artsy-craftsy" community stands in contrast to the Navy-dominated northern half. Not to be outdone, Oceana's primary FCLP field, Fentress, is largely surrounded by suburban build-up which has led to ever increasing noise complaints from taxpayers who don't appreciate Navy jets flying circuits to the wee hours of the night. There have been incidents of aircraft taking ground fire- and being hit- at both locations and the tale of at least one case of an enraged citizen driving a truck onto Fentress and chasing the LSOs one fine evening in the 1980s.

Disestablishment of the home squadron at NAS Cecil Field continued while Attack Det 66 was at sea. On 1 October the detachment's members gathered on the flight deck and held a somewhat less-than-solemn ceremony to commemorate the formal disestablishment of their home unit; the det, however, remained on board for the entire cruise. Eventually the Navy's *Prowlers* and *Hornets* assumed the AGM-88 mission for the air wings, but for the rapidly dwindling light attack community "Attack Det 66" provided some welcome late-career visibility for the *Corsair II* and its personnel.

The *Punchers* got in a little bit of fun on 21 October when they participated in a SinkEx against a decommissioned Italian Navy frigate. Operating in the eastern Med at the time, a VA-75 *Intruder* with skipper Cmdr. Al Harms and Lt.Cmdr. Keith Naumann pumped a *Harpoon* – sans warhead – *into* the target hulk. One of the *Kennedy* battle group's surface combatants then fired a second similar missile into the target ship. Working with other air wing aircraft, VA-75 crews subsequently sent the frigate to the bottom using a mix of laser-guided and conventional munitions.

Four days later the deployment took a major downturn for all concerned when a VS-22 *Hoover* failed to return for recovery. The wing quickly mobilized, launched, and scattered to search for the missing *Checkmate* S-3A. Tragically, at one point during the operation a -533 *Intruder* impacted the water while trying to get photos of a Greek freighter, killing the crew of Capts. Tim Morrison and Russ Schindelheim. A second *Night Hawks* A-6 went down later in the cruise on 17 January 1987; again the cause of the crash was an impact with the surface, and again the crew – Capt. Mike Gonzalez and 1st Lt. Bob Cox – died. Gonzalez was rescued the following day after a terrible storm at sea but passed on due to exposure and hypothermia before they could get him to sickbay.

On 28 January the ship and wing received notification they would be extended on deployment due to several hiccups in the vicinity of Lebanon. *Nimitz* – which arrived in January with CVW-8 and VA-35 as *Big John's* relief – instead operated alongside *Kennedy* for nearly two months. Once finally released CV-67 turned for home, returning to Norfolk on 3 March 1987. With their return to Cherry Point VMA(AW)-533 had completed two Med cruises in slightly over three years. As for Al Harms' *Sunday Punchers*, during the deployment they qualified 25 *Kennedy* Centurions, while the skipper tallied his 2500th *Intruder* hour. Also setting personal milestones were Lt.Cmdr. Keith Naumann, with 1,500 hours, and Lt.Cmdrs. Bill West and Joe O'Donnell, with 1,000 hours each in the A-6.

As the *Night Hawks* and *Punchers* wrapped up in the Med the Navy finally let their AirPac counterparts in *Ranger* out of the barn. After over two years of assignment to CVW-2 and two short "surge op" deployments the *Green Knights* of VMA(AW)-121 finally got their turn with a full deployment in July 1987. The last time -121 had gone out on a carrier was in 1960 when the squadron – flying A4D-2s – joined CVG-15 in *Coral Sea* for a WestPac. Now, working with the *Swordsmen* of VA-145, they were going out again as part of a regular carrier deployment. After two years of training and preparation, the "surge ops" approach had proven both unpopular and expensive, particularly due to high transit times, cost, and stress on the personnel.

Ranger departed NAS North Island on 14 July heading for WestPac and the IO via Operation *ReadiEx 87* off Hawaii. The cruise "only" lasted five and one-half months, but there was a lengthy period in the North Arabian Sea, including one stretch of 95 straight days. Once in the Indian Ocean the big carrier participated in operations with French counterpart the *Clemenceau*, giving the two *Intruder* squadrons a chance to play with the French air wing's *Crusaders* and *Super Etendards*. On the down side, the *Knights* dumped a plane on 23 October when BuNo 155469 suffered a dual-engine flameout. Fortunately, the crew was recovered without further incident. Otherwise the carrier, her air wing, and battle group spent a lot of time escorting tankers through the Persian Gulf. Notably, throughout the proceedings *Ranger* served as flag for both *Commander, Carrier Group One* and *Commander, Battle Group Echo*; the force totaled 17 ships, including Battleship *Missouri* (BB-63).

Led by Cmdr. Steven "Axel" Hazelrig – who relieved Ken Bixler on 2 November – the *Swordsmen* had an outstanding deployment, culminating in their winning the "Golden Tailhook" award for best cruise landing grades; pilot Lt. Brad "Coz" Cosgrove took the award of Top Tailhooker for the excursion. The Marines of VMA(AW)-121 also notched several notables, including the first ever launch of a *Harpoon* by a Fleet Marine Force aircraft. Capt.'s J.P. "Blotto" Stevens and Steve "Nick" Nicholson did the honors in plugging the target hulk. In addition, the skipper, Lt.Col. Joe "Java" Weston, rang up his 4,000th *Intruder* flight hour.

Ranger completed her cruise on 30 December 1987, with the *Swordsmen* returning to Whidbey Island and the *Green Knights* going home to MAG-11 at MCAS El Toro. As with their AirLant/FMFLant counterparts, they had proven adept in air wing operations with two squadrons of A-6s, and like their east coast brethren would get another chance in approximately a year. In fact, for the Navy overall operations with two squadrons of *Intruders* in select air wings was proving to be a pretty good idea, even with the oncoming flood of FA-18 *Hornets*. However, the Marine A-6 community remained a small one, and if the Navy hoped to expand its medium attack presence for any period of time, it would need to stand up several more squadrons. Amazingly enough that is exactly what happened.

One Going, Several Coming

At the end of 1986 the Navy had 15 carriers in service. The oldest, *Midway*, dated her commissioning to 10 September 1945 and remained forward deployed in Japan. Her sister *Coral Sea* wasn't much younger, with a commissioning date of 1 October 1947. While the "Ageless Warrior" had performed sterling recent service with the FA-18/A-6E-equipped CVW-13, the Navy knew her days were drawing to a close.

As for the big decks, *Forrestal* still soldiered on at 31 years of age; *Sara* had completed SLEP at Philadelphia Naval Shipyard in 1983 and was replaced by *Independence*, which entered the yard in February 1985; and *Kitty Hawk* and *Constellation* were scheduled to follow. The newest *Nimitz*-class nuclear carrier, *Theodore Roosevelt*, was commissioned on 25 October 1986 and was undergoing sea trials while the keels went down for *Abraham Lincoln* and *George Washington* on 3 November 1984 and 25 August 1986, respectively. The Navy proposed a third and fourth carrier – *John C. Stennis* (CVN-74) and the as-yet unnamed CVN-75 – under the FY88 shipbuilding budget, but budget limitations and a recalcitrant Congress delayed their funding. Indeed the cool, considerate gentlemen of Capitol Hill concurrently attempted to force the early retirement of *Midway*, which suffered from handling problems following the installation of "stabilizing" blisters. The Senate effort failed, but work on *Stennis* did not commence at Newport News until 1991, long after both Reagan and Lehman left office.

That left 12 available decks with four more coming, and their arrival would allow the retirement of the four oldest carriers, or at least the ones in the worst condition. To equip these carriers the Navy concurrently worked at providing enough air wings in and around rearranging squadron assignments. Entering 1987 the service had 13 operational carrier air wings: AirLant's CVW-1 (tailcode **AB**), CVW-3 (**AC**), CVW-6 (**AE**), CVW-7 (**AG**), CVW-8 (**AJ**), CVW-13 (**AK**) and CVW-17 (**AA**); and AirPac's CVW-2 (**NE**), CVW-5 (**NF**), CVW-9 (**NG**), CVW-11 (**NH**), CVW-14 (**NK**) and CVW-15 (**NL**). Adding to the mix were the two reserve wings, CVWRs -20 (**AF**) and -30 (**ND**). Of this illustrious group two wings had dual squadrons of *Intruders* assigned: *Kennedy's* CVW-3 and *Ranger's* CVW-2. However, each of these wings used one squadron of Marine A-6s, and John Lehman wanted more of a Navy *Intruder* presence on the decks. To achieve that end he directed the establishment of more A-6 squadrons, along with one more CVW. The Navy figured with one carrier always in overhaul or otherwise available, it would need that fourteenth wing. If circumstances like a national emergency directed manning and equipping a fifteenth deck one of the two reserve CVWs could theoretically fill in.

Hence, on 1 November 1986 *Carrier Air Wing 10* stood up at NAS Miramar, in the process gaining the famous "NM" tailcode from the former CVW-19. Its predecessor was an AirLant unit established as Carrier Air Group 10 on 1 May 1952, and eventually assigned the "AK" tailcode. Redesignated CVW-10 on 20 December 1963, the wing made three SEA deployments in *Intrepid* before disestablishing on 20 November 1969. Cmdr. Paul "Paco" Campbell led the new edition of the wing, which was penciled in for assignment to *Nimitz* when it replaced *Kitty Hawk* in the Pacific Fleet. The Navy also established six new squadrons for the wing: VFs -191 and -194; VAW-111; VS-35; HS-16; and VA-185. Two FA-18 units and a squadron of *Prowlers* were supposed to follow.

VA-185's squadron's establishment ceremony was held at NAS Whidbey Island on 1 December 1986, marking it the first of three new A-6 operators to stand up in the coming year, and marking the largest number of A-6 squadron activations since the first half of 1968 when VA-145, VA-176, and VMA(AW)-332 all switched over to the *Intruder*.

With no lineage or similarly-numbered predecessors, the men of the -185 set to work in preparation for operations with CVW-10 under skipper Cmdr. William J. Magnan and XO Michael J. "Crash" McCamish. For their name, they too chose *Nighthawks*, and for their emblem they used a diving eagle. However, much to everyone's surprise, VA-185's assignment to CVW-10 was short lived. In mid-1987

The *Green Knights* of VMA (AW)-121 were assigned to CVW-2 in 1986 and spent over three years with the Air Wing. NE400 is shown during workups at Fallon in February 1986. (Michael Grove)

The *Nighthawks* of VA-185 were established at Whidbey Island in December 1986. After only nine months in CONUS the squadron moved to join VA-115 as the second *Intruder* squadron in Japan. Already wearing the "NF" of CVW-5, 402 is on the Whidbey ramp roughly six weeks before their overseas move. (Rick Morgan)

the squadron was told it was going to move to Japan and join CVW-5 as the second A-6 operator in *Midway*. The first thing on many minds was what to do with their Whidbey property. The move brought that wing in line with the current "*Coral Sea* Air Wing" configuration, now sporting three FA-18 operators, two each *Intruders*, and one each VAW, VAQ, and HS.

On the East Coast *Coral Sea* had gained her second *Intruder* squadron on 12 September 1986 with the assignment of VA-65. In joining VA-55 in CVW-13 the *Tigers* continued SecNav's plan – approved on 1 September 1986 – of increasing all-weather attack assets per wing while reducing the total number of support personnel. Due to catapult and hangar deck limitations, neither of the surviving *Midway*-class ships gained a fighter squadron, making do instead with three *Hornet* units. Nominally, each A-6 squadron received eight aircraft with some shared maintenance operations. The *Nighthawks* packed up, and on 13 September 1987 formally reported to CVW-5 at Atsugi. By that date two additional A-6 squadrons had stood up; although no one could know it at the time, they were the last.

The first came on 6 March 1987 when the VA-36 designator returned to service in ceremonies at NAS Oceana. The squadron adopted the *Roadrunner* nickname of its predecessor, a long-time A-4 operator out of Cecil Field, but chose a more accurate rendition of the speedy southwestern fowl vice the earlier cartoon version. Officially the new *Intruder* operator was the second Navy attack squadron to bear the VA-36 designation, however – in one of those weird periods Naval Aviation periodically goes through – there *had* been a third, for less than 24 hours. On 1 July 1955 VA-36 established at NAS Jacksonville with F2H-4s and immediately redesignated as VF-102. Concurrently, the existing VF-102, an F9F-5 operator, was redesignated as VA-36. Following an initial WestPac deployment with *Air Task Group 201* in

The *Roadrunners* of VA-36 were established in March 1987 and assigned to CVW-8 and the new *Theodore Roosevelt* (CVN-71). AJ531 is seen here in June 1990. It would be lost with both of its crew in combat during *Desert Storm* on 2 February 1991. (Rick Morgan)

Bennington the *Roadrunners* ultimately transferred to CVG-3 and commenced a lengthy series of trips to the Med. The squadron made two Vietnam deployments – with CVW-9 in *Enterprise* over 1965-1966 and with CVW-10 in *Intrepid*, 1968-1969 – before retiring its C-model *Scooters* at Cecil on 1 August 1970.

Notably – and despite Navy policy – then NavAirLant Chief of Staff Capt. Fred Metz played a direct role in the selection of the VA-36 designator; he had also selected VA-55 when the previous MATWING-1 squadron stood up four years earlier. Metz made no bones about the fact he felt new units should have a historic background, no matter what the Navy thought. The fact that the previous *Roadrunners* had been Metz's first fleet squadron – prior to his transition to *Intruders* and combat tours with VA-85 – didn't hurt. Suitably equipped and led by Cmdr. T. Lamar Willis, the new VA-36 checked into *Carrier Air Wing 8* in September 1987 and started preparations for their first deployment.

About that time – and at the other end of the country – came the fleet's sixteenth and last *Intruder* squadron. At NAS Whidbey Island on 1 September 1987 the *Silver Foxes* of VA-155 established under Cmdr. Jack J. "Black Jack" Samar. Again, Metz played a direct role; by now he was a rear admiral and commanded the wing at Whidbey Island, and it was reportedly he who selected the VA-155 designator. However, the classic "Skull and Snake" patch did not survive from the predecessor *Corsair* operator, at least initially. Samar was a big fan of the National Football League's Los Angeles Raiders and had a new patch devised with the appropriate colors, while coming up with a motto, "Medium Attack in Silver and Black." The emblem featured a silver fox head with two very subtle pieces of shading on its face: a shamrock, for the skipper's alma mater Notre Dame, and an outline of Whidbey Island. Both were frequently lost in subsequent reproductions.

The new VA-155 was the third squadron to bear the number. The initial version stood up as VT-153 at NAS Sanford, FL, on 26 March 1945; it was redesignated as VA-16A on 15 November 1946, made one deployment in *Antietam* (CV-36), and redesignated again to VA-155 on 15 July 1948, prior to its 30 November 1949 disestablishment. The previous edition was originally a reserve AD outfit, VA-71E out of NAS Glenview. As VA-728 the squadron went into combat in Korea from *Antietam* with CVG-15 in 1951. The outfit redesignated as VA-155 on 4 February 1953 and proceeded through three more variants of the noble "Spad" before swapping out for A4D-2s in 1959. The *Foxes* made three Vietnam deployments with CVW-15 in *Coral Sea* and *Constellation* before moving to CVW-2 and *Ranger* in 1968. Following transition to A-7B *Corsair IIs* in 1970 they transferred to CVW-19/*Oriskany*, where they worked alongside the equally famous *Blue Tail Flies* of VA-153 and *Barn Owls* of VA-215. In 1976/77 they accompanied CVW-19 on the farewell cruise in *Franklin D. Roosevelt* and disestablished on 30 September 1977 at NAS Lemoore, having completed 19 deployments in 10 carriers over 31 years of service.

VA-155 was originally established to join the short-lived CVW-10. NM501 displays the air wing's code for the squadron's establishment ceremony, 1 September 1987. (Rick Morgan)

With the establishment of the "new" *Silver Foxes* at the Rock, CVW-10 finally gained a replacement for the departed VA-185. However the assignment didn't last long; the demise of the Navy's fourteenth air wing on June 1988 resulted in the disestablishment of all of its squadrons with the exception of VAQ-141 – established 1 July 1987, but never officially assigned to CVW-10 – and VA-155. With its wing's demise the *Foxes* transferred to CVW-17 with the intention of cruising in the Med with VA-85 in *Saratoga*. As such, it became the only AirPac A-6 squadron ever assigned to an AirLant wing. Part of the squadron joined a portion of its new Air Wing as it accompanied *Independence* for its post-SLEP trip around the Horn to San Diego from August to October 1988. With that trip complete plans changed again, and -155 found itself assigned to CVW-2, replacing VMA(AW)-121 and joining VA-145 as the second *Intruder* operator in *Ranger*.

However, this turnaround in wing and squadron fortunes was still some months in the future. Having played a direct role in the "re-estab-

VA-155 had KA-6Ds assigned for a short period prior to its assignment to CVW-2. NM514 was hit by a VA-128 aircraft while involved in night tanking training at El Centro in February 1988. The pilot was able to safely recover in spite of the complete loss of the rudder and the starboard horizontal stabilizer. The errant replacement pilot was able to land his *Intruder*, as well. (Bob Lawson)

The *Double Eagles* of VMAT (AW)-202 folded their tent at Cherry Point in September 1986, with Marine *Intruder* training moving to VA-128 in Whidbey. (David Brown)

lishment" of the *Silver Foxes*, Adm. Metz still had his sights set on the still expanding medium attack community at Whidbey; there were still several air wings that needed a second all-weather attack asset, and he had several historic numbers and names in mind.

In May 1987 Metz fired off a letter to *Commander, Naval Air Forces, U.S. Pacific Fleet* recommending the next six squadrons stand up as VAs -153, -93, -212, -164, -215, and -152. The stated purpose was:

> "... rather than creating a new squadron with no history, I recommend the recommissioning (sic) of an old attack squadron, providing the new squadron the rich heritage and tradition bequeathed by the combat tested squadrons listed in enclosed (1) (CMVWP ltr 312/1155, 21 May 1987)."

Despite the fact the Navy did not "recommission" squadrons – i.e., when a squadron disestablished, it and its lineage and combat honors went away for good – conceivably Metz's plan would have brought back the *Blue Tail Flies*, *Blue Blazers*, *Rampant Raiders*, *Ghost Riders*, *Barn Owls*, and *Wild Aces* as *Intruder* operators out of a crowded NAS Whidbey Island.[5]

However, it was not to be. The Reagan/Lehman buildup of the military in general, and Naval Aviation in particular, was just about tapped out, and there would be no additional A-6 squadrons. Indeed, the Marines had already started taking steps to reduce their *Intruder* component. On 30 Sept 1986 VMAT(AW)-202 at MCAS Cherry Point disestablished, ending 17 years' service as the Marines' A-6 RAG. The last skipper, Lt.Col. Randy Fridley, presided over the ceremonies, which saw only six A-6Es and one TC-4C remaining on the ramp at "Cheerless Pit." With the demise of the *Double Eagle* all Marine pilot, B/N, and maintenance replacement training transferred to VAQ-128 at Whidbey.

The five Fleet Marine A-6 squadrons would eventually turn in their trusty *Intruders* for other aircraft, while – surprisingly, for a community which had so recently grown – their Navy counterparts also started falling by the boards. Still, these events were a few years off, and there was still plenty of business in the interim to keep the crews entertained.

The IO Becomes Routine

Mid-1987 found several squadrons doing that deployment thing at several corners of the world. In the east VA-35 pulled out of Norfolk with CVW-8 in *Nimitz* on 30 December 1986; the cruise was particularly memorable, for it marked the transfer of CVN-68 to the Pacific Fleet. The big ship took the "scenic route" getting to its new homeport of Bremerton, heading from the Med across the South Atlantic, around Cape Horn, and up South America's Pacific Coast. She pulled into her new home on the Puget Sound in late July 1987, with the *Panthers* detaching and returning to Oceana on 26 July. Heading in the other direction was *Kitty Hawk*, which departed North Island on 3 January 1987 with CVW-9 and VA-165 onboard. *Nimitz* had gone to the left coast to balance CV-63's departure for Philadelphia Naval Shipyard and its turn in SLEP; *Kitty Hawk*, however, got to go the long way around. On 20 May the two carriers exchanged escorts in the Med, then proceeded to their new homeports.

The *Boomers* were in the midst of a particularly smashing period under Cmdrs. Robert T. Knowles, John C. Scrapper, and Don C. Brown. Scrapper relieved Bob Knowles on 31 January 1986, inheriting a squadron that had won the 1985 Whidbey *Intruder* Derby, been awarded Maintenance Squadron of the Year, and had also achieved the Battle E and Meritorious Unit Citation.

In turn, Brown relieved Scrapper on 1 August 1987, roughly a month after the completion of *Kitty Hawk's* around-the-world deployment. Under their tutelage VA-165 continued to build on its outstanding record as a master practitioner of all-weather attack: taking the 1986 and 1987 Battle Es; a second Meritorious Unit Commendation; the 1986 Safety S; winning the COMMATVAQWINGPAC *Intruder* Squadron of the Year award in both 1986 and 1987; the Norden Pickle Barrel Award for bombing excellence in both years; and again took the honors in the 1987 Intruder Derby. To top it off, the *Boomers* also won the 1987 McClusky Award as the top attack squadron in the Navy.

With Scrapper's departure Cmdr. William H "Otis" Shurtleff moved into the XO's slot, and the squadron and CVW-9 prepared for its next deployment in *Nimitz*. Unfortunately, during workups on 30 November 1987 the squadron lost an aircraft, albeit under unusual circumstances. *Nimitz* suffered a flight deck incident, and KA-6D BuNo 152950 managed to get exploded by 20mm rounds from another aircraft.

[5] AirLant had similar plans to increase their A-6 strength by several squadrons. VA-12, the *Ubangis* and a former A-4 *Skyhawk* unit, was slated to be the next one established at Oceana, after VA-55. The squadron was cancelled about 1987 when it became apparent that the Navy didn't have the aircrews or airframes to support the ambitious community growth plan.

"An early morning I will always remember," says Shurtleff. "An A-7 plane captain and at least one A-7 ordie were working on a gun on a *Corsair* parked on elevator 2. For some reason, the ordie bypassed the ground safeties on the gun and fired four 20mm HE rounds into the fully loaded KA-6D parked in front of the A-7.

"515 had flown three or four events that day and was one of our best tankers, so it had been spotted for the first launch, right behind the cat. The aircraft exploded in a fireball and the ship went to GQ. The flight deck conflagration was initially huge, and consumed not only the tanker, but the seven other aircraft parked nearby. Flight deck fire-fighting sprinklers with AFFF and hose teams of combined ship's company and air wing personnel put the fire out fairly quickly."

Shurtleff adds about a dozen Boomer maintenance guys had been working on 515 until a few minutes before the fire, and several of them were members of the first fire fighting crews on the scene:

> "Caught in the explosion were the A-7 plane captain and a second class petty officer – an AQ, I think – from the VAQ squadron, who was walking past the tanker to the EA-6B parked in front of 515. Just the wrong place, at the wrong moment. One of our young plane captains actually rushed into the fire with a CO2 bottle and extinguished the A-7 plane captain – who was on fire – and dragged him clear. This young man – the A-7 PC – was severely burned, and although he lived for nearly three days – long enough for him to be evacuated ... to the burn center – he died of his wounds. The EA-6B maintenance tech was killed in the fire."

Several Boomers received decorations for their efforts in fighting the fire, and the young squadron plane captain who attempted to save his *Corsair* counterpart received a higher award for heroism. As for the tanker, it was completely destroyed and was pushed overboard from elevator four the following day.

"It floated for a period of time, then sank," Shurtleff remembers:

> "A very spooky feeling, watching the aircraft in the water. A total of seven other aircraft suffered damage as a result of the fire: the EA-6B parked in front of the tanker, and six other A-7s. Seemed just to me since the A-7 had shot down our KA-6D, they should pay the highest price. It was one of the longest half-hours of my life until we learned that none of our guys had been hurt. Don Brown was CO at the time. I know he had to be twisting inside, but he remained the absolute essence of a calm and composed leader, an example to us all."

The *Boomers*, *Air Wing 9*, and *Nimitz* cleaned up, mourned their dead, and carried on as the professionals they were.

During this period the Navy continued the program initiated by John Lehman of regularly heading to the northern expanses of the Atlantic and Pacific, partly to give the Soviets something to think about, but also to give the flight crews, flight deck personnel, and ship's companies practical training in bad weather operations. During his tenure Lehman made no bones about the fact that if the balloon ever went up he expected his carrier battle groups to "go after the bear in his lair." The architect of the NorPac ops was Lehman protégé and CinCPac Adm. James "Ace" Lyons. First up was *Third Fleet*, which participated in Operation *Kernel Potlach 87/1* from 23 to 28 January 1987. The operations – which included winter amphibious assaults on Shemya and Adak – were performed under the watchful eye of CVW-15 in *Carl Vinson*.

The *Knightriders* of VA-52 had "enjoyed" a relatively uneventful cruise, departing ConUS on 12 August 1986, but they certainly made up for it by flying the Aleutians in winter. The deployment-ending swing up north marked the Navy's first ever by a carrier battle group in the Bering Sea, and others followed. As for Lyons, he would later comment via press release, "No longer will we permit the Soviets to operate with impunity in this important area."

From 2 March to 29 April the VA-145 *Swordsmen* got to see the Northern Pacific "up close and personal" while operating from *Ranger* with CVW-2. Later in the year VA-95 made the trip in *Enterprise*, deploying from NAS Alameda on 25 October and returning to the Bay Area on 24 November 1987. Notably, during the *Big E's* trip CVW-11 fielded Operation *Shooting Star*, which saw *Lizard* A-6s fly mock bombing runs towards Petropavlosk. The *Intruders* turned back for the carrier before they reached the 100-mile mark, but obviously the Soviets took little notice of the proceedings. To add insult to injury, Atlantic Fleet performed similar ops off the Soviet's west coast. The *Thunderbolts* of VA-176 made a NorLant excursion in *Forrestal* with CVW-6 from 28 August to 8 October 1987.

Back in the Pacific, VA-165 – now led by "Otis" Shurtlef – took its turn with *Nimitz* from June to July 1989. He recalls:

> "It was interesting, as we went north of the Aleutian Island chain and operated in the Bering Sea.
>
> "Weather patterns changed quickly, and was often rotten, with rain, and some of the densest fog I have ever seen. Long range strikes to targets in Alaska meant a lot of aircraft returned overhead with little spare gas. I recall launching in a tanker to give to the most needy, and the air boss was asked if he could see the bow. He replied that he couldn't even see the 'crunch pole,' which was just forward of elevator No. 1! We played at a concept called Fjord Ops, taken from Norway operations. We were supposed to hide in steep-sided inlets and launch and recover aircraft. The theory was long-range targeting radar would not see the CVN tucked so close to land and masked by high terrain.
>
> "I'm sure it worked, but using Adak – with several high mountains – made the evolution interesting. Wasn't really thrilled to be launched in the rain/low vis and have the boss radio laconically, 'caution, high terrain, one and a half miles on the bow.' Sure changed our departure procedure, I can tell you."

Still, a primary focus for the fleet remained the Persian Gulf and the Red Sea. With Iran and Iraq still going at it after seven long, bloody years, the situation in the Northern IO remained dangerous, and the point was hammered home on 17 May 1987 when the frigate *Stark* (FFG-31) collected two AM-39 *Exocet* missiles launched by an Iraqi *Mirage* F.1.

From its start the Iran-Iraq war had gone badly for both sides. Iraq had initially invaded Iran, but pulled back most of its troops by June 1982. The Iranians then attempted their own invasion, capturing Khorramshahr, but the offense deteriorated into human wave attacks against strong Iraqi positions. Repeated assaults against poison gas – the first use of the weapon in any battlefield since World War I – and electrified marshes only succeeded in running up the Shi'ite body count. For good measure the Iraqis also used chemical weapons on their own Kurdish minority in the northern part of their country. In 1983 the United States increased support for Iraq and its leader, Saddam Hussein, to the point of putting an ambassador back in the country and providing intelligence support. Hence, while not exactly allies, U.S. forces in the region started to look at the Iraqis as less of a threat than their Iranian opposition.

Iraq first started targeting Iran-bound tankers at the north end of the Gulf in 1981. In March 1984 the nation began going after Iran-bound tankers in earnest, as well as those of neutral nations, attacking 71 ships through the remainder of the year, primarily with air-launched *Exocets* acquired from France. Similar attempts to take out Iran's Kharg Island terminal, however, failed. On 13 May 1984 the Iranians fought back by attacking a Kuwaiti tanker near Bahrain; more attacks followed. At the end of the year the two nations agreed – with UN prodding – to knock off the attacks on civilian and neutral targets. Iraq quickly broke its "word" and resumed attacking shipping and oil terminals, and the Iranians followed suit.

On 1 November 1986 Kuwait announced it would seek third-party protection for its shipping. Notably, the Soviet Union made the initial response by agreeing to charter several tankers to Kuwait. The United States followed suit on 7 March, agreeing to reflag and rename 11 Kuwaiti tankers and place them under the direct protection of the U.S. Navy. Officially the U.S. would protect the Kuwaiti ships from Iranian attacks, but the Iraqis in particular continued to show a unique proclivity for launching, locking on a blip, and firing away without confirming what they were shooting at. That is apparently what happened with *Stark* in May 1987, although no one's ever really confirmed whether the attack was an accident or intentional.

In any event, *Stark's* crew – operating in international waters outside of the declared war zones – tracked the *Mirage* inbound without broadcasting any warnings to the aircraft or taking defensive action, and at 2019 the two missiles impacted port side forward. One *Exocet* failed to explode, but spread its fuel through the interior of the ship. The other did detonate, pushing burning fuel through the superstructure and all but gutting the ship. The crews' quarters were immediately incinerated, followed by the radar equipment room and combat information center. The crew was able to save their vessel – at the cost of 37 lives – but it was a near run thing, taking nearly 24 hours of damage control efforts. After a tow to Bahrain for immediate repairs *Stark* was able to return to the United States under her own power and was eventually rebuilt by Litton-Ingalls in Pascagoula at a cost of about $90 million. She did not return to service until August 1988.

The U.S. government eventually accepted the Iraqi's apology over the "mistake," but the Navy responded by increasing carrier battle group presence in the region. However, the official explanation for the influx of forces into the region – above and beyond escorting tankers – was the continued warlike behavior of the Iranians. In July 1987, the formal escort of the reflagged tankers began as Operation *Earnest Will*. Subsequently, carrier aviators went into the mix of warring nations, Revolutionary Guard forces, air attacks, and scattered neutrals. When *Stark* got hit VA-196 was in WestPac with CVW-14 in *Constellation*, having departed ConUS on 11 April. *Midway* added to the mix, making two of its standard short cruises – 9 January to 20 March 1987 and 23 April to 13 July – before doing a full-blown six-month deployment, departing Yokosuka on 15 October 1987. The latter trip took the ship to the Indian Ocean and marked the first deployment for CVW-5 with the newly arrived VA-185. *Ranger* also got into the act, cruising from 14 July to 29 December 1987 with VA-145 and VMA(AW)-121 onboard.

On the LantFlt side of the house *Coral Sea* deployed again to the Med on 29 September 1987. Making the trip this time with CVW-13 were VAs -55 and -65, with the *Tigers* moving over from CVW-7 in *Eisenhower* to join the *Warhorses*. The original "All Grumman Air Wing," CVW-3, remained active; it completed its second twin-*Intruder* deployment in March 1987 in *John F. Kennedy*. VMA(AW)-533's maintenance officer for the *cruise* was one Maj. John "Mooncap" Thornell, a veteran A-6 driver who had made two previous cruises with the *Sunday Punchers* as part of an exchange tour. Notably, during his tour with the Navy Thornell had also qualified as assistant CAG LSO with CVW-3. Back with -533, he eventually fleeted up to the XO's slot. On 4 August 1987, while performing a post-maintenance check flight of A-6E BuNo 161685, Mooncap and his B/N, squadron operations officer Maj. Jeb "Rebel" Stewart, unexpectedly also got to do a Martin-Baker check flight:

> "We were on a test-and-go on a hot humid afternoon," Thornell later recalled. "Launch was on runway 23 at Cherry Point – which heads to the town of Havelock – at 1300. Right at rotation, complete left side hydraulic failure and left firelight, with all of the left side annunciator lights. We couldn't get the gear up and had to get out of town! The gate guard saw flames covering the whole plane. About 10 seconds after the left fire light, we got a right side fire light. We made it to the trees out of town, about three miles. Jeb got out at about 600 feet and 90 degrees AOB (Angle of Bank). I got out at about 400+ feet and 140 degrees AOB. I went through the tops of the trees in the seat, and then the drogue and chute blossomed and hung me up in the trees; the plane crashed about 94 feet from me. Our flight time was 42 seconds!"

Stewart survived the ejection with no major damage other than a lost watch. The XO, however, did not fare as well; he had thoracic compression fractures, a broken sternum, a ruptured esophagus from where the oxygen regulator went into his chest, and a lacerated knee from the canopy glass. But – remarkably – he survived, and eventually battled his way back to reassume his duties with the *Night Hawks*.

"After extensive bouts with NAMI (Naval Aerospace Medical Institute at NAS Pensacola) I got up and went on the next cruise on *JFK*," said Thornell:

> "I later served as the last Marine A-6 squadron CO, while 'Rebel' Stewart retired as lieutenant colonel. He now works at Virginia Tech as an information technology specialist. Thanks to Jeb's experience, our dance card was filled at the exact same instant ... kinda neat to watch his boots go by as you turn to say 'EJECT!' It kind of confirms the decision."

Two additional losses occurred the following month, with the loss of three aircrew. VA-145 lost a bird on the 18th of September while operating off *Ranger* in the IO. Both crew, LtCdr Denny Seipel and Lt Healy, were killed in a mishap that appeared to involve collision with the water during night return to the ship. Five days later the *Knightriders* lost their Skipper. Cmdr. Lloyd Sledge's squadron was conducting a rare pre-dawn fly-off from *Carl Vinson*, which had just completed a work-up period off San Diego. Sledge and his B/N J. Morrison were launched from the ship for Whidbey Island and were climbing out off Los Angeles when their aircraft, 155674, suffered a freak complete electrical failure. As occasionally happened under such circumstances, the canopy, without electrical power, cracked open, dumping cabin pressure. Sledge, who was not wearing his oxygen mask, passed out before he could deploy the emergency generator (RAT). The B/N now found himself trying to fly the aircraft from the right seat in the dark with no instruments. It was not long before the *Intruder* was out of control, and the B/N did the only possible thing, which was to eject.[6] He was soon picked up by a Coast Guard helicopter and would return to flight duty. The *Knightrider's* popular CO apparently never revived and went in with his aircraft.

Praying Mantis

Operations in the Gulf continued through the end of 1987 and on into 1988 with sporadic violence. By the spring of the New Year, however, Navy *Intruder* crews once again found themselves "at the tip," albeit not quite at the same level as two years earlier with Libya. Overall, 1988 proved to be a year of incredible "ups" balanced by several "downs" that affected the entire medium attack community.

The level of both the violence and direct American involvement escalated. While American forces were escorting tankers through mine-infested waters on 10 August a VF-21 F-14 operating from *Connie* fired two *Sparrows* at a presumed Iranian F-4 that appeared to threaten a P-3. Neither missile hit anything. The same day the tanker *Texaco Caribbean* collected a mine while riding at anchor at Fujayrah; the explosion

[6] Amazingly, The *Intruder* was not built with 'command eject' which allowed the B/N to eject the pilot. The modification adding this feature came too late for this crew.

VA-95 introduced the *Skipper* laser guided missile to combat during *Praying Mantis* in 1989. NH501 sits with Mt. Baker in the background at Whidbey in December 1986. (Rick Morgan)

marked the first evidence of mines outside of the Gulf proper. Two weeks later on 24 August *Kidd* (DDG-993) fired warning shots at two Iranian dhows after they approached a convoy in Strait of Hormuz. Subsequently, *Jarrett* (FFG-33) warned off an Iranian warship when it approached the same convoy.

Early on the morning of 21 September Army MH-6 "Little Bird" helicopters from the *160th Airborne Regiment (Special Operations)* – operating from U.S. frigates – caught the freighter *Iran Ajr* literally in the act: it was laying mines about 50 miles northeast of Bahrain. They Army helos shot up the ship, killing three in the process. The following day Navy SEALs boarded the wrecked vessel and took 26 Iranians into custody along with a load of 10 mines. Now the U.S. had its "smoking gun," but there was more to come.

On 15 October, following additional confrontations in the Gulf, Iranian forces fired a CSS-N-2 *Silkworm* missile into an anchorage nine miles from Kuwait's Mina al Ahmadi oil terminal. The missile hit the tanker *Sungari* and damaged it, fortunately without loss of life. The following day another *Silkworm* launched from the Iranian position on the captured Iraqi territory of the Fao Peninsula and hit *Sea Isle City*, one of the reflagged Kuwaiti tankers. This time, 18 were injured including the master, a U.S. citizen. Three days later the Navy struck back, attacking the Iranian Rashadat oil platform while SEALs destroyed electronics and surveillance equipment on a second platform. *Kidd* – ironically, built for the Imperial Iranian Navy but retained for U.S. service after the fall of the Shah – *Leftwich* (DD-984), *John Young* (DD-973) and *Hoel* (DDG-13) did the honors, blasting the rig with 1,065 rounds of 5-inch fire after giving the occupants 20 minutes to pack up and get out of Dodge.

By now things were really getting crowded in the Gulf with ten western Navies, the Soviets and eight regional navies or coastal defense forces running around the area, keeping an eye out for Iranian mines and missiles and occasionally taking pot shots at the numerous armed small craft that hummed around the region. Finally on 14 April 1988 the situation took another downturn when another American combatant got hit; again, it was a *Perry*-class guided missile frigate.

The ship was *Samuel B. Roberts* (FFG-58) a 1986 product of the Bath Iron Works whose namesake was a valiant destroyer escort (DE-413) lost in the 25 October 1944 action off Samar during the Battle of Leyte Gulf. While escorting tankers *Roberts* hit a mine; the force of the explosion broke the gas turbine engines from their mounts and blew shrapnel throughout the engine room, opening up the hull, overhead and bulkheads. The main engine room exploded – with flames shooting out of the stack to a height of 150-feet – and then flooded. As with their counterparts in *Stark* the crew of *Roberts* fought valiantly to keep their stricken vessel afloat and they pulled it off. The frigate managed to stagger into Bahrain using her auxiliary thrusters; once there she was loaded onboard the Dutch heavy-lift ship *Mighty Servant 2* and returned to the U.S. for more extensive repairs. Fortunately, there were no fatalities due to the mining or subsequent damage control but 10 sailors were injured. *Samuel B. Roberts* finished repairs at Bath in late 1989 at an estimated total cost of $37.5 million.

Four days later, on 18 April, the Navy initiated Operation *Praying Mantis* with strikes against both Iranian surface combatants and oil platforms. In brief, during the two-day operation aircraft from CVW-11 in *Enterprise* went trolling for the ships while the destroyers and frigates engaged in more ShootEx's against oil platforms. For starters the blackshoes went after two oilrigs, targeting the Sirri and Sassan platforms. As before U.S. intelligence had evidence they were being used for surveillance and targeting duties while also providing cover for high-speed attack boats operated by the *Pasdaran*, or Revolutionary Guard.

Merrill (DD-976) and *Lynde McCormick* (DDG-8), working with *Trenton* (LPD-14), which was carrying Marine helos, led off at first light on 17 April. At 0755 the ships approached the Sassan platform and gave the inhabitants five minutes to evacuate. An Iranian 23mm gun briefly responded about 10 minutes later – in and around screams to higher headquarters in Farsi – but a direct hit by one of *Merrill's* 5-in 54s quickly silenced the gun. The tin cans allowed a tug to come in and pull the rest of the occupants off the rig and then Marine UH-1Ns and CH-46Es operating from *Trenton* inserted a raiding party made up of members of Marine Air-Ground Task Force 2-88. Once the grunts secured the platform demolition and intelligence-gathering parties went aboard, did their job and departed. Two hours later the Sassan platform disappeared in an explosion.

The group tasked with taking out the Sirri rig – *Wainwright* (CG-28), *Bagley* (FF-1069) and *Simpson* (FFG-56) – pretty much followed the same script except their first shots hit a compressed gas tank, turning the rig into an inferno. Hence, the SEAL team tasked with taking the platform and laying charges had to wait for another day. Adding to the festivities, a helicopter chartered by NBC for coverage of the events was reportedly fired on by the Iranians. However the shots were inaccurate and the helo escaped unscathed. The same could not be said for *Joshan*, a French-built *La Combattante II*-class missile boat that crossed paths with the Sirri group about three hours later. When the Iranian skipper declined to alter course or otherwise depart the area, the Navy surface action group fairly warned him, "Stop your engines and abandon ship; I intend to sink you."

Joshan's CO responded by uncorking a *Harpoon* which flew down the starboard side of *Wainwright* without apparently tracking. The U.S. warships responded by firing four SM-1 SAMs into the patrol boat, using the missile's backup surface-to-surface mode. All four hit and what remained of *Joshan* eventually succumbed to *Harpoons* as well as five-inch and 76mm gunfire. A short time later *Wainwright* used two SM-2ERs to hole an Iranian Air Force *Phantom* that had strayed too near; the F-4's pilot managed to get his stricken plane back to Bandar Abbas but his day was definitely done.

When it came to handling the larger units of the Iranian Navy the job fell primarily to *Carrier Air Wing 11* and its attack squadrons, notably VA-95. As recalled by one Lizard JO, "We were directed to specifically go after the *Sabalan* (F74), an (English-built) Vosper Thorneycroft *Samm*-class patrol frigate. This was to be a carefully planned 'eye for an eye' mission; reportedly, the ship's skipper was notorious for shooting up merchant ships, then coming up on the ship-to-ship (radio) and saying, 'Have a nice day.'"

The *Sabalan* and her three sisters were built in England in 1971-72 and were fairly typical small third-world frigates. Displacing only about 1500 tons fully loaded, (compared to American *Perry* at 3600t) the four Vospers, with only a single 4.5 inch gun, elderly *Seacat* SAMs and handheld IR missiles, were not exactly the Soviet combatants the U.S. Navy had been trained to fight. But they would do.

In order to find the ships *Enterprise* regularly launched SUCAP packages consisting of *Green Lizard* A-6Es under the escort of F-14A *Tomcats* from VF-114 and VF-213. One group passed over a clump of merchant ships off Bandar Abbas around 1100, looking for the *Sabalan*, but found nothing. Without a target the remainder of the War at Sea (WAS) package remained on the flight deck. About that time multiple Iranian Boghammers, small boats used by the Revolutionary Guards to launch quick attacks with small arms and RPGs, flushed from Abu Musa Island and started shooting up oil rigs in the Mubarak oil field off the United Arab Emirates. They also made a few passes on an American platform support boat, the *Willie Tide*, whose crew immediately yelled for help. Two A-6Es from VA-95 and an F-14A responded and – after receiving the okay from higher authority – the *Intruders* sank the lead Boghammer with several *Rockeye*. The other boats called it a day and hauled tail back to the beach.

A little later that afternoon intelligence resources indicated that several larger ships at Bandar Abbas were about to get underway. A few minutes later the ship launched a *Lizard Intruder* on a hunter-killer mission with DCAG-11, Capt "Bud" Langston, at the controls. The crew was specifically directed to find the *Sabalan*, positively identify it and then terminate its career. Once they got in the target area, things picked up quickly.

Off Larak Island the B/N picked up a large blip on the radar. The question was, "Is it the *Sabalan*?" Attempts were made through the frustratingly dense Gulf haze to identify it with the FLIR and by *Tomcats* using their TCS. The decision was finally made to make a close pass and positively identify the contact by eyeball. The aircraft flew parallel to the ship at low altitude at about 500 knots.

It was then the B/N noted 'lightbulb flashes' from the bridge wing. It was obvious that they were being fired upon. Langston started moving the plane around and dropped flares. A smoke trail was seen rising from the fantail, which they assumed was a handheld SAM.

After gaining permission, the crew set up for a *Harpoon* shot which hit the frigate dead amidships. They then rolled in with the LGB in a 30 to 40 degree dive. The bomb came off and although lasing was initially smooth the turret went stupid and the bomb landed in the water abeam the ship, port side amidships. By now the squadron XO, Cmdr. John Schork, was inbound from *Enterprise* with the rest of the WAS package – his A-6 and six A-7s. The initial aircraft then came around and scored with an AGM-123 *Skipper* in the first use of that weapon in combat.

Having done as much damage as they could, they departed the area and returned to Mother. As they hauled out, guided missile destroyer *Joseph Strauss* (DDG-16) pumped a *Harpoon* into the target, followed by another *Harpoon* strike and two additional *Skipper* shots by XO Schork and his B/N ... and then the CVW-11 A-7's came in with

The Iranian frigate *Sahand*, built in England by the Vosper Company in 1972, destroyed by the U.S. Navy in the Persian Gulf on 18 April 1988. The 1500-ton vessel would be the largest man-of-war sunk by *Intruders*. (U.S. Navy)

"Cruise patches" were routinely made by Air Wings during deployments, and found their way onto flight jackets fleet wide as a "been there, done that" mark of honor. CVW-11 made this one to commemorate their role in the Persian Gulf during the 1988 cruise.

Mk.83 dumb bombs, set up a wagon wheel and also started dropping. The ship by now was a flaming wreck. An additional strike that cycled in a few minutes later found an oil slick and not much else.

As it turned out, CVW-11 had actually sunk the sister ship *Sahand*. The *Sabalan*, the initial object of their attention, finally stood out that afternoon and was hit by another *Green Lizard* A-6 crewed by skipper Cmdr. Bill Miller, and Lt.Cmdr. Joe Nortz; the *Sabalan* was crippled and left Dead In the Water (DIW). Meanwhile, air wing planes were circling around and looking to continue the SINKEX when word came that the State Department had negotiated an end to the conflict, something like, 'We sank one of your ships, do you want us to sink another?'

In one of those miracles of modern communications, Secretary of Defense Frank Carlucci personally ordered the A-6s and A-7s to break off their attack on the badly damaged *Sabalan*, which was eventually towed back to port.

By sundown *Praying Mantis* was over and by all counts Iran had suffered an astonishing defeat at sea. "This is probably the largest U.S. naval battle since World War II," crowed *Enterprise* skipper Capt. R.J. "Rocky" Spane afterwards with some hyperbole "The combined efforts of *Enterprise* and CAG-11 in *Battle Group Foxtrot* again and again, through exercises – and today – the real thing, has proven to be an unbeatable team, a winner."

While some naval analysts and others with a sense of history may have cringed at the "largest ... naval battle" comment one thing was certain: the *Green Lizards* had become the first AirPac squadron to brave antiaircraft fire and sink an enemy vessel in open-ocean combat since Vietnam. For the *Intruder* community, that had prepared for years to take on the Soviet fleet, the roughly 1,500 ton Iranian frigate *Sahand* would be the largest warship they would ever sink.

The *Lizards*, after *Praying Mantis*, turned back east and started the long trip home. The *Green Lizards* finished the deployment, their tenth, on 3 July 1988.

Texaco: The *Intruder* as a Tanker

TANKER POSIT?!: The traditional call of a jet in need.
Of all forms of manned flight, carrier aviation is considered by many to be the most demanding. In order to carry out their combat mission aircrews must fly day or night, and in frequently frightful weather. The arrival of jet aircraft to carrier decks in the 1950s quickly led to the requirement that a means to refuel while airborne be developed. The first carrier based tankers were AD *Skyraiders* and AJ *Savages*, using the English-developed "probe and drogue" method. Before long A4D (A-4) *Skyhawks* were also carrying out tanker duties, using external, self-contained D-704 or Sergeant-Fletcher built "buddy stores." The A-6 series was plumbed to carry an external fuel store from inception, and in time the type proved to be what was probably the Navy's most versatile tanker.

"TEXACO OVERHEAD, ANGELS SIX": Ship's call to alert aircraft that gas was airborne.
Compared to both its predecessors and successors in the mission, the *Intruder* possessed both sufficient fuel give and performance to allow it to either loiter overhead the CV as a departure or recovery tanker, as well as accompany a strike package towards the target. Only the Douglas A-3 series could do both missions as well.

Prior to the arrival of the *Intruder* the AJ *Savage* series and A-1 *Skyraider* were also configured to carry refueling packages. The AJ used an internal hose-drum unit, while the *Skyraider* used a buddy store on centerline. The A-5 Vigilante and F-4 were also tested as tankers, the *Phantom* with a centerline buddy store, and the "Vigi" with an internal package located in the aircraft's linear bomb bay, but neither appears to have been used operationally in this role (and neither, as afterburner-equipped aircraft, made much sense in the mission).

The "typical" early 1980s air wing deployed with four or five KA-6Ds and ten to twelve A-6Es that could carry buddy stores. In addition, the two A-7 squadrons on board also pulled tanker duty with D-704 toting *Corsairs*. This allowed sufficient tankers to fuel the two fighter squadrons onboard – either F-4s or F-14s. This all changed with the arrival of the FA-18 *Hornet* to the fleet in 1984. Overnight the available tanker assets went from three squadrons to just the *Intruders*, while there were now four afterburner-equipped units to feed. The *Intruders* carried the load until the S-3 *Viking* was modified to carry the buddy store on its left wing. The last KA-6D deployed in 1993, with VA-165, while bombers carried buddy-stores for their air wings through to the end.

"401, HORNET BALL, 3.2, TRICK OR TREAT": Typical call of a jet which either traps on its next pass or will need a tanker or have to divert ("bingo") to a field.
During normal day carrier operations a tanker will hang out at 6,000 feet ("Angels 6") and await any customers that need fuel. Normal pattern is a left-hand orbit overhead "Mother" at 250kts with the store retracted. Launching aircraft that require fuel (usually fighter-types) will hit the tanker right after leaving the boat, typically taking 2 or 3,000 pounds before proceeding on mission. As a recovery tanker the duty A-6 may be asked to "hawk" a specific jet in the landing pattern with the goal of being above and to the right of the customer with the hose extended in case the aircraft on approach bolters (misses the wires) or is waved off. Flying the night tanker took a good deal of finesse and experience, and the mission was usually left to more senior pilots in the squadron. Even though a frequently thankless job, the presence of "Texaco" overhead was usually mandatory during night cyclic operations, and the tanker was almost always the last one aboard at the end of the night.

A tanker-configured "Pecker Head" A-6E from VA-36 drags a basket off Puerto Rico in November 1990. The *Intruder* was the fleet's standard tanker aircraft for over two decades. (Rick Morgan)

Above: Refueling from Air Force tankers became routine from the mid-1980s. The KC-135 had a flexible hose and basket assembly attached to the end of its "flying boom," which allowed Navy aircraft to refuel from it. A *Green Lizard* A-6E SWIP tanks off a Washington Air National Guard KC-135E over the Snake River in Idaho in April 1992. (Rick Morgan)

Right: Things do not always go as planned while tanking, as witnessed by this VA-196 KA-6D that returned to NAS Fallon wearing the basket from a USAF 96th Bomb Wing KC-135A in September 1984. The 135's notorious "hard basket," aka "The Iron Maiden," was considered to be much more difficult to deal with than the standard Navy "soft" tanker basket. The anonymous pilot in this incident went on to command an EA-6B Prowler squadron, as well as a Carrier Air Wing before retiring as a Captain. (Rick Morgan)

Below: A *Prowler* RAG instructor pilot demonstrates how not to plug from the *Intruder* off Whidbey in July 1989. The basket from the VA-52 KA-6D is well below the "electric's" nose – not the recommended method of tanking. (Rick Morgan)

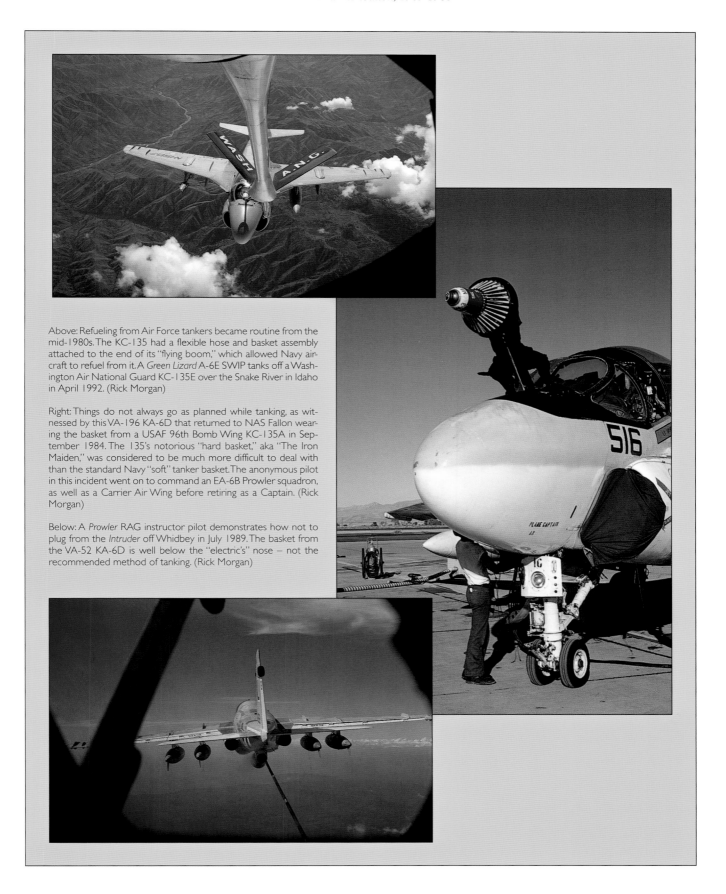

CHAPTER NINE

Cold War's End, 1988-1991

Downturn

With the conclusion of operations against Iran and the apparent resolution of the eight-year Iran-Iraq War, the U.S. Navy looked confidently into the future. In terms of personnel, equipment, weaponry, and newer systems coming on line – including the odd 40-year-old battleship – the sea service was enjoying good times. Still, after eight years of steady growth there was mounting evidence that the bloom was off the rose on several fronts.

Following the 22 October 1988 re-commissioning of *Wisconsin* (BB-64) four battleships prowled the seas, along with 15 carriers, 36 ballistic missile submarines, 94 fast attack submarines, 36 cruisers, 69 destroyers, and a staggering 115 frigates, plus 90 other surface and subsurface combatants. An additional four carriers and 70 or so other ships neared commissioning or were under construction. However, despite this wealth of combat power, it became more and more apparent in 1988 that John Lehman's promised 600 ship Navy was in serious trouble. Thus, Naval Aviation also faced problems.

Lehman himself had resigned effective 11 April 1987, ending a remarkable six-year term as Secretary of the Navy. His replacement was former combat Marine officer, novelist, and Assistant Secretary of Defense for Reserve Affairs James H. Webb. Webb was quickly confronted by the outcome of several years of "backend loading," in that the bills for many of the programs that his predecessor initiated were now coming due. Adding to the mix was new Secretary of Defense Frank Carlucci, who replaced Caspar "Cap" Weinberger in 1987. Later that year, on 8 December, President Reagan and Soviet Premier Mikhail Gorbachev signed the Intermediate Range Nuclear Forces (IRNF) agreement, which pulled the Army's MGM-31A *Pershing II* and Air Force's BGM-109 *Gryphon* cruise missiles out of Europe in return for the removal of Soviet SS-20 missiles. It was not ratified until 27 May 1988 and did not come into effect until 1 June, but by that date the new SecDef was looking for additional ways to cut the defense budget. The Army and Air Force made the case that without the protection afforded by the nuclear umbrella offered by IRBMs and GLCMs their conventional forces in Europe would require substantial enhancement. Where to get the money? No problem; it was taken out of the Navy's budget.

In the initial 1989 budget submission Deputy Secretary of Defense William H. Taft IV proposed cutting budgets across the board,

A division of *Swordsmen* are seen flying in close formation off Whidbey on 10 August 1988. 516 displays the tailpipe numbers used by a number of squadrons. The thought being that they would be more visible to aircraft attempting to rendezvous at night, the numbers being illuminated by the position lights mounted on the rear of the wing pylons. Squadron CO CDR Steve Hazelrigg leads the flight in NE500. He would be transferred to the Naval Air Test Center at Patuxent River, MD, and die in an A-6 mishap on 15 August 1990. (Rick Morgan)

The A-6F offered a substantial increase in capability over the previous versions. Five were built, and only three ever flown before the entire program was canceled. The first F-model is seen at Grumman's Peconic Field in New York while the program still had life. The larger intake "cheeks" from the new F404 turbofan engines are obvious, as are the two new wing stations, improved DECM antennas, and APU exhaust, which is located aft of the wing trailing edge. (Grumman)

with the Navy taking an 11.3% hit, the Air Force 9.3%, and the Army 8.1%. Carlucci's formal FY89 budget request, however, increased the Navy's share to $12.3 billion while reducing the Army and Air Force cuts to $2.1 billion and $500 million, respectively. In effect the Secretary of Defense had decided to use Navy money for more important things. Webb – who had attempted to keep several of the Navy's programs going and fully supported Lehman's 600-ship Navy – fought the cuts, but finally resigned in disgust on 22 February 1988; William L Ball replaced him, and the axe started to fall. Among the first aspects of Naval Aviation so affected was the fourteenth air wing, CVW-10, which was disbanded. The wing disestablished at Miramar on 1 June 1988, preceded by subordinate units VFA-161 on 1 April, VFs -191 and -194 and VAW-111 on 30 April, and VS-35 and HS-16 on 1 June. Up at Whidbey VAQ-141 reported to CVW-8, while VA-155 transferred to AirLant's CVW-17, effective 1 May 1988.

It would get worse over the following years, although at the time few probably could have foreseen what was coming down the pike, including an overall 22 percent defense budget cut by the end of 1990. In the interim all of the services furiously attempted to maintain funding for new weapons programs, such as the Navy's SSN-21, *Trident* D-5 missile, JVX (V-22) tilt-rotor program, and the *Burke*-class DDGs. The Air Force and Army clamored equally for the B-2, C-17, AH-64D *Apache*, Advanced Tactical Fighter, rail-garrison MX/*Peacekeeper*, and "*Midgetman*." Continued funding for President Reagan's Strategic Defense Initiative – nicknamed "Star Wars" by the press and other opponents – only added to the mix.

These rumblings were just that: distant thunder on the horizon. While the shutdown of a complete, new air wing shocked everyone, there were still plenty of carriers and A-6 squadrons to go around, and the medium attack community had much to look forward to, including more weapons, a new *Intruder* variant, and some sort of advanced technology follow-on. Even the reserves were about to get into the act. These were truly good times for the *Intruder* community.

We'll see your A-6F and Raise you an A-6G and SWIP

When first proposed by John Lehman, Naval Air Systems Command, and Grumman in the mid-1980s, the A-6F had the look of a world beater. The F-model fulfilled two major purposes: it upgraded the basic A-6 with the latest technology, weapons, sensors, and power plants – thus substantially increasing its already excellent capabilities – and also allowed carrier aviation to continue to take the fight downtown until the arrival of the *Intruder's* advanced technology replacement. Regrettably, in the end the A-6F was stillborn, and again it was a case of too many programs, not enough money, and a lack of high-level advocates.

The program started with great promise during Fiscal Year 1984 with an allocation of $1.4 million for concept development. Looking ahead a bit, NAVAIRSYSCOM foresaw spending $379.8 million for five Full-Scale Development aircraft over FY88 and 89, followed by 150 new-construction A-6Fs and 230 rebuilt Es.

The all-new F could justifiably be considered Lehman's pride and joy. The improved *Intruder* incorporated composite wings, two smokeless 10,080-pound non-afterburning F404 engines – the same engine designated for the FA-18C/D – and an AN/APQ-173 radar featuring sharper resolution, longer range, and additional modes. The radar tied into new digital computers, avionics, and other systems similar to those found in the *Hornet* and planned F-14D variant of the *Tomcat*. Inside the cockpit the pilot and B/N would work five multi-function displays – two on the left and three on the right – while the pilot gained a true Head-Up Display, or HUD. Using the new displays, the crew could employ a full range of targeting and navigation sensors, including the TRAM's laser/infrared displays, to place iron and guided weapons on target.

In addition, the aircraft would have an integral self-defense capability. To that end, Grumman added two new stations outboard of the wing fold (which had always been found on the EA-6A), allowing carriage of the AIM-9L *Sidewinder* or AIM-120 Advanced Medium Range Air-to-Air Missile (AMRAAM), with fire and forget capability. While the aircraft had long been able to carry the '*Winder*, the new stations meant that it could do now do so without sacrificing bomb capacity. Additional internal modifications included the installation of an auxiliary power unit, allowing engine start without a power cart, and an Aircraft-Mounted Accessory Drive (AMAD) on each engine. All engine-driven functions operated through the AMADs, greatly simplifying engine changes and other accessory work. Also, fuel lines were

A *Sunday Puncher* KA-6D in the groove for *Kennedy*, January 1990. (U.S. Navy, PH1 Michael Flynn)

rerouted through the tanks, and an onboard HALON fire-extinguishing system added, improving combat survivability.

The first example, BuNo 162183, flew from Bethpage on 26 August 1987. However, by that time SECNAV Lehman had departed the friendly confines of Washington D.C., and newer Navy leadership was taking a hard look at carrier air wings – both in number and makeup – and the future of medium attack. Notably, during the FY88 budget battles the Senate came down on the side of the AV-8B and the *Intruder's* follow-on, the Advanced Tactical Aircraft, by zeroing funding for the A-6F. Conversely the House canceled the improved *Harrier*, cut back on funding for the ATA, and restored funding for the A-6F. Plans to build the A-6F finally stalled when Congress refused to authorize procurement of the first 12 A-6Fs in favor of the ATA.

The fight over an improved Intruder continued for another three years. In 1990 the Navy and Grumman proposed an A-6G, which featured the APQ-173, improved avionics, and J52-P408 engines, which were already carried in EA-6Bs and the A-4M. Their approach called for 300 A-6Gs and up to 200 rebuilt A-6Es, which would have kept production lines at the Iron Works humming for several more years. However, it too was cast aside in favor of the "great white hope," the ATA program. The ATA – now formally designated the A-12 – reached FSD status on 23 December 1987 after several months of study by two major contractor teams. Notably, the team of General Dynamics-Fort Worth and McDonnell Douglas took the contract for development and eventual production, much to the surprise of the Grumman/Northrop team. The aircraft design and specific – *very* highly classified – were thought to incorporate the long-rumored "stealth" technology, an area in which GD and McAir had little known experience.

Still, on 13 January 1988 the Navy awarded a fixed-price incentive contract of $4.379 billion for the new plane's development, including some monies originally designated for the A-6F and A-6G. The move nailed the lid on the new model *Intruders*, but some money became available for systems and airframe upgrades to the trusty A-6E.

The chief upgrade was called "SWIP," for Systems/Weapons Improvement Program. The plan called for the rebuilding of 342 A-6Es to the SWIP standard; which would allow the Intruder to hold the line until the A-12A arrived during the early 1990s. A key feature of the SWIP birds was an all-new composite wing developed and produced by the Boeing Military Airplane Company at its former B-47/B-52 plant in Wichita, KS. Known almost universally, if incorrectly, as "plastic wings," the new sections were stronger and had split fuel tanks, among other improvements which made them less susceptible to combat damage. That alone marked a big improvement for the aging *Intruder*, which continued to suffer from wing fatigue problems. Above all, the SWIP re-wing program held the promise of finally ending questions as to the service life of the A-6's wing life, a problem that had plagued the type for years.

In addition, Grumman and the Navy planned to introduce a digital missile fire control system to allow employment of a wider range of guided weapons. These included the AGM-88 High-Speed Anti-Radiation Missile (*HARM*), the AGM-65D and F Imaging-Infrared (IIR) *Maverick*, the AIM-9L/M variants of the trusty *Sidewinder*, and the AGM-84E Stand-Off Land Attack Missile, or *SLAM*. Internally, the aircraft would upgrade to AN/ALR-67 and ALQ-126B ECM gear, and to top it off, the cockpits would be wired and modified for the use of night-vision goggles, or NVG.

In preparation for the program Boeing started building up its workforce in Kansas and stood by to receive *Intruders* on a ramp which had previously been the domain of SAC bombers.

Several of the improvements to the *Intruder* were already in test and evaluation by the time the SWIP program built up steam, and a few were in the fleet; in general, squadrons continued to get the various upgrades while Boeing worked up the re-winging program. For example, in September 1986 the *Tigers* of VA-65 became the first A-6 squadron to fully qualify with NVGs. Led by Cmdr. Robert Leitzel, they deployed the following September in *Coral Sea* with CVW-13. Roughly two years later, on 14 June 1988, *Sunday Puncher* CO Cmdr. John Meister and Ops Officer Lt. Cmdr. Rich Jaskot made the *Intruder's* first *HARM* launch. The event took place during a War At Sea (WAS) exercise as part of the work-ups for the squadron's August deployment in *John F. Kennedy* with CVW-3.

However, there were drawbacks to the upgrade program and, as is often the case, in several instances the squadrons themselves were forced to work out the bugs. Few squadrons received "all up" aircraft; Grumman and its sub-contractors generally delivered the new SWIP birds on a piecemeal basis, with the airframes containing some of the improvements but not all. Capt. Ron Alexander, who commanded VA-34 when the *Blasters* started receiving their new A-6E/SWIPs in late 1990, recalls:

> "The airplanes showed up without equipment, so we'd get the airplane with the new wing, but we wouldn't get the SWIP electronics, just the typical TRAM stuff. Then the stuff showed up and the boxes wouldn't even fit the airplane. The wiring was a little out of whack, too.

"I started flying the A-6 in 1976 and went through a couple of different iterations of mods, but this one was by far the worst, as far as getting the airplanes up and running. Everyone was really excited about getting new stuff and getting new weapons that they could use. The training wasn't an issue; it was just a case of everyone having to turn to and get the plane up and running. It was really painful; we really had a time of it."

As for the wings themselves, Boeing Military Airplane Company quickly ran into difficulty, primarily due to problems with the composite manufacturing process. In May 1988 BoMAC delivered the first composite wing set, which Grumman promptly installed in a new production A-6E. However, by November the Wichita plant was reportedly eight months behind schedule in its delivery of 179 wings. Still, the company reached a milestone on 3 April 1989 when the first re-winged A-6E, BuNo 155682, made its initial flight from the Wichita facility under the able hands of Steve Speight and Glenn Cermann. After initial tests the *Intruder* moved to Pax River for more exhaustive evaluation at the Naval Flight Test Center.

About the time SWIP hit the fleet Grumman was proposing yet another improved version. Called "Block 1A SWIP," it would be another modification of existing airframes and feature improved cockpit systems, including a Heads-Up Display (HUD) for the pilot and a multi-function display (MFD) for the B/N. There would also be new defensive countermeasures systems that would include an additional ALE-39 chaff/flare bucket (bringing capacity to 90 rounds) and a new towed decoy. A halon fire-suppression system would be installed, as would an improved computer. The wing's inboard leading edges and slats would also be modified, re-gaining some slow flight characteristics that had been lost as the type had gained weight over the years.

Airframe number M228 (155682) was identified as the test aircraft, with an induction date at Grumman set for January 1991. The first of 50 "production" aircraft were supposed to be delivered in December 1994, and Initial Operating Capability (IOC) was set for June 1995. Once again, the project was cancelled. SWIP, as it turned out, would be the final version of the A-6E to see fleet service.

Intruders in the Reserves

While John Lehman was gone as SecNav, several of his initiatives continued to bear fruit. Among these was the concept of "horizontal integration," i.e., the wholesale remaking of Reserve Air Wings (CVWR) -20 and -30 to more closely resemble their active-duty counterparts. The conversion of a single light attack squadron in each wing from A-7s to A-6s was part of the program.

The two wings dated to early 1970 and were a direct outcome of the ill-fated 1968 attempt to call multiple Naval Air Reserve squadrons to active duty. The Department of Defense activated six units at the end of January 1968 following the seizure of USS *Pueblo*; unlike the Air Force's call-up of multiple Air National Guard – which saw several F-100 units immediately move to Southeast Asia and into combat – the Navy's call-up flopped. As a result, in July 1968 the Naval Air Reserve commenced a major rework of its forces, emphasizing training, equipment, and the ability to seamlessly integrate with the active forces when required. On 1 April 1970 Carrier Air Wing Reserve 20 established at NAS Jacksonville; the same day, CVWR-30 stood up at NAS Alameda. Concurrently, NAVAIRES established Carrier Air Anti-Submarine

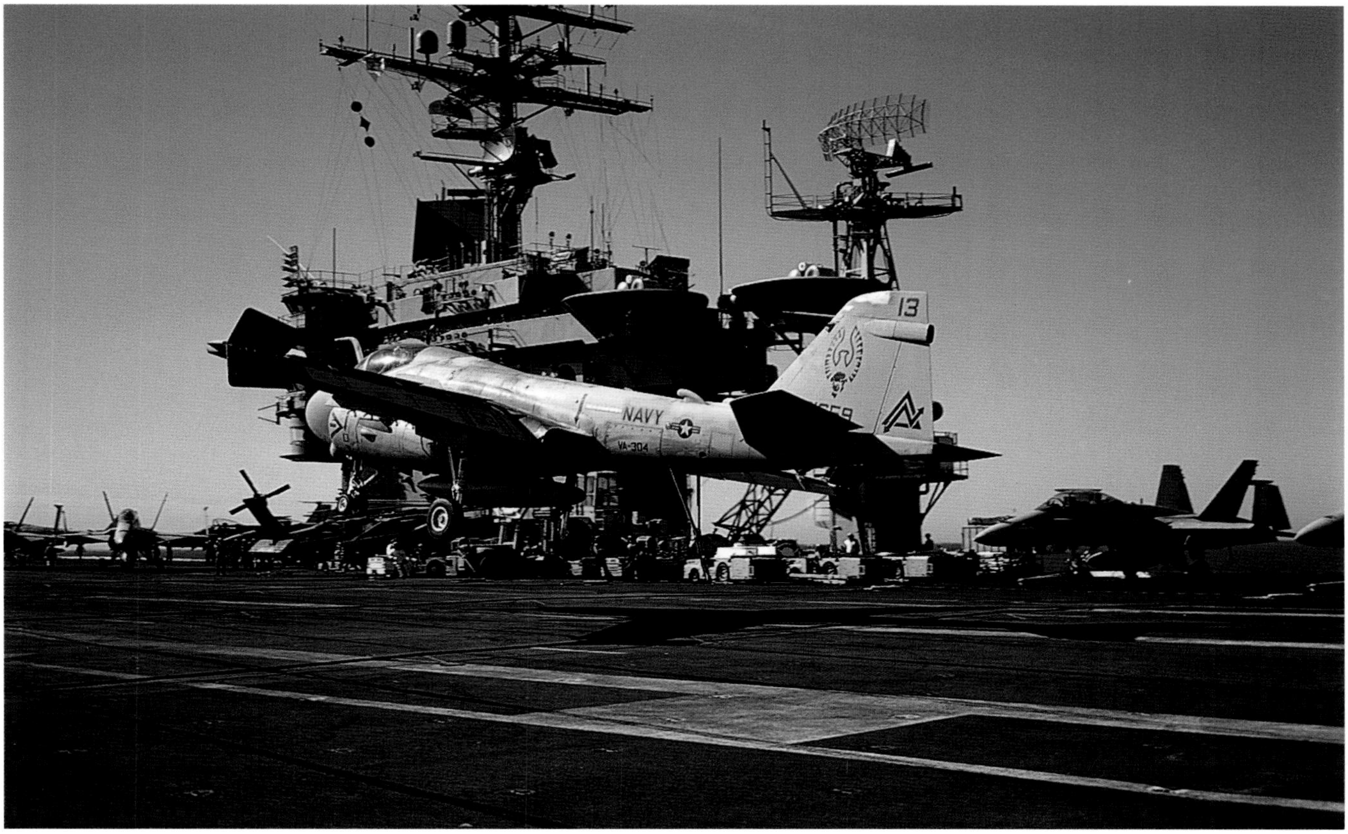

The Naval Reserve received their first *Intruders* in July 1988, with Alameda-based VA-304 doing the honors. VA-205, in Atlanta, would become their east coast counterpart two years later. The *Intruder* period in the Reserves lasted only seven years, as both squadrons were disestablished during the post *Desert Storm* drawdown. Here *Firebird* ND513 conducts CQ on *Carl Vinson* off San Diego on 31 May 1993. (Rick Morgan)

Warfare Group Reserve 70 at NAS Norfolk, CVSGR-80 at NAS North Island, 12 reserve patrol squadrons, and three reserve fleet logistics squadrons.

The reserves designed each new carrier air wing to emulate the active duty CVWs, with a full complement of fighter, attack, airborne early warning and electronic warfare squadrons. The hodge-podge of former reserve squadrons spread all over the continent with obsolete aircraft were re-organized into two coherent Carrier Air Wings, with organizations that roughly matched the regulars. Over the years the wings trained together, went to the boat together, and made weapons deployments to the traditional garden spots, like Fallon and Yuma. In due fashion they upgraded their equipment; the FitRons moved from remanufactured *Crusaders* to the mighty *Phantom*, while the light attack squadrons upgraded from A-4s to A-7As.

Time passed, and the units continued to train and regularly upgrade. In the late 1980s, however, both wings and their assigned squadrons took a major step forward in combat capability by transitioning to the same type of aircraft as their regular navy counterparts. Under Lehman's horizontal integration program the four fighter squadrons swapped out their F-4s for *Tomcats*, starting with VF-301 in late 1984, while four of the six attack squadrons (VAs -203, -204, - 303, and -305) upgraded to FA-18As between October 1985 and April 1991. The other two – VA-205 at NAS Atlanta and VA-304 at NAS Alameda – traded in their *Corsairs* for *Intruders*, giving both CVWRs their first all-weather attack capability. On 22 July 1988 the *Firebirds* of Alameda received their first KA-6D; their initial A-6E arrived on the ramp two weeks later, on 5 August. Two years later, on 22 August 1990, the *Green Falcons* of Atlanta received their first KA-6D and retired their remaining A-7Es. The squadron's first A-6E checked in on 17 November.

Both of the new A-6 squadrons set to work, bringing the all-weather attack capability to their respective wings while carrying out their normal commitments and annual training. For example, over 9-21 July 1989 VA-304 deployed to NAS Fallon for its annual two weeks AcDuTra with CVWR-30; the only wing player to miss the event was Whidbey Island's VAQ-309, which was transitioning from the EA-6A to the EA-6B. However, the *Cougars* of VAQ-139 – recently returned from a WestPac/IO cruise with CVW-14 in *Constellation* – stepped in and filled the dance card. The *Firebirds*, led by Cmdr. Mike "Dibs" Dibello, had by all accounts a smashing desert cruise, ringing up 106 sorties for over 212 hours during their first full training period with their wing.

Not that there wasn't some controversy involved with the Reserves and their A-6s. At the same time the SWIP aircraft were appearing in the fleet NavAirRes reportedly circulated a letter requesting right-off-the-line SWIP airframes for the two Reserve squadrons, prior to their arrival in the deploying fleet units. The stated requirement was based on the need that Reserve units never knew when they would be required for call-up, and therefore "required" the best equipment available so that they would be ready when the time came.

The reserve's request was met with derision at COMMATVAQWINGPAC, where they were working to get enough of the improved SWIP airframes to the units deploying to forward areas. One staff member, a former squadron CO, pointed out that the reserve's own monthly reports stated they were spending most of their flight time on day bombing, low levels, and cross country events. He was appalled that they should get the most capable airframes when they weren't carrying out more training in the critical A-6 specialty, night low level attack.

Whatever the reasoning, the Reserves would get SWIPS – but not immediately. Both A-6 squadrons continued to train and fly the aircraft for several years. However, neither was ever called up to active duty and, like their reserve carrier air wings, they never deployed. That particular honor fell to two little-known but equally important groups: the reserve Squadron Augmentation Units, or SAUs.

Created on 1 October 1983 to train and manage Selected Reserve (SelRes) personnel, VA-0686 at NAS Oceana and VA-0689 at NAS Whidbey Island quietly did their job for years while functioning as a ready source of trained aircrews and maintainers for the active duty squadrons. These units would normally come onto Oceana and Whidbey on weekends and use fleet or RAG aircraft to train on, while being available for call-up to augment deployed units as necessary.

Their schedules were varied, and the squadron members never deployed as a unit per se; an example would be VA-0686's operations in 1988, which saw participation in live ordnance drops in North Carolina in April while under the control of FACs from the Air Force's *507th Tactical Air Control Wing*, followed by a trip to Nellis AFB in July for Red Flag with VAs -65 and -55. Still they carried on, regularly providing personnel for the deployed *Intruder* squadrons and integrating flawlessly. The training paid off in short order; when the time came both SAUs were more than ready, while the reserve carrier air wings stayed home.

If it's Tuesday That Must be the *Aeronavale*

The *Green Lizards'* brief combat of late 1988 notwithstanding the last two years of the 1980s were marked by heavy operations but rare confrontation. As before, both AirLant and AirPac cycled carriers and carrier air wings through the usual stops; they showed the flag and kept an eye on the Soviets and the occasional hot spot while honing their power projection skills.

The *Warhorses* of VA-55 had undergone a particularly entertaining year, concluding their second deployment alongside VA-65 in CVW-13 on 28 March 1988. The cruise started on 29 September 1987; on 5 December, while anchored off Palma de Majorca, squadron XO Ralph "Benny" Suggs – attack pilot and LSO extraordinaire – relieved Cmdr. Craig "Gator" Chewning as *Warhorse One*. Throughout the remainder of the trip Suggs and XO John "Worm" Henson led the *'Horses* through several operations, including Exercise *Louisianne*, which saw sorties against French *Mirage F1s* and F-8FN *Crusaders*. A later operation off North Africa, *African Eagle*, had the Moroccans come out and play.

As is typical in naval aviation, it was not all fun and games, even during peacetime. One night the crew of Lt. Rob Spratt and Lt.Cmdr. Bud Hunsucker had their pucker factor escalate exponentially when they suffered a complete electrical failure. Hunsucker pulled out his PRC-90 survival radio and was able to contact *Caron* (DD-970), which relayed the news of the *Intruder* crew's predicament to an orbiting VAW-127 *Hawkeye*. The E-2 then vectored in an FA-18 for an assist; just as it arrived, the A-6's standby gyro tumbled, removing their last instrument reference. Under the *Hornet's* watchful guidance Spratt and Hunsucker managed to fly 150 miles to a civilian field, all the while monitoring their pitot-static instruments by flashlight. They managed to save their A-6, recovering without the assistance of speed brakes, flaps, AOA indicator, anti-skid, or nosewheel steering. It was just another day at the office.

Spratt – the *Warhorses'* LSO – later played a role in the recovery of another stricken *Intruder*. Lts. Russ Knight and his B/N were coming in for a recovery one day when their starboard main-mount failed to extend. After attempts to blow the gear down failed they turned for NAF Sigonella, some 400 miles away. Spratt proceeded them and was able to wave the field arrested pass, which occurred with minimal damage to the aircraft and crew.

On the slightly more upbeat side, on 2 December Capt. Bruce B. "B3" Bremner – skipper of *Coral Sea* and former CO of VA-85 – scored his 1,000th trap in a CVW-13 *Intruder*. Bremner was something of a fabled individual, having survived two and a half *Intruder* ejections in his days – a B/N jumped out on him once. He would eventually be promoted to Rear Admiral and take command of the wing at Whidbey. Following a quiet Christmas in Naples, *Coral Sea* spent three more months in Med before heading for the Tidewater barn. In its stead, *Dwight D. Eisenhower* checked in with Sixth Fleet.

The *Blue Blasters* of VA-34 – now led by Cmdr. "Butch" Bailey – departed with CVW-7 in *Ike* on 29 February 1988 and enjoyed a busy

Only one AirPac *Intruder* squadron was ever attached to an AirLant Air Wing. VA-155 joined CVW-17 in *Independence* while that carrier was transferred to the Pacific Fleet in the autumn of 1988. *Silver Fox* KA-6D AA514 taxis to "Indy's" cat 1 during that evolution. (U.S. Navy)

cruise, flying with anyone and everyone through a series of operations and exercises. During the course of their six month deployment they participated in *Dragon Hammer '88*, a two-week NATO exercise that allowed them to hassle with Royal Navy *Sea Harriers* operating from HMS *Illustrious*, Belgian F-16s, and Turkish Air Force F-5s and F-104s, plus the occasional odd Air Force B-52. Subsequent exercises put the *Blasters* up against *Aeronavale Super Etendards* from *Escadrille 14F* and F-8FN *Crusaders* from E.12F operating out of NAS Heyeres. Adding to the festivities, Cmdr. Bailey recorded his 1,000th trap during the deployment.

So did Cmdr. Ross "Rookie" Word, CO of VA-176, which arrived in *Forrestal* with CVW-6 during early May. His *Thunderbolts* also notched a notable cruise, which included 100 straight days of flight operations in the North Arabian Sea under Operation *Earnest Will*. As if an extended tour in the NAS was not enough fun, the ship and air wing returned to ConUS via the North Atlantic, where *Forrestal* participated in *Teamwork '88* with *Theodore Roosevelt*. That sojourn included operations with Brits, Norwegians, and French, which meant the skies were full of *Jaguars*, *Tornadoes*, *Phantoms*, *Buccaneers*, *Sea Harriers*, *Super Etendards*, and F-16s. Needless to say, the exercise also drew the duty Soviet *Bear Ds* to the area for a look. Following their return to the states, on 7 October 1988 the *'Bolts*, CVW-6, and *FID* all received the Meritorious Unit Citation for the previous year's participation in *Ocean Safari* and the recently concluded cruise. Rear Adm. Leighton "Snuffy" Smith made the formal presentation onboard *Forrestal* to the ship's CO, former VA-35 skipper Capt. John Pieno. VA-176 topped their year by taking the annual *Intruder* Derby at NAS Oceana.

The year continued for the other AirLant A-6 squadrons, with each taking its turn on station and rolling up the brownie points and personal bests. In August 1988 the *Sunday Punchers* went back to Sixth Fleet in *John F. Kennedy*, with the *Hawks* of VMA(AW)-533 joining them again. For both squadrons, getting to the Med had involved the usual hard work and long hours, but now came the payoff. *Big John's* 2 August 1988 departure from Norfolk marked the first deployment of the A-6E SWIP, nine of which were flown by VA-75. The *Punchers*, led by Cmdr. John Meister, also took along three "straight" A-6Es and four KA-6Ds, giving the wing a total of 24 bombers and four tankers.

Kennedy in-chopped to the Med on 17 August and, as with its predecessors, quickly found itself involved in a series of operations and exercises, starting with *National Week*. Following a short port call in Naples the CV-67/CVW-3 team headed to North Africa for ten days of *Sea Wind '88* with the Egyptian military, followed by 20 days of *Display Determination* in the western Med. During the latter op the *Hawks* and *Punchers* launched all 28 *Intruders*, executed a simulated long-range strike and mine drop, then returned for a 28-plane fly-over. *African Eagle* '88, a December exercise with the Royal Moroccan Air Force and U.S. Air Force, helped bring 1988 to a close. The New Year arrived, and with one month to go in cruise CVW-3's *Tomcats* grabbed national headlines. On 1 January 1989 two VF-32 *Swordsmen* crews shot down two Libyan Air Force MiG-23s with a mix of *Sparrows* and *Sidewinders* after the Libyan pilots got frisky. As had happened eight years before, Qadaffi's crews had tested the U.S. Navy fighter pilots and were found lacking in several vital skill areas.

The *Sunday Punchers* and *Night Hawks* concluded their second consecutive *JFK* cruise on 1 February 1989. For the Navy all-weather-attack types, the deployment marked a successful debut for the SWIP bird, but it got even better. On 22 October in Oceana, while the squadron was still out and about the Mediterranean, VA-75 darn near scored a sweep at the annual *Intruder* Ball. Ops Officer Lt.Cmdr. Rich Jaskot was named "Intruder of the Year," while Lt. Jim Sauger took "B/N of the Year" honors. AOC Charles Carter was tabbed as the top *Intruder* Ordnance Supervisor of 1988. In addition, VA-75 took home the year's Norden Pickle Barrel Trophy in recognition of their high bombing scores. As for the Marines, their third carrier deployment in six years was also one of note, as they recorded 2,300 hours and 1,200 trips for the cruise, along with sortie and boarding rates in the high 90s. The *Hawk's* XO, Lt.Col. John Thornell, came home to a personal achievement; at the annual Tailhook Association confab in Las Vegas "Mooncap" received Marine Aviator of the Year honors. Not bad for a bunch of gyrenes.

In mid-January, just prior to their departure from the Med, *Kennedy* had joined up with *Theodore Roosevelt* for one last exercise, *National Week*. The op lasted five days, incorporated over 25 ships, and gave the departing carrier and air wing a good opportunity to view *TR*, the newest deck in the fleet. Not only was the carrier a new one, but so was one of CVW-8's squadrons: VA-36, on its first deployment since standing up in March 1987.

September's *Teamwork '88* marked the first excursion for the new CVW-8 medium attack team of VA-35 and VA-36, and by all accounts it had gone well. Dubbed the "notional air wing" in some circles, Air Wing Eight had two squadrons each of *Tomcats*, *Hornets*, and *Intruders*. During the mini-cruise the *Panthers* and *Roadrunners* participated in a lengthy series of War at Sea exercises, including flight operations from within Vestfjord, Norway. With his squadron suitably primed and ready to go, *Roadrunner One* Cmdr. T. Lamar Willis turned command over to Cmdr. Dan Franken on 1 December 1988. Twenty-nine days later, *TR* and CVW-8 departed Hampton Roads for points east. They rolled home again six months later on 30 June 1989. Over the course of the deployment the two *Intruder* squadrons kept busy with stops in Spain, Egypt, Turkey, Morocco, Israel, France, Palma, and Monaco, in and around flinging ordnance at target complexes in several of those nations and Sardinia and Tunisia. The exercise package included *Phinia '89* with the French and April's *Dragoon Hammer*. While there were a few rough edges resulting from the operations with the new carrier, in all everyone who made the tour was pleased with the outcome.

For VA-85 with CVW-1 in *America* the most "fun" may have come during workups in March 1989. While training in the Caribbean the squadron got to flight-test several *Harpoons* and *Skipper* laser-guided weapons. On one op *Black Falcon* Lt.Cmdrs. Mark McNally and Ron Sites managed a shack (direct hit) on the target hulk with an AGM-84, while Lt.Cmdr. Bob Fuller and Lt. Craig Peterson achieved similar results with an AGM-123.

Getting back to work, CVW-17 deployed on 11 May 1989 with the prospect of six months in the Med and the Indian Ocean. Three weeks later – on 5 June and while in port at Palma – VA-85's Cmdr. James B. Stone, Jr., turned the reins over to his XO, Cmdr. Dean W. Ellerman, Jr. Notably, the *Falcons'* change of command was part of a three-way ceremony; during the same event VAQ-137 and VS-32 also welcomed new skippers. About a month later, while in the North Arabian Sea, veteran *Intruder* pilot and former VA-115 CO Capt. Robert R. Wittenberg relieved Capt. Jay L. Johnson as CAG-1. Otherwise the *Falcons* kept busy doing the usual *Intruder* stuff of visiting strange and exotic foreign lands and occasionally bombing them (in an approved and friendly fashion, of course). Later, during *National Week '89*, *America* prowled around the Gulf of Sidra in company with *Theodore Roosevelt*, but this time no Libyans came out to play. Following an extended tour in the IO VA-85 returned home to Oceana on 10 November 1989.

Over on the other side of the world Oceana's AirPac counterparts performed similar duties, cruising around the Western Pacific and Indian Ocean, showing the flag, and maintaining a forward presence. As 1988 ended, VA-52 was wrapping up a standard six-month deployment in *Carl Vinson* with CVW-15, while VA-165 was two months into their own cruise with CVW-9 in *Nimitz*.

The *Knightriders* started their trip on 15 June 1988, heading west via the Bering Sea and some operations with F-15s from Elmendorf AFB's *21st Tactical Fighter Wing*. The tempo remained brisk throughout the deployment and included target runs in Alaska, South Korea, Japan, Okinawa, the Philippines, Malaysia, Indonesia, Oman, Thailand, Somalia, and Guam. Even the French came out to play, providing counter-air F-8s and *Super Etendards* from *Clemenceau* (R-98), also on a WestPac/IO cruise. After spending three months on Gonzo Station *Nimitz* finally turned for home; VA-52 returned to Whidbey Island on 15 December, having scored several personal and unit bests, including skipper Rich Dodd's 1,000th trap.

Led by "Otis" Shurtleff, the *Boomers* of VA-165 hauled out of the Rock on 2 September 1988 for their turn in the Western Pacific and Indian Ocean. Things got ugly early, with *Nimitz* pulling 70 underway days in the Gulf of Oman that took the ship and assigned squadrons through the Christmas "holiday." Fortunately, as compared to the previous workups and deployment, this one was fairly uneventful, with one exception.

During a swing through the North Pacific a squadron A-6E was forced to divert to Shemya AFB at the extreme west end of the Aleutian Islands. Shemya was truly one of the garden spots of the U.S. Air Force, described by one officer as "25 knots of direct cross-wind in blowing snow and near 0-0 visibility," and that was on a *good* day. The base opened for operations in May 1943 and served during World War II as a staging point for bomber and patrol operations. Postwar VQ-1 operated from the field; by the late 1980s it was an important surveillance post for monitoring Soviet ICBM test ops, housing a huge AN/FPS-108 *Cobra Dane* phased-array radar, as well as RC-135S *Rivet Ball* operations of the *6th Strategic Reconnaissance Wing*.

Fortunately, the *Intruder* recovered without incident, scoring an "OK-1" using the field gear. According to *Boomer* B/N Rich "Simo" Simon, it was just another day at the office when flying in the Aleutians:

> "It was pretty interesting, because we were doing ops in the islands," he comments. "We'd go into the sounds and would start shooting aircraft as the ship left the sound. The weather was terrible; you'd get this scud layer at about 200 or 400 feet, you'd be coming down, it was clear, and then suddenly you'd land without ever seeing the ball. We saw the mast of the ship moving through the fog and just kinda came straight down.
>
> "We also did some low-levels in Alaska and then pulled into Anchorage. We were on our way up to our admin, going up the elevator in the hotel, and there was an older guy with his wife. We got to talking to him and learned he was in the *Boomers* when they were flying 'Spads'! We invited them into the admin and ended up spending the evening rolling dice with him."

After its ops in Alaska *Nimitz* continued to the Indian Ocean under a full EMCON plan that cut down on intercepts by the Soviets. Simon recalls it was kind of dicey sending the KA-6s out ahead to refuel the *Tomcats*, because the plane didn't have an INS or radar.

"We'd fly ahead, fill up the *Tomcats*, and then try to find our way back and hope the ship was there," he adds.

According to "Simo," once *Nimitz* arrived on station in the IO it never actually entered the Persian Gulf, but the crews prepared for any eventuality, including possible combat over Iran:

> "We'd typically fly missions right up to the mouth of the Gulf," he says. "It would vary; sometimes we'd be high and sometimes we'd be low, sometimes at day, sometimes at night, depending on what time we took the tankers through. I can remember doing some contingency planning on Iranian targets, but the interesting thing is we looked at 'back door' stuff. What we looked at doing was sending stuff up that would jam and shoot HARM in the front end while we would send the Intruders in the back door. There are some mountains to the north; we'd come in through there. Bandar Abbas has some jetties that almost look like crab pincers; you could put your offsets there and bomb from there."

However, things stayed quiet during *Nimitz's* tour of the region, and the remainder of VA-165's deployment went without incident. The *Boomers* concluded their deployment – the 18th in their history and 14th with A-6s – on 4 March 1989.

Other Whidbey squadrons proceeded with their deployment schedules, but the duty WestPac 911 force continued to be *Carrier Air Wing*

5 in *Midway* with VAs -115 and -185 assigned. Over the summer of 1988 the ship remained tied up at Yokosuka undergoing repair work, but that did not translate into time off for the air wing personnel. The two *Intruder* squadrons spent part of the summer undergoing a Medium Attack Advanced Readiness Program (MAAARP) at Osan Airbase, Korea, mixing it up with the *51st Tactical Fighter Wing*. In August the two squadrons resumed FCLPs in preparation for a return to the boat.

For the *Nighthawks* it had been a pretty successful first year as part of the CVW-5 team, but as can be expected, they spent their actual one-year anniversary, 10 September 1988, underway. In fact, between the fall of 1988 and the spring of 1990 *Midway* deployed a staggering six times, for varying periods. Occasionally though the *'Hawks* got the bennies. In May 1989 – in between two relatively short cruises totaling, oh, about 14 weeks – VA-185 brought Vice Adm. Mitsuo Kanasaki, Commander-in-Chief of the Japanese Maritime Self-Defense Force, aboard *Midway* for his first trap. With squadron Ops O Lt.Cmdr. Joe O'Donnell doing the piloting the admiral was greeted by Rear Adm. Lyle Bull, *Commander Battle Force Seventh Fleet*; *Midway* skipper Capt. B.J. Smith; and CAG-5 Capt. Dave Carroll. After a tour and a period of observing flight operations CinC JMSDF said his good-byes and returned to Atsugi via a shot off *Midway's* pointy end.

Next out of the barn was *Constellation*, once again deploying westbound with CVW-14 and VA-196 embarked. Now led by Cmdr. Bruce Stuckert, the *Main Battery* hauled out the first of December and made a quick transit to the Indian Ocean, with occasional stops in places like Diego Garcia and Karachi, Pakistan. Notably, on this cruise CVW-14's *Intruders* finally got an assist in the inflight refueling mission: the mighty *Sawbucks* of VS-37 became the first S-3A squadron to deploy with a tanker role. During the course of the deployment the *Hoovers* passed over 1,000,000-pounds of JP in and around their other missions of airborne ASW, bombing, and MineExs. VA-196 took advantage of the "break" to hone their skills, participating in weaponry practice off Diego Garcia, ops with the Thai and Singapore air forces, and exercises, such as *Beacon Flash* and *Busy Customer*. The payoff was port call to Perth, Australia, for some well-deserved liberty prior to their 1 June 1989 return home.

CVW-2 in *Ranger* also had an outstanding cruise with VA-145 and VMA(AW)-121 honing the tip of the spear. For the *Swordsmen*, the 24 February 1989 departure marked the AirPac introduction of the A-6E SWIP, and they made the most of it. By the end of the cruise on 24 August 1989, the squadron rolled up more than 2,600 hours and 1,350 traps, with 16 *Swordsmen* qualifying as *Ranger* centurions. *Green Knight* skipper Lt.Col. Pete "Nodak" Jacobs also set a record of 500 career traps, 200 of which came in *Ranger*. It was a remarkable accomplishment for a Marine pilot in the Cold War era, and marked how important the Marine A-6 squadrons had become to the carrier Navy. Equally notable, his squadron took the wing's Indian Ocean Bombing Derby for the second consecutive cruise, -121's last with CVW-2.

As for VA-95, they too made the WestPac rounds in late 1989, departing with CVW-11 in Enterprise on 17 September and returning on 15 March 1990. Fortunately, this deployment was a tad less "sporting" than the previous one. Led by Cmdr. John Schork, the proud *Lizards* went back to sea, having scored a hat trick, notching the Rear Adm. Clarence Wade McClusky Award, the NavAirPac Battle E, and their third Safety "S."

Even VA-155 got into the action, finally. Under the guidance of "Blackjack" Samar and Ron Zimmerman, the *Silver Foxes* took the "NM" of the dear departed CVW-10 off their tails, painted on the "AA" of CVW-17, and made their first deployment on 15 August 1988. The event was the transfer of the recently SLEP'd *Independence* to NAS North Island. AirLant threw together a mix of squadrons, aircraft, and personnel – including VF-103, VFA-131, and VS-30 – and sent the carrier south around the Horn. Commanded by former VA-65 skipper Capt. William R. "Buzz" Needham, *Indy* managed port calls in Rio de Janiero and Acapulco in and around operations with Brazilian, Argentine, Chilean, and Peruvian air force and navy units.

The trip ended with the *Indy* pulling into its new homeport on 8 October 1988, escorted by *Lake Champlain* (CG-57). The end of the short cruise also marked the end of VA-155's affiliation with CVW-17; on 1 October 1989 the *Foxes* formally transferred to CVW-2, taking VMA(AW)-121's place in that wing's lineup.

Yuma, Fallon ... and there go the Marines

In some ways, above and beyond the unique – if brief – assignment of Whidbey-based *Silver Foxes* to CVW-17, there was other evidence of consolidation or cooperation between the east and left coast A-6 communities. Much of it just made plain good common sense.

One example was the establishment of a joint permanent weapons detachment at NAF El Centro, CA, in September 1988. The two RAGS (VA-42 and VA-128) had been running regular weapons dets to El Centro, MCAS Yuma, and NAS Fallon for years, giving the replacements and instructors the chance to drop live ord and work on tactics

Even though the end was in sight for Marine *Intruders*, squadrons still deployed overseas to hone their skills. VMA (AW)-224 visited down under in 1989, deploying to RAAF Curtin, Western Australia, for operations with the Royal Australian Air Force. Nobody suspected that about a year later they would be in Bahrain gearing up for combat against Iraq. (J. Hunt)

and deliveries. Both squadrons started concentrating on dets to El Centro during the summer of 1987, and the following year formalized the arrangement through the establishment of the Medium Attack Weapons Detachment, officially assigned to the *Golden Intruders*.

Depending on whom you talked to, Naval Air Facility El Centro was either a great place for liberty or another one of those tours that had to be endured. The facility dated to the 23 July 1943 establishment of MCAS El Centro at a dusty site in the lower Imperial Valley, 10 miles north of Mexicali, Mexico, eight miles northwest of El Centro, and 128 miles due east of San Diego. Initially used for Marine fighter, bomber, and transport unit training, the base went into reduced operations at the end of World War II. On 1 May 1946 the Navy reopened El Centro as an auxiliary air station, and followed with the transfer of the Naval Parachute Unit to the installation in November 1947. Weapons training returned in the early 1950s with the establishment of Fleet Air Gunnery Unit, Pacific (FAGUPAC); in June 1956 the Navy and Marines held the first Fleet Air Gunnery Meet at the station. Over time the field also became the winter home of the *Blue Angels*, while the Navy continued to make improvements to the training ranges and target complexes, leading to its 1 July 1979 redesignation as a naval air facility.

What with clear skies, wide-open ranges, and a "temperate" climate, El Centro proved to be a great place for weapons training. Each RAG squadron put one class through MAWDet every other month. The training schedule included: division formation; air-to-air refueling; manual weapons delivery (in other words, the B/N let the pilot drop the bombs); war-at-sea tactics; close air support; deep strike interdiction; strike planning; defensive air combat maneuvering, or DACM; and of course, lots of live ordnance expenditures. The students completed their det training by planning and leading multi-aircraft coordinated strikes, resulting in a ground shaking "graduation exercise."

As for the fleet squadrons – and their air wings – they continued to make the sporadic weps det to the other two primary range complexes, with Fallon hosting most of their visits. According to "Simo" Simon, the live ordnance exercises were always the highlight of predeployment workups:

> "At Fallon we did the standard thing, buzzing cows and running up to Dixie Valley," he recalls. "We bombed this Army depot (Hawthorne Army Ammunition Depot, a former Navy facility) which had miles of bunkers stretching off into the desert. Our directions were something like 'hit the bunker four from the right, five deep,' which was different.
>
> "Another time we were at the beginning of workups. I was flying with Rob 'Levi' Perrettis; he got the name at an admiral's social when everyone wore suits except Levi, who showed up in jeans. It was a normal Fallon det, with A-7s in the wing, and we had a bunch of *Skippers* at the time. I don't know if they just wanted to get rid of them, but we shot the hell out of Wildcat (Bomb Range), up in Idaho. There was a (target) train, and we'd shoot up a bunch of boxcars. About 50 percent of the *Skippers* went off target, all over the place, but others would hit."

On another occasion and during another work-up cycle, Simon was flying with pilot Rick "Yak" Yasky when he had a couple more live ord experiences:

> "We were taxiing out there – there must've been something like 30 planes – and something happened, so we taxied back. As a result, I broke my regular checklist procedure. We got it fixed, we launched, we hit the target, and I started jumping up and down in my seat ... and realized, 'heyyy, I forgot to strap back in.' Now, Yak always carried a tape recorder, recording our conversations. At this point he pulled the recorder out and announced, 'Okay, I think we'll stop this now.'
>
> "Another time we flew this strike ... we launched off Mexico, met with some tankers off Los Angeles, and the tankers drove us all the way up to Lemoore. We then dropped down into Nevada and did a low level into Saylor Creek, Idaho, all at night. We launched at something like two or three in the morning, and when we came off the target the sun was coming up. We were beat, but it was gorgeous. We went up to high level all the way back to Whidbey. It was like an eight, eight-and-a-half hour flight; they almost had to pull us out of the cockpit. On the way back Yak said, 'Gee, I could sure use some coffee,' so I pulled out this thermos and said 'I hope you don't want sugar.' I then pulled out another one full of orange juice. He couldn't believe it."

"Simo" adds *Boomer* weapons deployments to Fallon were always colorful for another reason: the crews always went fully armed.

"We were big on guns," he comments:

> "We'd fly down with three or four guns in the cockpit, and we'd bring a van full of ammo. Then we'd line up in the desert and shoot rocks, lizards, whatever. One time this rabbit came out and the whole line trained on this rabbit as it ran for its life, right down to the end of our line ... and guess who was at the end of the line? The skipper. We almost shot him."

Back at El Centro the two fleet replacement squadrons – which included Marine *Intruder* crews under the auspices of VA-128 – continued flinging their own bombs around the desert under possibly more "controlled" circumstances, but there were changes in the wind, not

The *Green Knights* became the first Marine *Intruder* unit to transition to the FA-18D in December 1989. Thirteen months later, during *Desert Storm*, the squadron would introduce the type into combat. (Bob Lawson)

just for the new crews, but also for all Fleet Marine Force A-6 personnel. While the Navy worked with General Dynamics and McDonnell-Douglas to deliver the A-12 to the fleet a few years hence, the Marines made no bones about the fact they did not want the aircraft at whatever cost. Instead, the service started reducing its A-6E component by transitioning the squadrons to the new FA-18D.

An outgrowth of the FA-18B, the D-model *Hornet* was designed not only for strike, but also tactical air control missions. Among its systems were the AN/ARR-50 Thermal Imaging Navigation Set, or TINS (an acronym already well known among Navy and Marine flight crews, for other reasons), an improved HUD, NVGs, Airborne Self-Protection Jammer (ASPJ), Data Storage Set (DSS), color digital moving map, the *Hornet's* AN/APG-65 multi-mode radar, and provisions for a Weapons and Sensors Officer (WSO) in the back seat. Range and combat radius problems? The *Hornet* had them in spades, but with the Marines' emphasis on close air support and airborne tactical control they could live with it. The *Sharpshooters* of VMFAT-101 at MCAS El Toro had assumed the Marine FA-18 RAG duties in 1987 and prepared for the same role with the FA-18D.

While the first D-models didn't arrive in the FMF until the spring of 1990, the Marines redesignated its first *Intruder* squadron to fighter-attack (all-weather) in ceremonies at MCAS El Toro on 8 December 1989. The honoree was the *Green Knights*, now VMFA(AW)-121, led by Lt.Col. Gayle Adcock. The last -121 *Intruder* left the ramp on 21 December; five months later, on 11 May 1990, McDonnell-Douglas formally delivered the first FA-18D to the *Knights*.

While the Marines' push into *Hornetdom* undoubtedly left some scratching their heads and probably infuriated Navy supporters of the A-12 program – the last time something like this had happened was in the early 1970s when the USMC respectfully but firmly declined acquisition of the F-14A – the impending release of five squadrons-worth of A-6Es meant more airframes for the Navy and Naval Reserve. Everyone was confident they would do just fine holding the line until the arrival of the stealth attack aircraft, now scheduled for first flight around 1990. However, events around the world continued to change, and the medium attack communities – both Navy and Marine – were forced to change with them. As the saying goes, nothing goes out the window faster following first contact with the enemy than a carefully crafted plan.

Big Screen *Boomers*

Rumblings emanated from Eastern Europe and the Soviet Union in mid-1989 when Whidbey Island's VA-165 received one of the all-time good deals: the Navy selected the squadron for participation in the filming of the movie version of Stephen Coonts' *Flight of the Intruder*. After the movie "Top Gun" put F-14s and Miramar on the pop culture map, the *Intruder* community came up with a bumper sticker that read, "Fighter pukes make movies. Attack crews make history." Well, it was now the A-6's turn, and the Navy offered *Independence* and VA-165 to director John Milius, whose credentials also included such cruise classic "Man's Movies" as "The Wind and the Lion" and "Red Dawn." Mace Neufeld, known for "No Way Out," served as the producer.

The *Boomers* of VA-165 had a busy late-1980s, as shown by these two aircraft. NG501 was one of two A-6Es covered in a temporary desert scheme for a Red Flag exercise at Nellis AFB, NV. This aircraft would later join VA-155 and be shot down in *Desert Storm*. "NK502" was one of several squadron aircraft temporarily repainted to resemble 1972-era VA-196 *Intruders* for the movie "Flight of the Intruder." The bureau number is bogus (152792 actually belongs to a C-2A), and the white nose and rudder don't look quite right on an overall dull gray aircraft, but it does achieve the look needed for the movie. (Both pictures by Rick Morgan, 11 September 1989)

Conveniently, the *Boomers* were right in the middle of a one-year period between deployments, which made them available.

"We were selected because we were a squadron at Whidbey that didn't have anything to do at the time," says Rich "Simo" Simon. "It might have been partially because our air wing was switching to F-18s."

Among the things required by the movie production company, the squadron had to repaint their aircraft in correct, Vietnam-era markings for VA-196, Coonts' squadron from the Vietnam War. The normal fleet Tactical Paint Scheme (TPS) gray scheme had white added in some places and bogus VA-196 markings applied for the period required in the book.. According to skipper "Otis" Shurtleff, however, the *Boomers* did not really care, and everyone proceeded to have a large time.

"I could write page after page on the movie," he wrote:

> "This was the one really unique thing we did during my CO tour. From volunteering the squadron to Savannah, Georgia – just missing Hurricane Hugo – to almost a month in Hawaii, to the ops on the *Independence*, it was just about as much fun as a squadron could have."

Vietnam combat veteran and A-6 pilot Capt. Sam Sayers severed as a technical advisor, taking actors Danny Glover, Willem Dafoe, Brad Johnson – who played the protagonist and Coonts' alter-ego, Jake Grafton – Dann Florek, Tom Sizemore, and others under his "wing," teaching them the ropes and explaining all-weather attack aviation. Several of the actors went through aviation physiology, toured carriers, met with squadron and ship's personnel, and cobbed flights, all the while working at developing that peculiar "attack aviator" persona.

Simon was flying with Shurtleff at the time as the skipper's B/N. He comments:

> "First, we flew to Savannah; we must've taken four or six airplanes. We came in on the worst possible night; the hurricane that had taken out Charleston – I think it was Hugo – came through just a week before. I can still remember doing an *Intruder* approach into Savannah, and the skipper asks, 'Can you see anything?' I looked down, saw a 7-11, and could count the cars. Once we got out of the cockpit we asked 'Where are we going to? Charleston?' 'You're kidding! It's closed!' 'It's not in the NOTAMS (Notice to Airmen).' Their response was, 'That's because there's nobody there. It's gone.'"

In and around flying the movie's scenes out of Savannah Municipal – the Air Force's former Chatham AFB, with a "long" runway of 9003-feet – the -165 crews continually demonstrated the true "spirit of attack," as Simon continues:

> "We would normally carry a load of bombs with us, but Paramount didn't want us to drop them because they were paying for them. We'd have to come back in and grease a landing on a short runway with no arresting gear.
>
> "There was an old Civil War rice plantation to the south, and there's a snaky river that goes through there; that's where we did the filming. Otis and I were flying down this old muddy canal – we must've flown down it five or six times – and they had to keep adjusting the shot. We kept flying lower and lower each time, really scaring these birds. We were down at 30 or 50 feet above the canal, and they're blowing up dynamite charges. I'm looking off to the left, skipper's looking ahead watching these birds, and trying to keep from hitting a tree. I said, 'Skipper, the water's higher than we are.' He just nodded and said, 'I know, I know.'"

Other major filming took place in mid-October at Pearl Harbor, with additional scenes photographed on Kauai – the town of Hanapepe filled in as "Po City (Olongopo, Philippines)" – and the Pacific Missile Range Facility at Barking Sands. To add to the realism of the movie the production company even paid to have two privately owned A-1s flown over to Hawaii for the climatic crash, evasion, and rescue sequence. The portion of the movie took over a week to film and was done near Mount Waialeale.

"Otis was flying the camera plane," relates "Simo":

> "It's like a Learjet, and inside is a rail that the camera rode on, and there's a periscope out of the top and the bottom. He was going down the valley when they did that scene of the two planes

Coral Sea's 1989 deployment would be the last for the veteran carrier. CVW-13 had two *Intruder* squadrons assigned (VA-55 and VA-65), as well as three *Hornet* units. A *Warhorse* takes the shot from cat 2 while a *Fighting Tiger* waits its turn on cat 1. VA-55 would disestablish following the end of cruise. VA-65 would move to CVW-8. (U.S. Navy, PH3(AW) Wayne Edwards)

> going down the valley, when the doc (actor Dann Florek) gets sick. Otis tried to pull over the mountain, and the Lear didn't have the poop of the A-6, it didn't pull like on an A-6. They were on hot mike, and Otis said, 'God, I hope we make it.' That did a lot to instill a lot of confidence in everyone else."

The at-sea portion of the filming took ten days and nights; for most of the actors and production personnel, it was their first extended stay at sea on a man-o-war. Fortunately, *Indy's* crew – led by Capt. Tom Slater – meshed well with the movie company, and the cameramen were effectively given the run of the ship:

"The interaction with the actors was great," Simon adds:

> "I remember sitting at the duty desk as SDO, and Dick Russell brings in this guy in khakis; he's a lieutenant commander, and he's wearing a 'Tiger Cole' nametag, and I remember thinking, 'oh great, *another* O ... and this asshole needs a haircut!' My other recollection is we flew Dafoe off the ship; we flew him, Danny Glover, and Brad Johnson. I remember Brad Johnson at the club, we're all playing (the dice game) Klondike, and he was pulling out a wallet with $100 bills. We all rolled our eyes."

For the underway filming, the Navy and the movie's producers made a strong effort for realism by craning aboard and painting a static A-3 and F-4s, along with some visiting VA-122 *Corsairs* to give the feeling of a 1972 carrier air wing. The *Boomer Intruders* had their birds painted with the white noses and hi-visibility markings of the period, with stenciling for VA-196, CVW-14, and *Independence*.

Unfortunately, "Flight of the Intruder" was released in early 1991, and had nowhere near the impact of "Top Gun." The reviews ran the gamut from positive to awful, with one movie critic commenting that he spent the entire movie hoping the carrier would sink and take the cast with it. One former Navy pilot rated the final picture as a mixed bag, adding:

> "Willem Dafoe was cast as a dead-on-target Tiger Cole and Danny Glover played a good Skipper Camparelli. The leading man was largely unknown Brad Johnson who, although trying hard, seemed to be in the 'not quite right' category. The action and flying scenes were generally superb, although the movie's ending – which was substantially different from the book – seemed to draw a lot from Michener's 'The Bridges at Toko-Ri.' 'Flight of the Intruder' did do a pretty good job of portraying the frustration of the warfighters in 1972 and offered a hint of the beauties of Olongapo. Overall, the pluses made it a keeper, if not a classic."

Still, the filming provided a boost for the *Boomers* and the rest of the *Intruder* family, and gave attack aviators and maintainers everywhere something to cheer about. Coincidental with the release of the movie was the fact that the A-6 had reached its high water mark. 1990-91 would be the pinnacle of the Medium Attack community's history, with 24 squadrons (16 regular Navy, 4 Marine, 2 Naval Reserve, and 2 RAG) flying the type. Things would start changing in a hurry.

With what was going on in the rest of the world, something light and distracting like a movie was exactly what was needed, because the world was still a very dangerous place, as the political situation in the Soviet Union and Eastern Europe soon demonstrated. The world was changing faster than most had a chance to keep up with. The façade that had been the "Evil Empire" was collapsing. Mikhail Gorbachev had taken over the Soviet Union, while demonstrations and riots broke out in Eastern Europe, as the people went about throwing off the yoke of communism. Poland, Hungary, Romania, and the Baltic States all leaned westward, and on 3 October 1990, East Germany formally joined with the West to form a united *Deutschland*. The Cold War appeared to be coming to an end.

There was more to come over the following year. In the meantime, the Navy's carriers remained on patrol in the Med, with occasional forays into the North Atlantic. During the 12 month period where Poland, East Germany, Romania, and Hungary effectively fell out of the Warsaw Pact and the Communist orbit, the AirLant deployers – trained and ready as always for any contingency – included VA-55 and VA-65 with CVW-13 in *Coral Sea* and VA-85 with CVW-1 in *America*. Along with their Air Force and Army counterparts, they watched nervously, waited, and trained. However, a bigger problem was about to break out that would put *Intruders* – and indeed, the entire U.S. military – back into combat. By late 1990, as the first Soviet dominoes fell in Eastern Europe, America was well on its way to involvement in another shooting war, and this time it was in the Middle East.

The Proverbial Line In the Sand

Early on the morning of 2 August 1990 three Iraqi Republican Guard divisions crossed the border into Kuwait; within short order they conquered the small, oil-rich country. The world was stunned, including the neighboring Arab nations, although the signs of a possible invasion had been in evidence for some time.

The eight-year war with Iran had left Saddam Hussein's nation financially bereft. As is common with any dictatorship, Hussein decided the solution would be found through threats, intimidation, and possibly a new source of cold, hard – or, in this case, liquid – cash. In early 1990 the Iraqi leader told several Arab nations that his country needed $30 billion in order to rebuild and get back on a firm financial footing. The following May, during a meeting of the Arab League, Hussein specifically targeted Kuwait, demanding $27 billion. Referring to the nation as a "lost province" of Iraq, Hussein took particular exception to the smaller nation's rich ruling class, and its propensity for oil production above and beyond the levels approved by OPEC.

The United States quickly responded by freezing all Iraqi and Kuwaiti assets. However, the big concern was whether the Iraqi Army would stop at the Saudi-Kuwait border or continue into the Saudi Arabian Peninsula. Fortunately, indications were the Iraqis had reached the end of their supply tether, and in any event, were too busy looting, pillaging, and plundering their prostrate neighbor. That gave the Bush Administration some time to start efforts to a) protect Saudi Arabia (and its oil fields and government, which was generally friendly to the United States), and b) build some sort of coalition that could eventually force Iraq out of Kuwait.

In the near term, it fell to the U.S. Navy – through carrier aviation – to provide the first line of defense. On 2 August *Independence*, with CVW-14 and VA-196 onboard, was underway in the Indian Ocean. In the Mediterranean *Dwight D. Eisenhower* – with CVW-7 and VA-34 – was wrapping up a standard six-month deployment. Their schedules quickly went out the window; *Ike* quickly headed for the Suez Canal and a transit to the Gulf of Oman, while *Indy* headed for the North Arabian Sea from the vicinity of Diego Garcia. According to *Blue Blaster* CO Ron Alexander, the sudden turn was a big surprise, but the ship, air wing, and squadron responded in exemplary fashion:

> "We were at sea getting ready for a big NATO exercise off the coast of Israel and Lebanon – I can't recall the name of it – and I just remember it was a relief that we didn't have to go to the exercise," he says. "It was leading up to be *very* painful. I remember the planning for it and everything, and it was not going to be a lot of fun. The admiral got the news that Iraq had invaded Kuwait, and shortly thereafter we were told, 'okay, we're probably going to go down through the Suez into the Red Sea. We started trying to gather everything we could at that time, intelligence-type stuff, pictures, and there wasn't a lot."

The Blackburn *Buccaneer*

Perhaps the closest aircraft to the *Intruder* in terms of function was the Blackburn (later Hawker Siddeley) *Buccaneer*. Built as a twin-engined, two-seat (although tandem) low altitude carrier-based strike aircraft, the type would, in time, fly for three services and see combat in at least two conflicts.

The "Bucc" was roughly similar in performance to the A-6, a wee bit faster and farther ranging in the low-altitude environment that the Brits espoused, although it never did carry the equipment that would allow an "all weather" title like the A-6. Both types shared a not-quite-esthetic profile, a big, thick wing (complete with *Phantom*-like bleed-air boundary layer control on the "Bucc"), and even the same refueling probe configuration. There were differences, of course, as the English type featured an internal bomb bay, T-tail, and "petal" speedbrakes at the rear of the fuselage.

First flown in 1958, the "Bucc" was declared operational with 801 Squadron in 1962, and flew with distinction in the Royal Navy in four versions (S. – for Strike – Mk 1, 2, 2C and 2D), and for five squadrons until retired with the nation's last conventional carrier *Ark Royal* in 1978.

Export sales were approved, and in 1965 South Africa took the first of 16 S. Mk 50 airframes for use with its 24 Squadron. In spite of apartheid-inspired parts embargoes, the type saw combat from 1978 in the Namibian guerilla wars and remained in service until 1991.

The Royal Air Force (RAF) spent a good portion of the early '60s looking for a new strike aircraft, and looked seriously at both the supersonic TSR.2 and proposed General Dynamics F-111K. It canceled both projects and ordered the *Buccaneer* instead, buying new S. Mk2Bs and using former Royal Navy birds from 1969. Several squadrons were based in Germany as very low altitude NATO strike aircraft, while others were located in Scotland in the maritime role. Retirement started in the early 1980s, with the type being replaced by the *Tornado*. The RAF "Buccs" finally saw war in 1991, when the call came to send twelve to Saudi Arabia, where they participated in Operation *Granby* (the English name for Desert Storm). *Buccaneers* flew 226 sorties through the war without loss, largely serving as Pave Tack laser designators for *Tornados* dropping Laser Guided Bombs. *Granby* was the end of the road for the type in Her Majesty's service, though, and the last one ended squadron service in 1994.

Only 209 *Buccaneers* were built, as opposed to over 700 *Intruders*, but the "Bucc" was a world-class warplane that was remembered fondly by almost everyone who flew it.

A Royal Navy *Buccaneer* of 809 Squadron visits NAS Oceana on 14 July 1978. The unit was disbanded the following December, ending carrier-based *Buccaneer* operations. RAF "Buccs" would continue through 1994. (Mark Morgan)

The skipper recalls that – other than for a few chiefs who had served on Yankee Station – nobody in the squadron had combat experience ... not that it mattered:

> "They (the squadron) really stepped to it, and the fact that we were going to be late coming home probably never entered anybody's mind. Every plane was up; they worked off every gripe. The troops knew that there was a possibility that we'd be going into combat, and they wanted to send the best product they could."

Once on station, *Ike* and *Indy* joined the Navy's existing units in the region, assigned under *Commander Joint Task Force Middle East/Commander, Middle East Force* (COMMIDEASTFOR), Rear Adm. William M. Fogarty. Embarked in the command ship *LaSalle* (AGF-3), Fogarty could call upon a small surface force made up of one cruiser, a destroyer, and six frigates. None of these ships carried anything bigger than *Harpoon*, hence the two carriers, loaded for bear and ready for combat, formed the nation's thin blue line against further Iraqi aggression:

"That was obvious," Ron Alexander comments:

> "We knew that. We were put on alert just about nightly to be ready to go, to support any kind of movement that might be happening. We had to be there to support any of the ground troops if anything happened. We had *Hornets* (VFAs - 131 and -136) and two squadrons of F-14s (VFs -142- and -143) with us. As for the *Hornets*, we looked at what they were trying to carry – I think they were looking at two, maybe four bombs – and I said, 'Give me the tankers, we don't need these guys. We can carry the bombs and stay on station for a long time.' It was just an interesting time. We were about all they had that could be available."

Within short order, though, additional carriers and other units pulled into the line. *Independence* – with *Commander Cruiser-Destroyer Group*

One, Rear Adm. Joseph P. Reason, embarked – arrived first on 5 August. The following day Secretary of Defense Dick Cheney met with Saudi King Fahd and finally convinced the monarch that the kingdom needed the immediate support of the United States and its allies. The same day, the United Nations agreed to the imposition of economic sanctions against Iraq.

On 7 August, President George Bush formally announced that America's interests required an extensive military response to Iraq's invasion of Kuwait. Immediately, U.S. air and ground and additional naval forces started deploying to Saudi Arabia under the title of Operation *Desert Shield.*

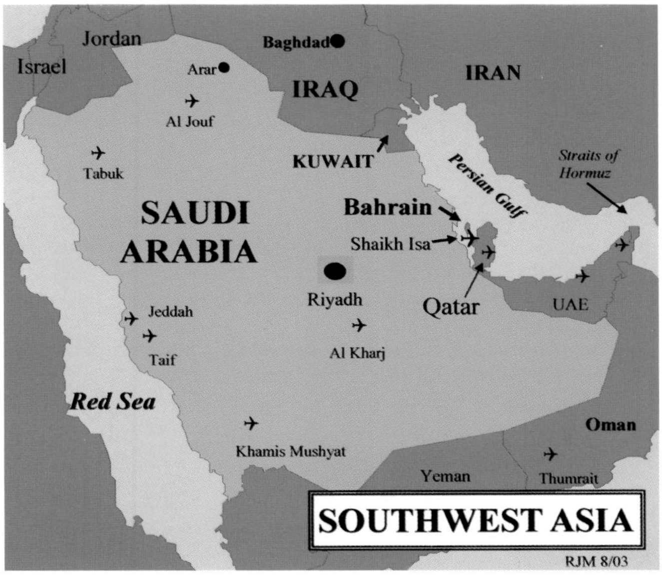

Intruders to the Gulf

Army and Air Force units started arriving in Saudi Arabia, initially under the cover provided by the two carriers in the region. The personnel of the *Independence* and *Eisenhower* battle groups knew they were "it" as far as providing striking power, at least until such time the United States could start putting heavy forces into Saudi Arabia. They took the role and worked on strike planning, albeit with a fair amount of higher-up-type confusion and minimal targeting or Order of Battle data.

"We flew quite a bit once we got there," comments Alexander:

> "We were in the Red Sea, and *Indy* was on the other side. It took some time to get down there, so as soon as they said we were going down we started trying to find what was there as far as training routes. There weren't any, really, so we built some and submitted them for approval. About two or three days before we left they finally got approved, so we didn't get to use them. *Sara* – who relieved us – did. KC-135s were the only thing we had for tanking. Every time we took off we went for the tanker, just for practice. We couldn't get any KC-10s; they were busy supporting the air bridge."

With the assumption of operational control by Commander-in-Chief *U.S. Central Command*, Gen. H. Norman Schwartzkopf, the buildup continued under various unified components. Commander, U.S. Seventh Fleet, Vice Adm. Henry Mauz, Jr., assumed the leadership of *Naval Forces Component, Central Command* (NAVCENT) while enroute from Yokosuka in *Blue Ridge* (LCC-19). Mauz initially had the two carrier battle groups to work with, along with the BG escort ships and units of the Middle East Force, but as with the other services, more was coming in a very big way. In late August *Saratoga* deployed from Mayport with CVW-17 and VA-35; about the same time *Wisconsin* (BB-64) headed east in the van of a Surface Action Group (SAG). A week later *John K. Kennedy* with CVW-3 and VA-75 headed out, followed in September by *Midway* (CVW-5/VAs -115 and -185) and *Missouri.*

"Big John" deployed with all of seven days notice. According to *Punchers* skipper Bob Besal, for a while there it looked like his squadron and *Kennedy* would miss the party, and no one in the squadron was happy with the prospect. Initially, the Navy told the squadron it wasn't going anywhere, but it would provide aircraft and crews to VA-35 in *Saratoga*:

> "The afternoon of the last Friday in July, the 27th ... I'm sitting in my office, feeling sorry for myself," he writes. "My 'one chance to deploy' while in command of the *Sunday Punchers* just evaporated: *JFK* will not sail for the North Atlantic NATO exercise due to 'budget constraints.' No steaming dollars for oil-fired ships. A nuke carrier – the *TR*? – will go alone instead of the dual carrier sail planned.
>
> "My two squadron intelligence officers, Ensigns Mark Elliott and Dan Brannick, stop in the office to give me an update. They've just returned from the Fleet Intel Center, which was abuzz with reports of Iraqi Republican Guard forces massed on the border with Kuwait. They give me the details of how many, why, and their best guess of when an invasion will occur. I'm interested, but it sure seems a long way from me being involved.
>
> "Tuesday, 31 July 1990: the ensigns swing by the office with an update. The IRG troops have pulled back from the border. The threat of invasion – according to the senior analysts – is defused ... but for some reason, the move doesn't sit well with Dan and Mark. They see it differently, and tell me that they believe Iraq will invade within 48 hours. I'm sort of impressed by their determination, and that they disagree so vocally with the sage leaders of the intel community. They're really sure that this is going to happen."

Two days later Iraq went into Kuwait. According to Besal, his two AIs instantly became stars, and were "looking pretty smug...and downright omniscient." He continues:

> "Tuesday, 7 August 1990: I'm summoned to COMMATWINGONE's office. My hopes for deploying to the Middle East are shot down. I'm directed by COMNAVAIRLANT to transfer four of my SWIP *Intruders* and five crews to VA-35 ASAP. I'm disappointed to say the least. VA-35 was sailing in *Saratoga* on schedule to relieve *Ike*. I call an AOM at 1000. While everybody knows something is 'up,' nobody was ready for the fractioning of the squadron. Many guys looked positively stunned. I announced my selection of those to go to sea with VA-35: pilots Lt.Cmdr. Ken Coburn, Lt. Les Makepeace, Lt. Spencer Butts, Lt. Mike Walsh, and Lt. Kurt Barich; and B/Ns Lt. Frank McCulloch (who's on leave in Mississippi), Lt. Tony Derossett, Lt. Al Misiaszek, Lt. Rick Mooday, and Lt. Jeff Hammann.
>
> "After the AOM several of those not chosen corner me in the office. I can honestly say I expected I might be confronted by some of those I selected as to why they shouldn't go ... nothing was further from the truth. Those not announced were demanding to know why they didn't make the cut, and volunteered to be next in line (I bit my tongue – I really wanted to say, 'You're in line behind me!'). The group was extremely talented. I kept all the department heads with me, as I felt I still had some semblance of a squadron to run, and I figured I needed their strong arms to keep the JOs under rein."

Afterwards Besal talked to the troops, announcing 35 maintenance personnel would also deploy with VA-35 under the capable guidance of

CWO2 Bill Shelton; they would fly to Rota and join *Sara* as she sailed past. After making the announcement squadron members continued to approach the skipper and volunteered to be the next in line. He recalls being "overjoyed at their response." On Wednesday, 8 August, VA-75 had its four best A-6Es – BuNos 162190, 162192, 162195, and 162197 – pumped, primed, and ready to go, and Besal completed his "data dump" with VA-35 CO Cmdr. Jim Anderson. They launched for *Saratoga* the following night:

"At 0130, I step out on my deck in the back yard and hear my four 'donations to the cause' take off from NAS Oceana. As they come into view above the trees, I watch them turn out to sea for night CarQuals and fly aboard *Sara*. I'm quite emotional and rather depressed," Besal recalls.

Two days later Besal says he attended the commissioning of a former staff mate onboard a submarine tender in Norfolk. When he got back to the squadron someone told him the commodore wanted to see him:

> "Not sure what to expect after the way events have unfolded ... was I ever wrong! He tells me the powers that be have decided to deploy *John F. Kennedy* next Wednesday, 15 August. I will pick up four 'unrestricted' (non-G-limited) jets over the weekend and transfer all four KA-6D tankers, two to VA-42 and two to VA-85. I will also gain four additional crews from VA-55, which is getting ready to stand down. After we catch up with *Sara*, we will get our crews and SWIP jets back. I am ecstatic. The squadron is mile high, too."

The *Sunday Punchers* quickly prepared for their short-fuse deployment. Over the weekend the squadron grabbed several crews from the SAU (VA-0686), and put in what Besal describes as a:

> "Weekend weapons det of intensive flying, including strikes with heavy inert bombs to Dare Target, opposed by Canadian F-18s. It was a super event and great preps for what lay ahead. Additionally, I sent a crew to Point Mugu for four days of intensive instruction on employment of the AGM-88E SLAM. Lt. Pete Kind and Lt. Charlie Giacomo returned to train a small team of *Sunday Punchers* – Team SLAM – and were also the first to fire the weapon in combat, on 18 January 1991 against the Al Quiam phosphate plant."

From there it was carquals and then out to the boat for the transit east, along with "a million strike planning sessions and briefs." Far from missing the dance, the *Sunday Punchers* were once again going to war.

Building the *Shield*

While United States Central Command was a descendent of the old Rapid Deployment Force – and thus designed specifically for this type of "come as you are" operation – it still took time to set up command and control for the various units that were arriving on line. As a result, there was a large amount of early confusion, as *Blue Blaster* CO Alexander recalls:

> "It was very interesting being there, because somewhere deep down in the bowels of the Pentagon and the JCS they might have had this stuff figured out, but it sure didn't seem like it. We couldn't talk to the Air Force guys that were there, and we couldn't talk to the headquarters they were starting to stand up. Coordinating and trying to get tankers was strictly by message traffic. By the time we left, though, there were starting to be communications back and forth."

Through all the preparations and deployments – planned or otherwise – the U.S. kept up the pressure on the diplomatic front, attempting to build a worldwide consensus against Iraq. On several occasions Hussein proved to be his own worst enemy, in and around the requisite chest thumping and calls for *Jihad*. On 8 August his government announced that Kuwait, the "long lost province," was now formally part of Iraq. The following day the United Nations vetoed the annexation claim.

In an interview on 10 August President Bush expressed his belief that Iraq would probably not invade Saudi Arabia, but added the buildup would continue. The President added he was "very encouraged" by the supportive actions taken by the Saudis and other members of the growing coalition. He did voice a warning, though, stating:

> "The troubling thing is, we're up against a man who is known for his brutality and irrationality, and who has taken a step that, though widely condemned, has still not been reversed."

Two days later the United States announced the implementation of an interdiction program against Iraqi shipping. The same day, elements of *Marine Aircraft Group 70* – the aviation arm of the *7th Marine Expeditionary Brigade, I Marine Expeditionary Force* – started preparing Shaik Isa Airbase in Bahrain for the arrival of Marine aircraft.

Shaik Isa, located on the island's south end, was already home for the country's small F-16 force, and was one of several largely unknown military bases that had been constructed in the region through the years by the U.S. in concert with local governments. Most had been built with ramp space and room to accept deployed American units "if the need ever arose," and many were used for that very purpose during *Desert Shield*. If not quite "secret" locations (the local population obviously knew about them) most were not depicted on aeronautical charts or mentioned in standard flight publications. Few, if any, Marine aviators had ever heard of Shaik Isa prior to deployment, and whole flights were told to fly to a specific latitude and longitude, call the tower on a specific frequency, and "look for a base to land at." It would become the Leatherneck's biggest fixed wing base of the coming war.

The initial deployment to Shaik Isa included the veteran *Bengals* of VMA(AW)-224; they were eventually joined by the *Night Hawks* of VMA(AW)-533 under Lt.Col. Beman Cummings, six fighter/attack squadrons flying *Hornets* (VMFAs -232, -235, -312, -314, -333, and -451), and VMAQ-2's *Playboys*.

Notably, the deployment of the *Bengals* contributed to the rapid "retirement" of another long-time Marine *Intruder* squadron. When -224 got the word to go it was short a couple of airframes; conveniently, VMA(AW)-242 was next in line to get the FA-18D. According to Lt. Cmdr. Tom Hickey, a career B/N who had served in Marine Air Weapons and Tactics Squadron 1 (MAWTS-1) at MCAS Yuma as an A-6 instructor from 1987 to 1992, the subsequent act of getting the *Intruders* to the *Bengals* did take a bit of doing:

> "In August 1990, VMA(AW)-242 flew to MCAS Cherry Point to deploy to Southwest Asia with VMA(AW)-224," he writes, "but the USMC aircraft had all-metal wings on their A-6s, and many were reaching their life limit and were downgraded to 3G aircraft. It was determined that 3G aircraft would not be safe for combat due to the probable onset of high-G maneuvers, so -242 was immediately stood down, and all of the aircraft were given to -224, which soon deployed to the Gulf. VMA(AW)-242 began an immediate transition to FA-18Ds."

Following this quick work by the Pentagon, the Marine hierarchy, and the two squadrons, -242 released its aircraft, and the *Bengals* – led by skipper Lt.Col. Bill Horn and XO Lt.Col. Elrath – were on their way to war. With its career in *Intruders* at an end, Lt.Col. Fred Cone's *Bats* shifted their attention to the *Hornet*.

One additional squadron also made the deployment to Bahrain: the *Green Knights* of VMFA(AW)-121, which had recently converted from *Intruders* to the D-model two-seat *Hornet*.

On 22 August 1990, President Bush authorized the call-up of 48,800 reservists of all services for active duty. Many members of the growing opposition movement to U.S. involvement in the desert had levied charges that the country was heading into another Vietnam; the reserve call-up – which was the largest by the Federal government since the Tet Offensive of early 1968 – showed that the leaders in Washington D.C. were *not* going to allow "another Vietnam." DOD subsequently increased the reserve authorization to 365,000, and the Navy, Air Force, Army, and Marines responded, calling up large numbers of their selected reservists. The personnel came from all over the United States and filled a number of roles, with heavy emphasis on medical and support units; indeed, with the recent reduction in the defense budget and the size of the active force, a lot of the military's capability had already transferred to the Guard and reserves.

Notably, among all of the Air Force, Marine, and Army Reserve units called up, the two reserve carrier air wings and their component squadrons – now equipped with *Tomcats*, *Intruders*, *Hornets*, and *Hawkeyes* – didn't get called up and never deployed into combat as units. However, while the *Green Falcons* and *Firebirds* never made it to the Gulf, the Selective Augment Units did. Cmdr. Thomas C. Stewart's group at NAS Oceana, VA-0686, ended up providing 18 aircrew and sailors for *Desert Shield* and *Desert Storm*. They worked in carrier, fleet command, and task force staffs, or on the decks of the carriers in the Persian Gulf and Red Sea. Four pilot-B/N teams reported to VA-42 and assumed replacement training duties, freeing several *Pawn* personnel for duty with combat squadrons. Others would go directly to fleet units to supplement squadrons in the Middle East.

On 22 August the *Saratoga* battle group arrived in the Red Sea and covered for *Dwight D. Eisenhower* and her supporting units. The *Blue Blasters* – armed and ready, but never called into combat – returned to Oceana on 12 September, five days after the *John F. Kennedy* battle group checked into the theater.

"I think we would've liked to have stuck around, but I think it probably would've gotten old by the time January-February rolled around," says Alexander:

> "August or September, we were ready to hang around. We went to the Med and had to hang around there for a couple of days as well, waiting for the Kennedy to come in. So we had two turnovers; we did the standard relief with *Saratoga* and did the same with *Kennedy*, except we had less to give, except airplanes. We ended up giving them all of our tankers."

The skipper says their return to Norfolk and Oceana was nothing out of the ordinary:

> "I guess the *Roosevelt* and *America* were out doing work-ups, but the piers weren't empty yet. That really showed in December, the first of the year when *everybody* left.
>
> As soon as we got home we gave the rest of our planes away, except for the ones that were G-limited," he concludes. "We gave them to -65 and -36, so we were down to about two airplanes. The SWIP airplanes were coming, and they didn't know how long that was going to take."

For XO Bob Besal, XO Cmdr. Kolin Jan, and the rest of the reunited VA-75 in *John F. Kennedy*, the schedule never let up:

"31 August 1990 – INCHOP Sixth Fleet," Besal writes:

> "CAG Hardin White hatches a plan to get KA-6Ds from returning *Ike*. We 'repatriate' Lts. Mike Walsh and Al Misiaszek, who have been getting fat while sitting in Sigonella awaiting transport to *Sara* in the Red Sea since 18 August (!). 01 September 1990: Turnover with *Ike*/VA-34 in Augusta Bay. I visited with CO Cmdr. Gene Neilsen and XO Cmdr. Ron Alexander to discuss operations they prepared for. As the emergency reaction force, *Ike* was stationed in the Red Sea almost immediately after Iraq occupied Kuwait. They were on the hook to blunt any Iraqi movement into Saudi Arabia. From the Kuwait-Saudi border, Iraqi armor was only eight hours from Riyadh, which made the Saudi royals so nervous that they left the capital and sought 'refuge' in Jeddah.
>
> "The *Blue Blasters* were nervous about 'blue on blue' with Saudi armor and posed the question of how to identify advancing Iraqi tanks from retreating Saudi tanks. I was told the response came from King Fahd: 'Bomb anything that is moving southwest ... we'll sort it out later.' It made for rather simple rules of engagement."

On 25 August the United Nations – after a great deal of debate and wringing of hands – authorized military interdiction in Kuwait. Roughly three weeks later, on 14 September, Iraqi troops in Kuwait City invaded several embassies and diplomatic missions, further infuriating the nations that were aligning against Iraq. While Jordan announced its support of Iraq, Hussein and his nation otherwise stood alone against the incredible armada forming in Saudi Arabia. However, the numbers did not prevent Hussein from waging a spirited propaganda war against the coalition in general, and the United States in particular. Back in the states, his efforts met with some support.

Opposition, Foreign and Domestic

On 8 November 1990 President Bush announced a substantial increase in the U.S.' commitment to Saudi Arabia, including additional carriers and Army and Air Force units from Europe. The purpose of the deployment was to give an "offensive option" to the coalition. There was no longer a question of just defending Saudi; now the intent was to forcibly evict Iraq from Kuwait, if diplomatic efforts failed.

The Marines tagged the *2nd Marine Expeditionary Force* and *5th Marine Expeditionary Brigade* for deployment, while the Navy prepared *Theodore Roosevelt* with CVW-8 and VAs -65 and -36 embarked, *Ranger* (CVW-2 with VAs -145 and -155), *Midway* (CVW-5 with VAs -115 and -185), and *America* (CVW-1 with VA-85) for deployment to the vicinity of Saudi and possible combat action. In numbers of carriers, ships, and squadrons it was the largest Navy combat deployment in decades.

Through all the diplomatic conniptions the air wings and squadrons continued to train, train, train. By now there were enough forces in the region to allow planning of strikes from just about every point of the compass. The targeting folks had already worked up a first night's list, as relayed by VA-75's Bob Besal:

> "There were several options for us to fly across Turkey into northern Iraq against targets there. Also reviewed first edition *Desert Storm* ATO (Air Tasking Order) and Master Attack Plan. Our first night targets were 'impressive:' Ministry of Defense headquarters, Baath Party headquarters, Military Intelligence headquarters, and the Government Control Building, all in downtown Baghdad. We will attack them in conjunction with *Sara*/VA-35. 11-12 September 1990: Inport Alexandria, Egypt. Deputy CAG Butch Bailey returns from a liaison trip to Riyadh. Tells me that VA-75 is 'in the spotlight;' we're on the hook for 11 up bombers with six hours notice."

On the 14th *Kennedy* transited the Suez Canal to the Red Sea. The following day it joined with *Saratoga*, and VA-35 returned the four "lost" *Sunday Puncher* crews and SWIP aircraft. Three of the borrowed VA-55 crews went home, while VA-75's four non-SWIP air-

Ranger's CVW-2 was equipped as a so-called "All Grumman Air Wing" for *Desert Storm*, with two large *Intruder* squadrons making up the ship's entire attack capability. The bow is clobbered with at least 17 VA-145 and VA-155 *Intruders* visible in this pre-war picture, as well as VF-1 and -2 *Tomcats* and VAQ-131 *Prowlers*. The single S-3B present proves, of course, that there were also Lockheed products on board. (U.S. Navy, via Steve Barnes)

craft went to VA-35. Otherwise, the two air wings – and every other military unit in the theater – continued to train.

"I received intel reports from ONI (The Office of Naval Intelligence, specifically the portion called *Spear*, which was made up of tactical aviators who evaluated intel from the flyer's viewpoint) – our former Deputy CAG, Capt. 'Carlos' Johnson – who told us of photographs of 'ten thousand gun barrels' pointed skyward in Baghdad," Besal comments:

> "I was concerned ... until the start of *Desert Storm* we continued to train to the most challenging mission – low-level night attacks. It kept our 'sharp edge.' It kept the adrenaline pumping and the crews focused. But I felt if we were to succeed, we needed a 'new way of doing business.' We had to get out of the AAA envelope, especially considering the quantity. That much lead flying was bound to hit something if it was in range. Next, we had not had an opportunity to find an urban complex target at extremely low altitude, much less attack one. I likened the task to something like flying through Richmond or Washington D.C. at rooftop level. There's no place to train for that; not many cities will permit you to rage tactically through their sky. The magnitude of the task – even if not being shot at – was daunting.
>
> "No one except the XO, Cmdr. Kolin Jan, had ever seen a missile fired at him (Jan participated in *El Dorado Canyon* on the Al Azzizyah Barracks strike while a department head in VA-34). It's easy to tell someone how to maneuver to defeat a missile, but doing it, while navigating at low-level and then hitting the target ... well, that's easier said than done by all of us rookies."

Standard doctrine for the *Intruder* called for low and fast. Besal and a few others in the community, however, had another idea:

> "I called a meeting of my division leaders – my pilot, Lt.Cmdr. Mike 'Ziggy' Steinmetz, Lt.Cmdr. Ken Coburn, Lt.Cmdr. Joe Kilkenny, Lt. Les Makepeace, Lt. John Zinck, Lt. Kurt Barich, and Lt. Kurt Barnard – in my stateroom to lay out the concept and rationale for a high-altitude attack. While I was hoping for a 'buy-in' from all, I was open to other ideas.
>
> "Essentially, I believed that we should ingress at high altitude to take several tactical advantages:
>
> • First and foremost, higher altitude could negate much of the AAA threat.
> • Second, it would provide more room for maneuvering to avoid SAMs.
> • Third, it would give us a greater radar horizon, and therefore more time for target identification.
> • Fourth, since we were all carrying laser-guided bombs and needed to designate the target until bomb impact, higher altitude delivery gave us a 'God's-eye view' of the targets in the city. It would also make tracking the target easier, because the turret slew rate required to keep the laser spot on the target would be slower.
> • Fifth, we would likely get good fuel economy from winds aloft while inbound. The only reason for us to go low would be if we were forced down by enemy fighters. And given what I saw to be our own fighter support, I really thought it unlikely. After the attack, we would egress low, which could help to 'sort out the picture for the fighters'; we'd be out of the way.

"I got a thumbs up from all my wingmen for the plan. Selling it uphill to some other 'elder, wiser *Intruders*' was a different story. After listening to the whole plan CAG Hardin White blessed it, and then I went to Rear Adm. Riley Mixson, the Red Sea Battle Force commander, who also concurred. Since we'd be attacking in concert with VA-35, I visited *Sara* to sell the high-altitude idea. It was not eagerly accepted by some of those who'd put in a lot of time planning already, and not really 'approved' by older community members – the *Sara's* CO and CRUDESGRU Chief of Staff – who thought that 'low is the only way, and we've always done it that way'. Skipper Jim Anderson bought the plan; as he was senior to me, his division would go through the target area first and the *Punchers* second.

"Sometime before kickoff, the planning cell in Riyadh received guidance that no 'non-stealth' aircraft were to go to Baghdad. Our first night targets were changed to facilities at the Al Taqaddum airfield, about 45 miles northwest of Baghdad center. VA-75 was tasked with two bunkered hangars used for MiG-29 assembly. We would strike each with two planes carrying four GBU-10 2000-pound laser-guided bombs apiece."

As the year wound down talk and speculation continued to run rampant. While additional troops, aircraft, and equipment went into the region, military experts and other analysts ran the possible American casualty count up near 30,000, with as many as 16,000 wounded. Army commanders later projected a 10 percent casualty rate for any ground invasion of Kuwait; Army, Marine, and Navy field and afloat hospital units were appropriately bolstered, primarily by medical personnel. The Iraqi government did its part, claiming that any attack into Kuwait or Iraq would result only in "rivers of Americans' blood." In a later speech delivered just before the coalition air attack, Saddam Hussein further dismissed the United States, announcing:

> "The Americans will come here to perform acrobatics like Rambo movies. But they will find here real people to fight them. They will see how the Iraqis – men, youngsters, and women – will fight them should they attempt to land anywhere in Iraq."

Finally, on 29 November 1990 the United Nations Security Council authorized offensive action if Iraq did not withdraw from Kuwait. If he did not release Kuwait by midnight, Eastern Standard Time in the United States on Tuesday, January 15, Hussein's "real people" would get their chance.

Happy New Year!

As 1990 rolled into 1991, the final military components arrived in the vicinity of Kuwait and Iraq. To the north in Turkey, U.S. Air Forces in Europe (USAFE) placed a composite wing they named *Proven Force*, with a wide mix of strike, air superiority, and electronic warfare aircraft. The task force would come in handy for attacks into Northern Iraq. Army and Marine ground forces were in place in Saudi Arabia, trained and ready to go. Several carriers were operating on both sides of the Saudi peninsula, with more coming as fast as they could.

There was one last stab at diplomacy. On 9 January 1991 Secretary of State James Baker and Iraqi foreign minister Tarik Azziz met in Geneva in an attempt to resolve the international dispute, but six hours of haggling led to nothing. Three days later the Congress of the United States of America voted approval for offensive operations in Kuwait. That same day United Nations Secretary General Javier Perez de Cuellar left Baghdad after meetings with Hussein, admitting he too had accomplished nothing. In any event, the coalition buildup was over, and after five months of non-stop building and preparation, the numbers of the forces gathered in Saudi Arabia, Turkey, Bahrain, and other locations was breathtaking. For the United States alone the totals included some 245,000 Army troops, 75,000 Marines, 60,000 Navy personnel, and 45,000 airmen flying/operating/maintaining/ preparing some 4000 tanks, 1700 helicopters, and 1800 aircraft.

On 15 January 1991 *Midway* and *Ranger* were on station in the Persian Gulf, cocked and ready to go, and with *Theodore Roosevelt* due to join the group imminently. In the Red Sea, *America*, *John F. Kennedy*, and *Saratoga* were in similar stages of readiness, along with herds of Marines in amphibs and the naval forces of some 15 other nations. For the personnel of *Sara* the preparations had taken a tragic turn following their arrival in the region. On 22 December 1990, in the port of Haifa, Israel, a liberty boat returning to the carrier capsized and sank, drowning 20 members of the crew; the remains of the twenty-first, AO3 Anthony J. Fleming of Buffalo, N.Y., were never found. Ship's and CVW-17 personnel grieved, picked up the pieces, and returned to their patrol station a couple of weeks later.

Marine *Intruder* aviators were in place with MAG-11 (which had replaced MAG-70), Shaik Isa, Bahrain. Also ashore was a large chunk of the U.S. Air Force, U.S. Army, and U.S. Marine Corps serving with representatives from 19 different nations. All watched and waited as the 15 January deadline passed with more rhetoric from Hussein and not much else. Iraq had fortified Kuwait, placing anti-tank traps and ditches along the border with Saudi Arabia, while parking 10 divisions along the Kuwait coast, just in case those two brigades of Marines lurking offshore decided to hit the beach. As the clock ticked down Hussein confidently promised his people the "mother of all battles" was about to take place. However, no one knew exactly what form the coalition's attack would take.

They got their answer promptly at 0240 local on Thursday 17 January when the skies lit up over Baghdad, as Iraq came under massive air and missile attack. In a remarkable example of modern technology –

Training for the fleet – a *Green Pawn* A-6E TRAM begins its takeoff roll at Oceana in February 1989 with a load of "blue death" Mk.76 bombs. (David Brown)

and a ratings coup – CNN carried the initial bombardment live while its reporters in Baghdad huddled under tables, occasionally peering out the windows of their hotel. Viewers world-wide watched as Iraqi tracers lit up the night and buildings started blowing up.

At a White House press conference held at 6:40 p.m. in Washington, D.C., presidential Press Secretary Marlin Fitzwater announced, "The liberation of Kuwait has begun." The clan had been called in and was now delivering the eviction notice to Saddam Hussein.

Intruder Paint

Not counting the occasional specially applied camouflage trials bird (such as green during Vietnam, or brown camo for *Red Flag* exercises or *Desert Storm*), the Grumman *Intruder* had only two standard paint schemes applied at the factory or depot throughout its life.

When the A-6 hit the fleet it was painted in a Light Gull Grey over Insignia White scheme (using Federal Spec. paint numbers 36440 and 17875) that had been the standard for carrier aircraft since February 1955. This scheme would remain the norm for the type through the first 20 years. The *Intruder's* signature massive black radome stood out in combat, and by the late '60s was left a 'natural' fiberglass color or, in time, painted gray and/or white.

Through Vietnam almost all squadrons put unit markings on their aircraft in full color, with the *Intruder's* tail proving a suitably broad canvas for spectacular artwork. VA-35's classy *Black Panther* motif and VA-65's dramatic orange stripes are but two examples of the lengths some units would go to establish an *esprit de corps* with their airframes. The problem is, these schemes also made these aircraft more visible – to both friend and foe, as well as proving a high-maintenance item for the squadrons' corrosion shops.

In 1980 AIRLANT began to evaluate a new aircraft paint scheme to replace the 25-year-old gull gray and white standard, with the goal being to come up with a design that would help reduce visual acquisition ranges in combat. By 1981 a formal "Tactical Paint Scheme" (TPS) had been developed, and it was adopted as the new standard. The design varied slightly between aircraft types, and on the *Intruder* it was specified as two slightly different shades of dull gray (F.S. numbers 36302, 36375), with all markings being applied in either the opposing color or a third similar gray. KA-6Ds were specifically directed to stay in the older gull gray and white motif in order to help aerial identification of the tankers (although this did not stop a couple of units from repainting KAs in their hangars in dull gray anyway).

The colorful squadron markings of old were, at least initially, deemphasized or banned outright, depending on command policy and how closely higher authority enforced the regs. Some squadrons advocated absolute minimal markings for reasons of "operational security." With the adoption of TPS, carrier aviation entered an era of gray that was bemoaned by aviation photographers worldwide.

As applied by the depot and when seen in good light, TPS could be sharp looking. By the mid-80s most aircrews agreed that TPS was an excellent paint scheme that provided better concealment in most conditions, and made the aircraft look like real warriors. The problem was they didn't stay that way for long.

When exposed to a salt spray and sun environment the closely matched shades of gray quickly faded into a single color. Safety markings and side numbers became hard to see or just vanished. Most ancillary markings were therefore quickly repainted in dark gray or black for practical reasons.

Maintenance troops worked their hearts out trying to keep up corrosion work, only to see their birds degenerate into globs of dissimilar, dull tones. What some aviation magazines called "complex cloud camouflage", was actually a typical TPS post cruise look. Supply stocks being what they are, maintainers frequently found themselves forced to use whatever paint was available, in other grays, or even occasionally light blues. Not allowed to conduct major re-paints at squadron level (at least not officially), and limited to small area patch work only, many TPS aircraft returned from deployment looking like they were ready for the boneyard. Dirt and bootblack tended to adhere to the dull paint, and it became generally accepted that "if an aircraft doesn't come back from cruise looking like 'S**T' the squadron wasn't doing corrosion work correctly." At least one west coast squadron reported having an A-6E refused a static display spot at an Air Force airshow because its warplane looked so bad. "My plane isn't dirty. It's painted that way" was the retort.

Improvements in paint quality and squadron-level application methods would help the looks of aircraft over time. By *Desert Storm* a balance had generally been achieved between tactical requirements and squadron markings, as well. Most units adopted reasonable squadron motif – in gray – that remained true to the intent of the paint scheme, while also allowing pride in ownership. Most air wings allowed squadrons to paint up one "Easter egg" in color as well – usually side number 500, the "CAG bird" – and nose art even showed up on some aircraft as well. This is the look the *Intruder* carried to its end.

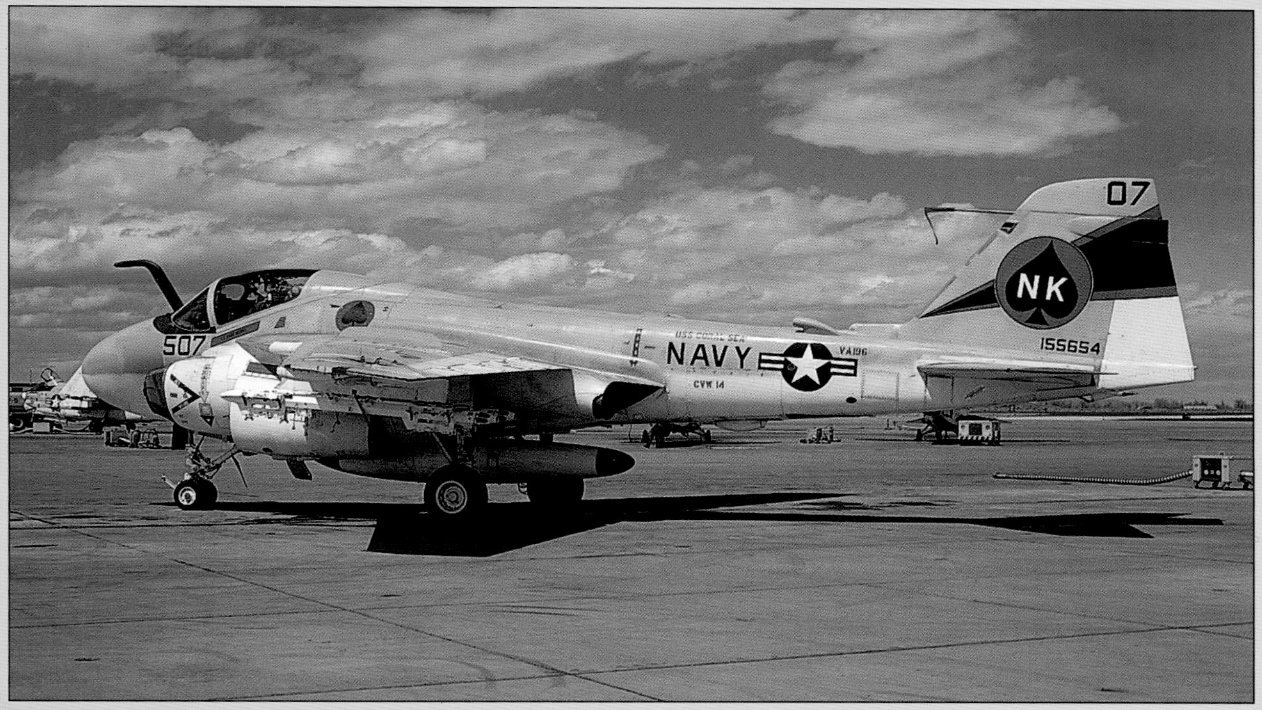

Extremes in paint. VA-196 A-6E displays the gull gray and white scheme, as well as the squadron's flashy and high maintenance tail markings. NK507 is at Fallon in May 1981 (Mike Grove). On the other extreme is VMA(AW)-533's aircraft in fresh TPS at Pensacola in June 1983. AA511, from VMA(AW)-533, displays "by the book" TPS, with very subtle markings and a nose number that is already starting to vanish into the base color. This aircraft would suffer an inflight fire only 13 days later, crashing near Emporia, VA, with the loss of the B/N. (Rick Morgan)

CHAPTER TEN

Desert Storm, 1991

As of the early morning of 17 January 1991, there were five carriers on station and another leaving a rooster-tail to get into the impending war. Now six years after the arrival of the FA-18 *Hornet*, the *Intruder* was still the primary attack aircraft and the undisputed "heavy hauler" of the fleet. In terms of total units involved, there were as many A-6 squadrons at sea as there were *Hornet*.

U.S. Naval Forces Central Command (NAVCENT) had recently come under the leadership of VADM Stan Arthur, a former A-4 pilot who had flown hundreds of combat missions in Vietnam. By all accounts, he was the right man for the job. In the Red Sea, Task Group 155/Battle Group Yankee steamed under the command of RADM Riley Mixson. *Saratoga*, with CVW-17/VA-35 onboard, and *John F. Kennedy* (CVW-3/VA-75 embarked) were preparing to make their first combat launches for what was now being called *Operation Desert Storm*. Nearby, *America*, (CVW-1/VA-85), having just arrived from its Atlantic crossing, was directed to be ready for the second day of operations. The delay was actually due as much to a lack of Air Force tankers in the "day one" game plan as for anything else.

On the Persian Gulf side, "The One Carriers" (CV-41, CV-61, and CVN-71) would employ the final "dual A-6" Air Wings, although all with different configurations. *Midway* (CVW-5, VA-115, -185) was flag for Task Force 154/Battle Group Zulu. *Ranger* carried CVW-2 (VA-145, 155). As of the 15th of January *Theodore Roosevelt* was actually just rounding the tip of the Arabian Peninsula, and it would be the only nuclear powered carrier to see combat in *Desert Storm*. *TR* carried CVW-8, with VA-36 and VA-65.

The war started at 0239 on the 17th, as Army, Navy, and Air Force aircraft poured into Iraq behind waves of cruise missiles launched from

A rack of Mk.82 *Snakeyes* sits in front of a VA-65 *Intruder* during *Desert Storm*, 7 March 1991. (Rick Morgan)

Navy ships, submarines, and Air Force B-52s. At Central Command's Saudi headquarters the staff watched and waited for signs that the war had in fact begun. CNN provided the answer to the world.

When correspondent Peter Arnett's voice disappeared eight seconds after Time on Target a big cheer went up in Riyadh. *Desert Storm*, America's largest war since Vietnam, had begun.

For Saddam Hussein and his military the first night proved to be a bloody mess; for the coalition forces, the assault marked a successful start to their efforts to free Kuwait. And for Navy attack aviators, it marked the first commitment to major, extended overland combat since the Vietnam War.

Intruders into the Storm

The first night of Operation *Desert Storm* included full participation by four carriers. *Ranger*, *Kennedy*, *Saratoga*, and *Midway* each contributed strikes in the opening waves, while *America* and *Roosevelt* would follow shortly. In Bahrain, both of the Marine *Intruder* squadrons launched aircraft the first night and continued operations thereafter.

The three separate locations quickly led to three largely distinct operating tempos, each driven by unique requirements and mission demands.

The Red Sea carriers operated about 450 miles from their closest targets in Iraq and over 650 from Baghdad. They would generally fly two or three major strikes a day and require significant coalition tanker support both to and from target. Their targets would be all over Iraq, and run from airfields and lines of communication in the west to leadership sites within site of the capital, and even as far north as Mosul. Missions could exceed five hours in duration, and crews would typically end the war with fewer sorties and more hours than their Persian Gulf counterparts.

The decision to move carriers into the Persian Gulf had not come lightly, due to the potential threat from Iran, as well as dire "fish in a barrel" predictions about risking carriers in the tight confines of the Gulf. Tanker support again drove the decision to move into the restricted waters. When the decision came the coalition moved in with everything it had though, and in the end they proved they could safely operate four carriers in the Gulf while dodging 100 other surface units and the odd mine.

The Gulf carriers largely flew cyclic operations, as distances (50 to 100 miles from Kuwait) allowed single or double cycled sorties, frequently without tanking, as far as the *Intruders* were concerned. As opposed to the Red Sea air wings, the Persian Gulf wings flew mostly battlefield preparation missions in Kuwait and southeastern Iraq. They would also deal with the Iraqi Navy, as well as hit significant interdiction and industrial targets around Basrah and the Shat al Arab waterway, which ran between Iraq and Iran. Flying into Iranian airspace was a source of some concern during the war, and it usually led to threatening calls on guard frequency from Iranian controllers, either to aircrew that had blundered across the border or, in the reported case of one carrier, use of the "neutral" country as a convenient method of egress from Iraq.

Opposing the coalition flyers, Iraq relied on a dense, well-linked system of Soviet and European – primarily French – radars, SAMs, and gun systems, tied together by the French designed *Kari* command and control system. Iraqi field units, such as the Republican Guard divisions, added to the mix with a large number of highly capable tracked, mobile, and shoulder-fired SAM systems.

The "strategic" SAMs included the old reliable SA-2 *Guideline* and SA-3 *Goa* missile systems. The mobile missiles were another concern, particularly the SA-6 *Gainful*, which had gained a fearsome reputation during the 1973 Arab-Israeli war. Short range SA-8s and SA-9s were also present, as were an unknown number of U.S.-built HAWK missiles that had been captured from the Kuwaitis.

The air threat was not to be taken lightly, either. The Iraqi Air Force had hundreds of fighter aircraft of Soviet and French makes,

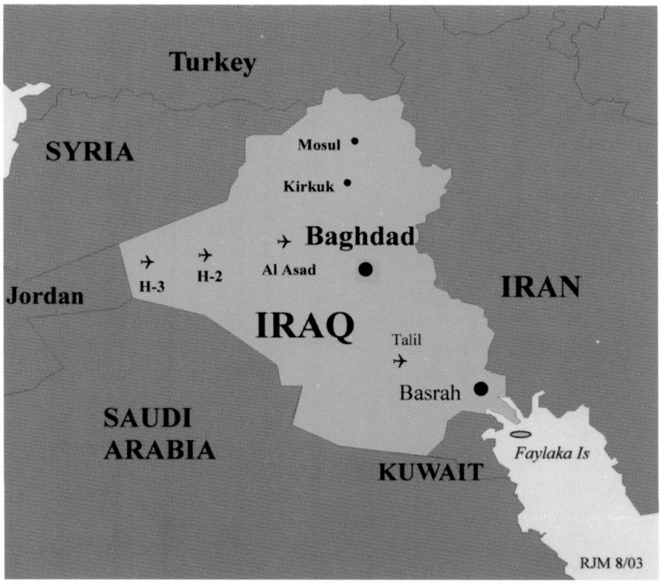

along with suitably "combat tested" veteran pilots, at least as far as many of the intelligence people and media were concerned.

Throw in a heavy mix of shoulder-fired weapons and trailer/tracked and self-propelled guns – plus thousands of troops firing their AKs at jet noise – and it is no wonder the skies were lit up "like the Fourth of July."

War from the West

Two carriers launched strikes from the Red Sea on the first night – *Saratoga* (CVW-17/VA-35) and *John F. Kennedy* (CVW-3/VA-75) – attacking airfields and other targets in western Iraq. *Kennedy*, as the only remaining Vietnam-era carrier not to have seen duty in Southeast Asia, was seeing its first extended combat. On the second day *America's* VA-85 made its first strikes, the carrier having arrived on station only two days previously after a direct transit from Norfolk.

The *Black Panthers* of VA-35, under the command of Cdr. Jim Andersen, saw the whole range of combat – the highs of successfully attacking targets in heavily defended areas to the lows of losing aircraft and shipmates. The "open back door" provided by the lightly defended Iraqi/Saudi border allowed the squadron to range throughout Iraq and deliver a reported 1.7 million pounds of ordnance throughout the war.

After looking at the threat from radar-directed SAMs and fighters the *Panthers*, or "Rayguns," from their radio callsign, chose to start the war at low altitudes. Their experience as the first *Intruder* squadron to qualify with Night Vision Devices (NVDs), as well as years of training in the all-weather low altitude environment had led to confidence in their tactics and equipment.

The first night's strikes were on the MiG base at Al Taqadum and other locations deep in central Iraq. While all of the *Intruders* made it back to the carrier, the *Sunliners* of VFA-81 lost an FA-18C to a MiG-25. Pilot Lt Cmdr Scott Speicher was quickly declared KIA.[1] The first morning strikes for CVW-17 took them to the airfields in Western Iraq, which still bore their old RAF names of H-2 and H-3. The targets were plastered, while *Sara's Hornets* bagged two MiG-21s in the attack, exacting a small measure of revenge for Speicher's loss.

[1] The wreckage of Speicher's *Hornet* was found in 1993. He has since been declared to be MIA as evidence arose that he might still be alive in Iraq. As late as mid-2003 the DOD has been unable to determine exactly what happened to the aviator.

The first *Intruder* lost in *Desert Storm* was from VA-35, which had 161668 shot down on the second night of the war. The crew of Bob Wetzel and Jeff Zaun were made POWs. The aircraft is seen here at Fallon during workups in May 1985 when the *Black Panthers* were assigned to CVW-8. (Michael Grove)

On the second night's strikes a four-ship of *Intruders* approached H-3 at below 500 feet and were met by a wall of fire. The second aircraft in trail, AA510 (BuNo 161668), took a fatal hit, and Lts. Bob Wetzel and Jeff Zaun were captured. Dash-three's bombs went well off target, while number four in the formation, AA502 (158539), was hit by AAA and a French-built *Roland* SAM, but managed to stagger into Al Jouf, Saudi Arabia.[2]

The second night losses led to no small amount of concern, as the controlling USAF E-3 AWACS aircraft on station initially reported Wetzel/Zaun's aircraft as a "Blue on Blue" kill by an Air Wing 17 F-14A+ using an AIM-54 *Phoenix* missile. Lt. Gen. Chuck Horner, USAF, the senior aviator in theater, was furious when he heard the report, and it took a quick investigation by Admiral Arthur's staff to convince him that it was, in fact, enemy action that had brought down "Raygun 510." None-the-less, at least two Air Force general officers remained convinced that the Navy had covered a fratricide incident up, and the story persists many years later.

On the night of 21 January the *Panthers* attacked the MiG base at Al Asad and had what was identified as a MiG-25 "Foxbat" fly through the formation on egress, this minutes after an F-14A+ from stable mate VF-103 had been shot down to the south.[3]

By the sixth day of the war CVW-17 had sustained three of the four Navy aircraft losses and had another one badly damaged. Air Wing personnel compartmentalized their grief and continued their work as the professionals they were. Low altitude work by the Navy was largely over, though. Vice Admiral Arthur sent a message out to the force stating that initial analysis had shown that AAA and tactical SAMs were more dangerous than the upper level threat, which was being handled by defense suppression aircraft. From now on *Intruders* would use medium to high altitudes in their operations.

The use of combat video was featured in daily press briefings and routinely made it to nightly news broadcasts. Newer generation aircraft, like the F-15E, F-117A, and FA-18 utilized commercially compatible tape systems that were easily converted for release to an eager media. The *Intruder*, which used an older and unique tape recording system, proved much harder to turn over for use. Nonetheless, a VA-35 crew was featured, anonymously, in the daily brief given to the press by CincCent, Gen. Norman Schwartzkopf, in Riyadh. While bombing a bridge over the Euphrates River with LGBs one night, B/N Lcdr. John Snedecker noticed what appeared to be a civilian car pass under his laser spot and continue off the structure. Paraphrasing General Schwartzkopf, "he was the luckiest man in the world, as now in his rear view mirror" – the bomb went off, right on target. Snedecker would end the war with 17 missions, considered typical for Red Sea flyers.

The two other two Red Sea A-6 squadrons, VAs -75 and -85, had elected to start the war at medium altitudes, counting on the ability of the EA-6B's jamming and *HARM* to keep the SAMs at bay. Pre-war briefings from groups like the Navy's SPEAR, a Washington-based intelligence outfit that used aviators to generate warrior-oriented threat information, convinced some that going low was not the wise course of action.

Among those who had come to that conclusion was *Sunday Punchers* skipper Cmdr. Bob Besal, who told his squadron "think high." Besal, who had relieved Cmdr. John Meister in September 1989, had worked

[2] Contrary to many reports that the damaged aircraft was left in Saudi Arabia, 158539 was dismantled and flown out in a C-5 and subsequently rebuilt by Grumman into a SWIP. It ended up with VA-304 and was stricken in 1994.

[3] The *Tomcat* was escorting a VAQ-132 EA-6B which was providing jamming and HARM support for the strike. Although officially recorded as an SA-2 loss, the *Prowler* crew reported seeing a MiG-25 *Foxbat* behind the *Tomcat* immediately after it was hit. The *Prowler*, now sans escort, ran south to Saudi Arabia at over .9 mach. The bandit was last seen headed north in full afterburner and may have been the one that flew through the *Intruder* formation.

VA-35 had a particularly rough evening on the second night of *Desert Storm*, losing one and having another seriously damaged. This close up shows the damage behind the right intake area of AA502, which was able to land at Al Jouf, Saudi Arabia, after being hit over Iraq. The airframe would later be flown back to the U.S., repaired, and returned to the fleet. (U.S. Navy)

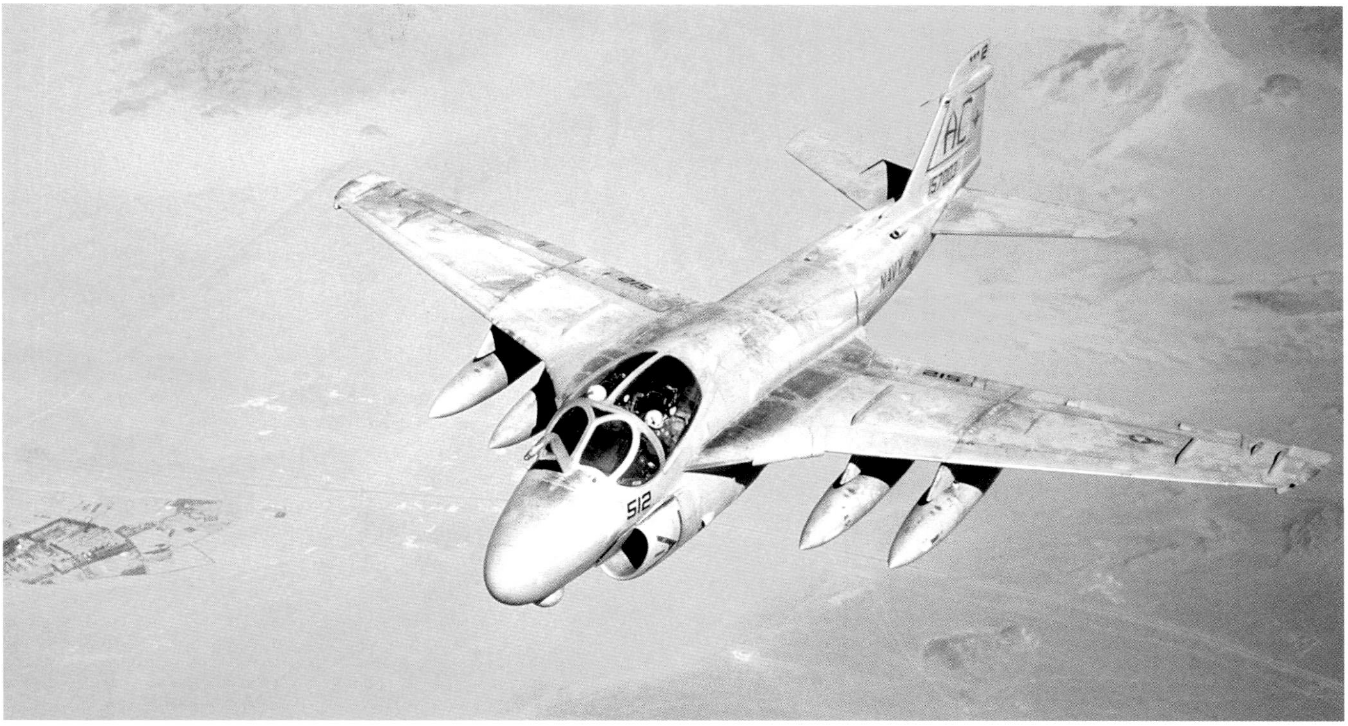

High over northwest Saudi Arabia, land once fought over by Lawrence of Arabia, a VA-75 A-6E comes off an Air Force KC-135 during *Desert Storm*. Tanker configured AC512 will now act as a "hose multiplier," refueling other CVW-3 aircraft before they press into Iraq. (U.S. Navy)

closely with XO Kolin Jan to make sure the *Punchers* were ready for the storm:

> "We sort of 'pioneered' the high altitude tactics used in *Desert Storm*," he comments. "This plan was met with some skepticism by the old hands – and some youngsters, as well – in the community, usually seniors with Vietnam combat time who figured they had a lock on the way to do business. I figured that the 10,000 gun barrels in Baghdad (that's the official intel we got) could get lucky with a target at low level.
>
> "Fortunately, I had a CAG who supported me as I sold the concept upstream. I am proud to say that we brought every crew and every plane home without a scratch ... and we went to all the 'hard targets.'"

VA-75's war was a hectic one, with the squadron delivering over 800 tons of ordnance, including AGM-84E *SLAM* (Standoff Land Attack Missile), which was based on the *Harpoon*. The successful employment of *SLAM* during the first night's combat marked the first use of the new weapon by a fleet squadron. As is, the *Punchers* would fly a total of 2150 combat hours in 280 sorties.

"Our CAG was Hayden Wright, and the DCAG was Butch Bailey, so we had an 'all A-6 front office,'' says Pete Frano, a squadron pilot:

> "CAG regularly flew with us; we led a lot, and there wasn't a guy in the air wing who wasn't ready to go.
>
> "VA-75 basically led everything that first night. We had E-2s and S-3s up near the border, and everyone was pretty anxious to go. On the long missions we got real proficient at tanking; one of the things we did that was different was we regularly flew with buddy stores. We'd take off, tank off the KC-135, then use the buddy stores to refuel the A-6s. That effectively doubled the number of hoses available and worked very well throughout the war."

Onboard *America*, the *Black Falcons* started the war on the 19th and entered a cycle with the other two carriers, conducting deep strikes throughout Iraq. The *Falcons* would hit targets well into the north of Iraq during the war, over 300 miles above the Saudi border – far enough that on at least one occasion their *Prowler* escort reported being intercepted by Air Force F-15s operating out of Incirlik, Turkey, under *Operation Proven Force*, which was attacking Iraq from the north. CVW-1 would move, with their ship, to the Persian Gulf in mid-February, and become the only carrier to work from both Task Forces.

War From the East

In the Persian Gulf the last three "dual A-6 wings" were fully involved under Task Force 154/Battle Group Zulu's flag, which was flown off *Midway*. On *Ranger* was the last of the so-called "All Grumman Air Wings," and embarked VA-145 and VA-155 with no light attack assets. It was led by career A-6 driver and former VA-85 skipper Rear Adm. Ron "Zap" Zlatoper. *Theodore Roosevelt* (CVN-71) carried VA-36 and -65 into combat; Capt. "Fox" Fallon's (former Skipper of both VA-65 and MATWING ONE) CVW-8, the Navy's largest, also had two *Hornet* and *Tomcat* squadrons on the only nuke boat to see action in *Desert Storm*.

Midway, with CAG-5 being the high-time A-6 B/N, Capt. Jim Burin, had VA-115 and VA-185 onboard, along with three FA-18A squadrons. CVW-5 ended the war without a loss. The wing's two small squadrons – VA-115 and VA-185, with nine aircraft each – flew hard throughout the war. *Eagle One*, Cmdr. Terry J. "T-Tom" Toms, recalls their initial activity:

> "The first night we went in low over Shaibah Airfield – actually, we were sending four airplanes to Shaibah, near Basrah, and two to El Jabar in Kuwait. We went up north – Bernie Satterwhite from -185 had the lead of the second wave, with his airplanes – and I was leading the third section with -115 airplanes. At the mouth

With its flak screens up in the cockpit, a *Silver Fox* heads out for a target during *Desert Storm* with a load of Mk.82 500 pound bombs. (U.S. Navy)

of the Al Faw peninsula we had to go out about seventy miles and around the backside of Basrah, and come back down on the airfield.

"The first thing I noticed was the river valley (the Tigris) was flashing. We're in the fog, and I was at about 350 feet, and it was really lighting up all around me. I thought, now that's kinda strange, because I was sure I had my lights off. I put my head up to about 400 feet, like a turtle sticking his head out of the pond, and realized people were shooting at us. We broke out of there and saw our first real heavy tracer fire. Nothing really sophisticated, just everyone shooting everything they had, kind of straight up to where the noise was. And as number three in the stream, that was *not* the place to be. Dash one just woke them up, dash two showed them where we were going, and all the bullets started to show up about the time I came along.

"The guys from Vietnam I'd talked to said, 'Look for the dark spots.' Well, looking through that big bug-eyed windshield that first night, there *weren't* any dark spots!"

Despite the reception, the *Intruders* pressed on and smashed their targets. Toms continues:

"I remember doing that one night with 10 APAMS; each canister carried 750 bomblets capable of penetrating four inches of armor, and we came in and dropped 10 apiece from 15,000-feet ... and I thought, what a lousy way to spend your evening camping out – if you're an Iraqi – as the airplanes are really too high to hear.

"You're just down there hunkered down in the sand for yet another night, and all of a sudden three football fields worth of bomblets tumble across your campground. That's gotta be depressing."

The two *Midway* squadrons launched right about 0200 and recovered 3+45 later with no losses. Among their group was Lt. Joe Kuzmik, the El Dorado Canyon veteran who was now serving as *Eagle* Admin Officer; as the "cagey combat vet," he was flying with a rookie pilot, Lt. Monty "Buckethead" Greene.

"I was assigned to the Al Jaber strike as a wingee to CAG Burin," Kuzmik states:

"We were carrying six Mk.83s on two MERs, as well as an ALQ-167 jamming pod. We launched and found out quickly that CAG had problems and couldn't go. It was going to be just us on this one!

"We coasted in low – about 300-feet – and I couldn't believe the amount of AAA that was coming up at us. I thought Libya was bad, but the Iraqis were even shooting small arms from boats off the coast. They must've been firing at the noise, as most of it was going well behind us – I was now glad CAG wasn't in front of us! It was about 3:40 a.m., the war had been going on for several hours, and I thought about how stupid it was to be flying this low with all of this lead in the air around us. I also realized that almost none was guided, but appeared to be randomly shot in the air.

"It was spectacular, but of little real danger, unless we ran into it. We hit the target and dropped our bombs on the money, with some great FLIR footage to show later."

Midway, with only two cats and three wires, had a reputation of launching and recovering planes as well and as fast as the super carriers. "*Midway* Magic" they called it. *Desert Storm* would prove a fitting end to the ship's career.

Bulletproof

Going into *Desert Storm*, *Ranger's* two squadrons truly felt they had something to prove. The "All-Grumman Air Wing" was visited by a senior rear admiral who had come aboard prior to the war and slighted them in an Air Wing All Officer Meeting (AOM) by saying that he was not sure he could use more *Intruders*, and would have preferred two more squadrons of *Hornets*. The *Swordsmen* and *Silver Foxes* took it as a personal challenge and set about to prove the Admiral wrong.[4]

With only two *Intruder* squadrons for attack assets, the *Ranger* became a primary night player for TF154. On the first night squadrons went low to attack the Um Qasr Naval Base, with excellent results. On the second night, however, intense AAA hit a *Silver Fox* that was trying to mine the approaches to the base. NE404 went down, with the crew of Lts. William "TC" Costen and Charles "Tuna" Turner being listed as KIA. From then on the wing went in at medium altitudes.

"We flew the first night against coast targets, hence the strike went in at high altitude with no major SAM threat," says J.R. Haley, a former VA-52 BN who went through pilot transition and saw combat with VA-145 from the left seat:

"The other strikes, going inland, went low in respect for the missiles. I flew the second night on a mining mission to Um Qasr

[4] The same flag officer gave an almost identical speech to aircrew onboard *Theodore Roosevelt* right after the war started. According to one member of CVW-8 "He insulted my ship, admiral, CAG and the Atlantic Fleet in general. If this was his idea of motivation, he sure had an odd way to go about it."

as number three of four aircraft. We went in very low, about 200-feet, dropped the mines as planned, and made a 160-degree turn to the briefed egress point, only to see a wall of flak in front, made a turn back to the west to avoid the guns, and headed for the alternate point."

According to Haley, he then saw a fireball, "Like a large, petroleum-fed explosion – not ordnance," at his ten o'clock. He says he didn't think anything of it until they got back on deck, when he realized it was probably VA-155's Costen and Turner – the second aircraft in the raid – hitting the ground:

"The approximate position was correct if they had egressed as planned ... not sure if AAA or jinking into the ground caused their loss; it could have easily been both. There were no calls and no beepers."

Cmdr. Larry Munns was VA-155's XO at the time. The operation called for laying Mk.36 *Destructor* mines at the mouth of the Khawr Az-Zubayr River; he comments:

"There were a lot of emotions and reactions. For some, it was really easy to get motivated and psyched. You do something over and over again and training and are ready to go ... but somewhere along the line you realize you'd just as soon the operations would end quickly.

"The squadron carried on. It was only the second day of a 43-day war – of course, we didn't know how long it would last. We did what aviators do best, compartmentalized and carried on."

"That was tough," recalls *Silver Fox* Damon Duncan:

"It reminded us of Vietnam, although in this case, four aircraft went out and three came back. Lt. Costen was my branch officer ... but we all picked up, kept servicing the planes, and kept going."

"We were finally able to hold a memorial ceremony after the flying was over, in March."

Munns adds:

"There was some speculation as to their disposition; it wasn't until later that the wreckage was found and their deaths confirmed.

"My experience in 20-odd years of cruising was that we brought everyone home. You get used to it. NATOPS was in full swing, just didn't expect to lose anyone. No one had been shooting at us for years."

Over in the other squadron, former enlisted avionics tech Lt. Cmdr. "Tugg" Thomson was flying as one of the most experienced B/Ns. With prior tours with VA-115, VX-5, and as a RAG instructor, he was also highly experienced with the SWIP aircraft. Looking back on *Desert Storm*, he recalls he was just about set to retire when the fleet got rushed to the IO, and was surprisingly given the option of staying at Whidbey:

"The skipper called me and said I could stay behind," says Thomson:

"We all knew there was going to be a war. He asked if I wanted to go and I looked at him, 'Are *you kidding?* I've spent 17 years flying this thing and you think I'm going to miss it?' But then I had to go home and 'splain to Lucy,' my wife, JoAnn.

"Our call sign was Ironworks; we were the all Grumman air wing, and we thought we were bulletproof. We deployed, spent New Year's Eve at Cubi Point – most of which I remember – and there was testosterone flying everywhere. We were going to kill them all and let God sort it out."

Thomson says the gravity of the impending situation finally sank in when everyone noted the intensity of the drills, the weapons loading, and the Marines practicing with their M16s from the flight deck:

"Gee, no kidding, we're not at Boardman anymore, we're not talking to the range at China Lake. Then the bravado of the event began to fade. People started to think, am I or am I not the man who could do this? I did not fear getting shot down, I did not fear getting killed ... what I feared was my family getting along without me. It was then I realized it was easier to go than to be left behind.

"The closer we got to the Straits of Hormuz, the quieter the ship got. I got made the tactics officer; part of my job was to murder board the strikes. There were six packages, VF-1 and VF-2 led one each, and (another one) was led by a guy named 'Ogre' O'Neill (Skipper of VAQ-131). The name fit, but he was a good guy to fly strikes with.

"Going low? There were two reasons: the element of surprise and terrain masking. However, I realized a) our element of surprise was zero, and b) in Southern Iraq, it's like trying to terrain mask in a parking lot. I said, 'let's fly high.' 'Uh, now, we can't do that, we've always done things low.' 'There are two advantages to flying high, you don't have to worry about the planet as much, and you can see the target a *long* way off.'"

With his SWIP background, Thomson regularly flew one of the two *HARM* planes, regularly going on strikes with Rick Price and J.R. Haley. The first mission particularly stands out:

"We all blasted off with an EA-6B, first light, 0100 brief and 0400 go," he continues. "If you've ever been to a strike brief, everyone's normally goofing off. I went into the *Silver Foxes'* ready room, they had the lead, led by Cmdr. Sweigert, and it was *quiet*. Still don't know if we're going to go.

"About five minutes later Admiral Zlatoper walked in, looked at us for a minute, and said, 'Gentlemen, the *Tomahawks* are in the air. God bless and God speed,' and he left and we got real quiet. None of us had been in combat before; none of us knew it was survivable.

"Off we went ... we weren't talking, except for occasional talk with *Red Crown* and the ship. Rick Price and I got over Iraq, looked down, and it looked like they were celebrating the Fourth of July with a vengeance. The A-6 is not designed for human use when you're on your way to the target. There's about five switches to push-pull, so we went in with the master arm on, launched the first *HARM* – WHOOOSSSSHHH! – 'Holy cow, *what are we doing!?*' Did that four times, 1+45, back to the ship."

The flight had a little fun on the way back, although at the time it didn't seem too funny. It turned out to be one of those "combat encounters" that all aviators experience at one time or another:

"We called off the target, checked in, dash one, dash two, dash three, dash four ... Dash Four is several octaves higher, said his ALR was screaming and he had a really bad indication at his six. We didn't know the Iraqis were averse to flying at night, just heard they had some neat stuff.

"Rick Martin came screaming out of there – you know you're in trouble when the *pilot* is doing all the talking – two *Tomcats* got a bead and rolled in on whoever was chasing Rick and Bud Abbott. They were at 100-feet, you could hear the plane rattling ... the problem is, flying over the Persian Gulf at 100-feet, they've got some really high oil rigs in the area, good chance to become a fireball.

"Two F-15s rolled in, and just about the time we thought there was going to be a missile in the air, we got a call ... it was an EF-111 who was just as scared as the plane they were following, yelling, 'DON'T SHOOT! DON'T SHOOT!' And you could her two *Tomkitty* and two *Eagle* drivers go, 'Damn! There goes my Silver Star!'"

In the meantime there was a war to fight, and operations continued unabated. The second wave of strikes departed at sunrise, including attacks on targets in both Iraq and occupied Kuwait. The Marines operating out of Shaikh Isa went after Iraqi aircraft where they could find them, making runs on the bases at Tallil, Sh'aybah, Al-Qumoah, and Ar-Rumaylah. By nightfall of the first day Iraq's C2 nodes, air defense assets, and Air Force were in shambles.

The following night Navy A-6Es, Air Force F-111s, and RAF *Tornadoes* continued pounding the Iraqi airfields. Between the *Intruders*, *Aardvarks*, F-15E *Strike Eagles*, and F-117As, the coalition's attack planners could call on over 200 aircraft equipped with laser designators for precision weapons.

During the second day the *Intruders* flew a number of Armed Surface Reconnaissance (ASR) missions, primarily to keep a watch out for Iraq's small – yet missile-equipped – Navy, and dispatch them in a friendly fashion if necessary. VA-145 CO Cmdr. Denby "Heel" Starling and B/N Lt. Chris "Snax" Eagle were trolling around on one of these hops when an oilrig service boat made the grievous error of hosing off a few shots in their general direction. Employing their standard ASR load out of two Mk.82s, two *Rockeyes*, and a *Skipper*, the crew turned and planted both 500-pounders into the boat; Starling and Eagle were turning for their second attack when the boat's crew jumped into the water.

In preparation for the conflict the Iraqi military had manned and armed several oilrigs; during the course of combat several *Swordsmen* and other *Intruder* crews flew anti-oil rig strikes. On one mission VA-145's Lts. Ron "Lobo" Wolfe and Rich "Dim" Witte plastered one platform that was serving as a patrol boat base. On the first pass they knocked down the AAA and small arms, and two passes later the crew dumped six Mk.82s onto the rig, destroying it. Later in the war, on 23 January, Lts. Charlie Giles and Tom "Pickles" Vlasic chased an Iraqi patrol boat under another one of the ubiquitous rigs; if the boat's skipper thought he was safe, he was wrong, because the A-6 crew sent in a *Skipper* under the rig, blasting the craft.

"Hiding under oil platforms is not a tactic that works well against *Skipper,"* pilot Giles commented later.

While the Allies had successfully – even spectacularly – knocked down a substantial portion of Iraq's defenses the first two nights, the threat still existed. For the few remaining old hands with combat experience, it must have been "*deja vu* all over again," as their gear lit up with SA-2 indicators. Yet, much of the enemy's equipment was brand new, and it came from a wide range of suppliers.

However, while the crews were concentrating on negotiating the gauntlet of missiles, AAA, and small arms, coalition planners were furiously looking for another type of rocket: its name was "Scud."

Scuds and Boats

As the air war progressed through its first week bad weather settled in, making aircraft like the A-6 even more invaluable. However, the *Intruder's* crews also found themselves getting thrown into the famous – and mostly ineffective – "Scud Hunt." On the second day of the war Iraq launched several of the medium-range ballistic rockets at Israel, dropping two into Tel Aviv and Haifa and setting off chemical alarms.

Otherwise, the air war continued at its initial heavy pace. Coalition aircraft prowled the skies over Iraq and Kuwait, pounding hardened aircraft shelters and Iraqi forces on the ground. On those rare occasions when the Iraqi Air Force came up to play, its aircraft were quickly shot down; within short order Iraqi MiGs and other aircraft would start running for Iran and presumed sanctuary.

On the 18th, the Navy executed its first anti-surface warfare strike against the Iraqi Navy, with aircraft from *Ranger* and *Midway* hammering Iraqi gunboats and missile craft, as well as oil platform service boats. In *Ranger*, Rear Adm. Zlatoper served as the Gulf's *Zulu Si-*

Intruders excelled in the Armed Surface Recon (ASR) mission during *Desert Storm*. *Fighting Tiger* 504 taxis aft from the bow on *Theodore Roosevelt* on 22 January 1991, armed for boat patrol with a single Mk.20 *Rockeye* and a *Skipper II* laser guided missile showing. The seemingly light loadout was typical, and allowed recovery within weight limits that did not force the jettison of ordnance. Note the raised flak screen on the B/N's side and the ever-present NATOPS pocket checklist in the right quarterpanel. (Rick Morgan)

CVW-8 painted two *Intruders* in temporary camouflage for *Desert Storm.* VA-65's AJ503 is seen on 31 January 1991 as it receives routine maintenance prior to its next event. Both aircraft had the brown paint removed when it became obvious that the war was going to be fought at higher altitudes. (Rick Morgan)

erra, in charge of the anti-surface and anti-ship effort; while the Iraqi Navy did not have many large combatants, it did have a lot of little ones ranging from 1200-ton Warsaw-pact built *Polnocny* landing ships to *Osas* and captured Kuwaiti TNC-45 missile boats. Innumerable small armed boats supported the oilrigs, which themselves were heavily armed. And all were fair game for the Navy and supporting Army helicopter units.

On several occasions aircraft had taken fire from rigs parked in the Ad-Dawrah field off the Kuwaiti coast; naturally, the allies responded. A few days into the war a small force led by *Nicholas* (FFG-47) and Free Kuwait patrol boat *Istiqlal* (P-5702) rushed under a hail of fire from Army OH-58Ds and Royal Navy *Lynxes*, peppering nine individual oil platforms. A SEAL team went aboard one, took prisoners, and confirmed the presence of shoulder-fired missiles and communications equipment.

Over the second and third days of the air war coalition aircraft continued attacks against all forms of targets, with the A-6s carrying a heavy load. One dark morning over 80 night-capable aircraft performed another round of pounding on airfields and command posts, with *Intruders* joining in with F-117s, F-15Es, F-111Es, and *Tornado* GR.1s of the Royal Air Force and Italian Air Force. From Turkey, *Proven Force* aircraft swarmed into Northern Iraq, continuing the devastation; concurrently, Navy aircraft delivered the message on naval installations at locations such as Um Qasr, Shlaybah, and Ahmad Al-Jabir. Throw in the odd *Tomahawk* and occasional gaggles of B-52s thundering over the combat zone to pound the Republican Guard, and things got pretty hectic.

On the plus side, the same day USS *Theodore Roosevelt* – with CVW-8 embarked and flying the flag of Rear Adm. Davy Frost – had come roaring into the Gulf in a lather. The ship had departed Norfolk on a high-speed transit on 27 December and was more than ready to join in. Herb Coon's VA-65 *Tigers* and Ladd Webb's VA-36 *Roadrunners* turned in a blistering performance, even after starting three days late.

Lt. Cmdr. Don "DQ" Quinn was VA-65's admin officer at the time; he says the attitude among the squadron and wing was direct: "Don't start the war without us!"

"We'd had pretty much normal workups," he says:

> "We concentrated heavily on high-altitude bombing, as it seemed plausible we'd be doing that. The ship went hauling across the ocean; unfortunately, on the way over we lost a (VAQ-141) Prowler with something like 25 hours on it. Half the wing hadn't CQ'd yet, but that didn't slow us down; we got in some while waiting to go through the Suez Canal, did 32 knots into the Red Sea, and away we go."

Initially the two *Carrier Air Wing Eight* squadrons concentrated on surface surveillance and SurfCAP missions, flying with *Skippers*, *Zunis*, Mk.82 LBGs, *Rockeyes*, and the like. They still got their share of excitement, as DQ continues:

> "We saw quite a few SAMS, quite a lot of indications. We were flying with the old tried and true ALR-45/50 (RHAW gear), and there was a lot of AAA, particularly ZSU-23s north of Kuwait City. One of the squadrons called a *lot* of SAM launches; it might've had something to do with their number of awards ... they had (night vision) goggles, we didn't.
>
> "We had this one hop ... it was on a day when the ship was supposed to do a five-hour unrep. Lt. Jeff Martin and I were assigned as SurfCAP alert. We were getting tired, looking forward to having the afternoon off ... the aircraft was buried on the bow, and there was no way we were going to fly, particularly since there were three other carriers in the area. 'We won't launch.'
>
> "Launch?! YGBSM! The ship called the launch, cleared out five rows of aircraft, we manned up and were told 1500 launch, 2000 Charlie. I called back twice, 'Are you sure?' They were; we were to rely on the KC-135s at one of the tanker tracks."

Quinn and his pilot Martin launched and headed out to "check on some things;" as there was no specific assigned target:

> "We were just orbiting up there along with a playmate from VA-36," he continues. "We saw a large *Boghammer* boat – or at least its wake – moving from Kuwait City to Falafal Island. We got permission, rolled in with *Zunis*, pulled off, rolled in with LGB Mk.82s, off target. Pulled a 5-G left hand turn, uphill, pretty high – about 13,000-feet – there was a lot of AAA out there, but we

thought we were safe. In hind site, the second attack was a mistake; there was a lot of AAA and SAMS – they looked like Smoky SAMS (training rockets used to simulate SAMS in the U.S.) – and they were obviously shooting something big, 85mm, 57mm.

"THUMP! Right wing, BIG hole, about six-feet across. From where I sat it didn't look that bad, maybe 18-inches or so. We lost our combined hydraulics immediately and jettisoned everything. The MER on station five was fused to the parent rack, still armed, still there ... I could see the Mk.82. We eventually landed at Shaikh Isa in Bahrain, EOD came out, took a look, walked up to us ... 'Oh, by the way ...' They ended up unscrewing the mechanical fuze from the bomb."

Quinn says the aircraft – AJ503 (BuNo 155620) – was known as "Christine" by the squadron due to all sorts of spurious problems. The canopy slid open once in mid-strike, during an attack on the refinery and command buildings at Aisa Baier. Still, the maintenance crews patched the shrapnel damaged portions of the tail and fuselage, replaced the outboard wing, and "Christine" was flying again only 11 days later.[5]

Through the conclusion of the war CVW-8 would claim the destruction of 22 Iraqi boats, with the loss of one A-6: on 2 February VA-36's Lt.Cmdr. Barry Cooke and Lt. Pat Conner failed to return to the carrier. As last heard, the crew was working alone at low altitude, apparently dueling with an Iraqi boat off Kuwait City. Although nobody saw the incident, they were presumed lost to either AAA or an IR SAM. Conner had only recently joined the *Roadrunners* from VA-55. Although CVW-8 would lose two *Hornets* and one pilot in operational mishaps, AJ531 would be the wing's only combat loss during *Desert Storm*.

Still, the *Tigers*, *Roadrunners*, and other *Intruder* squadrons would continue to take the fight to the enemy, in and around increasing oil slicks and clouds of oil smoke, courtesy of Saddam Hussein. On 22 January four A-6s managed to disable an Iraqi T-43 minelayer; that evening, a P-3C detected and tracked an Iraqi tanker that was apparently involved in electronic warfare activities. A covey of A-6s from VA-115 and VA-185 hit the tanker just as it released a hovercraft, which immediately took cover underneath the piers of the Mina Al-Bakr oil terminal. Flushing the hovercraft, the *Intruders* quickly sank it with *Rockeye*.

On the 23rd the Iraqis started dumping oil into the gulf and attempted to ignite it; they failed, for the most part. The following day *Intruders* from *TR* participated in the sinking of a *Zhuk*-class patrol boat and another minelayer, while other A-6s worked over the Umm Qasr Naval Base with some FA-18s. Near Qurah Island a third minelayer tried to evade the prowling Navy aircraft and hit one of its own mines; Army OH-58Ds operating from USS *Curts* (FFG-38) went in to rescue the 22 Iraqi navymen and drew fire from the island, so the helos and frigate responded.

What ensued was a six-hour operation that resulted in the first recapturing of Kuwaiti territory. After knocking down the opposition the helicopters from USS *Leftwich* (DD-984) put SEALS on the island. They captured 67 Iraqis and collected important intelligence on minefields in the area.

With one small component of Kuwait now free operations continued apace, with the A-6 personnel continuing a mix of overland strikes and surface ops. VA-145's "Tugg" Thomson got in a few missions:

"I found myself off Basrah and Bubayan doing ASR – *really neat* – with Mk.82s and *Rockeye*. The latter is a really great weapon, depending upon which end you're on. We also flew with *Skippers*, a 1000-pound LGB with a *Shrike* motor basically screwed in the end. If it hits you, it's definitely going to ring your clock.

"We found this trawler we were looking for, went by it, all those sparklers ... I turned to the pilot, 'You know, I think they're shooting at us.' We lined up for a bow-to-stern *Rockeye* pass, but both dudded, landing 50-feet in front. We couldn't shoot the *Skipper*, the laser was down ... we came in with two VT-fuzed Mk.82s, set up broadside – I'd noticed their two quad-50s were on both sides, able to shoot aft – so I thought I'd cut my losses.

"The first one dudded – by now I'm ready to shoot our gunner – and the second one landed in the water about 25-feet away. By the way, this was the first time I'd ever been shot at, and I found I was *mad*, not scared ... how DARE they!? So, as the bomb hit the shooting stopped, just like turning off a light."

With roughly one week's operations completed, there was a brief – very brief – pause in the action. It was becoming apparent to most that there would be no quick political solution to the conflict.

"We got three days into the air war and no one told us to stop, so we did a seven day plan ... and then just kept going," he comments. "Hussein wasn't giving any signs of capitulating, and we suddenly realized we were going to be there for a long time."

Send in the Marines

While their naval *Intruder* brethren were going downtown and sweeping the seas, the Marine A-6 crews of MAG-11 at Shaikh Isa, Bahrain were also doing their part. As of yet, all of the coalition ground forces were still on Saudi soil, so there wasn't much call for the Marines' *raison d'être*. Close Air Support. Therefore, they got to fly a wide range of missions at which the noble Intruder excelled, attacking installations and ground targets while generally contributing to a lack of sleep among the Iraqis.

On the first night of *Desert Storm* crews from VMA(AW)-224 and -533 went after Iraqi Army targets inside Kuwait, relying on their buds in VMAQ-2 to keep the SA-6s beaten down. Operating 10 aircraft each, they then spread out over the following days, working over Southern Iraq and Kuwait, including ops against armored vehicles, tanks, artillery, and rocket batteries assigned to the Republican Guards.

Alongside the Marine Intruders was a former "family member": VMFA(AW)-121, with their new FA-18Ds. Former RF-4 RSO and A-6 B/N Paul "Henry" Bless was with the squadron and recalls the unique circumstances that went with the squadron's combat introduction of the "Fast FAC" version of the *Hornet*:

"We'd gone through two COs," he says:

"Our first CO was kind of a Jekyl and Hyde kind of guy, a great American until he had a couple of drinks in him (he was relieved), and Lt. Col. Steve 'Muggo" Mugg, our XO and a former A-6 B/N, took us to war.

"Conditions weren't bad for us. We lived indoors, didn't live in tents. Since Bahrain's a small country they went through their barracks and told their enlisted troops to move home with their parents. There were a few senior enlisted guys who stayed in one room at the end of the hall, and we pretty much lived indoors. We ate a lot of hot meals, had a bus service that drove us from billeting to the airfield ... BIG airfield, long and concrete, but we laid down a lot of matting.

"We ended up strictly doing the FAC mission; we didn't drop one bomb, despite what anybody says. The reason was we didn't have any reconnaissance. The Marines' had stood down (VMFP-3), and we had Air Force recce available, but it would take two weeks to get the info. Two weeks when you're fighting mechanized forces is basically worthless.

[5] The aircraft also was one of two in CVW-8 which had desert camouflage applied (VA-36's AJ533, 161667 being the other). Both had the special paint removed by 9 February when it became obvious that operations were going to be at higher altitudes.

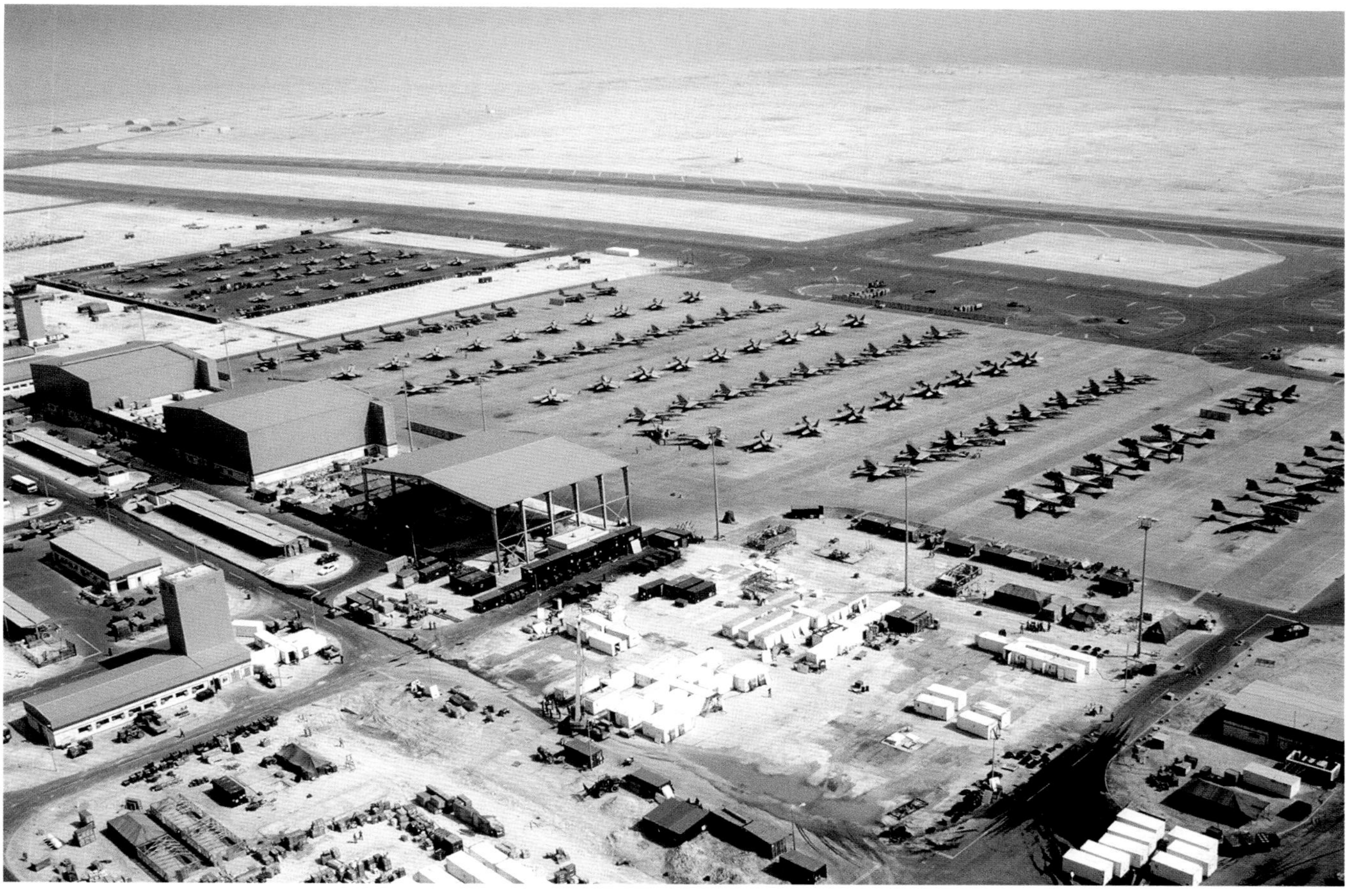

Boomtown. Marine *Intruders* were based at Shaikh Isa, Bahrain, during *Desert Storm*. The obscure base located on the island's southern end had been built for a small Bahraini F-16 force. At the start of the war the field also housed Air Force F-4Gs and RF-4Cs, as well as 1st MAW's *Intruders, Hornets,* and *Prowlers.* (U.S. Navy)

> "We'd end up going up ahead and determining if the targets were legitimate," Bless concludes. "If they weren't, we'd find something else and mark the target. That was our FAC mission."

The newly re-equipped *Green Knights* would spend the war doing the FAC and Tactical Air Coordinator-Airborne (TACA) missions, launching with full loads of 20mm ammunition and 2.75-inch white phosphorus spotting rockets. In and around regular trips to the KC-130T tankers, they applied themselves by calling in strikes and providing BDA.

As for the two other squadrons, which were still flying A-6s, they ended up ushering the *Intruder* out of Marine combat history in valiant fashion. Retired Marine Master Sgt. Al Parks was Assistant Ordnance Chief for the *Hawks* of -533; he recalls a lot of personalities and a lot of weapons to choose from:

"The standard weapons load were 10 aircraft, two flights per day (or more), starting at approximately 1700 to 0600," Parks writes:

> "The aircraft mix was 12 Mk.82 conventional/retarded, one 500-pound LGB/GBU, one *Shrike*, and 22 Mk.82s or Mk. 20s. All the aircraft had a mix of chaff and flares; on special missions they carried four 1000-pound or four 2000-pound bombs or LGBs. We flew every type of ordnance the A-6 was capable of carrying except special weapons (nukes) and *Harpoon/SLAM*, as our mission was to take Kuwait City and support the Marine ground forces.
>
> "The skipper for -224 was Lt.Col. Bill Horne, XO Lt.Col. McElrath, and AMO was Maj. Marty Post. The -533 CO was Lt.Col Beman Cummings, XO Lt.Col. Langdon, and AMO was Lt.Col. Life Larsen."

Among "Pig" Horne's pilots in VMA(AW)-224 was young Marine Captain and Eastern Washington University graduate Tom Uryga. He mentions that in February 1990 he was on his way to Cherry Point but was not sure to which squadron he was supposed to go; -332 was short of pilots and was about to go to Iwakuni, but he ended up joining one of his best friends in -224. Less than a year later he and his friends were seeing the elephant:

"We flew, with 10 airplanes, something like 450 combat missions," Uryga comments:

> "We didn't take a scratch. We were among two A-6 squadrons and five or six *Hornet* squadrons, and those first cheesy FA-18Ds that showed up. CNN was there; they were trying to pump up the F-18Ds big time.
>
> "We had a good squadron and a great bunch of guys. Bill Horne, our CO, actually flew F-4s in Vietnam. When we got over to the Gulf Gen. (1st MAW Commander, Royal) Moore flew with us, Charlie Carr flew with us, every other Tom, Dick, and Harry that ever sat in an A-6 showed up to fly with us. We even had these rocket scientists from MAWTS down in Yuma; we anointed them the 'Jedi.' They came up with all this stuff, but

some of their specific employments didn't work in the real world. We were probably the first people that ever did high-altitude steep-dive releases of *Rockeye*.

"In the first part of the Gulf War," he continues, "We started out with deep strikes on Basrah, Shaibah, and Tallil. We'd use a mix, since the weather was shitty, 2000-pound LGBs and 2000-pound 'steel nose cone.' We'd try to pickle the LGBs right in the middle of the basket. The first night of the war we came back with better BDA than the F-117s had."

Uryga says the *Bengals* initially were going to take some -242 personnel with them, but ended up taking just three aircraft; the emphasis was on gaining A-6s that did not have G-limitations:

"We did end up with some 3G limited jets, it was pretty much done by g-meter counts and no apparent damage. It was a big issue because of the weapons loads we were going to carry. We were working up, August to October, 120 degrees and 100 percent humidity, and you'd have some very nasty 7000-foot takeoff rolls. Oh God, I had one, it was slightly ... it was a two percent uphill runway and you always took off northbound. It was uphill and it was hotter than ... I think it was an 11,000-foot strip, and I had one that was like a 9000-foot takeoff roll."

According to Uryga, the squadron had a mix of new guys and some who had been there a long time, adding nothing seemed to bother the old hands:

"My first mission was with a guy named Rob McCarthy, an RAF Exchange officer and *Tornado* weps officer. He had absolutely no fear of death; he flew in the air show demo team in *Tornados* where they do the 100-foot inverted supersonic pass. Nothing scared Rob.

"First night we flew, we're heading into Iraq, and we're heading into Shibah Airfield. This was the first time we actually flew with the lights out and actually flew with the green formation lights ... every one was different with the lights. Some lights are burned out, you couldn't' tell what was what, you thought you were rendezvousing on Battlestar Galactica because of all the different lights.

"Off to the left side, as we're crossing the Gulf, you could see the SAM launches and AAA in Kuwait City, huge, holy shit, look at this stuff. Rob's got his head in the boot, doing updates. We're getting up there over the salt marshes, little stuff – red tracers – were arcing below you, bigger stuff, white tracers, reach up higher, about 20K feet. Then you have some big AAA, big white bursts up at our altitude, but it was less frequent. I think the red stuff was ZSUs ... it was disturbing enough, but you knew they weren't all radar guided because you didn't have a lot of radar warning about everything.

"About this time Rob pulled his head out and said, in typical British fashion, 'I say Tom, it appears those blokes are shooting at us.' 'No shit, Rob, put your head back in the boot.'"

The two Marine *Intruder* squadrons continued to carry the load through the initial phases of the air campaign, and then shifted to the mission they do best – support of Marines on the ground – once the invasion actually began. Uryga adds that as a newly promoted Captain and the -224 S-4, or "embark officer," he gained a new appreciation for expeditionary operations and "jointness" at remote Shaikh Isa.

"I had a lot of adventures in getting the squadron to the Gulf," he continues:

"First was the 'Tampon Crisis' (we took girls with us), where the wing considered female hygiene supplies as big an issue as MOPP gear. Then there was the ammo crisis.

"I issued 30,000 rounds of .223 for the M-16s to our marines (air crew, officers, and senior NCOs had 9mm), and the Wing Embark Officer wouldn't let me put them on the transport planes to the gulf with the ammo (the Air Force was willing). He claimed we could get all we needed from the MPS ships. Bullshit! MPS was for the division. I had to reclaim it all, dump it in a big box, which I put in the trunk of my car, and took it to my neighbor – a gunnery sergeant sniper who was with group supply – who took it to the rifle range. I made a lot of friends with that donation. I followed the squadron about 10 days later with the rest of the 'rear-party,' who made sure everything we needed was sent.

"We had Air Force RF-4s in from Germany, and the Reno Air Guard at our base (as well as F-4G *Wild Weasels*). Each 20 Air Force planes came with aircrew and over 1600 support weenies, compared to our squadron's 10 planes and 235 aircrew and support, plus augment corpsmen, flight surgeons, armorers, etc. Talk about fat staffing ... Col. Reich, our group CO, had six or seven USAF counterparts, all non-aviators, to contend with over issues like Marines peeing in USAF toilets."

Still, the Marines persevered and performed their mission; indeed, after the first 48 hours of combat operations, the men of VMA(AW)-224 proudly adopted the motto, "We own the Night." According to

The *Bengals* were one of two Marine *Intruder* units deployed to *Desert Storm*. WK507 appeared at Kelly AFB, TX, only a few months before deploying to Southwest Asia. (Phillip Friddell)

F-111: The *Intruder's* Air Force Counterpart

The Air Force's closest counterpart to the A-6 was the General Dynamics F-111, an aircraft that went through years of development, controversy, and turmoil prior to complete vindication. Oddly, the 111 would fly over three decades without an officially approved nickname, but was known up until its last day of service by the unofficial moniker "Aardvark" (as in the long-nosed, earth-burrowing creature). It would take about 30 frequently rocky years before the F-111 erased all doubts as to its legitimacy and was granted a legal name.

The F-111 arose from the TFX ("Tactical Fighter, eXperimental") program, where the McNamara DOD decided that Air Force low altitude/high speed all weather strike aircraft and Navy fleet air defense requirements could be met by the same airframe. The Navy, never enthusiastic about the design, spiked the F-111B in spite of successful carrier qualification. Luckily for the service, Grumman learned from the type by applying the swing-wing technology and AWG-9 radar/Phoenix missile system to what became the F-14 (One can only imagine the fun carrier fighter squadrons would have had over North Vietnam in 1972 if the F-111B had been the fleet's primary VF type, as apparently intended by the McNamara Pentagon).

Compared to the *Intruder's* maximum load of 28 500-pound bombs, the "Aardvark" could carry "only" 24, but it could haul them much faster and a lot further (reportedly the 111 was designed to be able to cross the Atlantic Ocean without tanker support). The "F for fighter" designation was a misnomer, as the type was an attack aircraft in function, if not name. Low altitude/high-speed solo penetrations of enemy air defenses at night and in foul weather were the type's forte, not chasing MiGs.

The F-111A was introduced by the Air Force in 1966, and six aircraft were sent in March 1968 to South East Asia as operation *Combat Lancer* for what in effect was a combat operational evaluation. Within two weeks a pair of the birds and their crews were missing. Two replacement aircraft were sent from Nellis AFB, and one of these was lost with two men only days after arrival in theater. The type was soon withdrawn from combat under a dark cloud.

Four years later, the *474th Tactical Fighter Wing* reintroduced the type to combat during *Linebacker I* and *II*, mostly flying solo night strikes up north. Although considered very effective, six aircraft were lost, and the F-111 remained a "high interest" aircraft throughout its second combat deployment, as every loss came under intense political and media scrutiny.

By the end of production GD had made 531 F-111s for the Air Force in five basic models (TAC's A, D, E, and F, and SAC's FB-111A). Systems improvements in the type pushed the frontiers of digital technology and precision delivery techniques, frequently leading to serious support and maintenance complications, particularly in the D model. With the E and F versions, however, the Air Force had a superb weapon system as good as, if not better than, any comparable aircraft in the world. Foreign sales were approved, with Australia buying 24 F-111Cs, while a potentially lucrative "F-111K" order to the RAF fell through.

By 1975, the type toiled as the only original Air Force tactical design among ramps full of Navy originated F-4s and A-7s. In fact, the "Aardvark" equipped only four wings in Tactical Air Command (TAC) and United States Air Forces Europe (USAFE), with each wing flying a different version (one each with A/D/E/F). The 111's Rodney Dangerfield "don't get no respect" position was reinforced in the 1980s, as it was obscured by glamorous new F-15s and 16s. In the Strategic Air Command (SAC), two wings of FB-111A "Switch Blades" quietly carried out nuclear alert duty alongside more numerous B-52s until retired in July 1991, with some being modified as G models for TAC use, and others going down-under to augment the Aussie fleet. From 1981 42 A models were modified for the Electronic Combat mission as EF-111A *Ravens*.

In spite of its rocky early years, by the late 1980s the type's supporters and aircrews had developed the 111 into a weapon system of enormous potential that would be realized late in its career in two separate actions. In *El Dorado Canyon* 13 United Kingdom-based F-111Fs carried out a 14 hour/2800 mile flight to bomb Libya, marking the longest mission by tactical strike aircraft in aviation history. During *Desert Storm*, the type was arguably the most important bomb dropper in the Allied order of battle, flying about 4000 sorties *without loss*.

Like the *Intruder*, the entire Air Force F-111 force was targeted for accelerated retirement immediately after its *Desert Storm* success. Also like the A-6, critics maintained that the type was being put out to pasture without adequate replacement. Nonetheless, before the last F-111F was flown to the boneyard for storage on 29 July 1996, the bomber version was finally given a legitimate and officially recognized nickname: *Aardvark*. It was as if the Air Force had finally acknowledged its red headed stepchild and provided a fitting end to a valiant and frequently unappreciated airframe.

The General Dynamics F-111 was the closest thing the Air Force had to the *Intruder*, at least in terms of mission, if not performance. Following a difficult combat debut in Vietnam the type proved itself beyond all question in Libya and Iraq. (U.S. Air Force, via Marty Isham)

Many *Intruders* returned home from *Desert Storm* with elaborate scoreboards displaying their success in the Gulf War. VA-145's NE501, an A-6E SWIP, displays 56 mission markings, 12 radars (presumably representing HARM shots), and three boats. The small symbol below *Swordsman* XO Rick "Stump" Cassara's name denotes his status as a "boat ace." NAS Whidbey Island, 6 June 1991. (Rick Morgan)

Intruders from CVW-2 destroyed the thermal power plant at Az Zubayr on 7 February 1991. VF-2 TARPS imagery confirmed the level of destruction left in their wake. (U.S. Navy, via Steve Barnes)

Uryga, the *Bengals* alone ultimately dropped over two million pounds of ordnance, including Mk.84 slicks and LGBs; Mk.83s ("A few; most went to the *Hornets*, as they were range/payload limited"); Mk.82 slicks and LGBs; *Rockeye*; a few each APAM and *Gator*; FAEs ("They completely disintegrated the Winnebago-like command vehicles the Iraqis used in Kuwait"); Napalm ("700-foot laydown ... that's *low*!"); *Skipper*; seismic sensors ("They went 20-feet into the sand"); and letter bombs.

"Once the inventory sheet indicated a B61," he recalls, in reference to the air-deliverable nuke. "I don't know if it was really there, but it disappeared from the list *very* quickly."

The Battle of Bubayan

On 25 January the Iraqis began pumping millions of gallons of crude oil directly from the wellheads and oilrigs into the Persian Gulf. Two Air Force F-111Fs "capped" the wells by bombing the manifolds with 2000-pound guided bombs, ending this particular annoyance.

Otherwise, aerial operations continued against Iraqi forces, including attacks on lines of communication and bridges. On 27 January the focus of many of the attacks shifted directly to Republican Guard targets; that same day Central Command declared it had attained complete air superiority over the Kuwait Theater of Operations. The coalition maintained a barrier air patrol between Iraq and Iran, shooting down several Iraqi aircraft that were attempting to escape.

Two days later Iraqi forces made their one and only incursion into Saudi Arabia by attacking the small, abandoned town of Al-Kafji. While not a major full-scale battle, the assault may have served as an attempt to give the allies a public relations "bloody nose." The Iraqi forces that took the town included battalion-level armored and mechanized infantry units; they ran up against A-6s dropping *Rockeye* – directed by VMFA(AW)-121 crews – and strong resistance on the ground by U.S. Marines and Saudi National Guard armor units. Over the course of the engagement the Iraqis lost numerous tanks and armored personnel carriers while attempting to dodge a constant hail of accurate ordnance; the allies lost one AC-130 gunship in return, along with several casualties on the ground, but coalition forces decimated the Iraqi attack by the 31st, in the process taking several hundred POWs.

The night of the 29th also saw the commencement of the single largest anti-surface operation of the Gulf War. Known afterwards as "The Battle of Bubayan," the 24-hour operation resulted in the complete destruction of the small Iraqi Navy.

The event kicked off when VA-145 and VA-155 *Intruders* on a surveillance mission detected four unlit boats heading from the vicinity of Shatt-al-Arab towards Iran. Cmdrs. Rick "Kato" Noble and Rick "Stump" Cassara were in the *Swordsman* aircraft, alongside the *Silver Foxes'* Lt.JG Greg "Davy" Crockett and Lt. Dan "Snake" Kalal. They had just thumped a *Silkworm* missile site at Ras al Qulayah when they stumbled across the hostiles.

Operating under the control of an orbiting E-2C, the A-6s first went down to positively identify the four craft; they turned up an FPB-57 and two TNC-45s, undoubtedly captured Kuwaiti equipment complete with missiles. *Swordsman* XO Cassara made the determination the craft were hostile and – following an exchange with the orbiting *Hawkeye* – went in for the attack. Cassara and Noble took out the lead vessel following a direct hit by a 500-lb LGB, then turned on the second boat and also rendered it dead in the water. The *Swordsmen* then lased the third boat, which the -155 crew proceeded to destroy, with a little friendly assistance from an FA-18 that had arrived.

By this point the fourth boat – later identified as an *Osa II* missile boat – was departing for Iran and presumed sanctuary at a high rate of speed. Cassara and Noble went after it but were unable to track the vessel due to FLIR problems, so they pulled off. A pair of Canadian CF-18s that were configured for combat air patrol then came in and strafed the boat; the Osa made it to Iran, but its superstructure was swiss-cheesed.

Over the following day another group of 20-plus combatants flushed from Az-Zubayr and Umm Qasr and attempted to make the run to Iran. Coalition aircrews destroyed the majority, including three *Polnocnys* (980 ton Polish-built landing ships), four TNC-45s, two *Osas*, and a *T-45* minesweeper. Coalition aircraft operating from *Ranger*, *Roosevelt*, and *Midway*, directed by E-2Cs, P-3Cs, and the occasional helicopter, succeeded in destroying multiple Iraqi ships, while sending several to the bottom; ultimately, one *Osa II* and a *Polnocny* managed to escape to Iran without any damage. Combatants that failed to clear the Iraqi piers were dispatched on the spot.

VA-115's Joe Kuzmik participated in the 'SinkEx', recalling:

> "We were directed on a boat that the E-2 held right off Kuwait City's harbor. We went in to investigate right after a *Ranger* strike cleared the area and found one small boat pulling a larger one. FLIR proved it to be an *Osa*-class missile boat behind a smaller patrol boat, which was towing it for whatever reason.

The *Black Falcons* started *Desert Storm* from the Red Sea off *America*, then moved with the ship to the Persian Gulf to finish the war. A VA-85 A-6E loaded with Mk.83 1000-pound bombs takes fuel from a 2nd Bomb Wing KC-10A *Extender* over the North Arabian Gulf while a *Roosevelt*-based F-14A from VF-41 observes. Air Force "big wing" tanking was crucial for the prosecution of the war. KC-10s, which had a standard Navy "soft" drogue, were preferred by almost all Navy pilots over the KC-135s, with their "hard baskets." (Rick Morgan)

"We put a single GBU-12 (500-lb LGB) through the *Osa's* stern, and its detonation was rapidly followed by each *Styx* surface-to-surface missile canister blowing up in succession. When the FLIR imagery came back after being whited-out, the *Osa* was nothing but debris, and the patrol boat in front was on fire. The footage of this attack later showed up on the Discovery Channel."

Following the Battle of Bubayan, the Iraqi Navy was combat ineffective; those remaining boats that cleared the ports were quickly taken under fire. On 2 February *Intruders* and other aircraft hammered the Al Kalia naval facility while sinking two more missile patrol craft. Regrettably, the same day VA-36's Cooke and Conner were downed near Faylaka Island. It was the third and last *Intruder* loss of the Gulf War.

On 8 February *Naval Forces U.S. Central Command* announced it had complete and total control of the northern Persian Gulf; for all intents and purposes the Iraqi Navy no longer existed. A few weeks later the Hoover community would add their own conclusion to the proceedings. On 20 February the crew of a VS-32 operating from *America* were vectored into a surface target by *Valley Forge* (CG-50). Using three Mk.82s – and a D-704 buddy store that was unintentionally dropped – they scored a kill on an Iraqi gunboat, the first S-3 to ever sink a ship in combat.

With the navy out of the picture, the carriers redoubled their efforts against strategic and tactical targets within Iraq and Kuwait. All sorts of installations remained under the gun, including hardened command and control installations and airfields. *Intruders* and other Red Sea-assigned carrier aircraft hammered the SAM production facility at Al-Falliyah, along with targets in the vicinity of Samarra. Marine A-6s participated in attacks on Iraqi armored and mechanized units, on occasion going after individual tanks and armored fighting vehicles.

The pounding continued for several weeks. During this period *America* departed the Red Sea to join the carrier forces in the Persian Gulf; it detached on 7 February and made quick transit to be present for the upcoming ground offensive.

Capt. Dave Polatty – a former VA-42 instructor and skipper of VA-115 – was serving in *America* as DCAG-1. He comments:

"We started initially in the Red Sea, did the first half from there, and then went to the Gulf for the remainder. It was pretty interesting.

"I told CAG I'd fly every night (from the Red Sea). You would take a package of 12, go to the two KC-135s off the coast – probably the roughest part of the mission – rotate on and off to the border, kick off, sneak into Iraq, hit the target, come back, and hit the tankers."

"We took off one night for some SCUD busting," Polatty continues:

"Not very successful,[6] so someone said after the second day, 'Let's go where they're storing the SCUDS.' We went west of Baghdad, about 150 miles, didn't realize how long it was. We planned to take four A-6s with 22 Mk.82s each, straight path through the clouds; (then the Iraqis) put this mobile SAM on the exit route. I'm strike lead, so we called to VAQ-137 and asked for a spare *Prowler*.

"I tell you, As soon as the first bombs hit the place lit up like the Fourth of July. You could see it from 200 miles away. On the way out an SA-3 lit me up. The *Rook* (EA-6B) turned into it (shot a HARM) and took it down.

"That's the flexibility of Naval Aviation; you *never* want to go into combat without that kind of capability."

On the other side of the war the Navy kept up the pounding. On 7 February a mixed flight of VA-145 and -155 crews went after the Az Zubayr electric powerplant; the crews included Lt. Cmdrs. "Tugg" Thomson and "Tag" Price; J.R. Haley and Bud Abbott; VA-155 Lt.Cmdrs. Steve "Boots" Barnes and Dan "Dewey" Dewispelaere; and Lt.Cmdr. John "JJ" Jackson and Lt. Marcus "Hitch" Hitchock. They completely leveled the plant, destroying yet another component

[6] Official post-war analysis stated that not a single mobile SCUD launcher was apparently ever hit, this in spite of numerous aircrew who swear they destroyed such targets. Some may have been luckless semi-trucks, but the true success of the SCUD-hunts still raises heated arguments in some quarters.

of Iraq's infrastructure, although Thomson recalls these and other missions still tended to be "sporting."

"We had a lot of 'conga lines,'" he says:

> "You know what a conga line is? Four aircraft come back, Dash One and Dash Two are high-five'n, whereas Dash Three and Four, the crews' eyes are as big as saucers and their seat pans are stuck to their butts. They had to go to sickbay to get them removed.
>
> "Why? We learned by this point that the Iraqis aren't shooting unless shot at. So the first two aircraft drop, the Iraqis come out of their porta-johns, regain their seats at their weapons, and they are mad. So the last two planes get hosed. We had to figure a way to get around that; eventually cut back to a two-plane conga line, rolling from 20,000-feet, about 30 seconds apart. That's when we started going high."

All lessons learned were applied by all aircrews. The following day, on 8 February, VA-145's skipper, Denby Starling, led another multi-plane raid with VA-155 XO Larry Munns on the Al Basrah refinery. The joint was burning mightily by the time four *Intruders* departed the scene.

America reported to *Task Force 154* on 15 February, giving the Navy four carriers in the Persian Gulf to support the coming liberation of Kuwait. CVW-1 went back to combat operations, and *Midway* departed for Bahrain and replacement of the flight deck non-skid, which was down to bare-metal in places. *Midway* returned to the fight on 21 February.

Once in the Gulf the *Black Falcons* suffered their only loss of the war, although it was the result of a flight deck mishap and not combat. On the 15th BuNo 155602 lost its brakes while coming out of the wires and the crew safely ejected; the A-6 ended up hung up on the angle by its main mounts. With the remainder of the wing-orbiting overhead, the flight deck crew removed the *Intruder's* ALQ-176 pod and pushed the luckless aircraft over the side. Officially, the event was declared an "act of salvage" and not a mishap; they had a war to fight. Providentially, the loss was the only one suffered by *America* and *Air Wing One* during *Desert Storm*.

Such was the air war in the middle of February 1991. Iraq's ability to wage war in the air, on land, and at sea was being systematically reduced and rendered ineffective, all in preparation for the ground assault that everyone knew would come. Iraq still managed to get in the occasional counter punch; on 18 February, while operating in the North Persian Gulf, *Tripoli* (LPH-10) and *Princeton* (CG-59) ran afoul of mines.

Tripoli – ironically operating in the mine countermeasure role – suffered a 16 by 20-foot hole in her bow and was forced to limp into port. The cruiser *Princeton* actually found two mines that cracked its superstructure, buckled its decks, and started fires. Her crew managed to save the damaged ship and, following relief by another cruiser, *Beaufort* (ATS-2) towed her to safety.

All the aerial combat, all the preparations, all of the strikes and targeting, and all of the diplomatic maneuvering finally came to a head on Thursday, 24 February 1991. At 0400 local, United States, British, French, and other coalition forces crossed the line and invaded Kuwait.

Into Kuwait

The ground war – executed under the title of Operation *Desert Saber* – lasted a grand total of four days, or about 100 hours. On the right side of the line, the 1st and 2nd Marine Divisions and the 1st Brigade of the Army's 2nd Armor Division moved into Kuwait to take on the Iraqi occupiers. On the left flank were a brigade of the 101st Airborne Division, accompanied by the French 6th Light Armored Division, while in the center the Egyptian 3rd Mechanized Division led the way. While the 101st Airborne moved north towards the Tigris and Euphrates Valleys, VII Corps headed west out of the Wadi al Batan, led by the 1st Infantry Division.

It was the largest single ground and air offensive anywhere in the world since World War II. In some places the battered Iraqi Army put up a determined defense, but it could not rely on assistance from the Iraqi Air Force; like the Iraqi Navy, it no longer existed. All of the allied aviation assets were heavily involved, striking point targets or responding to calls for assistance by the ground forces.

"Once in Kuwait we did 'battlefield air interdiction,'" -224's Tom Uryga writes:

> "Basically, drawing the map into 'kill boxes' that we patrolled. The Air Force calls this CAS, by the way ... what the Marine Corps calls CAS, the Air Force calls crazy.
>
> "The B-52s flew 10-hour missions from Diego Garcia. None had the external racks they used in Vietnam. They carried only 56 500-lb bombs. Two A-6s did more than a B-52. They just salvoed the entire load in one pass. Their BDA pictures showed mile-long strips of closely spaced craters, usually laterally displaced from their targets (largely due to 100 knot high-altitude winds). Carpet-bombing made a lot of noise but didn't hit much.[7]
>
> "The typical *Hornet* load was four or six Mk.82 or *Rockeye*. Without tankers, they had about 10 minutes time-on-station in Kuwait. They made one pass, dropped two bombs, and went feet-wet to a tanker, gassed up, came in, and dropped the rest, went home, and got credit for two combat missions on one flight. Pretty cheesy way to earn air medals. *No* (Marine) A-6 ever saw a tanker, yet we had over 40 minutes TOS (Time On Station), even in northern Kuwait."

The *Intruder* was a hero to more than a few Marines on the ground during the war. One Forward Air Controller (FAC) attached to the infantry was an EA-6B NFO who stated:

> "Our first three choices for CAS aircraft were, no.1, the *Intruder*, no.2, *Intruder*, and no.3, *Intruder*. It had plenty of bombs and enough gas to stick around. The *Hornets* had less of each, and as for the *Harriers*, if we didn't have a target ready for them when they checked in they almost always went right back to the tanker"

On the 25th NAVCENT ordered additional raids in the vicinity of Al-Faw and Faylaka Island, primarily to keep the Iraqi ground forces in the area from shifting west to meet the allied threat; a substantial component of Iraq's ground forces were in the area, primarily in response to the threat of an amphibious landing by another marine division lurking offshore. The raids included *Intruder* strikes in the area around Bubiyan Island, which – in association with the other raids – did the job of keeping the Iraqis confused and pinned down.

Within 24 hours Iraqis were retreating back to their homeland or surrendering; the numbers were staggering, putting a severe strain on coalition forces who were trying to fight a war. On the 26th Iraqi forces attempted to flee Kuwait City using tanks, trucks, busses, pickup trucks, and whatever else they could load their gear and contraband in. The road north to Al Basrah became a virtual parking lot and a major killing ground; it would forever be remembered as "The Highway of Death." *Intruder* crews directly contributed to the deadly traffic jam, knocking down multiple bridges and bottling up the fleeing enemy; on one mission the VA-145 team of Starling and Eagle de-

[7] It should be pointed out that B-52s also flew from England and Saudi Arabia and that their crews view of their effectiveness would undoubtedly illicit a serious and professional exchange of opinions at the bar.

stroyed the Shuyukh Bridge over the Euphrates using two 2000-lb LGBs.

The Iraqi collapse happened so fast, it was hard to keep track.

"I launched on two hops on the 'Highway to Hell," says "Tugg" Thomson: "and the FEBA (Forward Edge of the Battle Area) had moved miles by the time I checked in with the FAC on the second trip. The A-6s were first on the scene, then the *Hornets*." Uryga recalls:

> "In the morning some Air Force A-10s showed up. The Army claims they did it (destroyed the retreating Iraqi Army) with dismounted infantry throwing grenades, but they were not in the area, or we wouldn't have been dropping.[8]
>
> "I flew three missions that night, went into the clouds at 7 or 8K, coming out around 2000-feet and dropping *Rockeye* down to the 1200-foot level. Default fusing was 1.2 or four seconds canister opening (COT); we had all weaponeered for four or eight seconds to allow high altitude bombing, so the lowest was 1200-foot level, or 1400-feet in 10-degree dive with a four second COT. If we'd had 1.2-second COT available we could have done 200-foot level lay-downs."

Crews reported the weather was abysmal, with Uryga describing clouds at 7000-feet and lower ceilings around 2500-feet. He adds there was an airfield about 12 miles to the west, still active with SA-8s and *Rolands*:

> "There was still stuff occupied to the east, so we were actually taking fire from both sides.
>
> "Our load out was normally 22 *Rockeyes*; -533 flew a few times with 28 – they took off the centerline – but had no time on station. We'd go up, drop, hot fuel, hot rearm, then go back. I did three missions that night until they finally kicked me out of the airplane. We flew exclusively at night ... we *owned* the night.
>
> "We're talking directly with Lt. Gen. Moore, he's at the Tac Center."

Uryga continues:

> "He calls my pilot, 'Wheels, this is Hunter, how bad's the weather up there? Can we send the *Hornets* up?' They wanted to come up and play. We went through in 30-minute cycles with -533; every 30 minutes there were four more Marine A-6s in the airspace over Kuwait. We just started laying down over the highway, pickling four or six at a time, coming back for another run, coming back for another run.
>
> "We dropped everything in the inventory except nukes and chemicals. We supposedly exhausted something like 40 or 60 percent of the entire U.S. stock of *Rockeyes*. I dropped 2000-pound bombs that were stenciled 'Philadelphia Naval Arsenal, 1957;' they were not thermally coated, but they all worked."

On the night of 27 February the war finally took what could be considered a "personal" turn; Air Force bombers dropped 5000-pound penetrator bombs on a command bunker northwest of Baghdad with the express purpose of catching and killing Saddam Hussein. The bombs worked as advertised, destroying the facility, but Hussein escaped. The following night *Ranger* launched its last combat strike of the war, led by VA-155's Lt. Marc Hitchcock and Lt.Cmdr. Steve "Boots" Barnes. The eight aircraft went 120 miles into Iraq and thumped the still retreating Iraqis.

The roads leading from Kuwait north to Iraq became killing grounds, as Allied airpower destroyed everything that moved. This scene is typical, with vehicles of all descriptions being present. Most, if not all, of the civilian cars and trucks were looted from Kuwait, and were being used by Iraqi Army units to escape from the rapidly advancing Allied Army. For most, it did not work. (U.S. Navy)

Storm's End

Within 24 hours, President Bush had declared an end to hostilities in Iraq. Kuwait was free, and Iraq was in no position to argue against the UN's cease-fire terms.

Central Command's post-war assessments determined Iraq's failed invasion of Kuwait cost them 36 fixed-wing aircraft and six helicopters lost in air-to-air engagements; 68 fixed and 13 rotary-wing aircraft destroyed on the ground; 137 aircraft effectively lost to Iran; 3700 tanks; 2400 armored vehicles; 2600 artillery pieces; and 19 ships sunk. A total of 42 Iraqi divisions ended the war in a combat-ineffective status, with total casualties estimated at "somewhere" between 25,000 and 100,000. The coalition forces eventually released over 70,000 POWs to the Saudi government. In return, United States' forces lost 124 servicemen killed in action and another 207 killed in accidents or by friendly fire; having provided the majority of the combat forces in the theater, the Americans suffered the majority of the losses.

The numbers racked up by the two Marine and nine Navy *Intruder* squadrons were staggering. VA-155, operating from *Ranger*, ended the war with 635 combat sorties for 1388 green ink hours, while dropping 2.9 million pounds of ordnance. The *Silver Foxes* claimed numerous Iraqi naval vessels, patrol boats, headquarters facilities, ammunition storage sites, AAA batteries, SAM sites, a thermal power plant, tanks, armored vehicles ... and one white Chevy Impala.

VA-145, also off *Ranger*, flew 498 combat sorties totaling 2150 combat hours, destroying 33 tanks, one bridge, 48 artillery pieces, 41 naval vessels, 20 ammunition storage bunkers, three chemical weapons storage facilities, and seven command and control sites. The squadron also expended 33 *HARMs* – the most by any *Intruder* unit[9] – and 11 AGM-123 *Skippers*. The following year the *Swordsmen* would claim both the Battle E and the McClusky Awards, the latter giving them bragging rights as the premier *Intruder* squadron in the fleet (Although open to argument from the other units involved).

[8] The "Highway of Death" probably got the attention of about every tactical aircraft in theater, all of which claimed their share of the carnage. The wreckage left on the road would soon be obvious through press photos of the area. One of the authors flew his last event of the war, in an EA-6B, in support of VA-36 against a similar target further north. During debrief the *Intruder's* B/N, LtCdr Bobby Goodman, of the Syrian prisoner event, remarked that while the radar picked up a lot of steel in the area, nothing was moving. We had little idea at the time how right he was.

[9] VA-75 was the other SWIP equipped *Intruder* unit in theater and shot HARM as well.

The *Panthers* return home to Oceana after *Desert Storm*, 27 March 1991. (Tom Kaminski)

VA-75, operating from *Kennedy*, recorded 280 sorties and 2150 hours, while *Saratoga's* VA-35 reported dropping 1.7 million pounds of bombs. VA-65, in *Roosevelt*, dumped 1.2 million pounds of ordnance while recording an impressive 97 percent sortie rate. "T-Tom's" Eagles of VA-115 contributed 723,000 pounds of ordnance, while Bernie Satterwhite's VA-185 – both flying from *Midway* – turned in a score of 457 missions, 940 combat flight hours, and 720,000 pounds of heavy metal on enemy targets. The two Marine squadrons, VMA (AW)s -224 and -533, operated a grand total of 20 aircraft while logging 795 combat sorties, all without a loss. The *Bengals* alone flew 422 missions and dispensed 2.3 million pounds of weaponry.

"Early on, when we did training operations with aircraft launching out of Saudi Arabia, it quickly became apparent that we had to get the carriers into the Persian Gulf, or it never would've worked otherwise. Tactically, the big lesson was you can never fight this war with last war's tactics. All the A-6s went in low at night the first couple of nights and never had to again. We were able to roll back their SAMs and C2, but we couldn't do much about the small arms and AAA," adds the irrepressible "Tugg" Thomson. "We had two regrets: we didn't get Saddam Hussein, and we didn't get Peter Arnett, either."

According to statistics in the DOD's authoritative *Gulf War Air Power Survey*, Navy *Intruders* flew 4824 combat sorties during *Desert Storm*, more than any type of tailhook aircraft. The 96 airframes involved averaged just over 50 missions apiece through the war, as opposed to 89 Navy *Hornets* that averaged fewer than 50 per jet. Not half bad for a "high maintenance" airframe in its fourth decade of combat service.

In 1998, when the Navy finally came out with its official history of the war, "Shield and Sword," it put *Intruders* on the cover.

The *Intruder* had shone once again; however, with *Desert Storm* the *Intruder* had flown its last hurrah. It would now take less than six years to go from hero to extinction.

CHAPTER ELEVEN

The End, 1991-1997

With the end of Operation *Desert Storm* and the successful liberation of Kuwait the Navy's *Intruder* squadrons resumed a "normal" peacetime deployment schedule. However, circumstances in the Middle East and elsewhere in the world conspired to keep anything from returning to normal, at least for the foreseeable future. On one hand, all of the U.S. military services continued to maintain a sizable presence in the Gulf region in order to keep an eye on Iraq while enforcing post-war diplomatic sanctions against the defeated nation. On the other hand, the Soviet Union continued to implode; that had a major, lasting impact on the U.S. defense establishment as a whole, and the *Intruder* community in particular.

By March 1991 most of the carriers involved in *Desert Storm* had headed home for s well-deserved heroes' welcome. *Theodore Roosevelt* and CVW-8 delayed a few months to cover the re-established Sixth Fleet commitment and take part in *Operation Provide Comfort*. Launching from off Incirlik, Turkey, the wing's aircraft flew, with Air Force tanker support, across the country to enter Iraq from the north, keeping a protective umbrella over the Kurds and a wary eye on the Iraqis. They would remain in the Med until relieved by *Forrestal*, with CVW-6 and William Ballard's VA-176, in May.

Nimitz, with CVW-9 and "Too Tall" Indorf's VA-165 embarked, became the duty carrier in the Persian Gulf in March, replacing *Ranger*. Too late for full-blown combat, the nuke carrier would establish the routine that would soon lead to *Operation Southern Watch*, which entailed regular patrols, and an occasional skirmish, over Iraq.[1] The *Boomers* returned to Whidbey in August, with Indorf turning over squadron command to Bob "Tails" Taylor.

Elsewhere in the world things were changing in a big way. The end of the "Evil Empire" in 1991, the new Clinton administration elected in 1992, and a general feeling in the country that there may be no enemies left led to a massive reduction of U.S. military power through the 1990s. For the Navy it meant the loss of several carriers, two airwings, and the inevitable end of the Attack Community. Part of this course was laid by the lack of a real replacement for the *Intruder*.

Avenger's Demise

While the Soviet Union imploded, events occurred back in the states that had a direct and highly negative impact on the Navy's medium attack community. On 7 January 1991 – before the air war against Iraq even kicked off – Secretary of Defense Dick Cheney determined Naval Air Systems Command, General Dynamics, and McDonnell-Douglas had sold him a bill of goods on the health of the A-12. Following day-long meetings in Washington, D.C. he canceled the program.

On 26 April 1990 the Secretary had informed Congress that the A-12 program was on time, on track, and at its expected cost levels. He stated the first flight of the unique stealth attack plane would take place no later than March 1991. However, the manufacturers did not share Cheney's optimism, or at the very least were having problems conveying the program's serious difficulties. Despite being something of a "black" program, several observers felt the aircraft was grossly overweight – to the tune of 8000 pounds – would not meet contracted performance requirements, was having problems with its F412 engines, and also suffered difficulties with its avionics and computer systems. Whether the program managers at Naval Air Systems Command covered up the bad information or just refused to believe it is still a subject of conjecture.

Finally, on 1 June 1990 the two prime contractors delivered the bad news: the plane was going to cost a lot more than expected, exceeding the contract's agreed-upon price; the A-12 would not be able to meet all of the performance specifications; and the first flight of the prototype would not take place until September. In response Secretary of the Navy H. Lawrence Garrett III ordered a full investigation of the troubled program, which ultimately determined the Navy and the contractors had withheld much of the negative information about the plane. On 10 December Garrett announced the impending retirement of Vice Adm. Richard C. Gentz, Commander, Naval Air Systems Command; several other officers directly involved in the oversight of the A-12 program also retired, as did John Betti, the Undersecretary of Defense for Acquisition and DoD's point man for the A-12 program, resigned. Dick Cheney then gave SecNav Garrett until 4 January 1991 to "show cause" why the program should continue.

The Navy and the contractor team failed to make their case. Cheney himself said:

> "No one can tell me exactly how much more it will cost to keep this program going."

Some estimates put the plane at $100 million per copy upon delivery – a substantial increase over 1989's estimated unit cost of $87 million – while defense critics put the price at more like $150 million per plane.

In any event, the program was untenable and died a nasty death on 7 January. The following day 3500 engineers and other aerospace workers went out the door at General Dynamics-Fort Worth, with similar

[1] At the time, of course, no one could've imagined that "OSW", or its *Northern Watch* cousin, would last until the demise of the Hussein regime during *Operation Iraqi Freedom*, in 2003.

numbers abruptly departing McAir's plant at Lambert Field in St. Louis. It was the largest program cancellation in the history of the Pentagon.

The Navy scrambled to come up with an adequate replacement for the *Intruder*. The first proposal – designated "AX" for Advanced Attack Aircraft, and with an intent of utilizing much of the A-12's technology – was effectively stillborn, despite the efforts of partners McDonnell Douglas/LTV, General Dynamics Fort Worth/McDonnell, and Douglas/Northrop. Startlingly, during this period General Dynamics bailed out of the military aircraft industry completely, selling its Fort Worth operation to Lockheed. One month ago the "Only one AX Team can do this" ads had General Dynamics listed as team leader; the following month the ads listed Lockheed Fort Worth.

Other proposals, including the A/F-X and the JAF – Joint Attack Fighter – received some interest and funding, but didn't go anywhere. The Navy later participated in the Joint Attack Strike Technology (JAST) program that pushed systems commonality between the Navy, Air Force, and Marine Corps. This program also did not pan out beyond technology demonstrations, but three years afterwards the services seriously pushed its descendent, the Joint Strike Fighter (JSF). In the end the Navy went for a larger, enhanced variant of the *Hornet* designated FA-18E/F. Cheney approved, and the Navy moved forward, but in the interim the service "gapped" the medium attack mission, making it official: there would be no formal replacement for the A-6.

Within the *Intruder* community the reaction was typically shock and disbelief. During an all-officer brief at Whidbey in late 1992 a Vice Admiral from D.C. delivered a blunt message to the *Intruder* crowd: "The *Hornet* is the future of Naval Air. Either get on board or get out."

For the first time in most everyone's memory, the Navy voluntarily relinquished one of the primary missions of carrier aviation. Squadrons were already falling by the boards; VA-55 had shut down prior to *Desert Storm* on 1 January 1991 with the disestablishment of CVW-13, and the short-lived *Nighthawks* of VA-185 followed on 30 August, leaving the *Eagles* of VA-115 as the sole *Intruder* operator with CVW-5 in *Midway*. In the words of VA-155's *Desert Storm* XO, Cmdr. Larry Munns, it quickly became apparent to everyone:

> "... they weren't decommissioning aircraft to maintain air wing size. They'd decided to kill the entire community."

Crunch Time

The sword fell mightily during 1992. Over the course of 33 years – dating to the first YA2F-1, BuNo 147864 – the Iron Works had turned out over 700 *Intruders*. On 31 January Grumman rolled out the last-ever production A-6E, BuNo 164385, and started to secure the Calverton, NY, production line. Unfortunately, the last *Intruder* did not survive long in the fleet. On 1 April 1992 Grumman formally delivered the aircraft to VA-145 in a ceremony at Whidbey Island. About 18 months later, while on deployment with VA-95 in the Persian Gulf, BuNo 164385 suffered a midair that sent it and sister BuNo 161682 to the bottom of the ocean. Fortunately, all four *Lizards* successfully ejected and were recovered with minimal injuries. They would be the last of 16 *Intruders* lost in mid-air collisions.

The A-12 *Avenger II*, or "The Flying Dorito." At one time the presumed future of Medium Attack, General Dynamic's A-12 was shrouded in secrecy for most of its time, as well as the subject of a great deal of low voiced speculation at both Oceana and Whidbey. When the design was unveiled at the 1990 Tailhook Reunion reactions ranged from enthusiasm to disbelief. In the end, cost overruns would doom the aircraft, and it was canceled by then-Secretary of Defense Dick Cheney in January 1991. (via *Naval Aviation News*)

The last *Intruder's* demise mirrored Grumman's prospects. While the company's E-2C *Hawkeye* still filtered out of the factory in small numbers, the F-14 community was going through its own contractions, while the FA-18 assumed primacy in carrier aviation. Within a couple of years the Navy would cut back its dedicated fighter community, with each air wing reducing to a single *Tomcat* squadron and consolidating the survivors at NAS Oceana. The VF squadrons adopted a truly multi-mission capability, performing fleet air defense, tactical reconnaissance, and the occasional precision strike missions with skill and aplomb, but their days were numbered too, and everyone knew it. Grumman itself – the proudest and most famous name in naval aviation history, and the home of the *Wildcat*, *Hellcat*, *Bearcat*, *Tomcat*, *Guardian*, *Albatross*, *Panther*, *Cougar*, *Tiger*, *Tracker*, *Hawkeye*, *Prowler*, and *Intruder*, among others – disappeared. On 18 May 1994 the Iron Works formally became a subsidiary of the new Northrop Grumman Corporation; later that year Northrop Grumman acquired much of the old Vought Corporation, another long-time carrier aircraft manufacturer.

As for Grumman's *Intruder*, the end came with startling swiftness, as squadron after squadron disappeared into naval aviation's history books. On 1 October 1992 the last *Thunderbolt One*, Cmdr. Lee Hawks, presided over the disestablishment of VA-176 in a ceremony at

The A-6 and Foreign Sales

Despite its success with the Navy and Marine Corps, the *Intruder* was never sold to a foreign nation. Within the A-6 community the story has been told frequently that the capabilities of the airframe were considered "too good" to sell overseas, that its nuclear capability in particular kept it on a reported "do not export" list.

In fact, the United States sold more than a few nuke capable aircraft overseas – the F-4, A-4, and even A-1 were all capable of carrying special weapons. As for the A-6 having a bomb system being too accurate for export, the F-111's sale to Australia and offers to the RAF make that claim suspect. In all probability, it was likely the A-6's homely looks and sub-sonic performance that had more to with lack of sales. Hot, pointed-nose, "go-fast" aircraft that looked good at commercial airshows probably had more to do with the lack of overseas sales than any conscious decision not to sell it. Through the '60s and '70s supersonic aircraft are what sold, including F-5s, F-4s, F-16s, F-18s, and F-15s. On the low end market, former Navy A-4s and A-7s were rebuilt and delivered to other nations. Usually overlooked is that the A-10 *Thunderbolt II* (aka, "Wart Hog") has also never been sold overseas – once again, a slow, visually oppressed aircraft that wasn't as sexy as other aircraft available to small nations on limited budgets. Nonetheless, it was reported in 1998 that France looked at resurrecting mothballed A-6Es for potential use as tankers off their new nuclear powered carrier *Charles de Gaulle*. In the end it was not to be.

VA-185 was the shortest-lived A-6 unit, flying less than five years, yet they still compiled a remarkable record. Here a diamond of *Nighthawks* turn north over Saratoga Passage near their Whidbey home prior to their September 1987 move to Japan. (Rick Morgan)

NAS Oceana. The squadron had quite a background, dating to 1955 when it stood up as a *Spad* outfit flying from *Randolph* (CVA-15) with Air Task Group 202. Thirty-seven years, two wars, two major international crises, and 26 deployments in nine carriers later – and, lest we forget, one MiG kill in Vietnam – the *'Bolts* went out with a bang, making a final deployment in *Forrestal* with participation in Operations *Provide Comfort* – the U.S.' support program for Kurdish refugees in Northern Iraq – and *Display Determination*. While always something of the quiet man of AirLant medium attack, VA-176 still rang up an impressive record, including an unheard of three Battle Es between 1988 and 1990. Now the *Thunderbolts* were gone, although their name and emblem lived an additional two years through VA-42, which had lost the *Green Pawns* title.

According to the recently retired Lt. Denny Franklin – who returned to VA-176 following his commissioning and a tour with HC-8 at NAS Norfolk – the partying went into the night, with the old hands occasionally scaring the junior squadron members with remembered names and stories.

"We were kind of the bastard stepchild of the wing," he comments:

> "We were 'F Troop,' every day. But the best thing about -176 was the camaraderie. Tim Beard was a real supporter of the enlisted guys, and the chiefs liked him; Mike Currie was a little more stern, but I got along with him good. JB (Dadson) did his tour in -176, and we won the Battle E Award. When I was the Maintenance Materiel Control Officer we won it three times in a row; nobody had ever done that before.
>
> "I fingered Beard on the flight deck once," he continued, laughing. "We had permission to keep the plane running after it came back because the CSD had died. The flight deck LPO told him to shut it down; I'd already given him signals to keep it turning. He opened the canopy, took his helmet off, and threw it at me. We laugh about it now ... we'd had something like 275 consecutive flights without missing a hop, this blew it. But we never lost an airplane. The second time around I took the guys who'd gone out before, went on cruise, and brought them all back and won the Battle E, proved we weren't F Troop. We had something to prove. We recruited maybe 100 guys who were with us before and brought them all back in there and fine-tuned everything, put everyone in the right spot, where they wanted to be, and not where the squadron wanted them, and it worked. I had a rule: as soon as a chief showed up he had a coffee cup and a flight jacket with a squadron patch within a couple of hours. We were the only ones who did that. It made them feel good. Worked them like a yard dog, but I told them I would.
>
> "We went to the ceremony and they did a real neat thing. We went to the O'Club, and they opened up bottles of tequila and passed them around to the old timers. There was a lot of bitterness; we'd won so many Battle Es and couldn't understand, we finally had done something and they did this to us first."

The squadron's demise was a component of the mass retirement of *Forrestal* and her wing, CVW-6. In the post-*Desert Storm* world the Navy had too many carriers and too many air wings, and – as the saying goes – sacrifices had to be made. The wing dated to its establishment on 1 January 1943 as Carrier Air Group 17. It saw combat in *Hornet* before going through several redesignations, becoming CVG-6 on 27 July 1948 and CVW-6 on 20 December 1963. CVW-6 disestablished on 1 April 1992; along with VA-176, the wing's VFA-132 *Privateers*, VAQ-133 *Wizards*, and VS-28 *Gamblers* also went into oblivion. The *Steeljaws* of VAW-122 assumed a counter-narcotics mission and managed to survive until 31 March 1996.

As for the *Forrestal* – the Navy's first "supercarrier" – she decommissioned 11 September 1993 at the Philadelphia Naval Shipyard, three

VA-176 was never assigned to any air wing other than CVW-6. They would leave the Navy's rolls, along with that organization, in April 1992. KA-6D AE514 is at Oceana in November 1988. (David Brown)

weeks short of her 38th birthday. The Navy killed a plan to make her the dedicated training carrier at NAS Pensacola; despite redesignating as AVT-59 *FID* instead retired, following 21 operational deployments, including a first cruise that put her right in the middle of the 1956 Suez Crisis and one brief trip to Vietnam.

The next carrier to retire was *Midway*, and as a direct result the *Eagles* of VA-115 had a particularly busy year. On 17 April 1991 *Midway* and CVW-5 returned to Yokosuka flush from their exceptional combat performance in *Desert Storm* and the celebrations began. On 1 July Cmdr. Terry Toms departed and Cmdr. James D. Kelly fleeted up to the skipper's spot. Two months later Cmdr. Bernie Satterwhite read the orders disestablishing VA-185, and VA-115 immediately became the single largest A-6 squadron in the Navy with 16 aircraft – including several SWIP birds, delivered from Whidbey by VA-196 – 42 officers, and over 240 enlisted.

As if that didn't make things interesting enough, in mid-August *Midway* pulled out for Pearl Harbor and a SwapEx with the inbound *Independence*; *Indy* was moving to Japan, while *Midway* was heading for the barn and a well deserved retirement. The aircraft and personnel from CVW-5 completed their move to the new deck by 28 August and turned back towards Yokosuka, while the proud and historic – if somewhat elderly and occasionally unstable – *Midway* continued to San Diego for decommissioning. *Midway's* final turn in the spotlight came in ceremonies at NAS North Island on 11 April 1992, ending 47 years of outstanding service to the nation. COMNAVAIRPAC, Vice Adm. Edward R. Kohn, Jr., presided at the ceremonies with *Midway* skipper Capt. Larry L. Ernst and XO Cmdr. John Schork – a former VA-95 commanding officer – assisting. Afterwards the carrier went to the reserve fleet at Bremerton, ending four-plus decades of service to the nation.

The *Eagles* continued to train with their new deck in and around showing the ropes to the wing's newest squadrons, VF-154 and VF-21, from Miramar, which gave CVW-5 its first dedicated fighter capability since the conversions of VFs -151 and -161 to *Hornets* in 1986. Otherwise, the VA-115 crews found time to participate in exercises, such as *Valiant Blitz* with Korea and *AnnualEx* with the Japanese Maritime Self Defense Force.

Meanwhile, *Kitty Hawk* finished her SLEP at the Philadelphia Naval Shipyard and headed back to the barn at Naval Station San Diego via the Horn. The two-month mini-deployment found Capt. F.G. "Wigs" Ludwig's CVW-15 hard at work; for VA-52 it marked the first time at sea in some 14 months, and the squadron made the most of it. The *Knightriders* managed to get in some true quality time flying low-levels in exotic locales like Venezuela and Chile, while hassling with the local air forces, including F-5s and *Mirages* of the *Fuerza Aérea de Chile*. Everyone had fun while the carrier passed Argentina; the *Gringo Gaucho III* exercise saw *Commando de Aviación Naval Super Etendards* and S-2s come out for a round of touch and goes. After operating off of *Veinticinco de Mayo's* small, 48-year-old deck, the Argentine Navy pilots probably had a fun time working off *Kitty Hawk's* roomy deck. The *Knightriders* did their part, assisting with the training while sharpening their own skills in preparation for an upcoming deployment.

Perhaps VA-95 "enjoyed" the most auspicious cruise of the post-*Desert Storm* year while operating with CVW-11 in *Abraham Lincoln*. Led by Cmdrs. Randy "Dirt" Dearth and Graham "Buck" Gordon, the *Lizards* made the standard tour of WestPac through the Indian Ocean, and on into the Persian Gulf. However, things got hairy on 9 July.

For years crews had referred to the EA-6B as the "*Intruder* Station Wagon"; on that date, Lts Mark "Master" Baden and Keith "Yogi" Gallagher got to try out an "*Intruder* Convertible," in an incident not unlike that of VA-42's Jim Brooks in July 1963. The two young crewmen were in the middle of a tanking hop when Gallagher's ejection seat suddenly fired, albeit partially.

About halfway through the flight they had noticed a fuel transfer problem with one of the drop tanks, and the override on the tank pressurization didn't make any difference, so they reviewed their NATOPS check lists and agreed to try a series of positive and negative Gs. Baden had just pushed the stick forward on the first sequence when – BOOM! – the cockpit depressurized. Glancing right, he saw

Following the end of *Desert Storm*, *Intruder* squadrons still went into the Persian Gulf and flew missions over Iraq as part of *Operation Southern Watch*. Onboard *Independence*, a pilot from VA-115 waits to start. (U.S. Navy, PH2 Deloach)

Gallagher's legs; he looked up and his BN was halfway out of the aircraft, mask and helmet gone, and head wildly whipping around. Gallagher's parachute deployed, dragging the B/N halfway through the shattered canopy and into the slipstream. Baden immediately slowed the plan and declared an emergency, all the while fighting control problems caused by the chute's entanglement in the plane's aft control surfaces. The Air Boss on *Lincoln* immediately cleared the deck and directed the pilot to come on in. Remarkably, Baden got the bird on deck within six minutes, and the deck crew was able to safely remove the unconscious Gallagher. One observer on *Kitty Hawk's* deck later recalled:

> "The amazed excitement this created for the flight deck crash and salvage crew was unbelievable! If I recall, this happened because of a fuel transfer problem. For reasons of medical fatigue (I think!) the clamp, or whatever, at the bottom of the BN's seat broke or came loose, and the seat traveled up the rail enough to partially fire. The parachute deployed and ... draped itself around the horizontal stab. The pilot did a tremendous job at maintaining a low enough airspeed to maintain flight and prevent the BN from being fully thrown into the canopy. We all marveled that the BN survived with relatively minor injuries!"

Afterwards AirPac named Baden its "Pro of the Week," and on 21 November *ComCarGru Three*, Rear Adm. Timothy W. Wright, presented Baden with the Air Medal. As for Gallagher, he got banged up quite a bit and had to wear a sling for a while, but survived and eventually returned to flight status. The unconventional recovery marked his 100th trap in *Lincoln*, and it also occurred on his 26th birthday; hell of a way to celebrate.

Adios *Tigers*

The year 1993 was a particularly tough year for the *Intruder* community – and concurrently, the fighter community – as several squadrons dropped from the rolls of attack aviation. Sweeping changes in the AirPac side of the house resulted in the 26 February 1993 disestablishment of Medium Attack/Tactical Electronic Warfare Wing, Pacific (MATVAQWINGPAC) – the host outfit at NAS Whidbey Island – and creation of Attack Wing, Pacific and Tactical Electronic Warfare Wing, Pacific. Capt. Baker R. "Whiskey Bob" Hamilton, a former CO of VAQ-142, assumed command of VAQWINGPAC, while Capt. Bernis H. "Butch" Bailey, a former skipper of VA-34 at Oceana, took ATKWINGPAC. Medium Attack Wing One (MATWINGONE) at NAS Oceana went through its own redesignation on 1 September 1993 and became Attack Wing Atlantic Fleet. (ATKWINGLANT).

By that date NAS Oceana was short another A-6 squadron. On 31 March 1993 the veteran *Tigers* of VA-65 formally stepped down, ending 48 years of exemplary service as a torpedo and attack squadron. Late of CVW-13 and *Coral Sea*, the squadron had transferred to CVW-8 in time for *Desert Storm*, and during its last deployment returned to the skies of Iraq for Operation *Provide Comfort*.

In 1992 the squadron was surprisingly removed from the "Teddy Roosevelt" in a massive and experimental reorganization of CVW-8. The *Tigers*, and stablemate VF-41, were beached to make space for a composite Marine helicopter unit (HMH-362) that joined the wing, along with VMFA-312 and its *Hornets*. As part of a "Special Marine Air Ground Task Force" the ship also carried a company of Marine infantry onboard to determine if they could partially replace a fully dedicated amphibious group. Although the subsequent Med cruise was said to go well, the general opinion was that rotary and fixed wing aircraft do not mix well, and the marriage created more operational problems than they solved.

It was the end of the line for the *Tigers*, however. The squadron had been a fixture at NAS Oceana for 42 years, notable for its pre-Tactical Paint Scheme (TPS) orange aircraft markings, bright orange flight suits, and "Cupcake" callsign. VA-65 had made the highly successful third *Intruder* combat cruise to Vietnam under Bob Mandeville, and over the years had turned out several other naval aviation notables, such as Tom Shanahan, Paul Hollandsworth, George Strohsahl, Herb Brown, Joe Prueher, Bob House, and Bill "Fox" Fallon. Yet the final *Tiger One*, Cmdr. James K. Stark, Jr., brought it all to a close in ceremonies held 26 March 1993.

Semper Fi

Only Three Marine *Intruder* squadrons remained at the end of *Desert Storm*. It would only take a matter of months until there were none.

The *Hawks* came home from the desert to Cherry Point, where they spent their last year as an *Intruder* outfit. The squadron was redesignated VMFA(AW)-533 on 1 Oct. 1992, and to show that Marines do have a sense of humor, towards the end the men of VMA(AW)-533 had held a Kangaroo Court. During the proceedings the participants determined Commandant of the Marine Corps Gen. Carl Mundy had "killed the A-6 without providing for an adequate replacement" and fined him $4.00. Someone sent the fine chit to the CMC, who responded by signing it and sending a check for the prescribed amount. The framed chit and check are still on display in the *Hawk's* ready room.

VMA(AW)-224, the *Bengals*, had returned from Bahrain at the end of March 1991. In a great case of, "*yeah, but what have you done for us lately?*," the squadron was sent for one final UDP pump to Iwakuni only six months later, to return in March 1992. Their relief in Japan was

The *Tigers* of VA-65 established one of the more distinguished records in *Intruder* history. The fact they were beached for a Marine helo squadron made things all the more bizarre. AJ511 drops down from low holding during *Desert Storm*. (Rick Morgan)

The Marines relinquished their last *Intruders* in April 1993. The honor fell to the *Moonlighters* of VMA (AW)-332, who painted up their last aircraft in full tail markings for the occasion. (David Brown)

VMFA(AW)-121 with its FA-18D *Hornets*, making 224 both the first and last *Intruder* unit to participate in the Unit Deployment Program to 1st MAW. Finally, on 6 March 1993, they were redesignated VMFA(AW)-533 and became an FA-18D squadron.

To VMA(AW)-332 would go the honor of retiring the last Marine *Intruder*. On 20 April 1993 at Cherry Point BuNo 161681 – suitably marked EA500, with a startling full-color *Moonlighters* tail – flew the last official USMC *Intruder* hop, ending 29 years of A-6 operations in the Corps. On 16 June the squadron became an all weather fighter attack squadron, leaving the Marine Attack mission to the remaining *Harrier* units.

As an aside, one other proud fixture of medium attack aviation – as well as naval aviation as a whole – closed in late 1992: the famous "Crossroads of WestPac," NAS Cubi Point. When Mount Pinatubo erupted in June 1991 the Air Force quickly abandoned Clark Air Force Base, north of Manila, but by early 1992 it was obvious that all U.S. military operations on the island were coming to an end. The citizens of the Philippines demanded the removal of all American facilities and personnel, and the U.S. agreed. Hence, all military facilities on Luzon started shutting down, with some units shifting to other locations and others disestablished. The last ship to depart Naval Station Subic Bay was *Belleau Wood* (LHA-3), which pulled away from the pier 24 November 1992; the "last guy down the hill" when NAS Cubi Point closed was its final commanding officer, the colorful Capt. Bruce Wood.

NAS Cubi Point – and its famous officers' club – closed in October 1992. There would be no more Liberty in 'Po City, no more banca boats, no more cats and traps on the club's infamous "simulator," and no more Cubi Specials with a San Miguel chaser. The famous – including several admirals and such luminaries as Undersecretary of the Navy and future Senator John Warner – the infamous and even the unknown and occasional Air Force puke had ridden the catapult and added their names to the Wall of Fame. Now it was gone, but fortunately, as a result

A division of *Swordsmen* descend over the Strait of San Juan de Fuca on 4 August 1988. (Rick Morgan)

of the efforts of former VA-52 and -128 skipper "Boxman" Wood and several others, the majority of the Cubi O'Club made it to the states, where it is now preserved at the National Museum of Naval Aviation as the Cubi Bar Café.

Still, for everyone who had ever rotated through Cubi, either as part of the station complement or enroute to some other exotic place, it was truly the end of an era.

Swordsmen, Foxes ...

Two other squadrons disestablished in 1993: CVW-2's medium attack component of VAs -145 and -155. Again, the loss was painful, but at least both squadrons had a rip-snortin' last tour. Prior to their final trip the *Swordsmen* and *Silver Foxes* went through the standard workups, and in January 1992, while operating at NAS Fallon, VA-155 skipper Larry Munns and B/N Lt. Greg "Davey" Crockett notched the first-ever AGM-65 infrared *Maverick* shot by an operational squadron. They followed up with a combined -145/-155 all *Maverick* strike led by *Swordsman* skipper Cmdr. Rick Cassara and -155's Lt.Cmdr. Matt "Buddy" Storrs. The squadrons deployed with CVW-2 in *Ranger* on 1 August 1992 for what was planned as a standard five-month WestPac/IO deployment. The cruise quickly turned hectic; while inport in Pusan *Ranger* was yanked out and dispatched with all due haste to the Persian Gulf for Operation *Southern Watch*, the enforcement of a no-fly zone over southern Iraq. The ship relieved *Independence*, and the wing's squadrons ended up spending two and a half months on the line flying armed reconnaissance missions over Iraq and the Gulf. Then the situation deteriorated in Somalia, and *Ranger* headed south for Operation *Restore Hope*, rendering aid to U.S. units on the ground in that strife-torn country.

Led by Cmdrs. Munns and Jerry Nichelson, VA-155 made the most of the time left prior to their impending demise. The squadron managed to intersperse its operational requirements with regular low levels in all those exotic places that you get to hear about, particularly Kuwait, where the landscape was still heavily littered with ready-made targets formerly owned by the Iraqi Army. Notably, on 7 December 1992 Munns achieved a remarkable record: on the squadron's first mission in support of *Provide Comfort* he notched his 5,000th flight hour in the noble *Intruder*, the only pilot to ever do so. The squadrons returned home on 30 January 1993 with CVW-2's three Whidbey-based outfits – VA-145, VA-155, and VAQ-131 – making a spectacular homecoming by putting 27 Ironworks products over Ault Field.

VA-155 departed first, disestablishing on 30 April 1993. On 15 March ComNavAirPac announced the *Silver Foxes* had won the Battle E for 1992, while also garnering a Navy Unit Commendation. Their last four aircraft went to VA-128 on 15 April ... and with that, the *Foxes* were gone.

The *Swordsmen* hung around a while longer, even outlasting *Ranger*, which decommissioned on 10 July 1993 in ceremonies at North

The *Silver Foxes* ended almost six years of service in April 1993. Although assigned to three different air wings, the squadron only deployed overseas with CVW-2. NE402 is shown being shot from *Ranger's* cat number three while a VF-2 *Tomcat* waits its turn. (U.S. Navy via Steve Barnes)

Island. Her last ever arrested landing – number 330,683 – came on 11 March, courtesy of an F-14 from VF-124, the AirPac *Tomcat* RAG. Four months later *Ranger's* last skipper, Capt. Dennis McGinn, sent the 35-year-old carrier into retirement with 21 WestPac and nine combat deployments to her credit. For VA-145 it also marked the end of a 35-year association dating to May 1958, when the *Swordsmen* first put their *Spads* onboard the new CVA-61 for her delivery voyage around Cape Horn. The two saw combat together in Vietnam – both with A-1s and with A-6As – and also served together in *Desert Storm*. In the end VA-145 also went out swinging, making the around the Horn transfer of *Independence* to the Pacific Fleet from May to July 1993 along with other squadrons of CVW-2; on 4 July skipper Cmdr. David A. "Roy" Rogers recorded the 1000th arrested landing of his career. During the course of their final year the *Swordsmen* also learned they had effectively swept the board in AirPac attack, earning the 1991 McClusky Award, Arleigh Burke Fleet Trophy, Safety "S," and their sixth MATVAQWINGPAC annual bombing competition in a row.

On the down side, there was a sense of betrayal during – or at the least anger – over the squadron's fate. As late as the two-month transit in *Independence* the Navy had told VA-145 it would survive one more year in order to make the next cruise in Ranger with CVW-2. Instead, they got the word that due to "budgetary reasons" they would quickly join the roll of deceased units. Accordingly, VA-145 came up with a last patch to "commemorate" its impending demise: it featured aces and eights, the infamous "dead man's hand" held by Wild Bill Hickock the night he died in Deadwood. According to skipper "Roy" Rogers, the Navy had shot the *Swordsmen* in the back.

Cmdr. Rogers, the 37th and last commanding officer, presided over the *Swordsmen's* "last swash," held on 1 October 1993 at NAS Whidbey Island. From the date of its call to active duty on 20 July 1950 as VA-702 through the end, VA-145 made 28 deployments in such famous carriers as *Intrepid*, *Hornet*, *Randolph*, *Boxer*, *Kearsarge*, *Enterprise*, and of course *Ranger*. From their beginning in AD-2s to their end with the latest, most capable versions of the A-6E, the squadron lived and fought only one mission: *attack*.

Black Falcons, Roadrunners ...

When the budget axe resumed slashing in 1994, much of the emphasis fell on NAS Oceana. Three squadrons at NTU disestablished during the year, starting with one of the youngest outfits, VA-36.

Established only seven years previously as part of the Navy's two *Intruder* squadron air wing plans, the *Roadrunners* managed four deployments in *Theodore Roosevelt*, including combat in *Desert Storm*. The last cruise started on 11 March 1993 and turned out to be a busy one, with operations over Bosnia-Herzegovnia, followed by a run through the Suez Canal for participation in *Southern Watch*. *Roosevelt*

The *Roadrunners* of VA-36 expired on April Fool's Day 1994 after only seven years of service. Tanker configured AJ540 is shown coming aboard TR during *Desert Storm* on 28 January 1991. The *Buckeyes* of VA-85 lasted until September of the same year. AB500 is seen at Oceana in April 1994. (Rick Morgan and David Brown)

Having adopted the nickname and insignia of VA-176 in 1992, the former *Green Pawns* of VA-42 were disestablished in September 1994, leaving all *Intruder* RAG and FRAMP functions to VA-128 in Whidbey. A *Thunderbolt* KA-6D is seen on the Oceana ramp in April 1994. (Brian Rogers)

and CVW-8 returned to the states on 8 September, and VA-36 immediately started shutting down; as was commonplace, some of the personnel departed for other communities, while a few took retirement. The squadron's last skipper, B/N Cmdr. Mark Himler, read the orders on 1 April 1994, and that was that for the second squadron to bear the number VA-36.

Next on the list was VA-85, one of the long-time attack operators, and the second *Intruder* squadron to make a combat deployment to Vietnam. In February 1992 the squadron had been one of the first *Desert Storm* veterans to return to the region following an accelerated workup schedule with *America* and CVW-1; fortunately the wing, led by career *Intruder* pilot Capt. Paul D. Cash, was up to the job. Cash scored his own personal record early in the deployment when he joined the Grand Club on 21 March 1992; naturally, he made his 1000th trap in a *Black Falcon* A-6E.

The squadron was back at it the following year, still with CVW-1 in *America* under the leadership of Cmdrs. Bruce Weber and John Scheffler. Sometimes workups can be fun; VA-85's Doug Franks and Lt.Cmdr. Rick Keller demonstrated this fact during *Ocean Venture '93* when they dumped an IR *Maverick* into the hulk of the former *Rushmore* (LSD-14). The old amphib, a 1944 product of Newport News Shipbuilding, took additional hits from a *HARM* and two *Harpoons* before going down.

America pulled out on 11 August 1993, enroute to the Med with side trips to the Red Sea and Indian Ocean. Notably, CVW-1 deployed with only one fighter squadron, the VF-102 *Diamondbacks*. The other air wing "fighting" squadron VF-33 had fallen prey to budget cuts earlier in the year. It was a strong indication of the way things were going in carrier aviation, although the wing still deployed with only two strike/fighter squadrons, VFAs -82 and -86. After relieving CVW-8 in *Roosevelt* CVW-1 assumed the duty with Operations *Deny Flight* and *Provide Promise* – both involving flight over Bosnia – and participation in NATO's *Sharp Guard*. The wing's crews also managed to fit in *Dynamic Guard '93* with the air forces of Turkey, Germany, Italy, and France, as well as aircraft from HMS *Invincible* (R-05) and the French Navy's *Clemenceau* (R-98). The skies filled with *Intruders* mixing it up with combinations such as Armee de l'Aire *Mirage* 2000s and Luftwaffe *Tornados*.

Roughly halfway through the cruise XO John Scheffler fleeted up, relieving Bruce Weber on 29 September. The ceremony took place while *America* was at anchor off Corfu, Greece. The *Falcons* continued to provide the sharp end, participating in exercises with skill and aplomb, at times designating targets for the wing's *Hornets* and the odd *Sea Harrier*, while on other occasions flying war loads over Iraq. Ah, but all good things come to an end, and the squadron's last deployment – and prospects – ended with its return to Oceana on 5 February.

As can be expected, VA-85 went out in style, starting with the flyoff. The squadron really whooped it up later in the year when it took the 1994 MatWingLant *Intruder* Derby. During the course of the competition Lt. Rob Bassett and Lt.Cmdr. Steve Togliatti notched the best single day strike, XO Cmdr. Mike Steinmetz and Lt. Joe Gardiner recorded the best single strike, Lts. Bruce Shuttleworth and Carl Grooms won the best SWIP WST (simulator) strike, and the intelligence team of Ens. Jesus Romero and IS2 Brad Loschen was named best in the wing. At the annual *Intruder* Ball Attack Wing commodore Bernie Satterwhite had the pleasure of solidifying the *Black Falcon's* standing as the top attack squadron in the Atlantic Fleet. A few weeks later, on 30 September 1994, VA-85 retired at the top of its form with 43 years of proud heritage, a bag full of awards, and a strong combat record in Vietnam, Libya, and Iraq.

Rick Keller – who later served as VA-165's last XO – remembers the ending, while on a high note, was very bittersweet:

> "We went on deployment in August, had a change of command in September ... and I think the new skipper knew we were going away. There were rumors, but he didn't say one way or another until Christmas. I think not knowing hurt us.
>
> "There were two things different between the 1993 shutdown of the community and the final closeout in 1996: in 1993 the reservists were getting IRAD'ed (Involuntary Release from Active Duty), and only a few senior B/Ns were getting selected from transition to other communities. That's one of the reasons I jumped at a CAG ops job. In 1996 the situation improved, with everyone getting selected. We screened two out of four department heads (for command), probably as good as we could do."

Thunderbolts (again), *Falcons*, *Firebirds* ...

VA-85's departure further emptied an already sparse ramp at Oceana. On 30 September 1994 VA-42 – for 31 years the AirLant A-6 RAG – disestablished, leaving Whidbey's VA-128 as the last squadron where a medium attack crew or maintainer could train. The disestablishment of the *Thunderbolts* – historically the *Green Pawns* – left the east coast attack wing with only three squadrons: VAs -34, -35, and -75.

During this period the naval air reserve went through their own upheaval with a similar outcome, i.e., no more medium attack.

The reorganization saw the disestablishment of CVWR-30 at NAS Miramar, along with all of its squadrons, with the exception of VFC-13. At the other end of the country CVRW-20 realigned with one F-14 squadron – VF-201 – while retaining its two *Hornet* squadrons, VFAs -203 and -204. The results were the same for VAs -205 and -304; there was no place for them in the "new" Navy, and they went away.

Atlanta's *Green Falcons* would be the last reserve *Intruder* unit. AF507 is shown at NAS Fallon, NV, in May 1992. (Michael Grove)

Remarkably, VA-304 at NAS Alameda had started receiving SWIP *Intruders* in April 1993 and recorded the first reserve A-6 *HARM* shot on 15 July, with Lt.Cmdrs. Luke Ridenhouse and Greg Upright doing the honors. Within a few months, however, the *Golden Hawks* ferried its A-6s to the Aerospace Maintenance And Regeneration Center at Davis-Monthan AFB. Sadly, the *Hawks'* brief career with *Intruders* came to a tragic conclusion on 4 May 1994 when Lt.Cmdrs. Randall McNally II and Brian McMahon crashed into San Francisco Bay. According to witnesses in the tower *Firebird 507* was about a half mile south of NAS Alameda when it suddenly rolled and hit the water, killing both men. The Navy was only able to recover about half of the wreckage.

During its last year of existence the Falcons of VA-205 assumed an electronic attack role, augmenting Fleet Tactical Readiness Support Group (the former Fleet Electronic Warfare Support Group, or FEWSG) following the disestablishment of VAQs -33, -34, and -35. As a result, the squadron ran several EW dets around the country using its A-6s as simulated cruise missiles and the like. In May 1994 the squadron made its last weapons det to NAF El Centro, led by skipper Cmdr. Kent "Trash" White, XO Scott "Lips" Ruppert, and Officer-in-Charge Dave "Bags" Sandgren. On 3 August 1994 Dave Sandgren and Lt.Cmdr. Bob "Wog" Ayers delivered the last VA-205 A-6E to the "boneyard" in Tucson, where it joined a growing line of *Intruders*. Four months later, on 31 December 1994, both reserve A-6 squadrons disestablished.

That last day of 1994 saw a wholly different carrier Navy than the one that had existed one or two years previously. The service was down to 11 air wings, of which seven still had *Intruders* assigned: VA-75 with CVW-3; VA-115 with CVW-5; VA-34 with CVW-7; VA-165 with CVW-9; VA-95 with CVW-11; VA-196 with CVW-14; and VA-52 with CVW-15. The numbers continued to change in a negative direction, and as it turned out, the next medium attack "honoree" was one of the service's oldest, most storied squadrons of all.

Black Panthers ...

By early 1995 there was one word or phrase in carrier aviation, and that was "Strike Fighter"; the noble A-6 just did not fit in. No one talked about the loss of the long-range, all-weather capability from the carrier air wings that had already lost their *Intruder* squadrons, and those that were making similar transitions. "Power Projection Air Wing" was another popular term, referring to one squadron of F-14s and three squadrons of FA-18s, along with the standard VAQ, HS, VS, and VAW presence. To be sure, the F-14s had finally picked up an air-to-ground mission. When the *Tomcat* received the LANTIRN pod it immediately became the premiere precision weapon delivery platform in the air wing. Concurrently, the Navy and McDonnell Douglas (later Boeing) continued to push the development of the FA-18E/F as the future of Naval Aviation., with the eventual goal of having air wings made up entirely of *Hornets*. Another item that was not talked about much was the loss of air wing inflight refueling capability. While the FA-18s tested with buddy stores early in their development, the aircraft's design and flight characteristics – and already limited range – made the use of *Hornets* as tankers unsavory.[2] Instead, the S-3 community quietly shouldered the load; perhaps it was a coincidence, but at this time the *Viking* squadrons started their inexorable march away from their primary mission of anti-submarine warfare.

And just when the *Intruder* faithful thought it couldn't get worse for their community, it did. On 31 January 1995 the world famous VA-35 *Black Panthers* disestablished at NAS Oceana. At the time of its demise the squadron was the oldest attack squadron in the Navy, and one of the oldest squadrons period with over 60 years of active service and participation in four of the nation's wars.

Like the other fallen units, VA-35 went out in proud fashion, led by skipper Cmdr. John S. "Jack" Godlewski. They had deployed with CVW-17 in *Saratoga* on 12 January 1994; during the course of the six-month cruise the *Panthers* made the standard round of operational and training evolutions, including flights over several of the world's hot spots. On 6 December 1993 Godlewski made his 1000th trap; XO Marty "Mallard" Allard also joined the Grand Club, reaching his milestone on 17 April 1994. *Sara* pulled back into Naval Station Mayport on 24 June 1994; the last of her 344,664 arrested landings came the day before when Air Boss Capt. Mark Kikta landed a VS-27 S-3B. She decommissioned quickly, lowering her pennant on 20 August following 38 years of strong, loyal service.

Five months later VA-35 – the veteran of 30 carriers and 69 deployments with aircraft such as the Vought SB2U "Wind Indicator," SBD, the Curtis SB2C "Beast," and the mighty Douglas AD *Skyraider* – was dead. To many the squadron's demise was representative of the Navy's headlong rush away from the "VA" designator and dedicated attack mission, and a grievous lack of attention to tradition. In any event, the last official log entry by a Black Panther was succinct: "31 Jan. 1995: SQUADRON IS ELVIS."

Pilot and naval aviation historian/novelist Barrett Tillman was one of the few honored with the title of "honorary *Black Panther*." As with many who studied carrier aviation, he had trouble reconciling the Navy's decision to do away with VA-35. He later wrote:

---

[2] The FA-18E/F "Super Hornet" versions would be built from the factory ready to carry buddy stores and be used as such from their first deployment, in 2002. The use of the fleet's newest, most capable and expensive strike platforms as tankers was not considered much of an issue, not that the Navy had many options.

The *Black Panthers* of VA-35 ended over 60 years of service in the defense of America in January 1995. AA502 is shown at Oceana in September 1994. (Brian Rogers)

"To my knowledge, VA-35 was the senior attack squadron on Planet Earth, let alone in the U.S. Navy.Tracing an unbroken lineage from 1934, the *Panthers* had been bombing their peoples' enemies for 60 years when a callous, unfeeling bureaucracy ended a heritage precious to naval history: Midway, Guadalcanal, the Fast Carriers, Korea, Vietnam, Desert Shield. The *Panthers* deployed in carriers with single-digit hull numbers and with nuclear power: both *Saratogas*, both *Yorktowns*, *Leyte*, *Enterprise*, *Coral Sea*, *Nimitz*, and *America*. *Dauntlesses*, *Helldivers*, *Skyraiders*, *Intruders* – carrier killers, ship slayers, and bridge busters.

"But much newer, far less deserving squadrons were retained while the *Panthers* got the ax. It was all the more outrageous because four new *Hornet* squadrons were established in the wake of VA-35's demise – a crime against history for which no one is accountable. When the admirals convened to establish the navy's so-called 'core values,' history and tradition were pointedly omitted. None cared then; few care now.

"The *Panthers* belonged to another time, another country. They did things differently there, leaving us only one option: honor the memory."

Knightriders, Golden Intruders ...

The next squadron to step aside was another former reserve outfit that had found itself uprooted and shipped off to the Korean War: They were the proud *Knightriders* of VA-52, which first saw the light of day as VF-884 out of NAS Olathe, KS.

The squadron departed with CVW-15 in *Kitty Hawk* on 3 November 1992 as the scheduled relief for *Ranger* and CVW-2. The ship had a hell of a time on the trip over, dodging Typhoons Clara and Gay before finally drawing liberty in Hong Kong and Singapore. Then came the run to Somalia, where VA-52 crews sortied with NVGs while providing on-call close air support. Notably, no Somali warlords stirred up any trouble while carrier aircraft were overhead. The situation escalated right after Christmas following an aerial confrontation between an Air Force F-16C equipped with AIM-120 *AMRAAM* and an Iraqi MiG-25 (the Iraqi lost). *Kitty Hawk* went through the Straits of Hormuz on 1 January 1993, and within short order the *Knightriders* were doing that *Southern Watch* thing, armed and ready for any eventuality. Iraq continued its intransigence, and the U.S. responded, ordering up a combined-force strike package. The U.S. cancelled the planned 12 January 1993 strike at the last minute due to bad weather, but another raid the following night went in, making for the biggest air show over Iraq since *Desert Storm*.

A total of 116 Navy, Air Force, French, and Royal Air Force aircraft pounded the targets; VA-52 skipper Cmdr. Rick "Downtown" Hess led one of the three CVW-15 strikes in the van of 10 *Intruders*. All made it in, pickling two 2000-pound GBU-10s each and returning safely to the carrier with great FLIR video, much to the delight of Navy PAOs. On a subsequent *Southern Watch* hop that took place on 23 January, one of the squadron's crews took out an Iraqi AAA site with another LGB; it was probably the last weapon ever dropped by an *Intruder* in anger.

Things settled down after the dustup with Iraq, and later in the cruise Skipper Hess and XO Cmdr. James "Jingles" Engle rang up their 3000th flight hours, putting them in select company in the medium attack community. The squadron returned to Whidbey on 3 May 1993,

VA-52 was disestablished in March 1995, along with the majority of CVW-15. NL503 is depicted in May 1991. (Brian Rogers)

The *Golden Intruders* ended their days as the last A-6 RAG on 30 September 1996. NJ801 is bathed in a western sunset in January 1989 enroute to a night bombing event at Fallon as part of one of the periodic weapons contests held by the Medium Attack Weapons School Pacific Fleet (MAWSPAC). The aircraft, piloted by an instructor pilot and student B/N, carries a load of dreaded Mk.76 "Blue Death" practice bombs on its inboard MERs. (Rick Morgan)

made the standard turnaround, and went out for one last deployment on 24 June 1994. The cruise was a tad more relaxed than its predecessor; *Kitty Hawk* managed to grab a port call in Sasebo – marking a rare visit there by a carrier – and then headed to the waters off North Korea to keep an eye on that eternally contentious peninsula. In June 1994, while roughly halfway through the squadron's "Last Crusade," XO Cmdr. C. Rivers Cleveland moved up to the CO's spot, relieving Jim Engle. Cleveland ended up recording his 1000th trap and 4000th Intruder flight hour during the cruise; Lt.Cmdr. Rick Postera notched his 2000th, while Lts. Karl Klederer, Dave Mack, Chris Rollins, and Sul Ozerden all rolled 1000.

Carrier Air Wing 15 returned from its last deployment on 22 December 1994 and started shutting down, bringing the Navy down to 10 air wings for 12 carriers. The wing disestablished in a combined ceremony with VF-51, the famous VF-111 *Sundowners*, and VAW-114 on 31 March 1995, ending 40 years of service. As for the mighty *Knightriders*, they concluded operations in a ceremony at Whidbey on 31 March 1995.

It did not take long for additional changes to the West Coast *Intruder* lineup. On 20 September Naval Strike Warfare Center – aka "Strike U" – sent its last *Intruder* to the boneyard at Davis-Monthan. On 4 June 1996 Strike U would merge with "Top Gun" (the Naval Fighter Weapons School) and "Top Dome" (Carrier Airborne Early Warning Weapon School, both formerly of NAS Miramar) to form the Naval Strike and Air Warfare Center, or NSAWC. Eight days later the Medium Attack Weapons School (MAWSPAC) at Whidbey became a *Prowler* operation as the Electronic Combat Weapons School (ECWS), with Lt.Cmdr. Steve Kirby assuming command.

Finally, on 30 September the *Golden Intruders* of VA-128 joined the ranks of the fallen. The disestablishment concluded 32 years of medium attack training at Whidbey dating to its first establishment of the VAH-123 A-6 detachment. Redesignated a squadron on 1 September 1967, VA-128 was the larger of the two fleet replacement squadrons by 1988. Over the years its instructors trained over 1300 pilots, 1160 B/Ns, and 17,000 maintenance and support personnel.

VA-128 was not the only RAG to shut down during this period; with the reduction of the VF community all of the fighter squadrons had consolidated at NAS Oceana, with VF-101 surviving as the Tomcat RAG. All S-3 training shifted to VS-41 at NAS North Island, while P-3C training consolidated at NAS Jacksonville with VP-30. However, the disestablishment of VA-128 was the obvious final nail in the coffin of medium attack. In order to support the final two years of *Intruder* squadron deployments the Navy started shifting personnel around. Many Aviators, Naval Flight Officers, and maintainers quickly found themselves completing cruises and immediately moving over to another squadron, while others wrapped their A-6 careers and headed elsewhere.

As for the old *Intruder* area at the south end of the NAS Whidbey Island ramp, it started filling up with new tenants: the P-3 and EP-3 squadrons of *Patrol Wing 10*, inbound from NAS Moffett Field.

Green Lizards, Boomers ...

With the end of 1995 quickly approaching, the squadrons continued to drop as deployed air wings completed their cruises. VA-95 was the next to depart the scene at the relatively young age of 23. As with their predecessors, the *Green Lizards* did not hang their helmets, but instead wrapped up their operations in an upbeat, professional fashion.

The squadron's workups for its next-to-last cruise proved colorful. On 4 February 1993 *Lizard* Lts. Kimo "Kaulianuinamakeha" and Daryl "Salty" Martis made the first peacetime launch of an AGM-84E Standoff Land Attack Missile (*SLAM*) while operating from an aircraft carrier. During this period skipper Graham "Buck" Gordon, the pride of the University of New Mexico, recorded his 4000th flight hour in the A-6, while Lts. Paul "Cord" Bunge and Aaron "Cuds" Cudnohufsky notched their 1000th hours. In an unusual sidelight to the final workups, during April Russian Tu-95 *Bear Gs* managed an over flight of *Abraham Lincoln* during a simulated strike mission. The G was the *Bear* variant, with the 200-mile-range AS-4 *Kitchen* missile, designed specifically to take out carriers. The overflight was the last one of a U.S. Navy carrier battle group until September 1997, when *Nimitz* got tagged.

The cruise itself – which started on 18 June 1993 – proved not to be anything out of the ordinary. *Lincoln* and CVW-11 did the usual *Southern Watch* patrols, including one spectacular 1200-mile sortie that launched from the Indian Ocean. Two VA-95 **Intruders** joined two *Tomcats*, two *Hornets*, and a single VAQ-135 *Prowler* on the strike profile, refueling from Air Force KC-10s. For the most part Iraq behaved itself during *Lincoln's* tour, although on one occasion *Black Raven* EA-6Bs hosed off a couple of *HARMs* at threatening radar sites. Ultimately *Lincoln* spent 12 weeks on station in the Persian Gulf before heading to Somalia for the contretemps there; the longest single at-sea period was 59 days – not quite up to the levels of the Iran Hostage Crisis of the late 1970s, but still memorable for the participants. After a quick port call in Australia – always an air wing favorite – *Lincoln*, CVW-11, and VA-95 returned home on 15 December 1993.

Which left one more deployment for VA-95, the famous "Lizard's Last Romp" of 1995 led by Cmdr. Pieter "Dyke" VandenBergh and XO Lt.Cmdr. Terry "Krafty" Kraft. Things got off to a fiery start during workups in the SoCal OpArea; on 15 February 1995 BuNo 155586 –

A *Green Lizard* hangs out off Whidbey Island, 27 April 1993. (Rick Morgan)

configured as a tanker – caught fire on the deck of *Lincoln* while Lts. Don "Bun Bun" Parker and Steve "Meat" Gaze were in the cockpit. *Lizard* ADC McMullen immediately grabbed a hose, sent in a firefighting team, opened the burning *Intruder's* engine bay doors, and then pinned the external fuel tanks. AMH2 Lloyd Billups saw the smoke early and also led a fire team, while AME3 Buddy Thompson ran in and pinned the ejection seats, preventing their firing while the crew safely escaped. All survived without major injury, although the fried A-6 was a write off. In a later ceremony Rear Adm. Jay B. Yakely, *Commander, Carrier Group Three*, awarded Chief McMullen the Navy Commendation Medal and Petty Officers Billups and Thompson the Navy Achievement Medal.

Lincoln departed ConUS on 11 April 1995 and headed directly for the Persian Gulf. On 29 July VA-95 loaded and launched all 14 aircraft for a long-range, fully-loaded "fam hop" into Iraq. The *Lizard Intruders* pressed the mission to the 14 simulated targets and recovered with nary a hiccup following a magnificent 14-plane flyover. Prior to returning home on 9 October 1995, several squadron members put their names in the record books: skipper Vandenbergh and Lt.Cmdr. Steve "Caz" McCaslin with 3000 hours; Lt.Cmdrs. Marion "Doc" Watson and Thomas "Spicola" Mascolo with 2000; and Lt. Dave "Psycho" Bates with 1000. Capt. Bob Taylor, DCAG-11, recorded his 1000th trap just before the end of the cruise on 3 September. Still, as the saying goes, all good things come to an end. The true Lizard's Last Romp concluded with the squadron's disestablishment at Whidbey on 18 November 1995.

The induction of the mighty *Green Lizards* into the choir invisible left the Navy with five VA squadrons: VA-34 and VA-75 at Oceana; VA-165 and VA-196 at Whidbey; and VA-115 at Atsugi. By now there was speculation over the future disposition of the four, but the question was settled well before the end of 1996. As a result, VA-165 "earned" honors as third-from-last *Intruder* squadron in AirPac and the next to shut down.

The *Boomers* had a raucous year leading up to their final cruise and disestablishment, led by Cmdr. Mark S. Needler, who had relieved Cmdr. Ron Stites on 17 August 1995. Looking around quickly for an XO, the Navy settled on Lt.Cmdr. Rick Keller, who had been holding down a staff job with *Carrier Air Wing One*. He'd assumed he was going from that job directly (hopefully) to *Prowlers*; instead he got to take a little side trip:

> "As CAG ops I was flying *Tomcats* primarily," he comments. "We didn't have *Intruders* on that cruise. We were at sea, the list came out, the CAG called, 'You were selected for *Prowlers*.'
>
> "About nine that night I got a call from the detailer: 'Can you be in Whidbey Island in two weeks?' 'Well, I can stand on my head if you like, what do you need?' There were about five of us lieutenant commanders lined up, but I was the guy who had to relieve onboard, and I next went to VA-165 as XO."

The detailer wanted Keller to report to VA-128 in two weeks in an attempt to requalify him before the squadron closed up shop. Keller told the detailer he didn't need to go through -128, he was current:

> "'How can you be current?' 'Trust me, I've got 10 hours in the last 90 days in *Intruders*, that's all that matters.'"

"Boomer" 512, an A-6E with a load of Mk.76 practice bombs, orbits over a cloudbank, December 1993. (Rick Morgan)

Next stop, Whidbey Island. Needler relieved Stites, Keller reported as an O-4 executive officer, and the squadron went on from there. Keller says his first meeting with DCAG was:

> "... funny. The first time I went to a group meeting with the DCAG he asked, 'Where's your XO?' 'I *am* the XO.' I think Dee (Mewbourne, the last XO of VA-196) went through the same thing. Anyway, the DCAG reeled his jaw in and off we went."

The "*Boomers'* Last Ride" – VA-165's 23rd and last deployment – began on 1 December 1995. During the course of the cruise the squadron assisted with the enforcement of UN sanctions against Iraq while making a side trip to a point off the coast of Taiwan, a result of the China-Taiwan election crisis. All of the squadron's *Intruders* received special CVW-9 markings for this last deployment, and AMH2 Kevin Welch and AO3 Daniel Summers paid particular attention to NG501 – BuNo 159314 – giving it a rendition of VA-165's original A-6 markings, as well as a painting of "Puff the Magic Dragon" on the left side of the radome, complete with flames. AMS3 Kazuya Miyashita, AMHAA Jeremy Vanderpool, AMSAN Ruben Romero, AE3 Tina Brooks, AD2 David Kwapizewski, and AMH1 Giff Johnson assisted.

At one point during the cruise the squadron got all 12 *Intruders* airborne for a commemorative flyby of *Nimitz*; afterwards, CAG-9 and former VA-95 skipper Capt. James "Jocko" Worthington commented that he did not think any of the other *Air Wing Nine* squadrons would have been able to get all of their birds up for one sortie.

VA-165 did well during its final deployment, and everyone had a good time. Unfortunately, the good times never last; this was as true for the *Boomers* as for all the other squadrons that had preceded them into retirement.

"Operationally, the squadron did just great," says Keller:

> "But we shut down six months too early to get the Battle E. It was a typical cruise, standard, people weren't thinking about us leaving. Towards the end, as we came out of the Gulf of Taiwan – which turned out to be much ado about nothing – we began at that point to think, 'this is it.' We started doing neat things, such as getting all 12 aircraft up for the flight. They all came back 'up and up,' which was pretty neat. After we got back Needles went to European Command early, so I got to be the acting CO. That was pretty good, being the acting CO of an A-6 squadron for a month."

VA-165 returned to Whidbey for the last time in May 1996, ending an association with CVW-9 that dated to February 1990. Prior to disestablishment the squadron performed one last photo op with restored A-1H NX39606, flown by Naval Reserve Cmdr. Bram Arnold over Deception Pass in formation with the last NG500, BuNo 162202. In preparation for the flight personnel from VA-165 repainted the *Spad* as AH500 in the squadron's historic *Carrier Air Wing 16* colors.

At the ceremony – held on 26 July 1996 – noted *Intruder* bombardier/navigator Rear Adm. Herbert A. Browne served as guest speaker. At the time he was Deputy Chief of Staff, U.S. Pacific Fleet; more importantly, he was a former *Boomer*. Four days later VA-165 disestablished, officially leaving VA-196 as the last *Intruder* squadron on the Rock, although VA-165 lived on a while longer as a detachment assigned to the wing. Keller says the det – with Greg Smith as OinC – kept two jets, about three each pilots and B/Ns, and around 60 maintainers at Whidbey in support of VA-196 during its last deployment. Eventually, though, this last fragment of the squadron dissolved:

"I got to take an A-6 to the boneyard," Keller recalls. "We delivered two up and up jets. They could've flown a mission that day."

Blue Blasters *and* Eagles ...

The disestablishment of VA-165 left VAs -115 and -196 as the last AirPac *Intruder* squadrons. Their AirLant counterparts also numbered two: VA-34 and VA-75. Of the four squadrons Navy leadership decided to keep two, with the *Eagles* and *Blue Blasters* transitioning to FA-18s. Thus, it fell to the *Sunday Punchers* and the *Main Battery* to make the last A-6 cruises and then retire the dedicated attack community.

Among the last things VA-34 participated in was the 50th anniversary of the D-Day landings at Normandy, France. Several squadron aircraft were painted with ceremonial "invasion stripes" for the occasion. Two *Blasters* head for *GW*'s cats in June 1994. Note the "*Intruder* eyes" painted on the radomes, as well. (Rick Burgess)

The newest carrier to deploy with *Intruders* was the *George Washington* (CVN-73), which carried VA-34 and CVW-7 in 1996. Sharing the deck was a single large F-14B squadron (VF-143) and two *Hornet* units (VFA-131, VFA-136), as well as *Prowler*, *Hawkeye*, and *Viking* squadrons (VAQ-140, VAW-121, and VS-31). (U.S. Navy)

VA-34 – another "youngster" at 26 years of age – made two final deployments with CVW-7 before turning in their trusty A-6s. Cmdr. Rich Jaskot led the first tour, from 26 September 1991 to 2 April 1992, in *Dwight D. Eisenhower*, with trips to the Med and the Red Sea.

The *Blasters* suffered a major loss on 21 April 1993 while deployed to Nellis AFB. Squadron XO Cmdr. John Dolenti, with Maintenance Officer Lt. Cmdr. Marshall Atkins driving, were coming off target on the range during a night *Red Flag* event when they hit the empennage of their flight lead during rendezvous. Both were killed – the lead aircraft landed safely with tail damage. The mishap was believed to have been related to use of Night Vision Goggles and their lack of depth perception.

Cmdr. Charles Hautau commanded the squadron's last deployment in A-6s, which ran from 20 May to 18 November 1994; it also marked the first cruise for *George Washington*. The *Blasters* managed a "career highlight" as a unit on 6 June 1994 when they participated in the 50th anniversary commemorative activities at Normandy. The squadron's part included painting several A-6s with the famous black-white-black invasion stripes and performance of a missing man formation over Omaha, Utah, Gold, Juno, and Sword Beaches. Otherwise, the remainder of *Washington's* maiden deployment was spent in Operations *Sharp Guard* and *Southern Watch*, along with operations over Bosnia as part of *Deny Flight*. VA-34 returned home on 16 November 1994 and later received the AirLant Battle E for the year, while also taking the Hughes Trophy.

As the duty "alert air wing" in the Western Pacific CVW-5 – now embarked in *Independence* – made several passes by Iraq and the Western Indian Ocean through 1996. On 27 August 1992 the carrier was the first to respond to President Bush's establishment of the no-fly zone over Southern Iraq; later that afternoon CAG-5 Capt. Bud Langston led the first sortie by *Eagle Intruders*, all armed with *HARMs*, 1000-pound LGBs, and the occasional AIM-9. The tour passed uneventfully if long, hard flight operations under arduous conditions – 110 degree heat and the occasional sandstorm – can be considered uneventful.

VA-115 had the unfortunate distinction of losing the last two A-6s in operational mishaps. On 14 October 1994 a squadron A-6E hit the ground while on a low level on the Japanese island of Shikoku. Lts. Eric Hamm and John Dunne would be the last of 161 men killed while flying the *Intruder* in peacetime. About eighteen months later, on 3 June 1996 the Japanese destroyer *Yugiri* gunned down an *Eagle* A-6 while participating in the annual *RIMPAC* exercise. The incident occurred during a gunnery exercise while the *Intruder* towed a target sleeve; apparently the tin can's *Phalanx* Close-In-Weapon System picked the wrong target and worked as advertised, hitting the *Intruder* with multiple 20MM shells. Fortunately, the crew of Lts. Will Royster and Keith Douglas were able to eject, although Royster suffered multiple injuries that would in time end his Naval flying career. As for the hapless A-6, it was gone; according to SM1 Bryan Potter, who observed the accident from the signal bridge of *Bunker Hill* (CG-52), "We picked up a few pieces of the aircraft. That was it." It would be the last of 191 *Intruders* lost in operational mishaps.

The Japanese ships participating in *RIMPAC* suspended further live firing pending an investigation, which later determined *Yugiri's* gunnery officer had given the firing order before the ship had confirmed the A-6's position. The incident's irony was not lost on the *Eagles*, in that their squadron had been originally formed in October 1942 as Torpedo Squadron Eleven (VT-11) to fight, of course, the Japanese.

On 17 June 1996 the VA-115 catapulted its last six A-6Es from the deck of *Independence*, ending 21 years of forward deployed *Intruder* operations in Japan. The squadron redesignated as VFA-115 at NAS Lemoore on 30 September 1996; at the time of the ceremony – which featured both an A-6E flown down from Whidbey and the squadron's first FA-18C – the *Eagles* were the oldest surviving attack squadron in the naval service. The following day VA-34 redesignated into the strike-fighter business at NAS Cecil Field.

And Then There Were Two …

The demise of VA-165 and conversion of VAs -115 and -34 left two squadrons to carry the medium attack flag: VA-75 and VA-196, and both were deployed. It was highly appropriate that they were the last attack squadrons; the *Sunday Punchers* had been the first fleet A-6 squadron, and the *Main Battery* was the first *Intruder* squadron to arrive at NAS Whidbey Island in November 1966. Providentially, the two squadrons now went out together.

Both made a series of deployments leading up to their final cruise, starting with VA-75's departure for Med with CVW-3 on 7 October 1992. Led by Cmdrs. Tom Hagen and George Starnes, the squadron went through the usual workings, even managing a little fun by hosing

It was appropriate that the first Atlantic and Pacific Intruder squadrons would also end up being the last. On 27 September 1996 the "CAG birds" from each unit met over the Persian Gulf while their carriers turned over duties as the 5th Fleet carrier supporting *Operation Southern Watch*. (U.S. Navy, LCDR Steven Nakagawa)

off the odd *Sidewinder* at flares during defensive air combat maneuvering (DACM) training; talk about true strike-fighters. The deployment went quietly until late January 1993, when the Navy yanked *John F. Kennedy* and the air wing out of Naples and sent them to the Eastern Med due to increased tensions with Iraq. Things cooled off enough for the carrier to head to the Adriatic for participation in *Provide Promise*, and then everyone went home, returning to Tidewater on 7 April 1993. Sadly, on 23 July Lts. Joseph K. Rough and Paul A. Ambrogi paid the ultimate price, dying in a crash near Roanoke, VA. The squadron regrouped, and under G.K. Starnes sallied forth once again on 20 October 1994, still as part of CVW-3, but this time embarked in *Dwight D. Eisenhower*.

As for the *Main Battery*, they did one cruise with CVW-14 in *Carl Vinson* from 18 February to 15 August 1994. This deployment had its moments; the United States and several regional allies became annoyed with the North Korean government after the latter refused to allow inspections of its nuclear facilities. Hence, VA-196 and the rest of the air wing spent some "quality time" in the Sea of Japan before heading to the Persian Gulf for their turn with *Southern Watch*. When not buzzing Iraq with a full load – more often than not, as the lead of a strike/recce package – the crews found time to participate in Operation *Beacon Flash* with Oman, *Nautical Artist* with Saudi Arabia, and *Eastern Sailor* with Qatar.

Overall it proved to be a hell of a year for the *Main Battery*. Skipper Cmdr. Richard "Kato" Nobel notched his 1000th career trap on 24 June while *Vinson* was enroute to Australia; he ended up with the award as CVW-14 Top Hook for the deployment. Elsewhere in the squadron Lt. Cmdr. Brad "Schlep" Leppla recorded his 1000th hour in the A-6, while Lt.Cmdr. John "Sarge" Alexander reached 2000 *Intruder* hours. After VA-196 returned to Whidbey it managed to sweep the AtkWingPac *Intruder* Derby, placing first in the day single, night single, night section, WST, NATOPS exam, and intelligence categories. All that was left was to accept the accolades as "Best of the West"; new skipper Cmdr. Joe "Killer" Kilkenny – who fleeted up from XO on 26 August – did the honors at the annual *Intruder* Ball in October. Kilkenney himself reached 3000 A-6 hours during the year, along with Lt.Cmdr. Paul "Jack" Webb. Lt.Cmdrs. Gary "Po Daddy" Poe and Pete Davenport rang up 2000, and Lts. Scott "Duffer" Duffy and Bob "Reckless" Eckles hit 1000 hours each.

While the old heads were making the records, both squadrons featured a mix of young crews – some making their one and only deployments in the A-6 – along with some old friends, who up to recently had been with another squadron. As the end drew near for the community, more and more personnel found themselves moving from a disestablishing squadron to another and quickly deploying again. One of the B/Ns who joined VA-196 towards the end was Lt. Steve Nakagawa, the son of Capt. Gordon Nakagawa, the previous *Main Battery* skipper and Vietnam-era POW.

"I left Pax (Naval Test Pilot School at Patuxent River) and got orders to the last west coast A-6 squadron, VA-196," he comments:

> "I basically finagled the orders. They needed another B/N, and then-XO Cmdr. Dave Frederick and Jim Engler, the aviation detailer, got me in. I convinced them I was current, even though the RAG had shut down ... I didn't know where my career would go from there. I got to -196, became the admin officer – still a lieutenant – and had to grow up and be a department head. We had a really great cruise; I flew with Dan Dugan. He'd done the last VA-95 cruise, and we were a really good team. We agreed we flew as a great team; the crew coordination while in that plane was unlike any other plane I'd flown, particularly at night."

When it finally came time for the final A-6 deployments, VA-196 headed out first, departing the Rock for the last time on 14 May 1996 under the leadership of Cmdrs. Dave "Fredo" Frederick and Dee Mewbourne. *Carl Vinson* made the requisite stops in Yokosuka, Hong Kong, and Singapore before checking in with *Commander, Fifth Fleet* on 1 July. Within short order the squadron was once again participating in *Southern Watch* ops over Iraq, with an occasional break for exercises, such as *Rugged Nautilus*; notably, the Joint Forces Air Component Commander (JFACC) was *Commander, Carrier Group One* Rear Adm. Timothy R. Beard, a long-time *Intruder* pilot and former commanding officer of VA-176.

As for the *Sunday Punchers*, they left NAS Oceana on 28 June 1996 embarked with CVW-17 in *Enterprise*, which had just come out of overhaul. Over 30 years before VA-75 had recorded a lot of A-6 firsts; now it was participating in the "lasts," including the last *Intruder* Advanced Attack Readiness Program (AARP) in September 1995 and the last *Intruder* participation in *FleetEx* in May 1996. In between the squadron engaged in landscape modification projects at NAS Fallon, Hill AFB, and the Vieques complex, led by skipper Jim "Gigs" Gigliotti and XO Pete Frano. That latter had taken the scenic route getting to the executive officer's slot; in effect, he was hand picked, as were the other last *Intruder* XOs:

"I went to VA-42, then was a department head at VA-75, then to NAVPERS, and then went to VA-75 for the 'last dance,'" says Frano:

> "Gigliotti was one of the guys to screen for A-6 command, and they knew he and 'Fredo" would be the last COs. (Detailer) John Meister didn't want to take a department head from behind these guys and make him XO, so he selected four officers who had already screened for *Prowler* XO: Dee (Mewbourne); me; Krafty (Terry Kraft); and Rick Keller. It was a good decision at the time; Meister thought it was best, particularly if – God forbid – anything happened to the CO. 'Pokey' came out of the War College, went to CAG ops, then to VA-165. Terry got VA-95, Dee got VA-196, and VA-75 needed an XO ... I was available, and there you go."

Frano adds that -75's last deployment was a "classic Med summertime cruise." The carrier and squadrons were able to hit all the ports they had missed on previous deployments, such as Palma, Rhodes, and Haifa. *Enterprise* and CVW-17 departed ConUS on 28 June 1996 and immediately went into operation *Decisive Endeavor* operations over the Balkans, relieving CVW-7 – and VA-34, also making its last deployment – in *George Washington. Big E* pulled out of a port call in Rhodes, Greece, on schedule, but then came the call: make for the Persian Gulf ASAP and join *Vinson* for possible combat operations. Notably, her departure left the Med without a carrier for the first time in anyone's memory, a result of the retirement of four carriers in three years. The squadron arrived on station in good form, got the briefings, and started flying low-levels over Kuwait.

In late August Iraqi troops entered the city of Irbil in violation of UN restrictions. The United States decided an armed response was appropriate and set in motion Operation *Desert Strike* to teach Saddam Hussein another lesson. Once again it looked like carrier aviation was going to carry the day ... with the *Intruders* in the lead, of course. With the arrival of *Enterprise* – and VA-75 – planners worked up a strike involving both A-6 squadrons. The *Intruder* was born into combat, and it would have been highly appropriate if the aircraft departed the fleet while in combat. However, it was not meant to be. At the last minute the Clinton administration went with sea- and air-launched cruise missiles, with B-52s from Andersen AFB on Guam delivering the latter. The two air wings provided support, primarily command and control and fighter escorts by VFs -11, -31, and -103:

"We went to sleep thinking we'd do *Desert Strike*," recalls Steve Nakagawa:

> "We woke up the next morning – about 0400 – and learned it had been called off for cruise missiles. The next night Dan and I led an eight-plane mini-strike, the first raid after *Desert Strike* happened. We got north of Basrah, tracers lit up the sky, but we didn't get any green ink for it. CAG was stingy, said that we're not at war, so that's not really a combat mission. We were flying with NVGs that night; there were so many fires it kind of blotted everything out."

After things settled down – after all, according to CAG-14 this wasn't really combat – the two squadrons briefly worked with each other, and at one point the two last *Intruder* squadrons put several aircraft in the air for some commemorative "last ever" air-to-air A-6 shots. One highlight was a formation flight between full color *Flying Ace 500* (BuNo 164382) from VA-75 and her counterpart from VA-196, *Milestone 500* (BuNo 159579). The same day yielded the only real adventure of VA-75's cruise. While enroute back from a *Southern Watch* sortie AA501 – with Lts. Don "Cakeboy" Breen and Jason "Teefus" Jones – was unable to lower its nose gear due to a failed actuator. After running through every NATOPS procedure the recalcitrant gear still failed to show a green, so *Enterprise* rigged the barrier. As soon as the *Intruder* took the wire the nose gear knocked forward and locked down; damage was minimal, and the aircraft returned to flight status shortly afterwards.

While a bunch of great, historic photos of the last two A-6 squadrons resulted from the combined operations, it was apparent there was some tension between the two ready rooms. Several of the members of VA-196 felt that VA-75 was getting all the press as the "last A-6 squadron," ignoring the fact that the *Main Battery* was the other "last A-6 squadron" and also deployed. Some of the tension was apparently justified:

"We had more attention than VA-196, absolutely," Frano confirms:

> "First off, there were all the A-6 admirals there, people like Stan Bryant and Ralph Suggs. Just go through the admiral list; they're in the Norfolk area and were in their proximity. Hampton Roads is pretty big. We had all the local TV stations coming over to us for interviews, flights ... shoot, I think all of the anchors for the local major stations got to fly in the A-6 with us. You had it all there, AIRLANT, CINCLANT ... it was a Navy town, and a lot more responsive. In Seattle, this kind of coverage wouldn't have sold papers."

Fraternal squabbling between the squadrons not withstanding, they soon went their separate ways. *Enterprise* went back through the Suez Canal and resumed station in the Mediterranean, while *Vinson* turned for home, making a port call in Hobart, Tasmania. In Pearl Harbor on 7 November 1996 the ship participated in the CINCPACFLT change of command ceremony; the ceremony also marked the retirement of Adm. Ronald Z. Zlatoper, former commanding officer of VA-85. It turned out the admiral had flight time in one of -196's birds, so the squadron put his name on the canopy rail and let him make his last *Intruder* flight prior to retirement. "It was a fantastic flight," Adm. "Zap" commented afterwards: "There are no bad flights."

The *Main Battery* returned to NAS Whidbey Island on Veterans Day, 11 November 1996. They participated in a 17-plane flyoff from the boat, including the five aircraft of VAQ-139; fortunately, the clouds lifted a tad at NAS Whidbey Island for the squadrons' return. VA-196 totaled 8200 flight hours and 2700 traps since the start of workups – including over 2500 hours on deployment – and maintained a sortie rate above 95 percent. In the words of AVCM(AW) David Selby, the maintenance crews worked above and beyond everything that was expected of them:

> "We may be the last A-6s, but no one's quitting. We're going out in style."

Pretty good for an antique! Notably, the *Main Battery's* final return to Whidbey Island marked the 30th anniversary of the squadron's first arrival with A-6As. It was a fitting conclusion for the "Last Attack In WestPac."

"We got back and started cranking up for the disestablishment," says Nakagawa:

"We took two planes down to Tucson; I was with Dan Dugan, and it was my last A-6 flight. We arrived about 1700. I'd seen (the boneyard at) Davis-Monthan AFB before, but this was different, because we taxied through the gate and into the storage area. It was kind of a bittersweet deal. It hit me the last time I went down the ladder ... we taxied up, and all these guys are scurrying up to take over. They give you a pen and ask you to sign the plane.

"My father and I checked our log books," he continues. "We found out that we'd flown ten of the same aircraft and had an interesting connection to the *Enterprise*. Dad flew off the *Enterprise*, was shot down, came back as XO, and cruised with VA-196 in *Enterprise*. My first cruise with VA-196 was on *Enterprise*. I did two deck cert dets in *Enterprise*, and my first cruise was around-the-world to bring the ship to Atlantic Fleet. In fact, on the second deck cert I did a Mode 1 ACLS flying with Mark Kelley, who's now an astronaut. That afternoon I flew *Tomcat* traps with his twin brother, Scott Kelley, also an astronaut. I picked up lieutenant commander on cruise, department head screen, got *Prowlers* while on cruise, and got to stay at Whidbey. It was a good community and a good deal for NFOs."

The *Main Battery's* return to Whidbey left VA-75 still out on the boat, undeniably the last-ever deployed A-6 squadron. The *Punchers* finished the cruise quietly but professionally, remaining in the Persian Gulf until the end of November. During the final portion of the cruise Cmdr. Gigliotti and his pilot, Lt. Bruce "Sick Dog" Miller, recorded 3000 and 2000 hours in the *Intruder* on the same flight, while Lt. Al Armstrong rang up 1000 hours.

On 19 December 1996 VA-75 made the last-ever cruise flyoff by an *Intruder* squadron. At high noon an A-6 taxied to the catapult for the last time in the operational Navy. DCAG-17 Capt. Carlton "Bud" Jewett and skipper Jim Gigliotti did the honors in front of a large crowd, while Vice Adm. Dick "Sweetpea" Allen, a VA-75 veteran from the late 1960s and former CO of VA-85, served as honorary catapult officer. *Flying Ace 501* launched, and that was it ... the *Sunday Punchers* had been the first to deploy, the first to see combat, and were the last to leave the flight deck. The "Final Punch" was over.

Last CQ

New Years Day 1997 marked the beginning of the *Intruder's* 34th year in the Navy, but obviously the end of its remarkable run was drawing near. At Whidbey Island and Oceana both squadrons continued to work hard and get in some flight time, but the main emphasis now was on giving the A-6 a proper sendoff. However, despite the regular delivery of aircraft to the boneyard at Davis-Monthan, both squadrons managed last carrier qualifications; VA-75, much to the consternation of some on the west coast, managed *two* "last ever" CQs.

In late January *Enterprise* pulled out for a few days to do its ordnance offload and the deck was available, so VA-75 sent eight crews out to the boat. Rear Adm. Stan Bryant, *Commander Carrier Group Four*, managed some deck time while flying with pilot and squadron maintenance officer Lt.Cmdr. Tom "Space" Spacey. The crew's final trap put Admiral Bryan in the Grand Club. Afterwards the Navy public affairs played up the event as the "last ever" CQ by an *Intruder* squadron, emphasizing – once again – that VA-75 was the first and last A-6 outfit. However, VA-196 stepped up with their own "last ever" CQ; according to *Attack Wing, Pacific* commander Capt. Terry Toms, the event really threw the VA-75 and the east coast aviation "establishment" into a major tizzy.

The event took place over 12-13 February 1997 onboard *Carl Vinson* in the Strait of Juan de Fuca off Whidbey Island. The CQ took place after a burial at sea for a Navy veteran; in preparation for that ceremony and the last CQ the Navy flew out 26 VIPs. The group included eight retired *Intruder* squadron commanding officers representing both coasts: Capts. Butch Bailey (VA-34, also former COMATKWINGPAC), Jim McKenzie (VA-52, former CAG-11), John Pieguss (VA-196, VA-42), Dave Waggoner (VA-52, NAS Whidbey Island), Dave Williams (VA-145, VA-128, NAS Whidbey Island), Bruce Wood (VA-52, VA-128, NAS Cubi Point); Cmdrs. John McMahon (VA-95) and Don Sullivan (VA-52); and Mr. Bruce Van Tassell, head of Whidbey's Northrop-Grumman operation.

Prior to the arrival of the A-6s the guests and ship's company participated in the burial at sea of Lt.Cmdr. Gerald Ray Roberts, a member of VA-196 during the squadron's last combat cruise with *Spads*. Commander Roberts died on 2 December 1965 when North Vietnamese

The *Milestones* conducted the last carrier qualifications for a Pacific fleet *Intruder* unit in the Strait of Juan de Fuca onboard *Carl Vinson* on 12 February 1997. NK500 is over the ramp with numerous photographers present. Washington's San Juan Islands are in the background. (Mark Morgan).

gunners downed his A-1H; he was the third and last combat loss of the cruise for the *Main Battery*. The Vietnamese government returned his remains to the United States in 1993, and the Army's Central Investigative Laboratory was able to make a positive identification in October 1996. When the commander's remains arrived at Travis AFB in early February 1997 members of his family and several of his squadron mates from the 1965-1966 cruise in *Bon Homme Richard* were on hand to finally welcome him home.

Vinson chaplain Cmdr. Bob Stone presided over the service with Commander Robert's widow, Mrs. Claudette West; his daughter, Mrs. Helene Kephart; his niece, Ms. Leslie Rathman; the VIPs, and approximately one hundred ship's company personnel present. *Vinson's* Marine detachment provided the honor guard for the ceremony; KA-6D 151819, with canopy rail lettered for LCDR G.R. "Jay" Roberts, served as a backdrop. After Mrs. West dispersed the ashes into the Strait of San de Fuca, Mrs. Kephart tossed a memorial floral wreath into the waters. Three A-6Es from VA-196 – assisted by a VAQ-139 EA-6B – performed a flyby of the ship, with one *Intruder* pulling out to form the traditional missing man formation.

Following the ceremony it was time to return to flight quarters; the preparations were punctuated by a surprise fly-by by *Thunderbirds* F-16B No. 7, which made a low level (well, low for the Air Force) pass along the port side of *Vinson*. Then the final event began. CVW-14 LSO Lt.Cmdr. Ron "Lobo" Wolfe called the deck, assisted by VAQ-139 LSO Al Bradford. Over the following three hours the last three AirPac A-6Es – NK500 (159579), NK505 (164377), and NK515 (161102) – flew several cats and traps, interspersed with *Prowlers* from VAQ-139. After each cycle the aircraft returned to the air station – in clear view from the carrier's flight deck – hot fueled and seated, then came back for another round.

Vinson's CO, Capt. Dave Crocker, invited the squadron back the following day for what became the last ever AirPac *Intruder* flight deck operations. At the end of the day catapult one filled with spectators from the ship, while an A-6 flown by Terry Toms and Cmdr. Dave Frederick, the last commanding officer of VA-196, took its place on cat two. When the cat officer touched the deck a cheer went up among all assembled, and the *Intruder* roared off the flight deck. Perhaps trite, but it was truly the end of an era; the AirPac A-6 types took comfort in the fact that they – and not VA-75 – had actually gone to the boat last. Afterwards Toms commented he received a lot of calls from staff types and newspapers back in Tidewater of the "what on earth do you think you're doing?!" variety.

On Thursday, 27 February, Terry Toms and a crowd of several hundred dedicated the A-6 Memorial at naval air station's historic Seaplane Base, the original NAS Whidbey Island. Old friends and shipmates gathered, speeches were made, and more than a few tears flew as Toms read off the names of the AirPac *Intruder* pilots and B/Ns who paid the ultimate price. Everyone in attendance knew the names on the plaque. Toms stated, in part:

"Since 1967, generations of pilots and bombardier/navigators have left their homes, families, and friends in Oak Harbor to

Only two Navy *Intruder* squadrons would survive to convert to the *Hornet*, VA-34 and VA-115 becoming Strike Fighter (VFA) Squadrons in the process. AA200 is at Oceana, 30 October 1999 (Tom Kaminski). NK200 is shown at Whidbey Island, 25 July 1997. (Mark Morgan)

fly, train, and fight. Eighty-six did not return. The first fell in Vietnam while I was a high school student. The last in Japan a year and a half ago.

"Some now lie in the soil of their homes. Others remain with their *Intruders* eternally. Some rest beneath the waters of Puget Sound a few miles from where we stand, others thousands of miles away in the Pacific and Indian Oceans and foreign jungles.

"They fell in the cockpits of our beloved bomber at Whidbey Island, the mountains and plains of Washington and Oregon, the dark waters of faraway oceans, the karsts of Vietnam, and the bleak sands of Iraq. Some fell in bright sunshine or calm evenings on what was scheduled as a routine training flight. Some spent their final moments in skies filled with tracers and missiles as they fought their way alone towards an enemy target. Frequently, the only official epitaph is "did not return from night combat mission."

"It has often been our legacy to die alone and unseen. The heart of all weather attack, is two aviators in a single aircraft going to bad places on dark nights.

"I remember the pure joy of flying A-6s on a beautiful day in the northwest and sharing the feeling with many of these men that we were the luckiest guys on earth. For every hard moment, I remember a hundred great ones. And if I'm lucky enough to hang up my g-suit the final time and walk away, so too were my friends who had the rare opportunity to experience the great moments, however briefly. I remember them.

"If time can't ease the pain, neither can it lessen the pride, gratitude, and love we feel for these Whidbey Island *Intruder* aviators who are flying on God's wing today and forever. Mission complete, my friends. You're cleared to high holding. The rest of us will rejoin later."

The following day, at NAS Oceana and NAS Whidbey Island, the last two attack squadrons in Naval Aviation marked their disestablishment and the end of the Medium Attack line in the Navy.

The Last Dance

On 28 February 1997, in dual ceremonies at NTU and NUW, attack aviation passed on as a dedicated mission of naval aviation. The ceremonies were well attended, with approximately 1800 appearing in Virginia Beach and over 1000 in Oak Harbor.

At Whidbey VQ-1 provided their hangar for the ceremony; it was appropriate, as the structure had housed VAQ-128 and several fleet squadrons for 30 years. During the ceremony COMNAVAIRPAC Vice Adm. Brent Bennett presented VA-196 with the Battle E award. The last officer to receive Attack Wing Pacific's Mike Bouchard Junior Officer Leadership Award was Lt. Chris Moore, a *Main Battery* LSO and Line Division Officer. Notably, Mike Bouchard, Jr., was in the audience.

Rear Adm. Lyle Bull, the Navy Cross recipient and former VA-196 skipper, made the farewell speech for the squadron, while Capt. Jim McKenzie did the same for the wing. McKenzie pointed out that Terry Toms and Dave Frederick had not only made the last A-6 cats and traps, they had also flown the last A-6 bolter and got hung up in the gear. That brought an appreciative roar from the crowd, as did Bull's story of a earlier *Milestone* cruise where, in the course of an evening of frivolity at the Cubi Point O'Club, someone threw a bottle of champagne through a glass door. When the squadron received the bill for the door's replacement, they paid for two doors, with the comment, "Now you owe us."

One key moment came when the former VA-196 COs did a quick inspection of the squadron; some walked without difficulty, some walked with canes, but all walked proudly in a very quiet hangar. After observing the procession Capt. Toms commented:

> "I always thought the next time I saw this many -196 people in one place I'd go to prison."

"That brought a lot of laughs," said Capt. John Pieguss afterwards:

> "I was Lyle Bull's XO at VA-196 in the mid-1970s, and we were quite a group. We had a great time because Bull's attitude was, 'Don't take yourself too seriously, work hard and play hard. My CO/XO tour was a good one, one of the best, no major problems. It was the best mix of personnel I'd ever seen, from the wardroom to the chiefs down to the enlisted men, absolutely minimal problems. But we did gain something of a reputation; there was a CAG who'd say we were one big discipline problem."

The events on the east coast were similar, with attendees such as Secretary of the Navy John Dalton, *Commander-in-Chief U.S. Pacific Command* – and former *Tiger* skipper – Adm. Joe Prueher, and *Commander-In-Chief, US Atlantic Fleet* Adm. J. Paul Reason. The farewell and cocktail party was held in VA-42's old hangar and featured a mock ready room put together by *Puncher* Lt. Mark "Klinger" Columbo, with the able assistance of several *Intruder* vets, including his father-in-law, the irrepressible Capt. J.B. Dadson, USN (Retired). Afterwards everyone

The *Sunday Punchers* celebrated their over three decades in the *Intruder* by painting one aircraft in colors based on their initial, 1964 cruise in *Independence*. A-6E BuNo 157027/AG514 at NAS Oceana, VA, on 7 March 1997, just minutes before the crew boarded the aircraft for its final flight to the Aerospace Maintenance and Regeneration Center (AMARC) at Davis Monthan AFB, AZ. (Tom Kaminski)

The departure of the A-6 left the fleet with a serious deficiency in the area of precision weapons delivery. The F-14 *Tomcat* community started carrying bombs during the mid-1990s, and by the end of the decade "Bombcats" were carrying the USAF-developed LANTIRN system, which allowed IR imaging and precise laser targeting. The *Tomcat* community, which now included a good number of former *Intruder* crews, did a commendable job of picking up what was, in effect, the Medium Attack mission. The photo shows a VF-154 F-14A launching from *Kitty Hawk* in March 2003 with a LANTIRN pod and two LGBs. (U.S. Navy, PH3 T. Frantom)

went to the World Famous Oceana Officers Club to continue the festivities:

"The preparations started in July, the first month of cruise," recalls Pete Frano:

> "We coordinated with VA-196; Fredo spoke to Gigs and said, 'Let's do something together.' They wanted to do it in January because they were already losing people, we wanted to it in March, and we compromised on February. A bunch of the JOs got involved – it was classic – and every JO had a job, invitations, food, accommodations, dignitaries, etc. We talked about painting the jets, sent out the invitations, and Gigs sent a letter to Admiral Prueher. Once he decided to attend it got *very* big, and we looked at it this way: with this many people, lets turn it into a big party and reunion. Thursday night we had a cocktail party in the original A-6 hangar, old No. 111. Friday was the ceremony, and Friday night was the party at the club. We had the tables set up Thursday evening with the shirts, souvenirs, all the geedunk that wasn't going to be available anymore. A lot of old *Intruder* people came in and loaned us their old Vietnam era flight suits, helmets, things like that. All of the old admirals, the old A-6 guys, were hanging out in there."

At the bitter end – and not without tears, both for fallen comrades and for the end of an era and a community – the last fleet *Intruders* were ceremoniously towed out of their respective hangars to meet their eventual fate. At Whidbey the last A-6 went out of the hangar with a full dress escort, including bagpipes. The final parties at both locations, which went long into the night, were legendary:

> "We took over the Oceana O'Club, said 'this is what we want,' adds Frano. "Try as we might, we didn't drink them out of beer. The manager said it was the best farewell party she'd ever seen."

Afterwards, each squadron painted up one plane with the assistance of their station's AIMD and put them on display. VA-75 took BuNo 151579, did it up right, and parked it in the grass across the street from the officer's club. It sits there now, in honored company with several other aircraft in the NAS Oceana historic collection. At Whidbey two aircraft went on display; BuNo 152907 resides in Oak Harbor's waterfront park, marked for VA-128, while former KA-6D 149482 – with tanker package removed – sits just inside the main gate at the naval air station, loaded for bear with dummy Mk.82s and proudly bearing the motto, "Last Attack In WestPac, 1966-1997."

There were still a few items to sort out. VA-196 formally disestablished at the ceremony on 28 February 1997, with *Attack Wing, U.S. Pacific Fleet* following on 30 April. VA-75 did not officially go away until 31 March 1997, with *Attack Wing Atlantic* stepping down on 30 June 1997. At the CNO Fleet Squadron Awards ceremony held at the 1997 annual convention of the Association of Naval Aviation in San Diego, VA-75 received the last C. Wade McClusky Award ever presented to an attack squadron. The award, named for the famous commander of the *Enterprise* Air Group at the Battle of Midway, subsequently went to strike-fighter squadrons. And, on 12 March 1997 the last crews of VA-75 made the no-kidding, last-ever A-6 CQ. Once again a deck was available – in this case, *Eisenhower* – and the squadron still had two planes left. CAG-17 Capt. Jim Zortman – former skipper of VA-52 – and DCAG-17 Capt. Bud Jewett – former skipper of VA-34 – made the last trip.

Exactly one week later, VA-75's Jin Giggliotti, "Sluggo" Hullinger, "Space" Dacey, and Pete Frano departed for Davis-Monthan with the last two birds. They made the grand tour, with several stops before turning back for the Arizona desert:

"It was a 'farewell tour," says Frano:

> "We started the cross-country at sunset out of NAS Oceana; there was a little bit of water on the runway. It was nice, it was appropriate. We then went to NAS Pensacola for a night, went to Trader Jon's and gave him a squadron memento. Some of the local people knew we were coming through.
>
> "The next day to NAS Meridian, where Capt. Bernie Satterwhite was the base CO. It was hilarious; they didn't expect us! We flew in, the local high school band was playing, the mayor

was there, and the local NJROTC. The weather was bad and they weren't doing any flying that day. There were a lot of A-6 guys there as instructors; there were no A-6 RAGs, so all the A-6 guys were getting sent to Training Command. It was neat; we pulled up, parked the jets, the band's playing, and Gigliotti made a little speech. We had lunch, flew to Albuquerque for gas, then on to Miramar for gas and the night," Frano concludes. "The next day into Davis-Monthan we flew a low-level. It was gorgeous ... we came on in and parked the jets.

"The squadron officially shut down on 31 March, although we had people actually working there until June. Lt. Cmdr. Phil Hullinger – my pilot on the last cruise – was the last officer officially assigned to VA-75. He was a B/N who transitioned to pilot, and probably had 3000 hours in the A-6. He easily screened for command, but they took away his plane, so he retired."

It was a legendary end for two legendary squadrons. "Legendary" also sums up the A-6 *Intruder*, the people that designed it, flew it, maintained it, went to war in it, and lived and died in it. Capt. James McKenzie, career attack pilot, former commanding officer of VA-52, and former *Commander Carrier Air Wing 11*, summed it up best when he closed the proceedings at Whidbey with the following quote from Feodor Dostoyevsky's *The Brothers Karamazov*:

> "Even if we are occupied with more important things, if we attain to honor or fall into great misfortune, still let us remember how good it once was here when we were all together, united by a good and kind feeling that made us perhaps better than we are."

CHAPTER TWELVE

Final Voices

For what it's worth, B/Ns have ALWAYS been "high-tech," or whatever passed for it during the A-6's era. EA-6B ECMOs too.

Two-anchor versus one-anchor talk aside, the key driver for the second-seat role has mainly been system complexity. The A-6A/B/C models had both complexity and – by comparison to today's standards – an atrociously low mean time between failure (MBTF). It was a rare A-6A that came back from even an undemanding full-system-required hop with all the key black boxes in good shape.

One's reputation as a B/N of that era rested as much on the ability to prosecute the mission successfully with a degraded system, or even no system. This was difficult enough with two pairs of arms flailing about the cockpit and certainly beyond the single-handed capabilities of even the best stick. Yours truly sports a lower CEP from running manual range line attacks than his full-system CEP simply because he had more experience with degraded system operations.

My stateroom-mate during my 1973 cruise (a fellow B/N) became the radar instructor at the Lemoore A-7E RAG after his tour with us. I remember him saying that the A-7E, even with its much-improved system reliability over the A-6A, was almost too much for many folks, especially nuggets right out of the TRACOM. Keeping one's brain the proverbial five seconds ahead of the airframe (and the system) must be particularly challenging in such circumstances.

As long as we push the human factors envelope with demanding, often-fallible systems there WILL be a future for the two-anchor community.

— Mike "Owl" Kanze
A-6A, A-6B *Pat Arm*, KA-6D
VA-128, VA-95
Coral Sea Centurion; Veteran of many missions to Toby's Tavern, the Sand Point O' Club and the Greek Village; always a winner at Mom's - Fallon; sometime Liar.

Concerning the *Intruder*, in my case I was always in the right place at the right time. I enlisted in 1948, got selected for flight school and commissioning. In the deepest valley after World War II I was selected for regular. I was selected for command twice – I was in the CO's office at VAH-123 when the A-6 people called, otherwise I would've gone to VAH-2 as CO.

— Cmdr. Jerry Patterson, USN(Ret)

In the old days we said "Douglas for dependability, Lockheed for looks, and Grumman for shear strength." You know that famous fifty-foot obstacle at the end of the runway that's always part of an aircraft's performance specifications? Grumman designed the A-6 to go *through* the obstacle, not over it.

We knew we were ugly – things hanging out all over the place; you could hang another antenna anywhere on the plane and no one would know the difference. We were just going to bulldoze our way through the air and get the bombs on target.

— Capt. Dick Toft, USN(Ret)

I had three periods – actually, four – where I was associated with the A-6. The first was with VA-65 as lieutenant commander during the combat workups and Vietnam cruise in 1969. We were deployed most of the year. The second time was as XO/CO of VA-65 during the mid-1970s. The third period was at NavAir as the program director for the tactical aircraft programs. That time frame was 1987 to 1988 and that was during the time we were trying to develop the A-6F and A-6G.

The last period of time was 1988 to 1990 when I was the commander of PMTC and I was flying the aircraft again in test. It got to fly some at PMTC and thumb my noses at my friends in the F-18s, they'd have to go back for gas and we kept chugging along. We were basically chasing cruise missiles and the like. On one hop the missile launched, did circles then went all the way up to Fallon. Two chase packages, lead aircraft in the second package went down, didn't show up, so I stayed on it with my A-6. The instant it entered the restricted area at NAS Fallon I zoomed – single engine – and managed to land.

— Rear Adm. George Strohsahl, USN(Ret)

The A-6 was a good aircraft that never fully lived up to its potential. It was designed for a specific mission, which it never fulfilled; however, as other roles were applied, it did them well. In *Desert Storm* the *Intruder* did great but it was a very benign environment compared to some place like Hanoi during the Vietnam War.

The A-6F would've been great; its bombing accuracies were only limited by the ballistics of the bombs themselves. With the death of the A-12 however the handwriting was on the wall. The A-6 was going away; it was too expensive to maintain in the mix on the carrier, was old and hard to maintain. The Navy made a wise decision in retiring it.

One thing, however, should be remembered about the *Intruder*: no aircraft, before or since, integrated the pilot and NFO as well. The B/N and pilot achieved a rapport that you never found in any other community. This helped develop camaraderie, which enabled us to perform our mission.

— Capt. Daryl Kerr, USN(Ret)

During my career I was fortunate to command the *Black Falcons* of VA-85, the USS *El Paso* (LKA-117), and the USS *Saratoga* (CV-60). All three went away in 1994. I figure I must be getting old.

— Rear Adm. Charlie Hunter, USN(Ret)

Secretary of the Navy Garrett said back in '91 that he wasn't going to spend money on any aircraft that was designed before the advent of the hand held calculator; one AIRPAC guy said we just didn't need the A-6's mission anymore. Still, you talk to the guys who flew in *Desert Storm* and they'll tell you about all of the F/A-18s lined up behind the A-6s as they went into the targets. The A-6 could carry an unbelievable amount of stuff: five 2000-pound bombs at 500-knots, or *28* 500-pounders. You drop that kind of ordnance and you have shit falling all day.

I've known a lot of aircraft to fall out over time, generally because something has come along to replace them, or because the plane had aged to the point where it was dangerous or has no capability. This is the first time we've retired an aircraft and a mission without a replacement.

— Capt. James McKenzie, USN(Ret)

I flew my last A-6 hop in October 1992, 25 years to the day from my first flight in the aircraft. Afterwards I served as AirLant Chief of Staff for Shore Activities, and retired in the summer of 1993.

I think I was pretty fortunate, when you come down to it, because the longer you were with the *Intruder* and its mission, the more you thought of it. The crew concept was something; every time you flew with somebody it just got better and better. The only way to fly the mission is with two people working close.

I had the opportunity to fly with some of the finest people and worked with the most dedicated maintenance people. Those maintenance people had a special work ethic and bond. Other guys were out in liberty in four-section duty and we were port and starboard, working weekends. It brought us together.

— Capt. J.B. Dadson, USN(Ret)

I ended up with 2614.8 hours in the A-6. I requested them out of Training Command; as an NFO, I knew it was the *premier* NFO airplane. In the community itself the NFO is equal, unlike in the Air Force.

I had a quiet career: no glory, but it was fun, and we did a good job. I've been at Whidbey Island since 1983 – except for a tour at the Army War College – and I've been real happy the whole time. I'm fortunate to have spent my entire career in A-6s, in their prime.

— Lt. Cmdr. Gary "Po Daddy" Poe, USN

I feel fortunate to have worked on A-6s my entire career. It was one of the most difficult – and now the oldest – planes in the fleet to maintain, with all the different missions we were tasked with. If you can keep this plane flying – like we have, with full FMC rates and everything – then you can keep *anything* flying.

We had a lot of pride in the A-6 community. The camaraderie extended through the whole squadron, from the pilots and B/Ns to the maintenance types. We took care of each other. I haven't seen it like this in other communities; we *are* a family, maybe because we've been so isolated up here.

I hope the young guys going to the other communities can spread this pride and esprit de corps to their new squadrons and the fleet. It is a true warrior mentality.

— AVCM(AW) Dave Selby, USN

Fighters make movies, attack makes history. I'm not sure what EA-6Bs do, and I really don't know what *P-3s* do but they sure do win "Yard of the Month" a lot.

— MCPO Greg Troyer, USN(Ret).
At his retirement ceremony

I knew I wanted to fly A-6s after getting a KA-6D hop with VA-52 on my NROTC Second Class Cruise; reading *Flight of the Intruder* my junior year cemented my decision. Actually, what really convinced me was my first class cruise with the VP-45 *Pelicans*. Naaah, none of this maritime patrol stuff for me ... I liked the idea of going low and fast and dropping bombs.

I was truly impressed by the way we managed to keep our 18 1/2-year-old planes flying. The maintenance guys kept them going, full mission capable, and we consistently posted a high sortie completion rate while racking up the flight hours. Quite often our FMC rate was higher than the *Hornet* squadron's. Morale was high. Everyone wanted to make sure this last cruise went right, and look good doing it.

For a while it looked like the A-6 would go out as it had come in, in combat. We were just hours from launching an attack on Iraq under *Desert Strike*, but they sent the cruise missiles in instead. Fortunately, we didn't have to go, but we *were* at the tip of the spear and ready to go. I've never known a more professional group than those who have flown the *Intruder*. Flying *Prowlers* is going to be different, to say the least – it's a very challenging mission – but nothing can replace the thrill of prosecuting a target and putting bombs on it.

It's a bittersweet honor being the junior B/N in the last AIRPAC *Intruder* squadron. I regret that I won't be able to go back to VA-128 and share what I've learned. That's just not going to happen. Grumman could not have designed a better aircraft. Wherever we go, we will take the spirit and camaraderie of Medium Attack with us. Hopefully Naval Aviation will benefit.

— Guy Miller
The Junior AirPac B/N, VA-196, 1996-1997

The thing I liked best about the aircraft – sitting side by side – was the relationship developed with pilots. You get to the point where it's all trust, non-verbal communication. I've worked well with every pilot I've ever flown with one exception.

The best times in the plane? Any low-level, terrain clearance.

— Cmdr. Patrick "Pokey" Keller
The last *Boomer Two*

I was honored and excited to be selected as the last commanding officer of the last AIRPAC A-6 squadron, particularly when it became apparent that the A-6s were going away. I'd really been hoping to get this job; there were a lot of good guys in my year group that could've done it but ended up elsewhere. Out of the thirty-three year history of the A-6, I've put in eighteen; I'm fortunate to have spent over half of the *Intruder's* time in the fleet flying the aircraft.

I've met my career goal, squadron command. It's been every bit as fun and every bit as frustrating as everyone said. Some days you sit at your desk and laugh and some days you pound your desk, but the best part is to be in a position to look at the people that work for you and see what a good job they've done. We have a *lot* of *good* people here. It's been very gratifying.

On our last cruise, the two F-18 squadrons – 25 and 113 – worked pretty hard at becoming better attack pilots, and we helped train them in offset bombing and other attack procedures. They'll need to take up the slack in the all weather attack role; they're trying.

The effects of the loss of the all-weather long-range attack capability from the air wings will take time to shake out ... the dice are still tumbling. We may not need our attack skills for the next ten years; we'll see.

— Cmdr. Dave "Fredo" Frederick, USN
The last *Main Battery One*

The jets at the end were the best. We finally got the Sundstrand CSDs, SWIP, software upgrades, finally got software upgrades for the INS. The jets were the best they ever were, they were fantastic.

As for VA-75's last traps, the boat was looking for aircraft and the CO of *Enterprise* was Capt. Brown so it was appropriate. We got invited. Another thing going for us, Jim "Zoro" Zortran and Bud Jewett were CAG/DCAG-17. They were both A-6 guys.

— Cmdr. Pete Frano, USN
The last *Sunday Puncher Two*

It was a well-earned badge of honor to say you were a B/N. In the military aviation world, it wasn't necessary to say A-6 B/N. There was only one B/N.

A computer is replacing the B/N. It is quicker and more accurate in many ways. It is tireless and it doesn't require career development or command responsibility. When it breaks, it can be fixed as long as you don't mind aborting the mission. B/Ns had a bag of tricks to get the job done no matter what. They could kick the computer pedestal on the lower left side, hold the circuit breaker in while switching from off to standby, work out the formula for time of fall and milliseconds between pickles, use emergency jettison with the right breakers pulled to get the full drop tank off, use the FLIR to find the incoming aggressors, use the radar as a backup radar altimeter, hold a loose avionics box in on a cat shot or slap a pilot on the chest to tell him it's time to eject.

When the last B/N steps across the canopy rail and descends the boarding ladder, Naval Aviation will loose a unique member of the club until a new aircraft comes along that will justify the use of the proud title of Bombardier/Navigator.

— Capt. Bruce "Boxman" Woods, USN(Ret)
Bombardier/Navigator

A-6E Performance

Most impressive bomb load – On a single pass my plane dropped 22 Mk.82s and my wingman dropped 28. Fifty *Snakeye* 82s on a single pass is impressive.
Speed – My fastest level flight was a 100-foot fly-by of USS *Abraham Lincoln* in a light KA-6D, 565-kts indicated airspeed. I saw Mach 1.01 in a 40-degree dive out of 40,000-feet in a clean A-6 at military power. I could see the shock wave forming on the wing but I don't think we were supersonic. With the Mach tuck we could not pull out of the dive until we slowed down passing 15,000-feet, just like NATOPS says will happen.
Altitude – I have been at 45,000-feet, the plane controls are really mushy but it hardly burns any fuel this high. I routinely dropped bombs from 100-feet on the low side and have flown lower, as have most crews.
Range – Twice I have flown non-stop from Whidbey Island to Pensacola, Florida with no tanking. Another time I launched on a cross-country 200-miles west of Los Angeles and flew non-stop to Pensacola. That same launch, another A-6 flew from the same launch position and landed at Andrews AFB in Washington, D.C.
Tailslide – With Lt.Cmdr. Steve Speight, we were doing a single aircraft maneuvering. When we did our 90-degree nose up, zero airspeed stall, we stopped climbing at 25,000-feet. We fell straight backwards to 15,000-feet. Normally the heavy A-6 nose always immediately pointed straight down after any zero airspeed maneuver.

— Lt.Cmdr. Tom Hickey, USN(Ret)

Distinguished guests, ladies and gentlemen, family, friends, *Intruders*, welcome. Had I done my homework, there are several things I would've done differently. One rule is never try to follow CAPT (Jim) McKenzie when giving speeches. The second is don't allow your picture to be placed in the program directly across from a page titled, "Last Tour For An Aging Veteran."

Well, the final moments are approaching. I think most of us, during the past five years, have gone through those stages describing the road to mortality. I guess my anger and denial phases ran longer than most and I'm not far enough into the finality of acceptance. It's hard to explain the depth of emotion we feel for that 15 tons of metal, composite material and wiring sitting so lethally quiet in the corner. For many of us, the significant events of our adult lives beyond our families are woven about a core that is the *Intruder*.

If programs and ceremonies and media coverage were proportional to sheer numbers and contributions to putting *Intruders* over the target, then 90 percent of our attention would be focused on the heart of our community: the sailors, young and old, who maintained our big bird for 30 years. We pilots and bombardiers didn't own the aircraft, we simply borrowed it from those incredibly talented and dedicated professionals.

If aircrew sustained the majority of losses, so too did we get the attention and glory. Perhaps it's our boyish good looks and natural charm, but I suspect the attention fell on us because we were the ones who got to drive the sports car. Tom Cruise probably wasn't interested in the role of a 19-year-old plane captain hauling a hundred pounds of tie-down chains up and down a burning, screaming flight deck 18 hours a day, but he should have been because there are heroes out there on the nonskid. They were known by a variety of terms based on their particular craft: Tweets, Q's, BB Stackers, Metal Benders, Mechs, Crud Crew and Shooters.

The first model of our warhorse, at the leading edge of technology, could be a tad temperamental. We aircrew would dismount, fill out a stack of discrepancies the size of the Old Testament and stroll off for a nap. From then until we manned up again, our sailors practiced their trade, a mixture of craftsman, surgeon, mother and witch doctor. And during their precious few nonworking hours, our sailors tried to snatch some sleep in their coffin sized personal space in a 100-man berthing compartment.

Some of my best hours in the Navy were spent in the shops late at night listening to sailors and marveling at their dedication. I remember a group of young mechs changing an engine on a dark flight deck in record time the night before we entered the Gulf. I remember how every ordnanceman that was temporarily assigned to the galley and laundry suddenly appeared in their red jerseys the first night of *Desert Storm* to hump bombs onto our A-6s.

I remember the glow in a tired fire control technician's face when I told him how the weapon system he worked on all night had just dropped the last bridge in Basra into the river. I remember a gray haired chief flight deck coordinator whose patented hillbilly drawl was unmistakable on the maintenance radio during every launch as he proudly called "*Eagle 501* airborne." And I also remember our Grumman and NAESU reps, standing side by side with the Troops, spending years at sea dispensing their knowledge and skills to generations of sailors.

To all the *Intruder* sailors and chiefs and maintenance officers, from your aviators, I offer a profound "thanks, shipmate." For every hand that pushed the pickle button or held a slew stick to put bombs on target, there were hundreds of other unseen dirty scraped-up hands lying on top of ours at the release point. Nothing hanging on my uniform means as much to me as being called "Skipper" by you.

There's another group of unique individuals I need to talk about today. If there was a single facet of our community that allowed the A-6 to reign supreme in conditions where other aircraft stayed home, it was our bombardier/navigators.

For as long as there have been B/N's, they have been the best the Naval Flight Officer community has to offer. They've led the way for the NFOs at every stage. First to command squadrons, airwings, and carriers. First to flag rank, first to rise to Naval Air Atlantic and fleet command.

Several months ago, I spoke at a ceremony in Lemoore, California as my old squadron, VA-115, transitioned to F-18s. Now, Lemoore is a stronghold of single-seat pilots. Light attack and strike/fighter pilots believe that a computer and their neat heads up display, known as a HUD, can do anything a B/N can do. I know otherwise, so I'll repeat what I sad that day, titled "A B/N is better than a HUD because":

- A HUD won't loan you 50 bucks when you divert unexpectedly.
- A HUD won't bring you back from a night of liberty when you're temporarily confused due to bad local water. On the other hand, it won't gleefully recount your deeds to you the next morning.
- You can't share a particularly glorious Pacific sunset with your computer while holding over the ship.
- A HUD won't get you a date with its sibling or friend. Conversely, its sibling or friend won't turn out to be the date from hell.
- A computer can't sense that you're having a bad day or night and turn up its performance a couple of notches.
- A HUD, like a B/N, helps you get an OK3 wire on a dark night, but it doesn't care if you don't. And it won't celebrate an OK, commiserate on a no-grade and buy you a cheeseburger in the wardroom either way.
- When you're engrossed in trying to salvage an unsalvageable airplane, a computer won't pull the ejection handle and save your life.
- That incriminating HUD video tape tells the skipper what appears to have happened on your flight. A B/N verifies your version of what really happened.
- And most importantly in our line of work, when the computer is sick or the coordinates are wrong or the target doesn't look anything like the photos and twinkly things are flying past your canopy and the whole strike is on the verge of going to hell in a handbasket, a HUD won't say, "Relax, fly the jet, I'll fix this," and then make it all happen.

So from generations of *Intruder* drivers to our right-seaters, our scopes, our beenies, our best friends – thanks pal. We drove you to the party but you're the ones who made 'em dance.

And so, finally, it's time to say goodbye to the focal point that brought us together: our *Intruder*, the last carrier-based pure attack aircraft. This is the age of multi-role capability. Since the A-6 isn't a fighter, it's been declared expendable. But I've often thought, as I review the history of carrier air power, a single mission so dominates the application of that power, it would seem logical to put aircraft that best fill that mission at the top of our priorities.

Although air supremacy is critical to get bombers to the target, carrier-based aircraft have shot down less than 75 airplanes since 1945. In the past 24 years, that number is six. In that same period, covering Korea, Vietnam, and *Desert Storm*, we've flown hundreds of thousands, perhaps millions, of attack sorties.

The sword now passes to the *Hornet* and *Tomcat* communities. It's imperative that the "A" in "F and A" is a capital letter. As for our faithful warhorse, her time has come. The *Intruder* will always be a "she" to me, not an it.

Each aircraft has a personality, karma. You have your babies – the birds in each squadron that always treated you best. And, especially in the old days, you usually had one jet known as a Christine, as in Stephen King's demonic car, which is usually surrounded by perplexed electricians and chaplains.

Ironically, we're parking our jets in the desert when they're at the top of their game. The mid-life updates which brought new electrical systems and composite wings and a host of other changes made her among the most reliable aircraft on the ship. Maintenance man-hours per flight hour while deployed the last few years have averaged slightly above 20. Compare that to other types. It's not a number widely publicized.

But right or wrong, the time is here. If those of us who fixed and flew her seem overly emotional and attached to our *Intruders* ... well, we are. Many of us viewed the significant events of our adult lives beyond our families through those bug-eyed windscreens.

We saw the dark side of a dangerous world. The images are burned in permanently. A flash in the night ahead as you approach a pitching carrier deck, signifying someone who didn't make it and praying it's not an A-6. Taxiing out at Whidbey to start a long cruise and seeing your wife and children in the parking lot. Seeing an A-7 splash into the sea and desperately searching for a chute, looking at hundreds of lines of tracers at 200 feet and 500 miles per hour knowing you have to go through them to get to the target, and looking down at the Whidbey chapel as you perform a missing man fly-by.

But there were other moments – great, wonderful magical moments framed in that canopy. Rolling into the groove behind the ship for the best E-ticket ride in the world, working hard together during a stormy night on a Cascade low level, dropping your practice bomb, hearing bulls-eye and high-fiveing your B/N on the downwind.

Rolling and running down canyons in the clouds, a thousand miles from land. Making just the right move, putting the tailpipes of one of your *Tomcat* buddies squarely in your gunsight, seeing your wife and kids standing on the ramp seven months and a small war later. Seeing a grimy troubleshooter emerging from under your jet with a smile and thumbs up, mouthing the words "Good to go, sir."

And especially for us – pulling up from the final point of our favorite Cascade low level after 30 minutes of pure joy and awe among the glaciers and canyons and seeing Whidbey framed in that windscreen – the water, the mountains, home.

And this marvelous machine was our chariot to all those places. She took us to bad places on dark nights to strike at our nation's enemies. She took us to beautiful places on clear days to marvel why anyone would want to do anything else. And she brought us home to the best of her ability. A unique machine, tough, forgiving, honest – a thoroughbred to ride into battle.

Several weeks ago Dave Frederick and I were fortunate enough to make the last Whidbey A-6 carrier landing right out here in Puget Sound. In those final moments sitting on the catapult with my *Intruder* trembling at full power, I tried to take it all in – the cockpit, as familiar as the face of an old friend, the cat track and flight deck stretching out to the Olympic Mountains in this place we love, the hundreds of cheering sailors lining the bow. All the wonderful magic things about our profession were there in that moment.

And I locked eyes with the cat officer, fired off my last salute as a carrier attack pilot, and felt that tremendous surge as we were flung towards the sky this final time. And as we flew down the cat I screamed in my mask. It was an explosion of emotion for great days, dark nights, lost shipmates, blue water, gold wings, exploding bombs, and OK-3 wire, red tracers, best friends, and above all, for my A-6.

No matter what else I may do, if someone ever asks me what I did with my life, I'll puff up my chest and tell 'em, "By God, I flew *Intruders*."

Good-bye, Baby.

— CAPT Terry Toms, USN, COMATKWINGPAC
Presented at the VA-196/wing disestablishment ceremony

The Mission: Attack

1. The mission of the aircraft carrier is to put ordnance on target. Everything else is simply support for the attack mission.

2. You win the war by killing the bastards by the thousands, not one at a time at twenty-thousand feet.

3. In peacetime, DCM is something the attack pilot uses to rejoin off the range.

4. In wartime, DCM is something the attack pilot uses to turn and shoot the enemy in the face who's trying to stop the attack pilot before he destroys his high value target.

5. There is no such thing as "Defensive" DCM. I become offended when someone jumps me enroute to my target, and much offense is intended when I have to take the time to blow his face off.

6. In wartime our POWs were not released because the enemy sent representatives to sit smugly at "peace talks"; they were not released because domestic anti-war groups unwittingly played into the hands of the enemy and tied the hands of their countrymen at arms; they were not released because the enemy lost five aircraft to a select few called "Aces"; they were released because brave men took their bombers downtown and spoke personally to their captors in the only language the enemy understands: IRON BOMBS RAINING DOWN ON THEIR HEADS.

7. These lessons have been forged in blood and steel by all those attack pilots and bombardiers who have gone before you; back when happiness was flying *Spads*; back when jets were hard-lightin' and mean, and only quiche eatin' airline pukes flew fans; back when *Spads* roamed the valleys and spit death to those who would try to stop the *Skyhawks*; in an earlier time when the biggest Cadillac in town was called *BUFF* and when men took pride in decorating their leather flight jackets with "I've been there" patches, and the enemy hid every 1+45 because he knew the next cycle of the attack carrier was headed his way. Times change, technology changes, but the men in the cockpit must be the same brave warriors every age has counted upon in time of peril.

8. Finally, and this is the bottom line, real men fly attack because they understand the most fundamental law of wartime negotiations: YOU NEGOTIATE WITH THE ENEMY WITH YOUR KNEE IN HIS CHEST AND YOUR KNIFE AT HIS THROAT.

— From the VA-165 Disestablishment Program

The King is Dead: Long Live the King

The retirement of the A-6 earlier than perhaps required reminded some of sports superstars who leave at the pinnacle of their games. Like Mario Lemieux and Michael Jordon the A-6 was a hero that departed the arena with a lot of game left and with the crowd screaming for more. Unlike both of these sports legends, however, the *Intruder* wouldn't return.

Naval Aviation rarely seems to have the luxury of deciding where its next war will be fought. Four short years after the retirement of the A-6 *Intruder* the U.S. Navy found itself fighting a war in Afghanistan. Anyone who says they saw this one coming is telling a story, to put it mildly.

As always, carrier aviation answered the call, this time using F-14 *Tomcats* – another classic from the Grumman Ironworks – as well as the McDonnell-Douglas (now Boeing) FA-18 *Hornet* as the primary weapon delivery vehicles. And although they performed well what the air wings really needed was a bomb truck, an aircraft that was light on tanking requirements and could carry a large load, which could be delivered with extreme accuracy. This war begged for the *Intruder* but it wasn't available. The *Toms* and *Hornets* also needed a lot of tanking and the only asset available from the Navy's flight decks was the S-3B *Viking* – which, although it was game, has never been a really adequate strike tanker that could keep up with the bomb carriers. Again, the *Intruder* would have been a more appropriate airframe in this situation.

But let's face it. The *Intruder's* time had come. Technology, the budget, and airframe life all conspired against the great aircraft. Now let's get over the loss of the A-6. The king is dead. Long live the king.

It's been argued that the Navy ended the career of the *Intruder* before its time was up. Actually, there were a lot of good reasons to accelerate the type's retirement, operational cost being one of the big ones. The Navy has also been anxious to get itself down to one basic jet airframe on its carrier decks, a process that is ongoing and will soon lead to the demise of the *Tomcat* and even it appears the *Prowler* by about 2015.

So how is the *Hornet*, as A-6 replacement, designated heir, and potential hall of famer? As of this book's writing the *Hornet* is a well-established veteran of over 25 years of fleet service. As a land-based replacement for *Voodoos*, *Mirages*, *Phantoms*, and even MiGs, the F-18 series has done everything it was asked of it. There seem to be few complaints about the type from the Aussies, Canadians, Spaniards and Finns.

On the boat in spite of an impressive operational record, its critics still center on its problems with range, even if its fuel requirements have supposedly have never been any worse than the F-4s. Right or wrong, the type seems to be forever stuck with the image of being "short legged". The E/F versions feature greater range and more capability but it still strikes some observers as odd that the newest, most capable, most expensive aircraft we have is being touted for its role as a TANKER.The fact is, the *Hornet*, particularly the E/F version, is the carrier plane of the future, and a lot of good men, and a few good women will make it work in fine Navy tradition.

Beyond that the F-35 Joint Strike Fighter is several years down the road. Perhaps the JSF will be the last major manned combat aircraft we buy, as strike UAVs become more capable and alluring to the bean counters. (as they say, unmanned aircraft make lousy POWs). Then there's the cost of new aircraft, which has been rocketing up over the years. Maybe we will yet reach that day suggested by some wags when we spend the entire annual DOD budget on one aircraft, and then we can at long last, to paraphrase President Coolidge's famous, and perhaps apocryphal, quote, "buy just one, and let all those pilot fellows take turns flying it".

In the end, there's every reason to believe that the FA-18 will some day hold the same revered place of honor now held by the likes of the *Intruder*, *Hellcat, Skyhawk* and *Phantom*. The *Hornet* community should hope, and strive, for no less

Homecoming

On Friday, 28 April 2000 – fully three years after the final retirement of the mighty *Intruder* – the nation observed through various means the 25th anniversary of the fall of Vietnam. In a highly publicized trip Senator John McCain of Arizona – *Scooter* pilot, dedicated attack aviator, retired Navy Captain and a former POW – made a return visit to Vietnam.

The visit – McCain's first since his 1973 repatriation from the Hanoi Hilton – drew a substantial amount of press coverage. Several pundits and members of the press concentrated on specific aspects of the Vietnam War, such as the Tet Offensive and "Agent Orange." Above and beyond the facts of McCain's shoot down and his imprisonment few commentators even talked about the thousands of Navy, Air Force, Army and Marine Corps aviators who answered their nation's call and never came home. However, a few months earlier, two *Intruder* crewmen finally returned home; other than local press coverage, few noted their arrival.

On 9 December 1999 the government announced it had successfully recovered and identified the remains of Capt. Norman E. Eidsmoe of Rapid City, SD and Lt. Cmdr. Michael E. Dunn of Naperville, IL. Eidsmoe – known to his friends and squadron mates as "Buzz" – was a member of the original VAH-123 *Intruder* cadre. He died near Vinh on 26 January 1968 while assigned to VA-165, along with Mike Dunn, his B/N. Thirty-two years after deploying with their squadron in *Ranger*, both men finally came home. They received their final promotions posthumously; Eidsmoe's family – wife Betty, their five kids and numerous grandchildren who never met their grandfather – later buried him in Arlington National Cemetery while Dunn's family interred him in Illinois. Notably, Rear Adm. Lyle Bull's son Lt.Cmdr. Dell Bull led the missing man flyover at Arlington.

Subsequently, the Laotians returned the remains of Maj. Charles E. Finney of Saltillo, Miss. Finney, a member of VMA(AW)-533, disappeared while on a mission over Laos on the night of 17 March 1969. A joint U.S.-Laotian team turned up the remains between 1995 and 1999 after talking with local villagers and excavating the crash site in Savannakhet Province. Again, the Army's Central Identification Laboratory on Oahu made the final identification; Finney's family also interred him in Arlington National Cemetery, on 17 March 2000.

The announcement came towards the end of Secretary of Defense William Cohen's official visit to Vietnam. The government reported that with the return of Finney, 2,029 servicemen of all military services remained missing in action from the Vietnam War. Efforts to locate and identify them continue.

Mark Morgan, Issaquah, WA
Rick Morgan, Woodbridge VA
2004

Appendices

Appendix A: Intruder by the Numbers

INTRUDER Bureau Numbers

Shop #	BuNo	qty	version	
1-4	147864-7867	(4)	A2F-1	redesignated A-6A on 18 Sep 1962.
5-8	148615-8618	(4)	A2F-1	redesignated A-6A on 18 Sep 1962.
9-20	149475-9486	(12)	A-6A	
21-44	149935-9958	(24)	A-6A	
45-87	151558-1600	(43)	A-6A	151595-1600 conv. To EA-6A on line
88-135	151780-1827	(48)	A-6A	
136-199	152583-2646	(64)	A-6A	
200-263	152891-2954	(64)	A-6A	
264-311	154124-4171	(48)	A-6A	
312-452	155581-5721	(141)	A-6A	
453-488	156994-7029	(36)	A-6A	
E-1 – E-12	158041-8052	(12)	A-6E	
E-12 – E-24	158528-8539	(12)	A-6E	
E-25 – E-36	158787-8798	(12)	A-6E	
E-37 – E -48	159174-9185	(12)	A-6E	
E-49 – E-57	159309-9317	(12)	A-6E	
E-58 - E-70	159567-9579	(12)	A-6E	
E-71 – E-82	159895-9906	(12)	A-6E T/P	TRAM provisions
E-83 - E-93	160421-0431	(11)	A-6E T/P	
E-94, E-95	160993-0994	(2)	A-6E T/P	
E-96 – E-123	160995-1111	(28)	A-6E TRAM	TRAM
E-124 – E-129	161230-1235	(6)	A-6E TRAM	
E-130 – E-161	161659-1690	(32)	A-6E TRAM	
E-162 - E-164	162179-2181	(3)	A-6E TRAM	
E-165	162182	(1)	A-6E SWIP	Systems Weapons Improv.
E-166 – E-170	162183-2187	(5)	A-6F	
E-171, E172	162188-2189	(2)	A-6E TRAM	
E-173 – E-184	162190-2201	(12)	A-6E SWIP	
E-185 – E-195	162202-2212	(11)	A-6E SWIP/CW	Composite Wings
E-196 – E-205	164376-4385	(10)	A-6E SWIP/CW	

New Construction

A-6A	488	
A-6E	70	
A-6E T/P	25	"TRAM Provisions". Wired for TRAM, delivered without turret.
A-6E TRAM	71	"Target Recognition Attack Multisensor"
A-6E SWIP	34	"Systems & Weapons Improvement Program"
A-6F	5	
TOTAL	**693**	

In addition, 15 EA-6As (156979-6993) were built new on the Grumman line, for a grand total of **708** *Intruder* airframes.

Cancelled Intruders:

The Navy reportedly assigned an additional 217 bureau numbers to A-6s that were canceled.
A-6A: (76) 151601-1612, 152955-2964, 155137-155190.
A-6B: (54) 154046-4099
KA-6D: (20) 158053-8072
A-6E: (37) 161112-1114, 161236-1241, 161691-1694, 161886-1897, 162211-2222,
A-6F: (30) 163955-3984

***INTRUDER* modifications**

The A-6 went through a mind-numbing number of modifications through the years, particularly in the E-model, which had at least four major sub-variants. (A-6E, T/P ("Cains"), TRAM and SWIP). A-6As which went through the A-6E, "A-6E T/P" or "A-6E TRAM" modification were assigned new shop numbers M-1 through M-240. The following information is from official Grumman sources dated May, 1995.

NA-6A: 147866, 147867, 148617,
NEA-6A: 149935
YEA-6B (1) 148615
NEA-6B (2) 149479, 149481
EA-6A: (12) 147865, 148616, 148618, 149475, 149477, 149478, 151595-151600
A-6B: (19): 149944, 149949, 149955, 149957, 151558-1565, 151591, 151820, 152616, 152617, 155628-5630
A-6C: (12): 155647, 155648, 155653, 155660, 155662, 155667, 155670, 155674, 155676, 155681, 155684, 155688
KA-6D: 78 from A-6A: 149482, 149484, 149485, 149486, 149936, 149937, 149940, 149942, 149945, 149951, 149952, 149954, 151566, 151568, 151570, 151572, 151575, 151576, 151579, 151580, 151581, 151582, 151583, 151589, 151783, 151787, 151789, 151791, 151792, 151793, 151795, 151796, 151801, 151806, 151808, 151809, 151810, 151813, 151814, 151818, 151819, 151821, 151823, 151824, 151825, 151826, 151827, 152590, 152592, 152597, 152598, 152606, 152611, 152618, 152619, 152624, 152626, 152628, 152632, 152637, 152892, 152893, 152894, 152896, 152906, 152910, 152911, 152913, 152914, 152919, 152920, 152921, 152927, 152934, 152939, 154133, 154147, 155691.
KA-6D: 12 from A-6E (all A-6A previously modified to "E" standards): 152587, 154154, 155582, 155583, 155584, 155588, 155638, 155686, 155597, 155598, 155604, 155619.
A-6A to A-6E: 120
A-6A to A-6E T/P: 94
A-6A to A-6E TRAM: 26
A-6E AWG-21 (AFC-409, 1978):151593, 151782, 151812, 152929, 155588, 158539, 158792, 158795, 159185, 159574.
A-6E TRAM AWG-21: 154140, 158043, 158796, 158797, 159177, 159182, 159571, 159579.
A-6E and A-6E T/P to A-6E TRAM: 242
A-6E SWIP: 188 modified from all sources
A-6E SWIP/CW: 136 modified from A-6E TRAM and A-6E SWIP

<u>TC-4C *Academe*</u>
155722-155730 (9) TC-4C
As of 2003, seven TC-4Cs remain in storage at AMARC at Davis-Monthan AFB, AZ. 155722 was lost with VMA (AW)-202 on 16 October 1975. 155723 is on display at the National Museum of Naval Aviation, Pensacola FL.

Intruder Specifications
Zero fuel weight: 27,400 lbs (A-6E), 28,300 lbs (A-6E TRAM), 27,300 (KA-6D)
Maximum gross weight: 60,400 lbs (field), 58,600 lbs (catapult)
Maximum ship landing weight: 36,000 lbs
Length: 54 feet 9 inches (radome to end of rudder)
Height: 16 feet, 2 inches (ramp to top of tail, wings spread or folded),
21 feet, 11 inches (wing folding arc)
Wingspan: 53 feet (wings spread), 25 feet 4 inches (wings folded)

Engines: two J52-P8 engines, 8,700 pounds military rated thrust

Fuel System: (JP-5 @ 6.8 pounds/gallon)

Total:	2,344 gallons	(15,939 pounds)
Fuselage:	1,326 gallons	(9,016 pounds)
Wings:	1,018 gallons	(6,923 pounds)

Aero 1D external fuel tanks: 300 gallons, 2040 lbs each

Maximum speed limited by airframe (there is no 'red line' airspeed- the *Intruder* could fly as fast as the crew, and gravity, could coax it to go)
G limitations: -2.4 to + 6.5, in normal symmetrical flight. (note: these are the best possible values. Actual G limits varied widely depending on turbulence, asymmetric external store loads and aircraft weight)

Source: NATOPS flight manual, dtd 15 Mar 1983.

PERSONAL MILESTONES:
<u>5000 Intruder flight hours</u>

Larry Munns:	Top Naval Aviator.
Jim Burin:	Top Naval Flight Officer.

<u>4000 Intruder flight hours</u>

Marty "Mallard" Allard
Graham "Buck" Gordon
Gene Porter
Denby "Heel" Starling
Jim "Simo" Symonds
WC "Gator" Chewning
Harry "Bud" Jupin
Mike Reilly
George Starnes
TJ "T-Tom" Toms
Jack "Yukon" Goldlewski
Dave "Nickel" Nichols
Dave "Roy" Rogers
Donald "Sully" Sullivan
Bruce "Boxman" Wood

101 reached 3000 flight hours
296 reached 2000 flight hours
715 reached 1000 flight hours.

Source: Disestablishment program for VA-196, 28 Feb 1997.

Appendix B: Navy Units

<u>NAS Oceana, VA</u>
Commander, Medium Attack Wing One (ComMatWing ONE)
1 Oct 1971 to 2 Aug 1993.
Commander, Attack Wing, U.S. Atlantic Fleet (ComAtkWingLant)
2 Aug 1993 to 1 Apr 1997.

<u>NAS Whidbey Island, WA</u>
Commander Medium Attack, Tactical Electronic Warfare Wing, Pacific Fleet (ComMatVaqWingPac)
30 June, 1973 to 31 Jan 1993.
Commander Attack Wing, U.S. Pacific Fleet (ComAtkWingPac)
1 Feb 1993 to 30 Apr 1997.

Most carrier-based Navy squadrons report to two senior commands, usually a Carrier Air Wing (CVW) for operational control (OPCON) and either the CVW or a Functional Wing commander for support and administrative control (ADCON) while based ashore.[1] The Functional or Type Wing usually acts as a liaison between the squadron and base, coordinates intermediate level maintenance activities as well as performing mundane administrative and inspection functions. Historically such commands were given the title "Fleet Air", as in "Fleet Air Norfolk" and these would provide support for all aircraft at a specific base or even region.

This worked well when Carrier Air Wings were located with all of their squadrons but during the middle 1960s aircraft were based with same types to simplify parts and training support. During the early 1970s Fleet Air commands were also broken up along type lines, so that distinct A-6 activities became "Medium Attack Wings", as opposed to "Fighter Wings" (at Oceana and Miramar) or "Light Attack Wings" (at Cecil and Lemoore). In Whidbey the A-6 and EA-6 communities were combined into the cumbersome "Medium Attack, Tactical Electronic Warfare Wing, Pacific Fleet" (MatVaqWingPac, or "VacuumPac" in informal conversation).

In 1993 these Flag level commands (typically a one-star Admiral) were reduced to Captain billets, as the Navy stripped field flag commands to beef up its joint and Pentagon staffs. At this time the "Medium Attack" title was dropped as the light attack community (A-7s) had transitioned to Strike Fighters (FA-18). Both *Intruder* wings survived until the end of the community. At Whidbey the A-6s and EA-6s were separated to form two distinct wings, each commanded by its own Captain.

Navy Squadrons

The squadron is the basic organizational unit in Naval Aviation. Squadron titles are assigned along mission lines and assigned by higher authority, with the VA designation denoting a fixed-wing attack unit. The A-6 community was known Navy wide as "Medium" attack (casual designation "VAM"), denoting it's relative position between the formally recognized "Heavy Attack" (VAH) units flying A-3s and A-5s and the so-called "Light Attack" (commonly called "VAL") squadrons that operated A-4 and A-7 types.[2]

Unit heraldry and nicknames are adopted by the squadrons themselves, although insignias require approval through the office of the Chief of Naval Operations.

During the period the *Intruder* was in the fleet (1963-1998) official policy was that while unit designations could be repeated, actual lineages and honors only applied to squadrons that stayed in a continuous state of existence. For instance, while VA-34 was established as an A-6 unit in 1970 and adopted the insignia and nickname of the preceding A-4 unit with the same *Blue Blasters* title; it *could not* (at least officially) carry on the honors or history of the predecessor. It was allowed to carry on the "traditions" of the original unit however, for whatever that was worth. For official purposes, the new VA-34 only dates from 1970, although this distinction is frequently missed (or ignored) in unit histories. This policy has caused no small amount of confusion for aviation historians ever since.[3]

Key to unit listings:
Nickname: Title(s) the squadron was known by, either official or unofficial. These may change in time or can be confused with a squadron's assigned radio callsign, which in some cases ends up also becoming the nickname. Callsigns are given where known for disestablished squadrons.
From: When the squadron was officially established (not "commissioned" a term reserved for ships and officers) or what type of aircraft it transitioned from into the *Intruder*. Although usually given as a specific month or even date, actual time to transition may actually span several months.
To: Either the official date of disestablishment (Naval squadrons are not "decommissioned") or date of redesignation. In some cases the actual date of the formal ceremony is also given, where it was known to differ from the official "paper" date.

Homestations: Naval Air Station(s) the unit was based at between deployments.
Models flown: Versions of the *Intruder* used by the squadron. Atlantic Fleet squadrons also frequently operated a small number of A-6B models for short periods while deployed to the Mediterranean through the early to mid-1970s, but these may not show up in all listings.
Subordination: For replacement units only. The reporting senior command for the squadron.
Notes: Significant points and other information about the command, including combat and other relevant squadron details.
Deployments: Dates the unit is forward deployed on a carrier. This list does not include operational areas near the Continental U.S. or shakedown cruises on new carriers. Information includes carrier, carrier air wing, type aircraft, tailcode and modex, dates and location of deployment. Exact dates are not given as sea going squadrons frequently travel to their ship's location prior to the official date of departure of the carrier, and can arrive home before or after the ship returns to its own homeport.
Med: Mediterranean Sea, NL/NorLant: North Atlantic Ocean, SL: South Atlantic Ocean, RS: Red Sea, WP: Western Pacific, IO: Indian Ocean, Carib: Caribbean, PG: Persian Gulf. Horn: "Around the Horn", as in the tip of South America. These events were usually conducted as part of a carrier transfer between fleets. World: "Around the World", transferring between coasts the long way, which could be either east to west or vice versa. Significant combat operations are denoted by a **boldface.** These include Vietnam (South East Asia, or SEA) and operations such as *Desert Storm, Prairie Fire, El Dorado Canyon, Praying Mantis,* Grenada, and Lebanon.

Fleet Replacement Squadrons
Training units were popularly known as RAGs, for 'Replacement Air Groups", a term that had technically become obsolete in 1963 when the word 'Group' was replaced by 'Wing' in Navy structure. The title stills sees use however and its longevity is perhaps due to the fact that it rolls out of the mouth easier than the latter official title, which was Fleet Replacement Squadron, or FRS. The use of the term RAG generates additional confusion as it's been used freely to describe specific *squadrons* even though it more properly describes the dedicated Combat Readiness Air Groups/Wings, (RCVW-4 in AirLant and RCVW-12 for AirPac) which were both disestablished in 1970.

RAGs are responsible for the training of replacement aircrew, transitioning squadrons and maintenance personnel under FRAMP, the Fleet Replacement Maintenance Program, and were normally under the command of a senior Commander following a successful tour in a fleet unit although for a short period in the 1980s they were designated "Major Commands" and had Captains assigned as commanding officers. Frequently overlooked in histories, RAGs were absolutely vital in both role and function.

VA-42
Nicknames: *Green Pawns, Thunderbolts* (from 19 Oct. 1992).
From: Transitioned from the A-1 *Skyraider* from Feb. 1963.
To: Disestablished, 30 Sep. 1994. (23 Sep ceremony)
Home station: NAS Oceana, VA
Models Flown: A-6A, B, C, E, KA-6D, TC-4C. **Losses**: 14 operational
Subordination: Replacement Carrier Air Wing FOUR (RCVW-4) Feb. 1963 to 30 Apr. 1970.
Commander, Fleet Air Norfolk (ComFairNorfolk): 1 May, 1970 to 30 Sep, 1971
Commander, Medium Attack Wing ONE (ComMatWing ONE): 1 Oct, 1971 to 2 Aug 93
Commander, AttackWingAtlanticFleet (ComAtkWingLant) 2 Aug 93 to 30 Sep 94.
Notes: VA-42 was the East Coast A-6 replacement squadron for over 32 years. In 1992 the unit abandoned the *Green Pawn* name and insignia, which had been with the unit since original establishment as an F4U *Corsair* outfit in 1950. In its place the *Thunderbolt* title and insignia was adopted from the recently disestablished VA-176. Through the years the squadron used the A-6A and E as its primary training aircraft and the B, C and KA-6D versions in a limited fashion. The Grumman TC-4C *Academe* arrived in 1968, and remained until the end. A few T-28s and A-1s were retained until the completion of propeller instrument training in March, 1964. During the 1980s a permanent detachment was maintained at MCAS El Centro, CA in conjunction with VA-128. Functioned as the A-6 NATOPS Model Manager throughout its tenure, and as such was responsible for standardized training in the type for all Navy and Marine squadrons. Assigned to Replacement Carrier Air Wing-FOUR (RCVW-4) from the start of *Intruder* training until the wing's disestablishment on 1 Jun 1970. Aircraft carried the "AD" tailcode throughout its life, representing AirLant replacement units.

VAH-123
Nickname: *Pros*
From: Established from Heavy Attack Training Unit, Pacific Fleet (HatTuPac) on 15 Jun, 1957.
Intruders were assigned from August, 1966.
To: *Intruder* detachment became VA-128 on 1 Sep, 1967. (see notes)
Home Station: NAS Whidbey Island, WA.
Models Flown: A-6A **Losses**: None
Subordination: Commander, Replacement Carrier Air Wing TWELVE (RCVW-12)
Notes: "Heavy-123" was the Pacific Fleet's A-3 *Skywarrior* replacement unit. It was the only Heavy Attack squadron to have *Intruders* assigned. A detachment was established in August, 1966 to begin west coast *In-*

truder training, the first A-6 arriving on the 17th of that month with a cadre of instructors going through training at VA-42. 123 was assigned to Replacement Carrier Air Wing 12, with its *Intruders* wearing an NJ-35X tailcode and modex. The squadron retained the *Intruder* replacement mission alongside its A-3 training role for about a year, when the detachment became the basis for VA-128 on 1 Sep, 1967. VAH-123 continued as the Navy's A-3 replacement squadron until disestablished 1 February 1971.

VA-128

Nickname: *Golden Intruders*, call sign "Phoenix", (frequently shortened on flight plans to "fenix")
From: Established 1 Sep. 1967
To: Disestablished 29 Sep. 1995
Home Station: NAS Whidbey Island, WA.
Models Flown: A-6A, E, KA-6D. TC-4C. (A-6B and C may have also been assigned for short periods)
Losses: 13 Operational
Subordination: Replacement Carrier Air Wing TWELVE (RCVW-12) 1967 to 30 Apr 1970.
Commander, Fleet Air Whidbey (ComFairWhidbey) 1 May, 1970 to 28 Feb. 1973.
Commander, Medium Attack Tactical Electronic Warfare Wing Pacific Fleet (ComMatVaqWingPac) 1 Mar 1973 to 21 Jan 1993
Commander Medium Attack Wing Pacific Fleet (ComMatWingPac)
Commander AttackWingPac (ComAtkWingPac): 21 Jan 1993 to 29 Sep 1995
Notes: The west coast replacement training squadron, formed from a detachment of VAH-123. First TC-4C assigned on 15 Mar, 1968. First A-6E arrived on 16 Dec. 1973. Assumed duties of training Marine A-6 crews from Oct. 1986 with the deactivation of VMAT-(AW)-202. The unit's insignia and traditions were adopted by VAQ-128, an EA-6B *Prowler* unit established at Whidbey on 9 Oct. 1997 with the name *Fighting Fenix*. It should be noted that the Whidbey-based *Prowler* RAG, VAQ-129, flew a very small number of A-6As while awaiting new EA-6Bs during the 1971-72 time frame.

FLEET SQUADRONS

VA-34

Nickname: *Blue Blasters*
From: Est. 17 Apr. 1970. (official date of 1 Jan)
To: Redesignated VFA-34 on 1 Oct 1996 and transitioned to FA-18C.
Homestation: NAS Oceana, VA
Models flown: A-6A,B,C,E. KA-6D. **Losses**: 0 combat, 9 operational
Notes: The *Blue Blasters* were the seventh, and last, of the Vietnam era *Intruder* units established at Oceana, Virginia. The official establishment date is recorded as 1 January, 1970, although formal proceedings were not conducted until the 17 April date. The unit adopted the insignia and traditions of a Cecil Field based A-4 unit that had been disestablished on 1 June, 1969. Not deployed to South East Asia, the *Blasters* would never see sustained combat operations, although they participated with distinction in Operations *El Dorado Canyon* and *Prairie Fire*. In the later the squadron was credited with sinking two Libyan missile boats. In 1988 the *Blasters* participated in Operation *Desert Shield* while a member of CVW-7. In 1994 the squadron participated in the 50th anniversary ceremonies of the D-Day Invasion of Normandy, France, painting aircraft with "invasion stripes" in the style of 1944. For its last *Intruder* deployment the squadron was involved in NATO peacekeeping activities in Bosnia, participating in *Operations Deny Flight* and *Decisive Endeavor*. Following what would be the only *Intruder* cruise in *George Washington*, VA-34 became one of only two Medium Attack squadrons to transition to the FA-18 *Hornet*.
Deployments:

9/70-3/71	John F.Kennedy	CVW-1	AB5xx	A-6A/B	Med/Norlant
12/71-10/72	John F.Kennedy	CVW-1	AB5xx	A-6A/B/C,KA	Med/NorLant
4/73-12/73	John F.Kennedy	CVW-1	AB5xx	A-6A/B/C,KA	Med/NorLant
6/75-1/76	John F.Kennedy	CVW-1	AB5xx	A-6E, KA	Med
9/76-11/76	John F.Kennedy	CVW-1	AB5xx	A-6E, KA	NorLant
1-8/77	John F.Kennedy	CVW-1	AB5xx	A-6E, KA	Med
6/78-2/79	John F.Kennedy	CVW-1	AB5xx	A-6E, KA	Med
8/80-3/81	John F.Kennedy	CVW-1	AB5xx	A-6E, KA	Med
8-10/82	America	CVW-1	AB5xx	A-6E TRAM, KA	Med/NorLant
12/82-6/83	America	CVW-1	AB5xx	A-6E TRAM, KA	Med/IO
4-11/84	America	CVW-1	AB5xx	A-6E TRAM, KA	Med/NorLant
8-10/85	America	CVW-1	AB5xx	A-6E TRAM, KA	NorLant
3-9/86	**America**	**CVW-1**	**AB5xx**	**A-6E TRAM, KA**	**Med/Libya**
2-8/88	Dwight D. Eisenhower	CVW-7	AG5xx	A-6E TRAM, KA	Med
3-9/90	Dwight D. Eisenhower	CVW-7	AG5xx	A-6E TRAM, KA	Med,PG, NL
5-11/94	Dwight D. Eisenhower	CVW-7	AG5xx	A-6E SWIP, KA	Med, RS, PG
1-7/96	George Washington	CVW-7	AG5xx	A-6E SWIP	Med, RS, PG

VA-35

Nickname: *(Black) Panthers* (callsign: "Raygun")
From: Transitioned from the A-1 *Skyraider* from July, 1965.
To: Disestablished 31 Jan. 1995. (24 Jan. ceremony)
Home Station: NAS Oceana, VA.
Models Flown: A-6A, B,C, E. KA-6D. **Losses**: 9 combat, 9 operational
Notes: One of the more illustrious squadrons in the history of Naval Aviation, the *Panthers* date back to 1934 when originally established as Bombing Squadron (VB) 3B. Combat honors include WWII (with SBDs and SB2Cs) and Korea (with AD *Skyraiders*). As an *Intruder* unit, VA-35 saw combat in Vietnam (four deployments) and *Desert Storm.* The unit carried out the first aerial mining missions since WWII on 26 Feb, 1967 at Song Ca, North Vietnam. In October, 1967 VA-35 became the first fleet *Intruder* unit to fire a live AIM-9 *Sidewinder* air-to-air missile. In 1979 it became the first fleet unit to accept A-6E TRAMs. Tankers from the squadron had a supporting role in the 1980 movie "The Final Countdown". VA-35 was the last AirLant *Intruder* squadron to deploy with a Pacific Fleet carrier, in 1970. When disestablished in 1995, the unit was the oldest continuously operational aviation command in the US Navy, after having flown off 30 different carrier decks and fought in four wars. The squadron's battle honors included Midway, Eastern Solomons, Guadalcanal, Iwo Jima, Korea, Vietnam and *Desert Storm.* For whatever reason, the Navy did not have a place for this distinguished unit in the early 1990s drawdown, preferring to retain several Strike Fighter squadrons with lineages only a fraction as significant.
Deployments:

11/66-7/67	**Enterprise**	**CVW-9**	**NG5xx**	**A-6A**	**WP, SEA**
1-7/68	**Enterprise**	**CVW-9**	**NG5xx**	**A-6A/B**	**WP,SEA**
9/69-7/70	**Coral Sea**	**CVW-15**	**NL5xx**	**A-6A**	**WP, SEA**
6-12/71	America	CVW-8	AJ5xx	A-6A/B/C, KA	Med
6/72-3/73	**America**	**CVW-8**	**AJ5xx**	**A-6A/C, KA**	**WP,SEA**
1-8/74	America	CVW-8	AJ5xx	A-6E, KA	Med
9,10/74	America	CVW-8	AJ5xx	A-6E, KA	NorLant
7-9/75	Nimitz	CVW-8	AJ5xx	A-6E, KA	NorLant
7/76-2/77	Nimitz	CVW-8	AJ5xx	A-6E, KA	Med
12/77-7/78	Nimitz	CVW-8	AJ5xx	A-6E, KA	Med, NorLant
9/79-5/80	Nimitz	CVW-8	AJ5xx	A-6E TRAM, KA	Med, SL, IO
8-10/80	Nimitz	CVW-8	AJ5xx	A-6E TRAM, KA	NL
8/81-2/82	Nimitz	CVW-8	AJ5xx	A-6E TRAM, KA	Med
11/82-5/83	Nimitz	CVW-8	AJ5xx	A-6E TRAM, KA	Med
3-10/85	Nimitz	CVW-8	AJ5xx	A-6E TRAM, KA	Med
8-10/86	Nimitz	CVW-8	AJ5xx	A-6E TRAM, KA	NL
12/86-7/87	Nimitz	CVW-8	AJ5xx	A-6E TRAM, KA	World
12/88-6/89	T. Roosevelt	CVW-8	AJ5xx	A-6E TRAM, KA	Med
8/90-3/91	**Saratoga**	**CVW-17**	**AA5xx**	**A-6E TRAM, KA**	**Med, RS**
5-11/92	Saratoga	CVW-17	AA5xx	A-6E SWIP, KA	Med
1-6/94	Saratoga	CVW-17	AA5xx	A-6E SWIP	Med

VA-36

Nickname: *Roadrunners* (callsign "Heartless")
From: Established 6 March, 1987.
To: Disestablished 1 April, 1994. (ceremony on 11 March, 1994)
Home Station: NAS Oceana, VA.
Models Flown: A-6E **Losses**: 1 combat, 0 operational
Notes: Formed during the *Intruder* build up of the late 'seventies, VA-36 deployed as half of a "dual *Intruder* wing" on three of its four cruises. The *Roadrunner* title was adopted from an earlier VA-36, a Cecil Field based A-4 *Skyhawk* unit that had disestablished on 1 Aug, 1970. Attempts to use the original outfit's insignia, based on the Warner Brothers' cartoon character, were officially disapproved, so the new unit designed a motif depicting a more realistic-looking bird grappling with a snake. During *Desert Storm* the squadron flew 578 sorties and dropped "over" 1.2 million pounds of ordnance, with the loss of one aircraft and crew. VA-36 flew only A-6E TRAM and SWIP (from 1992) aircraft throughout its short life.
Deployments:

8-10/88	T.Roosevelt	CVW-8	AJ5xx	A-6E TRAM	NorLant
12/88-6/89	T.Roosevelt	CVW-8	AJ53x	A-6E TRAM	Med
12/90-6/91	**T.Roosevelt**	**CVW-8**	**AJ53x**	**A-6E TRAM**	**Med, RS, PG**
3-9/93	T.Roosevelt	CVW-8	AJ5xx	A-6E SWIP	Med, RS

VA-52

Nickname: *Knightriders* (callsign: "Viceroy")
From: Transitioned from A-1 *Skyraider* from November, 1967.
To: Disestablished 31 Mar. 1995. (17 Mar. ceremony)
Home Station: NAS Whidbey Island, WA.
Models Flown: A-6A,B,C,E. KA-6D. **Losses**: 2 combat, 8 operational
Notes: The unit's somewhat unusual patch depicted a mace wielding knight astride a sea turtle, which led to the irreverent alternate nicknames of "Turtle Herders" or "Turtle Beaters". Nonetheless, the squadron built a strong record with three combat deployments to Vietnam in the A-6. VA-52 was the first Navy Squadron to be commanded by a Naval Flight Officer on 23 Nov, 1971, when CDR. Lenny Salo assumed the title. A-6E TRAMs arrived in 1982, SWIPs from 1989. Claimed distinction as first Whidbey unit qualified in night vision goggles (NVGs). On 19 Jan. 1993 *Knightrider* crews struck targets in Iraq, officially in response to anti-aircraft fire while flying a mission during Operation Southern Watch. This action may well have been the final combat action for the type.
Deployments:

9/68-4/69	**Coral Sea**	**CVW-15**	**NL4xx**	**A-6A**	**WP, SEA**
11/70-7/71	**Kitty Hawk**	**CVW-11**	**NH5xx**	**A-6A/B**	**WP, SEA**
2-11/72	**Kitty Hawk**	**CVW-11**	**NH5xx**	**A-6A/B, KA**	**WP, SEA**
11/73-7/74	Kitty Hawk	CVW-11	NH5xx	A-6A/B, KA	WP, IO
5-12/75	Kitty Hawk	CVW-11	NH5xx	A-6E, KA	WP
10/77-5/78	Kitty Hawk	CVW-11	NH5xx	A-6E, KA	WP
5/79-2/80	Kitty Hawk	CVW-11	NH5xx	A-6E. KA	WP, IO
4-11/81	Kitty Hawk	CVW-11	NH5xx	A-6E, KA	WP, IO
3-10/83	Carl Vinson	CVW-15	NL5xx	A-6E TRAM, KA	World
10/84-5/85	Carl Vinson	CVW-11	NH5xx	A-6E TRAM, KA	WP, IO
8/86-2/87	Carl Vinson	CVW-11	NH5xx	A-6E TRAM, KA	WP, IO
6-12/88	Carl Vinson	CVW-11	NH5xx	A-6E TRAM, KA	WP, IO
9-11/89	Carl Vinson	CVW-11	NH5xx	A-6E SWIP, KA	NP
2-7/90	Carl Vinson	CVW-11	NH5xx	A-6E SWIP, KA	WP, IO
10,11/91	Kitty Hawk	CVW-15	NL5xx	A-6E SWIP, KA	Horn
11/92-5/93	Kitty Hawk	CVW-15	NL5xx	A-6E SWIP, KA	WP,IO,PG
6-12/94	Kitty Hawk	CVW-15	NL5xx	A-6E SWIP	WP

VA-55

Nickname: *Warhorses*
From: Established 7 Oct. 1983.
To: Disestablished 1 Jan. 1991. (Ceremony 22 Feb)
Home Station: NAS Oceana, VA.
Models Flown: A-6E, KA-6D. **Losses**: 0 combat, 1 operational
Notes: Established as the first new AirLant *Intruder* unit during the mid-'80s build up. Adopted the nickname and traditions of a Lemoore-based A-4 squadron disestablished in 1975. VA-55 joined the newly formed CVW-13, and participated in Operations *Prairie Fire* and *El Dorado Canyon* against Libya on its first deployment. For the following two deployments VA-55 was joined in Air Wing 13 by VA-65. VA-55 was disestablished prior to the start of *Desert Storm* with many of its men transferring to other squadrons deployed to the combat zone.
Deployments:

10/85-5/86	**Coral Sea**	**CVW-13**	**AK5xx**	**A-6E TRAM, KA**	**Med**
9/87-3/88	Coral Sea	CVW-13	AK5xx	A-6E TRAM	Med
5-9/89	Coral Sea	CVW-13	AK5xx	A-6E SWIP	Med

VA-65

Nickname: *Tigers* (callsigns "Cupcake" and "Fighting Tiger")
From: Transitioned from A-1 *Skyraiders*, from early 1965.
To: Disestablished 31 March, 1992. (26 March ceremony)
Home Station: NAS Oceana, VA
Versions: A-6A, B, C, E, KA-6D **Losses**: 3 combat, 2 operational
Notes: Also known as the "Cupcakes", their official tactical callsign for many years and, according to some stories, the name of the circus tiger their insignia was derived from. The third *Intruder* unit in the Navy, made four extended combat deployments on four different decks. Its second Vietnam trip was cut short by the fire on *Forrestal*, after which a group of volunteers continued combat operations through a detachment with VA-196 in *Constellation*. A-6E arrived in 1972, A-6E TRAM in 1979. Participated in two "Dual A-6 Wings", with VA-55 in *Coral Sea* (1986-89) and with VA-36 in *Theodore Roosevelt* (1989-1991). Participated in Desert Storm, where it was credited with sinking 22 Iraqi Naval boats. The squadron was beached following *Desert Storm* in order to make room for Marine helicopters in an experimental fixed wing/rotary wing deployment for *Roosevelt*. It was disestablished within the year. The *Tigers* were famous for their peacetime use of orange flight suits well after everyone else had gone to the sage green.

Deployments:

5-12/66	**Constellation**	**CVW-15**	**NL4xx**	**A-6A**	**WP/SEA**
6-9/67	**Forrestal**	**CVW-17**	**AA5xx**	**A-6A**	**WP/SEA**
12/68-9/69	**Kitty Hawk**	**CVW-11**	**NH5xx**	**A-6A/B**	**WP/SEA**
6/70-1/71	Independence	CVW-7	AG5xx	A-6A	Med
9/71-3/72	Independence	CVW-7	AG5xx	A-6A, KA	Med
6/73-1/74	Independence	CVW-7	AG5xx	A-6E, KA	Med
7/74-1/75	Independence	CVW-7	AG5xx	A-6E, KA	Med
10/75-5/76	Independence	CVW-7	AG5xx	A-6E, KA	Med
3-10/77	Independence	CVW-7	AG5xx	A-6E, KA	Med
1-7/79	Eisenhower	CVW-7	AG5xx	A-6E, KA	Med
4-12/80	Eisenhower	CVW-7	AG5xx	A-6E TRAM, KA	IO
1-7/82	Eisenhower	CVW-7	AG5xx	A-6E TRAM, KA	Med
4-12/83	Eisenhower	CVW-7	AG5xx	A-6E TRAM, KA	Med
10/84-5/85	Eisenhower	CVW-7	AG5xx	A-6E TRAM, KA	Med
9/87-3/88	Coral Sea	CVW-13	AK51x	A-6E TRAM	Med
5-9/89	Coral Sea	CVW-13	AK51x	A-6E TRAM	Med
12/90-6/91	**T.Roosevelt**	**CVW-8**	**AJ 53x**	**A-6E TRAM**	**Med/PG**

VA-75

Nickname: *World Famous Sunday Punchers* (callsign "Flying Ace")
From: Transitioned from A-1 *Skyraiders* in 1963.
To: Disestablished 31 Mar. 1997. (28 Feb ceremony)
Home Station: NAS Oceana, VA.
Versions: A-6A,B,E KA-6D. **Losses**: 8 combat, 13 operational
Notes: The first and the last. The *Sunday Punchers* made the first deployment in the *Intruder*, going straight into combat in Vietnam. Over the next thirty-plus years the unit carried out 20 some odd cruises in six carriers. A-6E arrived in 1973, A-6E TRAMs in 1981. SWIP versions were assigned from late 1987. Participated in strikes in Lebanon from *Kennedy* on 4 Dec. 1983. The squadron participated with distinction in Operation *Desert Storm*, flying 498 sorties and dropping 1.6 million pounds of ordnance. The *Punchers* made the final *Intruder* cruise fly-off on 19 December, 1996. Although officially disestablished, final carrier ops for the type were conducted with former squadron aircraft and personnel off the Virginia Capes on 12 March 1997. Its final two aircraft were flown to the boneyard on 19 March. With its departure, VA-75 ended 53 years of service.
Deployments:

5-10/65	**Independence**	**CVW-7**	**AG5xx**	**A-6A**	**WP, SEA**
6/66-2/67	Independence	CVW-7	AG5xx	A-6A	Med
11/67-6/68	**Kitty Hawk**	**CVW-7**	**NG5xx**	**A-6A/B**	**WP, SEA**
7/69-1/70	Saratoga	CVW-3	AC5xx	A-6A	Med
6-11/70	Saratoga	CVW-3	AC5xx	A-6A/B	Med
6-10/71	Saratoga	CVW-3	AC5xx	A-6A/B, KA	Med
4/72-2/73	**Saratoga**	**CVW-3**	**AC5xx**	**A-6A/B, KA**	**WP, SEA**
9/74-3/75	Saratoga	CVW-3	AC5xx	A-6E, KA	Med
1-7/76	Saratoga	CVW-3	AC5xx	A-6E, KA	Med
7-12/77	Saratoga	CVW-3	AC5xx	A-6E, KA	Med
10/78-4/79	Saratoga	CVW-3	AC5xx	A-6E, KA	Med
3-8/80	Saratoga	CVW-3	AC5xx	A-6E, KA	Med
1-7/82	J.F. Kennedy	CVW-3	AC5xx	A-6E TRAM, KA	Med, IO
5-7/83	J.F. Kennedy	CVW-3	AC50x	A-6E TRAM, KA	NL
9/83-5/84	**J.F. Kennedy**	**CVW-3**	**AC5xx**	**A-6E TRAM, KA**	**Med**
8/86-3/87	J.F. Kennedy	CVW-3	AC5xx	A-6E TRAM, KA	Med
8/88-2/89	J.F. Kennedy	CVW-3	AC5xx	A-6E SWIP, KA	Med
8/90-3/91	**J.F. Kennedy**	**CVW-3**	**AC5xx**	**A-6E SWIP, KA**	**Med, RS**
10/92-4/93	J..F..Kennedy	CVW-3	AC5xx	A-6E SWIP, KA	Med
10/94-4/95	Eisenhower	CVW-3	AC5xx	A-6E SWIP	Med, RS, PG
6-12/96	Enterprise	CVW-17	AA5xx	A-6E SWIP	NL, Med, RS, PG

VA-85

Nickname: *Black Falcons* (callsign and alternate nickname, "Buckeye")
From: Transitioned from A-1 *Skyraiders*, from 1964.
To: Disestablished 30 Sep. 1994. (22 Sep. ceremony).
Home Stations: NAS Oceana, VA.
Versions: A-6A,B,E KA-6D. **Losses**: 12 combat, 16 operational
Notes: The Black Falcons were the second fleet A-6 squadron and were initially assigned to CVW-6 for a Med cruise in *Enterprise* before being reassigned to *Kitty Hawk* for war service. VA-85 was the second *Intruder* unit in combat, making two deployments with a Pacific Fleet carrier and air wing within 19 months. The squadron

flew four successive Vietnam cruises in less than five years, a record for Atlantic Fleet squadrons. The price however for this intense tempo was eleven *Intruders* and 13 men, three of which were commanding officers. Became the first unit to deploy with the A-6E. Participated in retaliatory strikes into Lebanon on 4 Dec. 1983, with the loss of one aircraft. During its 1985-86 deployment in Saratoga the unit was involved in *Prairie Fire* strikes against Libyan patrol boats. In 1990-91 the *Black Falcons* flew from America in Operation *Desert Storm*, flying combat missions from both the Red Sea and Persian Gulf. Transitioned to A-6E SWIP for its last deployment in 1993. At the time of its disestablishment VA-85 had completed 43 years of active service.

Deployments:

1964	(Enterprise)	CVW-6	AE4xx	A-6A	no cruise
10/65-6/66	**Kitty Hawk**	**CVW-11**	**NH8xx**	**A-6A**	**WP, SEA**
11/66-6/67	**Kitty Hawk**	**CVW-11**	**NH5xx**	**A-6A**	**WP, SEA**
4-12/68	**America**	**CVW-6**	**AE5xx**	**A-6A/B**	**WP, SEA**
8/69-5/70	**Constellation**	**CVW-14**	**NK5xx**	**A-6A/B**	**WP, SEA**
1-7/71	Forrestal	CVW-17	AA5xx	A-6A, KA	Med
9/72-7/73	Forrestal	CVW-17	AA5xx	A-6A, KA	Med
3-9/74	Forrestal	CVW-17	AA5xx	A-6E, KA	Med
3-9/75	Forrestal	CVW-17	AA5xx	A-6E, KA	Med
4-10/78	Forrestal	CVW-17	AA5xx	A-6E, KA	Med
11/79-5/80	Forrestal	CVW-17	AA5xx	A-6E, KA	Med
3/81-9/81	Forrestal	CVW-17	AA5xx	A-6E, KA	Med
6-11/82	Forrestal	CVW-17	AA5xx	A-6E TRAM, KA	Med
9/83-5/84	**J.F.Kennedy**	**CVW-3**	**AC54x,55x**	**A-6E TRAM, KA**	**Med**
8/85-4/86	**Saratoga**	**CVW-17**	**AA5xx**	**A-6E TRAM, KA**	**Med, IO**
6-11/87	Saratoga	CVW-17	AA5xx	A-6E TRAM, KA	Med
2-4/89	America	CVW-1	AB5xx	A-6E TRAM, KA	NL
12/90-4/91	**America**	**CVW-1**	**AB5xx**	**A-6E TRAM, KA**	**Med, RS, PG**
8-11/91	America	CVW-1	AB5xx	A-6E TRAM, KA	Med, RS, PG
8/93-2/94	America	CVW-1	AB5xx	A-6E SWIP	Med

VA-95

Nickname: *Green Lizards*
From: Established 1 Apr. 1972.
To: Disestablished 31 Oct. 1995. (18 Nov. ceremony)
Home Stations: NAS Whidbey Island, WA
Versions: A-6A,B,E, KA-6D **Losses**: 0 combat, 9 operational
Notes: VA-95 adopted the name of a former Pacific Fleet A-4 *Skyhawk* unit disestablished in 1970. While using the previous squadron's nickname (which had officially changed from the *Skyknights* in 1963) the *Intruder* unit designed a new patch featuring a trident-toting reptile over a rising sun. The squadron's unique title reportedly dates from the first unit's days in AD *Skyraiders* and an incident where their CAG blamed VA-95 for the presence of a large reptile in his stateroom bed one night.

The *Lizards* were the last of the Vietnam era *Intruder* squadrons to be established at Whidbey, and made one deployment in the combat area, supporting the clearing of minefields in Operation *End Sweep*. On their second cruise the unit participated in both Frequent Wind, the evacuation of Saigon and the re-capture of the *SS Mayaguez*. From 1979 to 1981 the unit made two deployments to the Med and/or IO with CVW-11 in *America*. The latter cruise also included A-6E TRAM aircraft. The *Lizards* played center stage in 1988's Operation *Praying Mantis* against Iranian Naval units and were largely responsible for the sinking of the light frigate *Sahand*, the largest man-of-war to be sunk by the A-6 type.

Deployments:

3-11/73	**Coral Sea**	**CVW-15**	**NL5xx**	**A-6A,B, KA**	**WP, SEA**
12/74-7/75	**Coral Sea**	**CVW-15**	**NL5xx**	**A-6A, KA**	**WP, SEA**
2-10/77	Coral Sea	CVW-15	NL5xx	A-6E, KA	WP
3-9/79	America	CVW-11	NH5xx	A-6E, KA	Med
4-11/81	America	CVW-11	NH5xx	A-6E TRAM, KA	Med, IO
9/82-4/83	Enterprise	CVW-11	NH5xx	A-6E TRAM, KA	NP, WP, IO
5-12/84	Enterprise	CVW-11	NH5xx	A-6E TRAM, KA	WP, IO
1-8/86	Enterprise	CVW-11	NH5xx	A-6E TRAM, KA	World
10-11/87	Enterprise	CVW-11	NH5xx	A-6E TRAM, KA	NP
1-7/88	**Enterprise**	**CVW-11**	**NH5xx**	**A-6E, KA**	**WP, IO, Iran**
9/89-3/90	Enterprise	CVW-11	NH5xx	A-6E, KA	World
9-11/90	A.Lincoln	CVW-11	NH5xx	A-6E SWIP, KA	Horn
5-11/91	A.Lincoln	CVW-11	NH5xx	A-6E SWIP, KA	WP, IO, PG
6-12/93	A.Lincoln	CVW-11	NH5xx	A-6E SWIP	WP, IO
4-10/95	A.Lincoln	CVW-11	NH5xx	A-6E SWIP	WP, IO, PG.

VA-115

Nicknames: *Arabs, Eagles*
From: Accepted A-6s from Jan. 1970 following 31 months of inactivity.
To: Redesignated VFA-115, transitioned to FA-18C 30 Sep, 1996
Home Stations: NAS Whidbey Island, WA, NAF Atsugi, JA (from Sep. 1973)
Versions: A-6A,B,E KA-6D **Losses**: 1 combat, 9 operational
Notes: VA-115 was an A-1 *Skyraider* unit assigned to CVW-5 when it returned to NAS Lemoore, CA from a cruise in *Bon Homme Richard* in July, 1967. In a move that appears to be unduplicated in the modern Navy, the unit was stripped of its aircraft and personnel, and the Commanding Officer of VA-125, an A-4 training unit, was named its CO. Why this was done has never really been explained, other than it kept a squadron technically on the books while the Navy decided what to do with it. In January, 1970 VA-115's title was moved to Whidbey Island, and the squadron opened for business again as the fifth Pacific Fleet *Intruder* unit. Initially assigned to CVW-16 (tailcode AH), 115 would re-join its old CVW-5 for its first war cruise in A-6s.

VA-115's historical nickname was *Arabs* (pronounced A-rab'), which was initially used by the *Intruder* unit. By 1979, however, political considerations led to the unit's adopting the less sensitive *Eagle* title. In 1973 the squadron moved with its air wing to Japan, where for the remainder of its existence it made numerous cruises throughout the Pacific and Indian Ocean areas as part of the Navy's unofficial "Foreign Legion". The squadron retired the final A-6As in the Navy in 1975, with at least some of the aircraft returning to CONUS on the decks of amphibious ships. From 1979 to 1984 VA-115 was the Pacific Fleet's repository of Standard ARM capable A-6Bs and AWG-21 equipped A-6Es. The squadron was reduced in size to nine aircraft in 1987 when VA-185 joined it in CVW-5, and as such it fought in *Desert Storm*. In 1991, VA-185 was disestablished, and 115 absorbed most of the command, raising its strength back to 15 aircraft. The *Eagles* lasted until 1996, when it returned to the US and was redesignated VFA-115. The squadron became the only Pacific Fleet *Intruder* unit to transition to another aircraft, and initially flew FA-18Cs with CVW-14. In 2000 it became the Navy's first fleet FA-18E unit.

Deployments:

(none)	(—)	CVW-16	AH5xx	A-6A	—
4-11/71	**Midway**	**CVW-5**	**NF5xx**	**A-6A, KA**	**WP, SEA**
4/72-3/73	**Midway**	**CVW-5**	**NF5xx**	**A-6A/B, KA**	**WP, SEA**
9/73	Midway	CVW-5	NF5xx	A-6A, KA	transfer to Japan
1973-1977	Midway	CVW-5	NF5xx	A-6A, KA	Ops throughout WestPac
1977-1980	Midway	CVW-5	NF5xx	A-6E, KA	Ops throughout WestPac
1980-10/90	Midway	CVW-5	NF5xx	A-6E TRAM, KA	Ops throughout WestPac
10/90-4/91	**Midway**	**CVW-5**	**NF5xx**	**A-6E TRAM, KA**	**WP, IO,PG**
4-10/92	Independence	CVW-5	NF5xx	A-6E SWIP	WP, IO , PG
11/93-3/74	Independence	CVW-5	NF5xx	A-6E SWIP	WP, IO, PG
8-11/95	Independence	CVW-5	NF5xx	A-6E SWIP	WP, IO, PG

Note: *Midway*, with CVW-5 and VA-115 embarked, was forward deployed to Yokosuka Japan from September, 1973 until 1992 when it was replaced by *Independence*. During that time it made numerous trips throughout the Western Pacific and Indian Ocean regions.

VA-145

Nickname: *Swordsmen* (callsigns: "Electron", later "Rustler")
From: Transitioned from A-1 *Skyraiders* from June, 1968.
To: Disestablished 1 Oct. 1993. (13 Oct ceremony)
Home Stations: NAS Whidbey Island, WA.
Versions: A-6A,B,C,E KA-6D. **Losses**: 0 combat, 10 operational
Notes: VA-145 was the third Whidbey based *Intruder* unit, and participated in combat in Vietnam (three cruises) and *Desert Storm*. The 1972 deployment to Vietnam was notable as the first combat use of laser guided bombs by Navy *Intruders*, using centerline mounted AVQ-10 *Pave Knife pods*. A-6E TRAMs arrived in Nov. 1981, the Pacific Fleet's first SWIPs in 1988. In 1984-85 the unit became the last location for Pacific Fleet AWG-21/ Standard ARM capable A-6Es. Participated in dual A-6 wing configurations with both the Marine (VMA-AW-121) and Navy (VA-155) squadrons. From 1986 to 1987 VA-145 accompanied *Ranger* and CVW-2 as they carried out "surge" operations to the Western and Northern Pacific operations areas.

Deployments:

1-7/69	**Enterprise**	**CVW-9**	**NG5xx**	**A-6A/B**	**WP, SEA**
10/70-6/71	**Ranger**	**CVW-2**	**NE5xx**	**A-6A/C**	**WP, SEA**
11/72-6/73	**Ranger**	**CVW-2**	**NE5xx**	**A-6A/C, KA**	**WP, SEA**
5-10/74	Ranger	CVW-2	NE5xx	A-6A, KA	WP
1-9/76	Ranger	CVW-2	NE5xx	A-6A, KA	WP,IO
2-9/79	Ranger	CVW-2	NE5xx	A-6E, KA	WP
9/80-5/81	Ranger	CVW-2	NE5xx	A-6E,KA	WP,IO
4-10/82	Ranger	CVW-2	NE5xx	A-6E TRAM, KA	WP, IO

1-8/84	Kitty Hawk	CVW-2	NE5xx	A-6E TRAM, KA	WP, IO
8-10/86	Ranger	CVW-2	NE5xx	A-6E TRAM, KA	WP surge
3,4/87	Ranger	CVW-2	NE5xx	A-6E TRAM, KA	WP surge
7-12/87	Ranger	CVW-2	NE5xx	A-6E TRAM	WP, IO
2/89-8/89	Ranger	CVW-2	NE5xx	A-6E SWIP	WP, IO
12/90-6/91	**Ranger**	**CVW-2**	**NE5xx**	**A-6E SWIP**	**WP, IO, PG**
8/92-1/93	Ranger	CVW-2	NE5xx	A-6E SWIP	WP, IO, PG
5-7/93	Constellation	CVW-2	NE5xx	A-6E SWIP	Horn

VA-155

Nickname: *Silver Foxes* (callsign: Vixen)
From : Established 1 Sep. 1987.
To: Disestablished 30 Apr. 1993. (27 Apr. ceremony)
Home Station: NAS Whidbey Island, WA
Versions: A-6E, KA-6D. **Losses**: 1 combat, 1 operational
Notes: The third unit so designated, VA-155 was directed to adopt the name and traditions of a previous *Silver Fox* squadron, an A-7 unit that had been disestablished at Lemoore in 1977. A new patch was designed for the unit featuring a fox's head, with a lightning bolt, a silhouette of Whidbey Island and a shamrock (the first CO, CDR Jack Samar, was a Notre Dame grad) shadowed into its face (these subtleties were lost in many renditions however) The unit's motto and colors, "Medium Attack in Silver and Black" also reflected the CO's affinity for the National Football League's (then) Los Angeles Raiders. 155 replaced VA-185 in the newly formed CVW-10, but never deployed with the short lived air wing. On 1 May, 1988, the *Foxes* became the only Whidbey Intruder squadron to ever be assigned to an east coast air wing, as it replaced a Marine squadron in CVW-17. The same year a portion of the unit joined the wing as it accompanied *Independence* around the Horn to its post-SLEP homeport of North Island, CA. In October, 1989 the *Foxes* moved to Pacific fleet CVW-2, and replaced another Marine unit by joining VA-145. The squadron's first extended deployment included combat operations in Operation *Desert Storm*, where it flew 647 sorties and dropped 2,289,910 pounds of ordnance at the cost of one aircraft and its crew. VA-155 was equipped with A-6E SWIP Intruders upon return to Whidbey, which it flew on its second, and final, cruise.
Deployments:

(none)	(—)	CVW-10	NM5xx	A-6E TRAM, KA	—
8-10/88	Independence	CVW-17	AA5xx	A-6E TRAM, KA	Horn
12/90-6/91	**Ranger**	**CVW-2**	**NE4xx**	**A-6E TRAM**	**WP, IO, PG**
8/92-1/93	Ranger	CVW-2	NE4xx	A-6E SWIP	WP, IO, PG

VA-165

Nickname: *Boomers*
From: Transitioned from A-1 *Skyraiders* in 1967.
To: Disestablished 30 Sep. 1996.
Home Stations: NAS Whidbey Island, WA
Versions: A-6A, B, C, E, KA-6D **Losses**: 2 combat, 6 operational
Notes: The *Boomers* made five combat deployments to Vietnam, and introduced the A-6C TRIM to war. In 1989 the squadron was picked to participate in the movie version of "Flight of the Intruder", with operations on *Independence*, and several aircraft painted in markings representing VA-196 in 1972. The 1990 cruise on *Constellation* involved only a portion of the unit as it accompanied the ship's post-SLEP transfer from Norfolk to North Island. The 1983 cruise in *Ranger* included a short period off Central America during political unpleasantries with the Nicaraguan Sandinistas. The 1993 deployment in *Nimitz* was the final cruise for the KA-6D type.
Deployments:

11/67-5/68	**Ranger**	**CVW-2**	**NE5xx**	**A-6A**	**WP, SEA**
10/68-5/69	**Ranger**	**CVW-2**	**NE5xx**	**A-6A**	**WP, SEA**
4-12/70	**America**	**CVW-9**	**NG5xx**	**A-6A/B/C**	**WP, SEA**
10/71-6/72	**Constellation**	**CVW-9**	**NG5xx**	**A-6A, KA**	**WP, SEA**
1-10/73	**Constellation**	**CVW-9**	**NG5xx**	**A-6A, KA**	**WP, SEA**
6-12/74	Constellation	CVW-9	NG5xx	A-6A, KA	WP, IO, PG
4-11/77	Constellation	CVW-9	NG5xx	A-6E, KA	WP
9/78-5/79	Constellation	CVW-9	NG5xx	A-6E, KA	WP, IO
2-10/80	Constellation	CVW-9	NG5xx	A-6E, KA	WP, IO
10/81-5/82	Constellation	CVW-9	NG5xx	A-6E TRAM, KA	WP, IO
7/83-2/84	Ranger	CVW-9	NG5xx	A-6E TRAM, KA	WP, IO
7-12/85	Kitty Hawk	CVW-9	NG5xx	A-6E TRAM, KA	WP, IO
1-6/87	Kitty Hawk	CVW-9	NG5xx	A-6E TRAM, KA	World
9/88-2/89	Nimitz	CVW-9	NG5xx	A-6E TRAM, KA	WP, IO
6,7/89	Nimitz	CVW-9	NG5xx	A-6E TRAM, KA	NP
2-4/90	Constellation	CVW-9	NG5xx	A-6E TRAM, KA	Horn
2-7/93	Nimitz	CVW-9	NG5xx	A-6E SWIP, KA	WP, IO, PG
11/95-5/96	Nimitz	CVW-9	NG5xx	A-6E SWIP	WP, IO, PG

VA-176

Nickname: *Thunderbolts*
From: Transitioned from A-1 *Skyraiders*, in early 1969.
To: Disestablished 30 Oct. 1992. (18 Sep. ceremony)
Home Station: NAS Oceana, VA.
Versions: A-6A,C,E KA-6D **Losses**: 0 combat, 10 operational
Notes: VA-176 was the fifth, and last, Atlantic Fleet *Intruder* squadron established during the Vietnam period. The squadron operated largely out of the Mediterranean with CVW-6 for the next 21 years, with limited combat operations during the Invasion of Grenada and strikes on Lebanon in 1983-84. The *Thunderbolts* were unique among the older A-6 units in that they made every deployment with the same Air Wing. They were also the only *Intruder* unit to deploy in *Franklin D. Roosevelt*. VA-176 was the first squadron to deploy with the KA-6D. 176 was the first Oceana *Intruder* unit to be disestablished during the post *Desert Storm* draw down, but its name and insignia lived on as the RAG, VA-42, adopted both in 1992.

Deployments:

1-7/70	F.D.Roosevelt	CVW-6	AE5xx	A-6A	Med
1-7/71	F.D.Roosevelt	CVW-6	AE5xx	A-6A, KA	Med
2-12/72	F.D.Roosevelt	CVW-6	AE5xx	A-6A/C KA	Med
9/73-3/74	F.D.Roosevelt	CVW-6	AE5xx	A-6A/C,KA	Med
1-7/75	F.D.Roosevelt	CVW-6	AE5xx	A-6A/C, KA	Med
4-10/76	America	CVW-6	AE5xx	A-6E, KA	Med
9/77-4/78	America	CVW-6	AE5xx	A-6E, KA	Med
6-12/79	Independence	CVW-6	AE5xx	A-6E, KA	Med
11/80-6/81	Independence	CVW-6	AE5xx	A-6E, KA	Med, IO, SL
6-12/82	Independence	CVW-6	AE5xx	A-6E TRAM, KA	Med
10/83-4/84	**Independence**	**CVW-6**	**AE5xx**	**A-6E TRAM, KA**	**Carib, Med, NL**
10/84-2/85	Independence	CVW-6	AE5xx	A-6E TRAM, KA	Med
6-11/86	Forrestal	CVW-6	AE5xx	A-6E TRAM, KA	Med
8-10/87	Forrestal	CVW-6	AE5xx	A-6E TRAM, KA	Med
4-10/88	Forrestal	CVW-6	AE5xx	A-6E TRAM, KA	Med, IO, NL
11/89-4/90	Forrestal	CVW-6	AE5xx	A-6E TRAM, KA	Med
5-12/91	Forrestal	CVW-6	AE5xx	A-6E SWIP, KA	Med

VA-185

Nickname: *Nighthawks*
From: Established 1 Dec. 1986.
To: 30 Aug. 1991. (6 Aug. ceremony)
Versions: A-6E, KA-6D. **Losses**: None
Home Stations: NAS Whidbey Island, WA, NAF Atsugi, JA (from 13 Sep1987).
Notes: The shortest lived of all *Intruder* squadrons; VA-185 was formed with the intent of being a charter member of CVW-10, but was instead transferred to Japan in September, 1987 to join CVW-5 in *Midway*. The *Nighthawks* became a small (nine aircraft) squadron, with seven bombers and two tankers normally assigned. The squadron participated in *Desert Storm*, but was disestablished soon afterwards, with VA-115 functionally absorbing the unit as CVW-5 became a single *Intruder* wing again. VA-185 flew over 14,000 hours in its four years and nine months, without loss of a single aircraft or aircrew in peace or war.
Deployments:

10/87-4/88	Midway	CVW-5	NF4xx	A-6E TRAM, KA	WP, IO
10,11/88	Midway	CVW-5	NF4xx	A-6E TRAM, KA	WP
1,2/89	Midway	CVW-5	NF4xx	A-6E TRAM, KA	WP
2-4/89	Midway	CVW-5	NF4xx	A-6E TRAM, KA	WP
5-7/89	Midway	CVW-5	NF4xx	A-6E TRAM, KA	WP
8-12/89	Midway	CVW-5	NF4xx	A-6E TRAM, KA	WP, IO
1-4/90	Midway	CVW-5	NF4xx	A-6E TRAM, KA	WP, IO,PG
10/90-4/91	**Midway**	**CVW-5**	**NF4xx**	**A-6E TRAM, KA**	**WP,IO,PG**

VA-196

Nicknames*:* *Main Battery*, (Callsign and alternate nickname: *Milestone).*
From: Transitioned from A-1 *Skyraiders* through 1966-67.
To: Disestablished: 28 Feb. 1997.
Home Station: NAS Whidbey Island, WA.
Types: A-6A,B,E KA-6D. **Losses**: 12 combat, 10 operational
Notes: VA-196's historical nickname, *The Main Battery* refers to the squadron being the primary weapon of the carrier air wing, and dates from its days in A-1 *Skyraiders*. The secondary nickname *Milestone* was the unit's official radio call sign, and used interchangeably with *Main Battery* for many years. In addition, the squadron frequently used "Devil" for an abbreviated radio callsign, based on its logo. 196 was the first Pacific Fleet *Intruder* unit, and the last to be disestablished. As the first West Coast *Intruder* unit, 196 actually went through

transition with VA-42 in Oceana. The squadron made five deployments to Vietnam while also losing more aircraft in South East Asia than any other *Intruder* unit. A-6E TRAMs arrived in 1984, SWIPs in 1990. In other major operations VA-196 was involved with include the aborted release of the Iranian hostages in 1982, and *Desert Shield* in 1990. Squadron member Steven Coontz immortalized the aircraft in his novel *The Flight of the Intruder*, which was made into a movie using VA-165 aircraft. VA-196 became the final Pacific Fleet *Intruder* unit, with its last carrier operations being conducted on *Carl Vinson* in the Strait of San Juan de Fuca in February, 1997.

Deployments:

4-12/67	**Constellation**	**CVW-14**	**NK4xx**	**A-6A**	**WP, SEA**
5/68-1/69	**Constellation**	**CVW-14**	**NK4xx**	**A-6A/B**	**WP, SEA**
10/69-6/70	**Ranger**	**CVW-2**	**NE5xx**	**A-6A**	**WP, SEA**
6/71-2/72	**Enterprise**	**CVW-14**	**NK5xx**	**A-6A/B, KA**	**WP, SEA**
9/72-6/73	**Enterprise**	**CVW-14**	**NK5xx**	**A-6A/B, KA**	**WP, SEA**
9/74-5/75	Enterprise	CVW-14	NK5xx	A-6A, KA	WP, IO
7/76-3/77	Enterprise	CVW-14	NK5xx	A-6E, KA	WP, IO
4-10/78	Enterprise	CVW-14	NK5xx	A-6E, KA	WP, IO
11/79-6/80	Coral Sea	CVW-14	NK5xx	A-6E, KA	WP, IO
8/81-3/82	Coral Sea	CVW-14	NK5xx	A-6E, KA	WP,IO
3-9/83	Coral Sea	CVW-14	NK5xx	A-6E, KA	WP,IO
2-8/85	Constellation	CVW-14	NK5xx	A-6E TRAM, KA	WP,IO
4-10/87	Constellation	CVW-14	NK5xx	A-6E TRAM, KA	WP,IO
12/88-6/89	Constellation	CVW-14	NK5xx	A-6E TRAM, KA	WP, IO
6-12/90	Independence	CVW-14	NK5xx	A-6E SWIP	WP,IO,PG
2-8/94	Carl Vinson	CVW-14	NK5xx	A-6E SWIP	WP,IO,PG
5-11/96	Carl Vinson	CVW-14	NK5xx	A-6E SWIP	WP,IO,PG

NAVAL RESERVE SQUADRONS

VA-205

Nickname: *Green Falcons.*
From: Transitioned from A-7E *Corsair II* from August, 1990.
To: Disestablished 31 Dec. 1995. (25 Sep. 1994 ceremony)
Home Station: NAS Atlanta, GA.
Versions: A-6E, KA-6D. **Losses**: None
Notes: Assigned to Reserve Carrier Air Wing Twenty (CVWR-20) with AF5XX markings. Carried out reserve duties from Atlanta for five years. Flew A-6E SWIP in 1993, Transferred all aircraft in August, 1994. Took up the electronic adversary mission in Oct. 1993 upon disestablishment of Key West-based VAQ-33. Was disestablished during period of regular Navy A-6 wind-down, and replaced in its reserve air wing by a Marine FA-18 squadron.
Deployments: None.

VA-304

Nickname: *Firebirds*
From: Transitioned from A-7E *Corsair II* from July, 1988.
To: Disestablished 31 Dec. 1994. (17 Sep. ceremony)
Home Station: NAS Alameda, CA
Versions: A-6E, KA-6D. **Losses**: 1 operational
Notes: Assigned to Reserve Carrier Air Wing 30, (CVWR-30) and initially marked with ND4XX markings, which later became the traditional *Intruder* 500 series numbers. Initially equipped with both A-6E and KA-6Ds, the tankers were retired in 1993 when A-6E SWIPs arrived. VA-304 flew occasional carrier qualifications on Pacific Fleet carriers and carried out other reserve duties out of its NAS Alameda home. The *Firebirds* were disestablished along with the remainder of its wing during the post-Desert Storm cutback period.
Deployments: None.

INTRUDER CARRIERS

The following list of carriers represents the flight decks from which the *Intruder* worked from and only covers the period from 1963 to 1995. Dates are given for when ships were redesignated from "Attack" carriers (CVA/CVAN) to "Multi-Purpose" (CV/CVN) with the assignment of anti-submarine assets to their air wings. Those ships not already done were redesignated as a group on 30 June, 1975.

Essex Class "27C" modification

Lexington	AVT-16

Although no *Intruder* unit ever deployed in a 27C *Essex* class carrier, *Lexington* was occasionally used for carrier qualifications by A-6 units, particularly the RAGs

Midway Class			
Midway	CVA-41	CV-41	30 Jun 1975
Franklin D. Roosevelt	CVA-42	CV-42	30 Jun 1975
Coral Sea	CVA-43	CV-43	30 Jun 1975
Forrestal Class			
Forrestal	CVA-59	CV-59	30 Jun 1975
Saratoga	CVA-60	CV-60	30 Jun 1972
Ranger	CVA-61	CV-61	30 Jun 1975
Independence	CVA-62	CV-62	28 Feb 1973
Kitty Hawk Class			
Kitty Hawk	CVA-63	CV-63	29 Apr 1973
Constellation	CVA-64	CV-64	30 Jun 1975
America	CVA-66	CV-66	30 Jun 1975
Enterprise Class			
Enterprise	CVAN-65	CVN-65	30 Jun 1975
Kennedy Class			
John F. Kennedy	CVA-67	CV-67	29 Apr 1973
Nimitz Class			
Nimitz	CVAN-68	CVN-68	30 Jun 1975
Dwight D. Eisenhower	CVN-69		
Carl Vinson	CVN-70		
Theodore Roosevelt	CVN-71		
Abraham Lincoln	CVN-72		
George Washington	CVN-73		

Notes

[1] Operational Control (OPCON) and Administrative Control (ADCON) of squadrons is a frequently confusing issue that varied widely depending on time and which coast was involved. It is beyond the scope of this book.

[2] There was one formally designated Navy light attack squadron, VAL-3, which flew OV-10 *Broncos* in Vietnam.

[3] The Navy changed this policy in 1998 under instruction OPNAVINST 5030.4E which covers the methods used by the Office of Naval History to track official lineages. Squadrons are now either "Established", "Redesignated", "Deactivated" or "Reactivated". They are no longer "Disestablished".

Appendix C: Marine Units

Marine Aircraft Wings: (MAW): The Marine Air Wing, usually commanded by a Major General, is the major aviation unit in the Fleet Marine Force and is normally paired with a specific Marine Division. Each MAW is made up of a number of Marine Air Groups (MAG), each MAG usually specializing on a specific mission or even type of aircraft. There are four MAWs, the 1st through 4th. The 1st MAW is associated with the Far East, being based in Iwakuni, Japan or Vietnam during the period the *Intruder* was used by the Corps. The 2d MAW is based in MCAS Cherry Point, NC, the 3rd MAW in MCAS El Toro, CA during the *Intruder's* tenure. The 4th MAW is a Reserve organization and never had VMA(AW) squadrons assigned.

Marine Aircraft Groups: (MAG): Each MAW has a number of MAGs, each Group being normally commanded by a Colonel. A MAG typically (but not always) controls a number of squadrons of like types or in the same mission area. The MAG performs certain administrative and intermediate maintenance functions for its squadrons. In addition, jet MAGs usually had a Headquarters & Maintenance Squadron (HAMS, or HAMRON) with TF-9Js, TA-4Fs or OA-4Ms assigned during the period of *Intruder* service for miscellaneous support or FAC duties.
MAGs that had *Intruder* squadrons assigned include, but are not limited to:

CONUS:
MAG-11: El Toro, deployed to Shaikh Isa, Bahrain 1990-91.
MAG-14: Cherry Point
MAG-13: El Toro   MAG-15: Iwakuni
MAG-24: Cherry Point
MAG-33: El Toro
MCCRTG-20: Cherry Point
MAG-70: Shaikh Isa, Bahrain (replaced by MAG-11)

Vietnam:
MAG-11: Da Nang RVN   7/65-5/71
MAG-12: Chu Lai, RVN   5/65-2/70
MAG-15: Nam Phong RTAFB 6/72-8/73

Squadrons: Marine *Intruder* units bore the title Marine All Weather Attack, as in VMA(AW). Squadrons are normally commanded by a Lieutenant Colonel with a Major or Lt Col. as second in command with the title Executive Officer (XO). Seven Marine squadrons flew the A-6 *Intruder* with all but the training squadron transitioning from the Douglas A-4 *Skyhawk* into the A-6. As opposed to their Navy counterparts, Marine squadrons maintain a historical lineage across periods of inactivity and are "activated" and "deactivated". A reactivated unit is considered to have all of the combat honors of its predecessors.

The five surviving Fleet Marine Force *Intruder* squadrons converted to the two-seat FA-18D, becoming All-weather Fighter Attack, VMFA(AW) squadrons. The single fleet Marine *Intruder* unit to be deactivated after Vietnam (225) was brought back as a *Hornet* unit in time.

Format:
Squadron designation, nickname(s) and assigned tailcode.
From: When squadron was redesignated as a VMA(AW) unit.
To: When squadron was redesignated or deactivated.
Types flown: The "A" and "E" models were standard for most squadrons although units that made carrier deployments were usually assigned KA-6Ds and, in one case, A-6Bs.
Carrier Cruises: Three Marine *Intruder* squadrons were assigned to Navy Carrier Air Wings and made extended deployments. Dates of deployment, ship, Air Wing, tailcode/side numbers and type aircraft are listed.
Combat: During the Vietnam war three Marine *Intruder* units were based from Da Nang and Chu Lai RVN as well as Nam Phong Thailand. One squadron saw combat from the carrier *Coral Sea*. Two squadrons were involved in *Desert Storm*, and flew from Shaikh Isa, Bahrain.
Homebase/UDP: The Marines initially based all of its *Intruders* at MCAS Cherry Point, NC, with three squadrons moving to Vietnam for combat duty. As the war wound on, two units were based at El Toro, CA and one was permanently stationed at MCAS Iwakuni, Japan. VMA (AW)-533 remained in the Far East until November, 1975 when it returned to Cherry Point to replace its A-6As with A-6Es. From 1975 to 1992 the Corps rotated units to Iwakuni and the 1st MAW for six to twelve month periods under what was called the Unit Deployment Program (UDP). UDP became the Marines' counterpart to Navy carrier deployments and was the primary event in a squadron's training cycle. During this period squadrons frequently left aircraft in Japan for the succeeding

units or swapped airframes with visiting Navy outfits. The program involved VMFA, VMA and VMA (AW) units under MAG-15 as well as their rotary wing counterparts, which deployed to Futenma, Okinawa.
Notes: "Nice to know" or miscellaneous information on the squadron.

Replacement Squadron

VMAT (AW)-202 *Double Eagles* (KC)
From: Activated 15 Jan 1968
To: Deactivated 30 Sep 1986
Types: A-6A, E, TC-4C **Losses**: 5 operational, 1 TC-4C
Homebase: MCAS Cherry Point, NC
Notes: The Marine A-6 replacement squadron. Subordinated to MAG-24 (1/68-4/68), MCCRTG-20 (4/68-12/75), MAG-14 (12/75-end). First commanding officer was LtCol J.K. Davis, and the first aircraft accepted was A-6A 151823. The training syllabus emphasized Marine missions, such as Close Air Support and beacon bombing with carrier qualifications not normally (if ever) conducted. The squadron also carried out basic airframe (non-mission) training for EA-6A aircrew in the VMCJ/VMAQ units. The *Double Eagle* insignia was approved on 27 Feb 1968. The squadron's first class, made up of five B/Ns, graduated 22 March, 1968. The first TC-4C (155727) arrived 20 March, 1968. Normal aircraft strength was nine A-6A, three TC-4C in 1969, twelve A-6E, two TC-4C in 1984, six A-6E, one TC-4C by 1986. Marine A-6 replacement duties were transferred to VA-128 at Whidbey Island upon the squadron's deactivation.

Fleet Marine Force (FMF) Units

VMA (AW)-121 *Green Knights* (VK)
From: Redesignated from VMA-121 on 14 Feb, 1969.
To: Redesignated VMFA(AW)-121 on 8 Dec. 1989.
Types: A-6A, E **Losses**: 0 combat, 6 operational
Homebases: MCAS Cherry Point, NC 1969 to 1977. MCAS El Toro, CA, 1978 -1989
UDP: 5/77-5/78, 10/82-4/83, 4/81-10/81.
Combat: None.
Carrier Deployments:

8-10/86	*Ranger*	CVW-2	NE 4xx	WestPac	A-6E TRAM
3,4/87	*Ranger*	CVW-2	NE 4xx	WestPac	A-6E TRAM
2-8/89	*Ranger*	CVW-2	NE 4xx	WestPac	A-6E TRAM

Notes: The sixth, and last, Fleet Marine *Intruder* squadron to be established and the first to transition to the *Hornet*. One of the most notable units in Marine aviation history, VMF-121 was the leading Marine fighter unit in World War II, being credited with 208 aerial victories, as well as being the squadron where the legendary Joe Foss won the Medal of Honor. The unit was awarded the Presidential Unit Citation (PUC) for its work at Guadalcanal during late 1942. During Korea, while flying AD *Skyraiders*, VMA-121 reportedly dropped more bomb tonnage than any other Marine squadron. VMA-121 flew A-4C and E *Skyhawks* from Chu Lai, Vietnam for two tours between 1966 and 1969. On Feb. 14 1969 the squadron was redesignated as a VMA(AW) and moved "on paper" to Cherry Point (without personnel) for *Intruder* transition. First A-6A accepted 24 Feb. 1969. Converted to A-6E during 1973.

Assigned to CVW-2/*Ranger* by 12/84, joining VA-145 in an "All Grumman Air Wing" with no light attack assets onboard. The ship and air wing were selected to participate in experimental Pacific Fleet "surge" ops where ship and air wing were kept at a high state or readiness for emergency use in the Pacific theater. From August 1986 to April 1987 the ship made two short "surge" deployments to WestPac. High transit times, costs and stress on personnel led to an end of the surge experiments and they went back to a more routine eighteen-month deployment cycle. Made full Western Pacific/Indian Ocean deployments in 1987 and 1989 before transferring back to 3rd MAW control at El Toro on 25 Sep 89. They were replaced in CVW-2 by VA-155. Upon return to Marine control, the *Green Knights* would become the first Marine *Intruder* unit to transition to the FA-18D.

VMA (AW)-224 *Bengals* (WT)
From: Redesignated from VMA-224 1 Nov, 1966
To: Redesignated VMFA (AW)-224 6 March, 1993
Types: A-6A,B,E, KA-6D **Losses**: 5 combat, 5 operational
Homebase: MCAS Cherry Point, NC 1966-1993.
UDP: 11/75-5/76, 5/78-5/79, 4/83-10/83, 10/84-4/85, 4/86-10/86, 10/87-4/88, 4/89-10/89, 10/91-3/92.
Combat: Vietnam (*Coral Sea*, 1971-72), Desert Storm (Shaikh Isa, Bahrain, 28 Aug 1990 to 26 March, 1991)
Carrier Deployments: (1)

11/71-7/72	*Coral Sea*	CVW-15	NL5xx	WestPac/Vietnam	A-6A,B, KA-6D

Notes: Unit was transferred "on paper" from Chu Lai, RVN to Cherry Point in 1966, with no personnel or aircraft making the move. First *Intruder* accepted 5 Nov. 1966. First Marine *Intruder* unit to make an extended deployment with a Navy Carrier Air Wing and the only Marine unit to fly A-6Bs. Saw extensive combat in Vietnam with CVW-15 off *Coral Sea* (CVA-43). Received the first Marine A-6Es in June, 1974 and made first UDP to Iwakuni with the type in November, 1975. Flew 422 combat sorties during Desert Storm while delivering 2.3 million pounds of ordnance. Made the final Marine A-6 UDP to Iwakuni, departing for Cherry Point on 20 March, 1992.

VMA (AW)-225 *Vagabonds, Vikings* (CE)
From: Redesignated from VMA-225 1 June, 1966.
To: Deactivated 23 June, 1972
Types: A-6A, E **Losses**: 2 combat, 1 operational
Homebases: MCAS Cherry Point: June, 1966 to Jan. 1969. Da Nang RVN: Feb 1969 to Jun. 1972. MCAS El Toro, CA: June, 1972.
UDP: (none)
Combat: Vietnam 1969-1972
Carrier Deployments: None.
Notes: Squadron was moved "on paper" from Chu Lai RVN in 1966 and was attached to VMA(AW)-533 for administrative reasons during a long, drawn-out rebuilding period. 225 resumed its own operations on 22 Oct 1966. Renamed the "Vikings" in October, 1968. Adopted the logo of the NFL's Minnesota Vikings on a background of "Carolina Blue". 225 arrived at Da Nang on 5 Feb, 1969 with twelve *Intruders*. Involved in early use of laser-guided bombs (LGBs) during 1970-71, using ground designation. Flew 13,000+ combat sorties in Vietnam. Squadron was "cadred" on 15 June 72, actually deactivated on 23 June. The *Vikings* would return on 1 June, 1991 as VMFA (AW)-225, flying FA-18D *Hornets*.

VMA (AW)- 242 *Batmen* (DT)
From: Redesignated from VMA-242 1 October, 1964.
To: Redesignated VMFA (AW)-242, 14 December, 1991.
Types: A-6A, E **Losses**: 4 combat, 9 operational
Combat: Vietnam.
Homebases: MCAS Cherry Point, NC 1964-1966. Da Nang, RVN 1966-1970. MCAS El Toro, CA 1970-1991.
UDP: 10/80-4/81, 4/82-10/82, 4/84-10/84, 10/85-4/86, 4/87-10/87, 10/88-4/89.
Carrier Deployments: None.
Notes: The first Marine *Intruder* unit. Served 47 consecutive months in Vietnam, which is believed to be a record for any Navy or Marine squadron. Lost three *Intruders* due to Viet Cong mortar attacks while in country, 30 Jan 1968 (2) and 27 July 1968 (1). Averaged 2000+ tons of bombs a month at the height of operations in Vietnam, with a high of 2700 tons dropped in January, 1969. Was the first A-6 unit attached to 3rd MAW at El Toro, in 1970. Initially assigned to MAG-33, transferred to MAG-13 in Dec. 1970. Squadron was functionally broken to a cadre status (non-operational) upon return and went through a very slow rebuild period due to shortages of people, aircraft and supplies. Transitioned to A-6E September, 1977 and A-6E TRAM in April, 1981. Made six UDP trips to MCAS Iwakuni between April, 1982 and 1989. Four squadron aircraft were transferred to VMA (AW)-224 for use in *Desert Shield/Storm* in August, 1990.

VMA (AW)-332 *Polkadots, Moonlighters* (EA)
From: Redesignated from VMA-332 20 August, 1968.
To: Redesignated VMFA-AW-332 16 June, 1993.
Types: A-6A, E **Losses**: 0 combat, 5 operational
Homebases: MCAS Cherry Point, NC.
UDP: 5/76-5/77, 5/79-5/80, 10/83-4/84, 4/85-10/85, 10/86-4/87, 10/89-4/90, 4/88-10/88, 1/91-10/91.
Combat: None.
Carrier Deployments: None.
Notes: The last A-4C unit at Cherry Point and the fifth of six Fleet Marine squadrons to get *Intruders*. Received A-6E from 1 Mar. 1975, A-6E TRAM from July, 1982. Participated in Unit Deployment Program to MCAS Iwakuni, Japan from 1976 through 1992. Original title of *Polka Dots* dates from Korea, reportedly when they accepted F4U-4 *Corsairs* from the *Checkerboards* of VMF-312 and repainted their nose markings into red dots on a white background. The Hat and Cane insignia is supposedly derived from the unit's '50's-era tailcode, which was "MR", and it's "gentlemanly" connotation. The *Moonlighters* name and insignia that came to use during the squadron's *Intruder* years was certainly easier to explain at the Cubi bar. The *Moonlighters* were the last Marine unit to turn in their A-6s and therefore the last formally designated All Weather Attack squadron in Corps history.

VMA (AW)-533 *Nighthawks* (ED)
From: Redesignated from VMA-533 1 July, 1965.
To: Redesignated VMFA-AW-533 1 Oct. 1992.
Types: A-6A, E, KA-6D **Losses**: 11 combat, 10 operational

Homebases: MCAS Cherry Point, NC 1965-1967, Chu Lai RVN 1967-1969, Iwakuni Japan 1969-1972, Nam Phong, Thailand 1972-1973, Iwakuni, Japan 1973-1975, MCAS Cherry Point, NC 1975-1992
UDP: 4/80-10/80, 10/81-4/82, 4/90-1/91.
Combat: Vietnam (Chu Lai, March 1967-Oct. 1969, Nam Phong Thailand June, 1972-Aug 1973), Desert Storm (Shaikh Isa Bahrain, Dec. 1990-Feb 1991)
Carrier Deployments:

4-10/84	*Saratoga*	CVW-17	AA 5XX	Med	A-6E TRAM, KA-6D
8/86-3/87	*JF Kennedy*	CVW-3	AC 54X/55X	Med	A-6E TRAM
8/88-2/89	*JF Kennedy*	CVW-3	AC 54X/55X	Med	A-6E TRAM

Notes: The *Nighthawks* were among the most combat experienced of all of the Marine *Intruder* units, having fought in Vietnam from two different bases as well as being heavily involved in *Desert Storm*. Following a two-year combat tour at Chu Lai the squadron moved permanently to MCAS Iwakuni, Japan, where it stayed for 6 years, with an almost one-year combat tour at Nam Phong RTAFB ("the Rose Garden") during the end stages of the Vietnam War. The squadron returned to North Carolina in 1975 where they received A-6Es. During the next 17 years the *Hawks* made three cruises to the Mediterranean, several UDPs to Iwakuni and saw combat in *Desert Storm*.

Appendix D: Intruder Combat Losses

Num	BUNO	Type	Squadron	Date	Location	d/n	Cause	Crew	Status
1	151584	A	VA-75	14-Jul-65	Laos	day	own ordnance	Boecker/Eaton	rcvd/rcvd
2	151577	A	VA-75	18-Jul-65	Laos	day	own ordnance	Denton/Tschudy	POW/POW
3	151585	A	VA-75	24-Jul-65	Laos	day	own ordnance	Bordone/Moffett	rcvd/rcvd
4	151588	A	VA-75	17-Sep-65	North Vietnam RP-4	night	poss AAA	Vogt/Barber	KIA/KIA
5	151781	A	VA-85	21-Dec-65	North Vietnam RP-6B	night	SA-2	Cartwright/Gold	KIA/MIA
6	151797	A	VA-85	18-Feb-66	North Vietnam RP-5	day	low pullout	Murray/Schroeffel	KIA/KIA
7	151794	A	VA-85	17-Apr-66	North Vietnam RP-2	day	AAA	Sayers/Hawkins	rcvd/rcvd
8	151798	A	VA-85	21-Apr-66	North Vietnam RP-3	night	AAA	Keller/Austin	KIA/KIA
9	151785	A	VA-85	22-Apr-66	North Vietnam RP-3	day	AAA	Weimorts/Nickerson	KIA/KIA
10	151788	A	VA-85	27-Apr-66	North Vietnam RP-3	day	AAA	Westerman/Weston	rcvd/rcvd
11	151816	A	VA-65	25-Jun-66	North Vietnam RP-3	day	AAA	Weber/Marik	rcvd/KIA
12	151822	A	VA-65	27-Aug-66	North Vietnam RP-3	day	AAA	Fellowes/Coker	POW/POW
13	151590	A	VA-85	19-Jan-67	North Vietnam RP-4	day	AAA	Brady/Yarbrough	POW/KIA
14	151587	A	VA-85	24-Mar-67	North Vietnam RP-6A	night	AAA	Ellison/Plowman	KIA/KIA
15	152609	A	VMA(AW)-242	17-Apr-67	North Vietnam RP-2	night	AAA	McGarvey/Carlton	KIA/KIA
16	152589	A	VA-85	24-Apr-67	North Vietnam RP-6B	night	AAA	Williams/Christian	rcvd/rcvd
17	152594	A	VA-35	19-May-67	North Vietnam RP-6B	day	SA-2	McDaniel/Patterson	POW/KIA
18	152625	A	VA-196	21-Aug-67	China	day	MiG	Bookley/Flynn	KIA/POW
19	152627	A	VA-196	21-Aug-67	China	day	MiG	Scott/Trembley	KIA/KIA
20	152638	A	VA-196	21-Aug-67	North Vietnam RP-6A	day	SA-2	Profilet/Hardman	POW/POW
21	152639	A	VMA(AW)-533	26-Aug-67	North Vietnam RP-6B	night	AAA	Bacik/Boggs	MIA/KIA
22	152601	A	VMA(AW)-242	30-Oct-67	North Vietnam RP-6A	night	AAA	Fanning/Kott	MIA/KIA
23	152629	A	VA-196	2-Nov-67	North Vietnam RP-6B	night	AAA	Morrow/Wright	KIA/KIA
24	152612	A	VMA(AW)-242	25-Nov-67	North Vietnam RP-6B	night	unknown	Abrams/Holdeman	KIA/KIA
25	152917	A	VA-75	31-Dec-67	North Vietnam RP-1	day	SA-2	Peace/Perisho	KIA/KIA
26	152636	A	VMA(AW)-533	18-Jan-68	North Vietnam RP-6A	night	unknown	Wallace/Murray	KIA/KIA
27	152901	A	VA-165	26-Jan-68	North Vietnam RP-3	dusk	unknown	Eidsmoe/Dunn	KIA/KIA
28	152588	A	VMA(AW)-533	30-Jan-68	Da Nang	night	Mortar attack	n/a	n/a
29	152644	A	VMA(AW)-533	24-Feb-68	North Vietnam RP-6A	night	SA-2	Marvel/Friese	POW/POW
30	152938	A	VA-35	28-Feb-68	off North Vietnam RP-4	night	unknown	Coons/Stegman	KIA/KIA
31	152944	A	VA-35	1-Mar-68	North Vietnam RP-6B	night	unknown	Scheurich/Lannom	KIA/KIA
32	152922	A	VA-75	6-Mar-68	North Vietnam RP-6B	night	unknown	Nelson/Mitchell	KIA/KIA
33	152940	A	VA-35	16-Mar-68	North Vietnam RP-6B	night	AAA	Shuman/Doss	POW/POW
34	154164	A	VMA(AW)-533	2-May-68	North Vietnam RP-1	day	AAA	Avery/Clem	KIA/KIA
35	152951	A	VA-35	13-May-68	North Vietnam RP-2	night	AAA	Brehner/Fardy	rcvd/rcvd
36	152949	A	VA-35	24-Jun-68	North Vietnam RP-2	night	AAA	Carpenter/Mobley	KIA/POW
37	154166	A	VMA(AW)-533	25-Jul-68	North Vietnam RP-1	day	AAA	Lawson/Brown	rcvd/POW
38	152595	A	VMA(AW)-533	27-Jul-68	Da Nang	night	Mortar attack	n/a	n/a
39	151561	B	VA-85	28-Aug-68	North Vietnam RP-3	night	SA-2	Duncan/Ashall	KIA/KIA
40	154127	A	VA-85	6-Sep-68	North Vietnam RP-2	night	AAA	Coskey/McKee	POW/Rcvd
41	154149	A	VA-196	30-Sep-68	North Vietnam RP-3	night	SA-2	VanRenselaar/Spinelli	KIA/KIA
42	154141	A	VA-52	13-Oct-68	North Vietnam RP-2	night	unknown	Orell/Hunt	KIA/KIA
43	154150	A	VA-196	18-Dec-68	Laos	day	AAA	Babcock/Meyer	KIA/KIA
44	154152	A	VA-196	19-Dec-68	Laos	night	AAA	Bouchard/Colyar	KIA/rcvd
45	152586	A	VMA(AW)-242	17-Jan-69	South Vietnam	night	AAA	Fickler/Kuhlman	KIA/MIA
46	154160	A	VMA(AW)-533	17-Mar-69	Laos	night	unknown	Armistead/Finney	KIA/KIA
47	155587	A	VA-65	3-Apr-69	Laos	day	AAA	Redden/Ricci	rcvd/rcvd
48	155611	A	VMA(AW)-225	21-Sep-69	South Vietnam	day	AAA	Busch/Hardgrave	rcvd/rcvd
49	155696	A	VMA(AW)-242	29-Sep-69	South Vietnam	night	unknown	Lono/Curran	KIA/KIA
50	155701	A	VMA(AW)-225	15-Nov-69	South Vietnam	night	AAA	Jessen/Tutor	rcvd/rcvd

Num	BUNO	Type	Squadron	Date	Location	d/n	Cause	Crew	Status
51	155613	A	VA-196	22-Nov-69	Laos	day	unknown	Richards/Deuter	rcvd/KIA
52	155607	A	VA-196	22-Nov-69	Laos	night	unknown	Collins/Quinn	KIA/KIA
53	152937	A	VA-196	2-Jan-70	Laos	day	AAA	Fryar/Brooks	KIA/KIA
54	155618	A	VA-196	6-Feb-70	Laos	dusk	AAA	Reese/Frazer	rcvd/rcvd
55	155677	A	VA-165	30-Dec-71	North Vietnam RP-2	day	SA-2	Holmes/Burton	KIA/rcvd
56	155652	A	VMA(AW)-224	9-Apr-72	Laos	dusk	AAA	Smith/Ketchie	rcvd/KIA
57	155709	A	VMA(AW)-224	3-May-72	North Vietnam RP-1	day	unknown	McDonald/Williams	KIA/KIA
58	155650	A	VMA(AW)-224	29-May-72	North Vietnam RP-6B	day	AAA	Schuyler/Ferracane	rcvd/rcvd
59	154145	A	VMA(AW)-224	11-Jun-72	North Vietnam RP-4	day	AAA	Wilson/Angus	KIA/POW
60	155690	A	VMA(AW)-533	7-Jul-72	South Vietnam	day	AAA	Kroboth/Robertson	POW/KIA
61	157018	A	VA-52	19-Aug-72	North Vietnam RP-6B	night	unknown	Lester/Mossman	KIA/KIA
62	155626	A	VA-75	6-Sep-72	North Vietnam RP-6B	day	SA-2	Lindland/Lerseth	KIA/POW
63	157028	A	VA-35	16-Sep-72	North Vietnam RP-6B	night	AAA	Donnelly/Buell	KIA/KIA
64	155700	A	VMA(AW)-533	11-Oct-72	North Vietnam RP-1	night	unknown	Peacock/Price	KIA/KIA
65	155594	A	VA-196	20-Dec-72	North Vietnam RP-6B	dusk	AAA	Nakagawa/Higdon	POW/POW
66	152946	A	VA-75	21-Dec-72	North Vietnam RP-6B	night	AAA	Graustein/Wade	KIA/KIA
67	155666	A	VMA(AW)-533	27-Dec-72	North Vietnam RP-1	night	unknown	Chipman/Forrester	KIA/KIA
68	155693	A	VA-115	9-Jan-73	North Vietnam RP-3	night	SA-2	McCormick/Clark	KIA/KIA
69	157007	A	VA-35	24-Jan-73	South Vietnam	day	AAA	Graf/Hatfield	rcvd/rcvd
70	152915	E	VA-85	4-Dec-83	Lebannon	Day	IR SAM	Lang/Goodman	KIA/POW
71	152928	E	VA-155	18-Jan-91	Desert Storm	night	AAA	Turner/Costen	KIA/KIA
72	161668	E	VA-35	18-Jan-91	Desert Storm	night	Roland SAM	Wetzel/Zaun	POW/POW
73	155632	E	VA-36	2-Feb-91	Desert Storm	day	AAA or IR SAM	Cooke/Conner	KIA/KIA

Notes: 73 Intruders were lost in combat, or about 11% of the production run. 86 men were either Killed in Action (KIA) or listed as Missing In Action (MIA) while flying the A-6. Triple A, meaning guns of all sizes, was the big killer in combat, accounting for over half of the losses. "Unknown" (lost/missing due to unknown reasons) is listed for 16 Intruders while radar directed SAMs shot down 11. Ten of these were SA-2s in Vietnam and one to what was believed to be a Roland in Desert Storm. Aircrew names and status used in this appendex are current with the official DPMO list dated 17 Nov 2003. It should be noted that several aviators originally listed as "MIA" are now listed "KIA" under a "presumptive finding of death"

Appendix E: Intruder Operational Losses

Num	BUNO	typ	Squadron	Date	Location	Details	Crew
1	149476	YA	NATC	18-Mar-63	Pax River	Impacted ground during trials. Pilot only one onboard.	F
2	149958	A	VA-75	21-May-64	Meridian MS	Electrical and hydraulic problems	2R
3	149938	A	VA-42	28-Aug-64	Columbia SC	Spin demo, lost control	2R
4	151780	A	VA-85	27-Jul-65	off San Diego	mid-air with 151786	1R/1F
5	151786	A	VA-85	27-Jul-65	off San Diego	mid-air with 151780	2R
6	151805	A	VMA(AW)-242	8-Dec-65	off Cp Hatteras	Hi-speed low over water, hit surface	2R
7	151567	A	VMA(AW)-242	28-Jan-66	Mattameuskeet NC	Stall and departure during maneuvers	2R
8	151586	A	VA-75	2-Mar-66	Oceana	Wing folded after takeoff	1R
9	151800	A	VA-85	15-May-66	Gulf of Tonkin	off CVA-63. Fuel starvation	2R
10	152602	A	VMA(AW)-242	20-Jun-66	Tangier Tgt MD	Night mid-air with 152605	2R
11	152605	A	VMA(AW)-242	20-Jun-66	Tangier Tgt MD	Night mid-air with 152602	2R
12	149949	A	VA-75	16-Jun-66	Mediterranean	off CVA-60. Hit water forward of boat after night cat shot	2F
13	149939	A	VA-42	27-Sep-66	Tangier Tgt MD	Hit ground while in bombing pattern	2F
14	149480	A	NATC	4-Oct-66	nr Pax River	Hit ground during test program	2R
15	149483	A	NATC	10-Nov-66	Pax River	Hit ground during test program	1R/1F
16	152608	A	VMA(AW)-242	23-Mar-67	Da Nang	Aborted take off due to USAF C-141 on rwy	2R
17	152900	A	NATC	28-Mar-67	Stumpy Pt Tgt. NC	Dual flameout during rocket run	2R
18	151803	A	VMA(AW)-224	25-May-67	Cherry Pt NC	Fire during hot fueling	2R
19	152613	A	VA-128	5-Sep-67	Whidbey Is WA	Dual flameout during FCLPs	2R
20	151799	A	VMA(AW)-225	4-Nov-67	Cherry Pt NC	Mid-air during tanking, hit by receiver acft	2R
21	152631	A	VMA(AW)-533	23-Feb-68	South Vietnam	Mid-air with F-8E in GCI pattern to Chu Lai	2R
22	152943	A	VA-35	12-Mar-68	Gulf of Tonkin	off CVAN-65. Hit water forward of boat after night cat shot	2F
23	152952	A	VA-85	30-Mar-68	nr Puerto Rico	off CVA-66, hit water fol. Attempt. rendezvous	2F
24	151817	A	VMA(AW)-224	18-Apr-68	nr Cherry Pt NC	Hit water at steep angle	2F
25	151578	A	VMA(AW)-242	1-May-68	Da Nang	Apparent control failure after launch	2R
26	151569	A	VMA(AW)-224	15-May-68	nr Topeka KS	Dual generator failure in IMC	2R
27	152903	A	VA-165	22-Jul-68	nr Whidbey Is WA	Dual flameout with hydraulics failing	2R
28	154139	A	VA-128	5-Aug-68	nr. Whidbey Is WA	Departed during aerobatics	1R/1F
29	151560	A	VA-196	20-Aug-68	Gulf of Tonkin	off CVAN-65, control problems	2R
30	154125	A	VA-128	27-Aug-68	Cypress IS WA	Hit trees during IMC descent	2R
31	151594	A	VMA(AW)-224	12-Feb-69	Blythe CA	Dual flameout	2R
32	152643	A	VA-35	7-Mar-69	Boyington VA	Day mid-air with 152897	1F/1R
33	152897	A	VA-35	7-Mar-69	Boyington VA	Daty mid-air with 152643	2F
34	152898	A	VA-35	3-Apr-69	Lauray VA	Impacted terrain during weather	2F
35	155663	A	VA-85	18-Apr-69	Olathe, KS	Lost engine on takeoff, ran off end of runway	2R
36	154143	A	VA-52	29-Jun-69	McChord AFB WA	Crashed on takeoff	2F
37	154153	A	VA-165	7-Jul-69	nr. Whidbey Is WA	Control problems in flight	2R
38	152633	A	VA-145	19-Aug-69	Boardman OR	Impacted ground in target area	2F
39	151574	A	VMAT(AW)-202	26-Aug-69	Kinston NC	Split S into ground	1F/1R
40	152926	A	VA-176	26-Sep-69	nr the Bahamas	off CVA-42, control problems on fly-off	2R
41	152646	A	VMA(AW)-332	19-Dec-69	Cherry Pt NC	Stall in landing pattern	2F
42	152891	A	VA-35	26-Dec-69	Gulf of Tonkin	Hit the water during Case I approach to CVA-43	2F
43	155639	A	VA-75	4-Jan-70	Mediterranean	off CVA-60, engine fire, loss of hydraulics	2R
44	151815	A	VA-176	16-Feb-70	Mediterranean	off CVA-42, hit water on wave off from single-engine landing	1F/1R
45	155605	A	VA-196	28-Feb-70	Gulf of Tonkin	off CVA-41, engine fire, hydraulic problems	2R
46	154165	A	VMA(AW)-533	4-Mar-70	Iwakuni JA	Lost control in IMC with flap/slat problems	2R
47	155641	A	VA-75	10-Apr-70	Atlantic Ocean	off CVA-60, catapult set 11,000 lbs light, aircraft hit water	2R
48	155614	A	VMA(AW)-533	22-Apr-70	South Korea	Out of control during bombing run.	2F
49	156998	A	VA-145	12-Jun-70	Boardman, OR	Right wing seperated during bombing pull-out.	1R/1F
50	155593	A	VA-176	21-Aug-70	Petersburg, VA	Hydraulic and control problems on FCF	2R

Num	BUNO	typ	Squadron	Date	Location	Details	Crew
51	152622	A	VA-128	17-Sep-70	Rosallia WA	Impacted ground during RBS run.	2F
52	157015	A	VA-35	20-Sep-70	nr Memphis TN	Flame out on approach	2R
53	155647	C	VA-145	8-Jan-71	Gulf of Tonkin	off CVA-61, into water after night cat shot	1F/1R
54	152899	A	VMA(AW)-121	18-Jan-71	Neuse River NC	Engine fire.	2R
55	156994	A	VA-145	24-Feb-71	Gulf of Tonkin	off CVA-61, Engine failure after night cat shot.	2R
56	154138	A	VA-176	24-Mar-71	Sigonella, IT	Landed with electrical failure, off runway, aircraft turned over	2R
57	155720	A	NATC	19-Apr-71	Pax River	Complete hydraulic failure	2R
58	155640	A	VA-42	29-Jul-71	Dare Cty NC	Engine fire.	2R
59	152598	KA	VA-115	12-Aug-71	Gulf of Tonkin	off CVA-41, massive fuel leak leading to fire	2R
60	152590	KA	VA-196	9-Oct-71	Gulf of Tonkin	off CVAN-65, instrument problems at night, stalled on launch	2R
61	151563	B	VA-42	15-Oct-71	Oceana, VA	Inadvertant pilot ejection, BN followed	2R
62	157022	A	VA-52	28-Oct-71	Boardman OR	Control problems in bombing pattern, out of control	2R
63	154157	A	VA-176	7-Apr-72	Mediterranean	off CVA-42, engine fire	2R
64	157008	A	VA-34	13-Apr-72	Mediterranean	off CVA-67, hit water during night bombing practice	2F
65	152597	KA	VA-115	2-May-72	Gulf of Tonkin	off CVA-41, Electrical fire in cockpit after launch.	2R
66	155609	A	VMA(AW)-242	18-Jul-72	29 Palms CA	Hit ground during high-speed pass over troops	2F
67	151559	B	VA-196	30-Jul-72	off San Diego	off CVAN-65. Aircraft pitched up after launch	2R
68	151571	A	VA-95	8-May-72	Boardman, OR	Hit ground on low-altitude/high-speed pass	2F
69	151580	KA	VA-85	9-Aug-72	Oceana VA	Crashed on short final.	2F
70	155705	A	VA-115	24-Oct-72	Gulf of Tonkin	On CVA-41, wheel failed, acft impacted pack of aircraft on deck	1R/1F
71	151806	KA	VA-196	31-Oct-71	Cubi Pt PI	Pitched up on takeoff, crashed	2F
72	152615	A	VA-128	15-Nov-72	Fallon NV	Impacted ground after 90 deg dive bomb run	1F/1R
73	155622	A	VA-75	28-Nov-72	Gulf of Tonkin	off CVA-60 cat shot, pilot's HSI lodged against stick, acft stalled.	1F/1R
74	154155	A	VMA(AW)-332	4-Dec-72	Pamlico Sd NC	Hit water during or after target run	2F
75	152604	A	VA-52	28-Mar-73	Boardman OR	Wing seperated from aircraft during bombing run	2F
76	158789	E	VA-42	15-May-73	Fentress VA	Fuel exhaustion	2R
77	155634	A	VA-52	30-May-73	Boardman OR	Flew into tree tops during low level	2R
78	152616	A	VA-34	3-Jul-73	Mediterranean	off CVA-67, acft lost control, possible fire	2R
79	155721	A	VA-128	19-Sep-73	Lakeview OR	Impacted ground during low level	2F
80	155603	A	VA-196	12-Nov-73	Astoria, Or	Aircraft went out of control while at 22,000 ft.	2R
81	157020	A	VA-52	8-Mar-74	Subic Bay PI	Dual engine flameout in the break at Cubi Point	2R
82	152637	KA	VA-145	12-May-74	off Hawaii	off CVA-61, mid air with F-4J at night while in tanking pattern	2R
83	151825	KA	VA-75	11-Oct-74	Mediterranean	off CVA-60. Flaps/Slats retracted on launch. Acft went in water	1R/1F
84	156999	A	VA-95	20-Nov-74	Whidbey Is WA	Aborted take off into water due to missing bolt on stabilizer	2F
85	155601	A	VA-196	14-Apr-75	Cubi Pt PI	Fire in flight.	2R
86	158048	E	VA-42	21-Apr-75	Virginia Bch VA	Fuel exhaustion while dealing with gear problems	2R
87	152640	E	VA-34	25-Apr-75	New River NC	Left wing seperated during bomb run. VA-75 acft on loan to 34	1F/1R
88	152892	KA	VA-34	10-May-75	Atlantic Ocean	off CV-67. Brakes on during launch, crew ejected	2R
89	152918	E	VA-85	25-Jun-75	Mediterranean	off CV-59. Night mid air with VAQ-134 EA-6B 158814	1F/1R
90	151824	KA	VA-115	20-Jul-75	Pacific Ocean	off CV-41. BN inadvertantly ejected during cat shot, pilot followed	1F/1R
91	149948	E	VA-35	23-Aug-75	Atlantic Ocean	off CVN-68. Mid-air with A-7E in tanker pattern	2F
92	158044	E	VA-42	31-Mar-76	Tangier Tgt MD	Impacted water following work at target	2F
93	159577	E	VA-95	6-Jun-76	Goldendale WA	Impacted ground during lead change on section low level	2F
94	152593	E	VX-5	30-Jun-76	China Lk CA	Mid-air collision with A-7E in landing pattern	2R
95	152914	KA	VA-52	17-Sep-76	Whidbey Is WA	Impacted water during simulated single-engine landing	2F
96	158534	E	VA-65	22-Sep-76	Fallon NV	Single engine due to bleed-air leak, unable to maintain altitude	2R
97	154130	A	VMAT(AW)-202	28-Oct-76	Cherry Pt NC	Loss of pitch control on takeoff	1F/1R
98	155631	E	VMA(AW)-533	24-Nov-76	Bluefield WV	Stalled aircraft at night while dealing with other problems	2R
99	155617	E	VMA(AW)-332	20-Jan-77	Luzon, PI	Impacted ridgeline while on low-level	2F
100	159309	E	VMA(AW)-332	2-Mar-77	Cubi Pt PI	Impacted water during night bombing	2F
101	155671	A	VA-128	7-Jul-77	nr Albuquerque NM	Runaway trim followed by dual engine flameout	2R
102	158049	E	VA-42	16-Aug-77	Oceana VA	Impacted short of runway on single-engine landing	1R/1F
103	151795	KA	VA-115	7-Dec-77	Pacific Ocean	off CV-41, Ramp strike on night landing	2R
104	151791	KA	VRF-31	18-Jan-78	Alameda, CA	Fuel starvation during approach: VA-75 aircraft on ferry flight	2R

Num	BUNO	typ	Squadron	Date	Location	Details	Crew
105	154151	E	VMA(AW)-224	1-Feb-78	Fallon NV	Impacted ground during night bomb run	1F/1R
106	149945	KA	VA-34	22-Feb-78	off Virginia	Lost at sea due to unknown problems	2F
107	158047	E	VA-85	29-Apr-78	off Italy	off CV-67, loss of control six minutes after cat shot	2R
108	155692	E	VA-42	20-Jul-78	El Centro CA	Hit by bombs from other aircraft, lost wing, hit ground on fire	2F
109	155645	E	VA-196	19-Sep-78	So China Sea	off CVN-65, fire in buddy store while tanking	2R
110	159313	E	VA-95	10-Oct-78	Whidbey Is WA	Impacted terrain in weather	2F
111	152947	E	VMA(AW)-332	16-Oct-78	Dare Cty NC	Impacted ground during bomb delivery	2F
112	151582	KA	VA-85	22-Feb-79	Roosevelt Rds PR	Out of control following takeoff	2F
113	157012	E	VRF-31	16-Mar-79	nr Grand Jct CO	Ferry flight to NARF Alameda, VA-34 aircraft. Flew into mountain	2F
114	160421	E	VA-176	17-May-79	Caribbean	off CV-62, Lost at sea during night bombing mission	2F
115	160428	E	VA-34	29-Aug-79	Lynchburg VA	Impacted terrain during low level	2F
116	157011	E	VA-35	6-Dec-79	Mediterranean	off CVN-68. Caught fire during bombing run.	2F
117	151566	KA	VA-165	12-Dec-79	off San Diego	off CV-64, impacted water following night cat shot	2F
118	152632	KA	VA-52	29-Dec-79	Indian Ocean	off CV-63, bad catapult shot	2F
119	160426	E	VA-196	8-Jan-80	of Philippines	off CV-43, pilot perceived soft cat shot and ejected. BN followed	2R
120	158050	E	VA-75	9-Jan-80	off Florida	off CV-60, drifted right on landing, impacted two F-4s, over side.	2F
121	152929	E	VA-75	14-Jan-80	off Florida	off CV-60, VDI box lodged against stick on cat shot, aircraft unflyable	2R
122	155675	E	VA-165	8-Feb-80	Mountain Home ID	Impacted terrain during low-level flight	2F
123	155624	E	VA-165	23-Jul-80	Indian Ocean	off CV-64, departed during aerobatics	2R
124	151895	E	VA-115	1-Oct-80	Indian Ocean	off CV-41. Lost aircraft on wave-off during single engine night landing	2R
125	151821	KA	VA-145	5-Nov-80	Indian Ocean	off CV-61, control and hydraulic problems	2R
126	155633	E	VMAT(AW)-202	2-Jan-81	Dare Cty NC	Impacted ground after bomb run	2F
127	158530	E	VA-85	6-Jul-81	off Sicily	off CV-59, dual engine failure due to fuel contamination	2R
128	155612	E	VMA(AW)-533	4-Feb-82	Iwakuni JA	Rolled right after takeoff and impacted ground	2R
129	152908	E	VMAT(AW)-202	12-Feb-82	nr Jacksonville FL	Impacted swamp during low-level flight	2R
130	154144	E	VA-65	22-May-83	Mediterranean	off CVN-69. Impacted water during mission	2F
131	154168	E	VMA(AW)-533	27-Jun-83	Emporia, VA	Fire while on low level	1R/1F
132	155680	E	VA-42	26-Aug-83	Pamlico Sd NC	Loss of control during maneuvering flight	2R
133	149941	E	VA-34	30-Mar-83	nr Kingman AZ	Impacted ground on low level	2F
134	155608	E	VMA(AW)-121	2-Oct-83	29 Palms CA	Wing seperated from aircraft during bombing run	2F
135	155660	E	VMA(AW)-242	18-Oct-83	Yuma, AZ	Uncontrolable roll on final to airport	2R
136	154132	E	VA-85	11-Jan-84	Mediterranean	off CV-67, nose gear failure on cat shot	2R
137	152934	KA	VA-75	12-Jul-84	off Florida	off CV-67, lost all control response while on descent	2R
138	152585	E	VMA(AW)-121	13-Jul-84	Beale AFB CA	Gear raised early during takeoff roll, aircraft settled on runway	2R
139	152894	KA	VA-115	15-Jul-84	nr Pusan, ROK	off CV-41, aircraft crashed while clearing range	2F
140	155667	E	VMAT(AW)-202	21-Nov-84	Pamlico Sd NC	Duel engine flameout	2R
141	158535	E	VMA(AW)-533	18-Dec-84	Avon Range, FL	Impacted ground during bombing mission	2F
142	158790	E	VA-75	27-Dec-84	Dare Cty NC	Impacted ground during low-level bombing run	2F
143	161673	E	VA-35	1-Apr-85	Panama	off CVN-67, failed to return from night low-level flight	2F
144	161684	E	VMA(AW)-121	22-Apr-85	Fallon NV	Impacted ground during night bombing practice	2F
145	154140	E	VA-176	10-May-85	Flandreau SD	Impacted ground at farm of BN's brother	2F
146	149954	KA	VA-85	1-Aug-85	Dare Cty NC	Departed flight during stall training	2R
147	155668	E	VA-176	11-Dec-85	Dare Cty NC	Impacted ground during night bombing practice	2F
148	161085	E	VA-42	3-Feb-86	Dare Cty NC	Impacted ground during night bombing practice	2F
149	151783	KA	VA-34	18-Feb-86	off Puerto Rico	off CV-66, massive FOD of stbd engine on cat stroke	2R
150	162181	E	VA-75	22-May-86	Oceana	Crashed on takeoff, killed civilian dependent in car	2F (+1)
151	159902	E	VMA(AW)-121	28-Jun-86	Pacific Ocean	off CV-61, total hydraulic failure	2R
152	159897	E	VMA(AW)-533	24-Oct-86	Mediterranean	off CV-67. Impacted water during SAR effort	2F
153	160994	E	VA-128	5-May-86	Boardman, OR	Impacted ground off target	2F
154	155688	E	VA-128	14-Jan-87	El Centro CA	Wing seperated from aircraft during bombing run	1R/1F
155	159905	E	VMA(AW)-533	17-Jan-87	Mediterranean	off CV-67, failed to return from flight	2F
156	155657	E	VA-42	12-May-87	Gulf of Mexico	on AVT-16, hook point failed on landing, went overboard	2R
157	161685	E	VA-85	20-May-87	Dare Cty NC	Impacted ground pulling off target	2F
158	151804	E	VMA(AW)-242	27-Jul-87	Kadena, Okinawa	Ejected from inverted spin	2R

Num	BUNO	typ	Squadron	Date	Location	Details	Crew
159	155625	E	VMA(AW)-533	4-Aug-87	Cherry Pt NC	Hydraulic failure and engine fire after takeoff	2R
160	161105	E	VA-145	18-Sep-87	Indian Ocean	off CV-61. Impacted water during night mission	2F
161	155674	E	VA-52	23-Sep-87	off Los Angeles	off CVN-70, total electrical failure during night climb out.	1F/1R
162	155649	E	VMA(AW)-121	23-Oct-87	Pacific Ocean	off CV-61, dual engine flameout	2R
163	158796	E	VA-145	14-Apr-88	nr Mt St Helen WA	Impacted ground on IR-344 night low level	2F
164	155637	E	VA-128	6-May-88	nr Mt St Helen WA	Impacted ground on IR-344 night low level	2F
165	152920	KA	VA-165	30-Nov-88	Pacific Ocean	on CVN-68. Hit by 20mm gunfire from A-7E undergoing maintenance	na
166	161109	E	VA-128	8-Aug-89	Whidbey Is WA	Impacted ground during airshow practice	2F
167	159572	E	VA-145	6-Nov-89	Whidbey Is WA	Total hydraulic failure on approach	2R
168	158798	E	VA-55	4-Dec-89	Dare Cty NC	Fuselage fire	2R
169	159174	E	VA-145	22-Jan-90	Whidbey Is WA	Pitched up out of control on take off	2R
170	152583	E	VA-196	24-Jul-90	Pacific Ocean	off CV-62, Lost at sea during night bombing mission	2F
171	152904	E	NATC	15-Aug-90	Pax River	Loss of control, aircraft impacted ground	1F/1R
172	155599	E	VA-176	6-Nov-90	Oceana VA	Engine fire immediately after takeoff	2R
173	152621	E	VA-42	7-Nov-90	El Centro CA	Engine fire after takeoff	2R
174	155602	E	VA-85	15-Feb-91	Persian Gulf	on CV-66, lost brakes after landing, hung up over side. Pushed into water	2R
175	159569	E	VA-42	9-May-91	nr Petersburg WV	Impacted ground during day low level	2F
176	155708	E	VA-176	19-Jul-91	Mediterranean	off CV-59, Impacted water	1F, 1R
177	159573	E	VA-128	19-Jul-91	off San Diego	on CV-61, landing mishap	2R
178	154148	E	VA-85	18-Sep-91	Mediterranean	off CV-66, lost engine on cat shot.	2R
179	151810	KA	VA-85	5-Dec-91	Mediterranean	off CV-66, engine failure on approach	2R
180	152620	E	VA-155	15-Oct-91	nr Rock Island WA	Impacted Columbia River during VR-1351 day low level flight	2F
181	155658	E	VA-128	17-Oct-91	El Centro CA	Lost wing on bomb run	1F 1R
182	159895	E	VA-95	22-Mar-93	Fallon NV	Crashed on takeoff	2R
183	155665	E	VA-95	14-Apr-93	Diamond, WA	Mid-air with Grumman Ag-Cat cropduster while on low level route	2R
184	158787	E	VA-34	21-Apr-93	Nellis NV	Mid-air with wingman during night rendezvous off target	2F
185	162196	E	VA-75	23-Jul-93	Roanoke VA	Hit ground on day low level	2F
186	164385	E	VA-95	8-Sep-93	Persian Gulf	off CV-63, mid-air with wingman, 161682	2R
187	161682	E	VA-95	8-Sep-93	Persian Gulf	off CV-63, mid-air with wingman, 164385	2R
188	155694	E	VA-304	5-Apr-94	Alameda, CA	Impacted water while in landing pattern	2F
189	161288	E	VA-115	14-Oct-94	Shikoku Japan	Hit ground on low level	2F
190	155586	E	VA-95	15-Feb-95	off San Diego	on CVN-72, fire on the catapult	2R
191	155704	E	VA-115	3-Jun-96	Pacific Ocean	off CV-62, Shot down by Japanese ship during exercise	2R

Appendix F: Intruders on Display

The following list of preserved A-6s includes all airframes known to the authors as of the end of 2003. More aircraft may yet come out of AMARC to add to this list. Not all of the *Intruders* on this list may, if fact, actually be on public display.

147865	MCAS Cherry Point, NC	
147867	Allegheny Arms & Armory Museum, Smithport PA	NA-6A
149482	NAS Whidbey Island WA: Ault Field main gate	KA-6D
151579	NAS Oceana VA	KA-6D
151782	USS *Midway* Museum, San Diego (planned)	A-6E
151826	Naval Aviation Schools Command, Pensacola	KA-6D
152599	Patriots Point, SC (USS Yorktown)	A-6E
152603	Richmond, IN	A-6E
152907	Beeksma Beach Park, Oak Harbor WA	A-6E
152910	Western Aviation Museum, Oakland CA	KA-6D
152923	NAS Norfolk, VA: Main gate	A-6E
152933	USS *Intrepid* Museum, New York, NY	A-6E
152935	Empire State Aviation Museum, Schenectady NY	A-6E
154131	Walker Field Park: Grand Junction CO	A-6E
154162	Palm Springs Air Museum, Palm Springs CA	A-6E
154167	National Air & Space Museum/Dulles (Udvar-Hazy)	A-6E
154170	MCAS Miramar, CA	A-6E
154171	Estrella Warbird Museum, Paso Robles CA	A-6E
155595	Pacific Coast Air Museum, Santa Rosa CA	A-6E
155610	National Museum of Naval Aviation, Pensacola FL	A-6E
155627	NAS Fallon, NV	A-6E
155629	Quonset Air Museum, RI	A-6E
155644	Chino, CA	A-6E
155648	NAS Atlanta, GA	A-6E
155713	Pima Air Museum, Tucson, AZ	A-6E
156997	NAS Patuxtent River, MD	A-6E
157001	NAVICP Philadelphia, PA	A-6E
157024	Defense General Supply Center: Richmond, VA	A-6E
158532	USS *Lexington*, Corpus Christi, TX	A-6E
158794	Museum of Flight, Seattle WA	A-6E
159901	NAF El Centro, CA	A-6E
161676	Lumley Aviation Center, Williamsport PA	A-6E
162182	Valiant Air Command, Titusville FL	A-6E
162184	Cradle of Aviation Museum, Garden Cy NY	A-6F
162185	USS *Intrepid* Museum, New York, NY	A-6F
162195	San Diego Aerospace Museum, CA	A-6E
162206	Oregon Air & Space Museum, Eugene OR	A-6E
155722	National Museum of Naval Aviation, Pensacola FL	TC-4C

In addition, over 60 *Intruders* were dumped at two locations into the Atlantic Ocean off Florida to form artificial reefs. Both are reportedly accessible to scuba divers.

Intruder Reef (29 degrees, 54 minutes North, 080 degrees, 48 minutes West)
A-6E: 149943, 149946, 149957, 151558, 151784, 152634, 152902, 154169, 155590, 158529, 158533, 158792, 158797, 159175, 159570, 159903, 160431, 161088, 161103, 161104, 161110, 161111, 161231, 161234, 161235, 161661, 161663, 161664, 161665, 161670, 161674, 161687, 161689.

Intruder Alley (29 degrees, 21 minutes North, 080 degrees, 64 minutes West)
A-6E: 149950, 149956, 151562, 151565, 151807, 152610, 152623, 152912, 154124, 154128, 154163, 155585, 155591, 155661, 155670, 155684, 158041, 158531, 158538, 158793, 159181, 160424, 160993, 161084, 161087, 161091, 161100, 161101, 161106, 161230, 161666, 161671, 161679, 161681, 161690, 162180.
KA-6D: 151820

Not unlike a lot of other Navy retirees, many *Intruders* headed for the deserts of Arizona when it was time to leave the service. A-6B 149955 is shown at MASDC in storage at Davis-Monthan AFB, AZ, in April 1976. She would be pulled out and rebuilt to "Echo" standards and return to what was now called AMARC in 1996. (John Kerr)

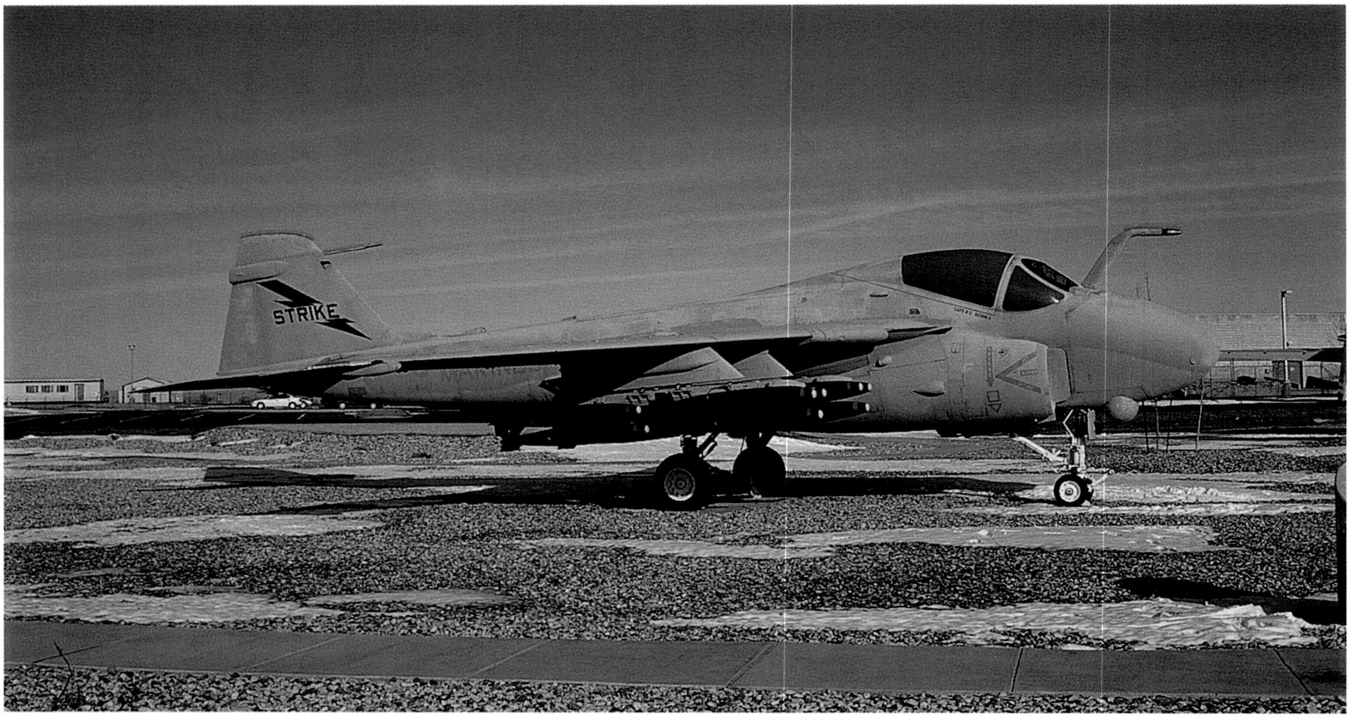

A-6E SWIP 155627 is typical of almost 40 *Intruders* now on display in the country, and is now located across from Strike "U" at NAS Fallon, NV. (Rick Morgan)

Quite a few *Intruders* continued to work for the Navy in a less-than-dignified fashion. Many ended up on bombing ranges as targets or, like this one, as a practice hulk on a carrier. 155644 is shown on *Abraham Lincoln*, NAVSTA Everett, WA, in August 1997. (Mark Morgan)

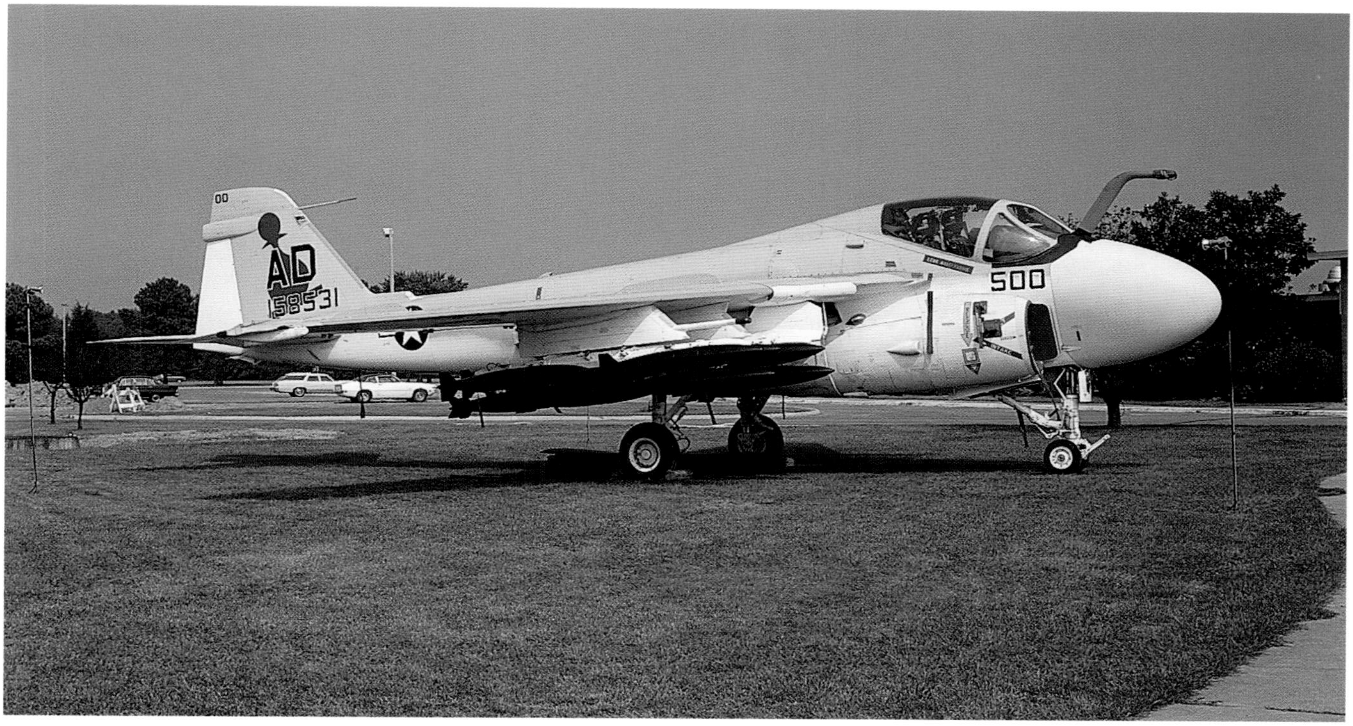

Contrary to appearances, AD500 has not been retired to display – it is a full up jet that was towed down the street and placed in front of the Oceana Officer's Club in September 1978 for the annual *Intruder* Ball. (Mark Morgan)

Intruder Reading

For an aircraft as important to the Navy as the A-6, there are surprisingly few books available on the *Intruder*. Most of the non-fiction efforts are softbound works written by aviation enthusiasts and frequently cater to model aircraft builders. Many are out of print. Here are some of the more notable efforts.

Fiction

Flight of the Intruder: Steven Coonts. 1986, Naval Institute Press. Well known novel of A-6s in Vietnam by a former VA-196 pilot. Probably the best aviation novel to come out of the war and made into a movie by Hollywood, starring Danny Glover, Willem Dafoe and Brad Johnson as the lead character, Jake Grafton.

First Person Non-fiction

Angles of Attack: An A-6 Intruder Pilot's War. Peter Hunt. Ballantine Books, 2002, 355 pages. Excellent paper-back book written by a pilot in VA-145 during Desert Storm. Proves beyond all doubt that while "fighter pukes make movies, attack crews make history".

Technical Non-Fiction

A-6 Intruder In Action (1), Aircraft no.20. Lou Drendel, Squadron/Signal Publication, 1975. 50 pages. Squadron's "In Action" series has been in production since the early 1970s and provides a large number of photos, drawings and first-person aircrew stories as well as the author's superb art work. Covers up to the A-6E's introduction as well as the EA-6A and B.

A-6 Intruder In Action (2), Joe Michaels, Squadron/Signal Publications, 1994. 50 pages. Second edition of Squadron's familiar series on the *Intruder*. This work covers some of the same territory of the earlier effort although it's updated and with new photographs and artwork, this time by Don Greer and Tom Tullis. Covers the EA-6A and B as well as the bombers.

A-6 Intruder in Detail & Scale, Bert Kinzey, Tab Books, 1987. Softbound. 72 pages. Nuts and bolts detail book oriented towards plastic model builders. Includes numerous pictures, and as a bonus an excellent piece on the Lebanon raids and Bobby Goodman's capture by the Syrians. Also covers a detailed review of scale plastic models available on the type.

Colors and Markings of the A-6 Intruder, Bert Kinzey, Tab Books, 1987. Softbound, 64 pages. Running mate of Mr. Kinzey's "Detail & Scale" work, this effort is a collection of aircraft pictures depicting paint schemes for Navy squadron and developmental birds. Excellent reference for modelers.

Grumman A-6A Intruder & EA-6 Prowler: Aircraft Profile No 252. Kurt Miska. 1973, Profile Publications Ltd. 24 pages. The Aircraft Profile series, out of England from the 1960s, was an excellent series of aircraft monographs that set the standard for aviation enthusiast publications for years. This edition, which went for the original price of $2, was one of the first good "books" written on the type and features technical descriptions, squadron use and other information. Color "profiles" featuring squadron markings were the hallmark of this series.

Grumman A-6 Intruder, Robert Dorr, Osprey Publishing, 1987. 199 pages. Part of Osprey's Air Combat Series, this work was probably the best single reference on the type available prior to the *Intruder's* retirement. Covers development and operations of the A-6 and EA-6 up to mid-eighties.

Grumman A-6 Intruder, Warbird Tech Series Vol. 33: Dennis Jenkins, Specialty Press, 2002, 104 pages. A "nuts and bolts" production, heavily illustrated with photographs and drawings out of NATOPS and technical manuals. Provides an amazing amount of in-depth information on all versions with the plastic modeler and obscure detail freaks in mind. More readable than NATOPS and a whole lot lighter.

Intruder, Lou Drendel, 1991: Squadron/Signal Publications. 64 pages, softbound. Part of the "Modern Military Aircraft" series, covers development and operations of the *Intruder*. In many ways, a follow-on to Mr. Drendel's "In Action" effort. Heavily illustrated, it includes several stories from Intruder crew as well as 1/72nd scale drawings and more of the author's excellent artwork.

World Air Power Journal, Vol 12, Spring 1993: A high-quality periodical that specializes in combat aircraft, WAPJ featured the *Intruder* in this issue with a 60 page spread written by noted author Robert Dorr. Well illustrated with mostly color photography as well as a cut away drawing of the type. Other features in this excellent magazine include articles on H-60 variants, RAF Phantoms from 56 Squadron and the Indian and Malaysian Air Forces.

NATOPS Flight Manual, Navy Model A-6E/A-6E TRAM/KA-6D Aircraft, (NAVAIR 01-85ADF-1). NAVAIRSYSCOM: Volume reviewed dated 15 March, 1983. Classic page-turner read by thousands within the Medium Attack community. Hard to put down prior to onset of sleep. Hundreds of pages of obscure minutia as well as gripping performance and systems diagrams. Contains everything you'd ever want to know about the *Intruder*. Don't miss the blood-curdling emergency procedures in section five.

Glossary

AAA:	Anti-Aircraft Artillery, generally ballistic ordnance fired at aircraft.
ADCON:	Administrative Control
AdMat:	Administrative & Material Inspection
AGL:	Above Ground Level: distance in feet, from the earth
AIMD:	Aviation Intermediate Maintenance Department
Air Boss:	Individual on the carrier in charge of the flight deck.
AIRLANT:	Naval Air Force, US Atlantic Fleet
AIRPAC:	Naval Air Force, US Pacific Fleet
Alpha Strike:	Term used to denote a major daylight strike by carrier aircraft into Vietnam
AMTI:	Airborne Moving Target Indicator
AOA	Angle of Attack
ARVN	Army of the Republic of Vietnam
ASW	Anti-Submarine Warfare
ASuW	Anti-Surface (ship) Warfare
ATKRON	Attack Squadron (see also VA)
AVCAD	Aviation Cadet
B/N	Bombardier/Navigator, an NFO specialty, title give to A-6 right seaters
BARCAP	Barrier Combat Air Patrol: Fighter orbits between a target and MiG bases
Battle "E"	Battle Efficiency award, given every 12 or 18 months to "top" squadron on each coast
BDA	Bomb Damage Assessment; a post-strike estimate of strike's effectiveness
Birdcage	In an A-6 aircraft, the extendible avionics platform in the rear fuselage
Blackshoe	Generic term for surface ship drivers in the Navy.
Boot	As in the Radar Boot, the black rubberized hood that covered the scope for daylight use
BOQ	Bachelor Officer's Quarters
Brownshoe	Generic term for aviators in the Navy
BuNo	Bureau Number- official serial number of each Navy airframe
BUPERS	Bureau of Personnel, the Navy's flesh peddlers, located near the Pentagon
Buster	Radio call telling aircrew to fly at maximum, non-afterburning speed
CAG	Air Wing Commander's title, from obsolete term "Commander Air Group"
CAINS	Carrier Airborne Inertial Navigation System
CarQuals	Carrier Qualification (CQ)
CAS	Close Air Support
CBU	Cluster Bomb Unit
CEP	Circular Error Probable: a diameter within which 50% of a group of bombs will fall
Chaff	Tiny aluminum strips ejected from an aircraft to confuse radar guided missiles
CINCLANT	Commander In Chief, US Atlantic Fleet
CINCPAC	Commander In Chief, US Pacific Fleet
Clag	Lousy weather
CO	Commanding Officer
CONUS	Continental United States (usually refers to the "lower 48")
CQ	Carrier Qualifications
CSD	Constant Speed Drive(s): engine mounted device connecting to auxiliaries
CVIC	Carrier Intelligence Center
DD, DDG	Naval ship designation for Destroyer and Guided Missile Destroyer
Desert Shield	Defense of Saudi Arabia and buildup to Desert Shield, 6 Aug 90-15 Jan 91
Desert Storm	Liberation of Kuwait, 15 Jan to 28 Feb 1991
Det	Detachment, a portion of a squadron separated for a specific mission
DIANE	Digital Integrated Attack and Navigation Equipment
Dirty Shirt	Wardroom No 1 in most carriers, where aviators usually eat. So-called because flight suits were allowed, as opposed to WR No 2, the "clean shirt", where more formal uniform is required.
DIW	Dead in the Water: a ship unable to propel itself
Dixie Station	Title of the patrol area for carrier operations off South Vietnam
DMZ	De-militarized Zone
DR	Dead Reckoning; navigation by time and compass alone
DRV	Democratic Republic of Vietnam
DST	Destructors, MK.80 series bombs fused as mines
E&E	Escape & Evasion: avoiding capture following being shot down behind lines
ECM	Electronic Countermeasures
ECMO	Electronic Countermeasures Officer, NFO specialty found in EA-6A and EA-6B aircraft
Eldorado Cyn	Operation name for strikes on Libya, 15 Apr 86
EMCON	Emissions Control, deliberate reduction of electronic usage, to deny enemy early warning
ESM	Electronic Support Measures
FAC	Forward Air Controller: air or ground-based individual that directs bombing
Fan Song	NATO code name for radar that controls Soviet-built SA-2 SAM
Feet dry	Navy term for an aircraft crossing the beach from the water
Feet wet	Opposite from 'feet dry', an aircraft heading out to sea

FF	Naval ship designation for a Frigate
FID	Popular name for the Forrestal (CVA-59) from ship's motto "First in Defense"
FITRON	Fighter Squadron (VF)
FLIR	Forward Looking Infrared
FMF	Fleet Marine Force
FNAEB	Field Naval Aviator Evaluation Board: formal proceedings where the service determines the suitability of a flyer to retain his wings.
FOD	Foreign Object Damage: damage to an aircraft engine due to ingestion of a hard object
Frag	Fragmentary order, usually part of a larger mission
FRAMP	Fleet Replacement Aviation Maintenance Personnel, enlisted/ground training portion of the RAG
FRBN	Fleet Replacement B/N
FRP	Fleet Replacement Pilot
FRS	Fleet Replacement Squadron (see also RAG)
FUBAR	from WWII, "F****d Up Beyond All Recognition"
Fuse plug	A device in the wheels that melts at high temps, releasing air from the tires and preventing a potential tire explosion.
GCA	Ground Controlled Approach
Green Ink	Signifies aircrew combat time, comes from ink color used in log books
GSE	Ground Support Equipment, also called "yellow gear"
GTMO	Naval Station Guantanamo Bay, Cuba
Guard Freq	Radio frequencies used for emergency transmissions, 243.0 mHz in UHF, 121.5 VHF.
H&MS	Headquarters & Maintenance Squadron, a Marine support unit assigned to the MAG
HATRON	Heavy Attack Squadron.
HE	High Explosive
Hot Dog	Radio call warning aircrew they are about to violate a border
IBN	Instructor B/N
IFR	Instrument Flight Rules
IMC	Instrument Meteorological Conditions
IO	Indian Ocean
IP	Initial Point: the navigation checkpoint prior to a target, or Instructor Pilot
IR	Infrared
Iron Hand	Navy term for anti-SAM missions.
Iron sight	Standard, non-system bombing where pilot uses 'old-fashioned' bomb sight
JBD	Jet Blast Deflector- device on carrier raised behind an aircraft prior to launch
Jink	Random movement of an aircraft to throw off AAA gunners
JO	Junior Officer, in the Navy the ranks LCDR and below
KIA	Killed in Action
LABS	Low Altitude Bombing System
LDO	Limited Duty Officer; pertains to professional maintenance types in a squadron
LLTV	Low Light Level TV
LSO	"Paddles", the pilots who monitor and judge all carrier approaches
MAG	Marine Aircraft Group
Marshal	In carrier aviation, the location where aircraft gather prior to recovery
MAU	Marine Amphibious Unit
MAW	Marine Air Wing
MCAF	Marine Corps Air Facility
MCALF	Marine Corps Auxiliary Landing Field
MCAS	Marine Corps Air Station
McClusky Award	Annual award given to "top" attack squadron in Navy
MCPO	Master Chief Petty Officer
Med	Mediterranean Sea, as in a "Med Cruise"
MEF	Marine Expeditionary Force
MER	Multiple Ejector Rack; an under wing device that can carry up to six bombs
MerShip	Merchant Ship
MIA	Missing in Action
MOA	Military Operating Area, an FAA sanctioned flight area reserved for military use
NAAS	Naval Auxiliary Air Station
NAF	Naval Air Facility
NAS	Naval Air Station
NATOPS	Naval Aviation Training & Operational Standardization
NAVAIR	Naval Air Systems Command, the group that develops and procures naval aircraft
NCO	Non-Commissioned Officer, typically a Petty Officer (Navy) or Sergeant (Marines)
NFO	Naval Flight Officer, non-pilot mission specialist aircrew
NKT	3-letter code for MCAS Cherry Point, NC
NTU	3-letter code for NAS Oceana, VA
Nugget	Slang term for a new, "rookie" Naval Aviator.
NUW	3-letter code for NAS Whidbey Island, WA
NVA	North Vietnamese Army
NVAF	North Vietnamese Air Force
NVG	Night Vision Goggles, helmet-mounted devices that give aircrew night sight
OinC	Officer in Charge
OOB	Order of Battle; intelligence assessment of an enemy's units and organization
OOD	Officer of the Deck, individual in charge of the bridge of a warship
OPCON	Operational Control
OPNAV	Chief of Naval Operation's staff
ORI	Operational Readiness Inspection, a pre-deployment wing/ship exercise
PAT/ARM	Passive Angle Tracking/Anti-Radiation Missile, system used with A-6B
PHD	Pilot's Horizontal Display, the A-6 scope the pilot's uses for basic flight instruments
PI	Philippine Islands
PIM	Point of Intended Movement: where the carrier says it is going, but rarely does
Plank Owner	An individual who is assigned to a ship or squadron when it is established
Pocket Money	Operation name for mining operations of North Vietnam in 1972
POL	Petroleum, Oils, Lubricants; shorthand for fuel-related targets
Poopy suit	Slang term for anti-exposure garment worn by aviators while flying over cold waters
POW	Prisoner of War
Praying Mantis	Operation name for retaliation strikes on Iranian military targets, 18 Apr 88
R&R	Rest & Relaxation
RadAlt	Radar altimeter
RAG	Replacement Air Group
RAN	Reconnaissance Attack Navigator, NFO position in the RA-5C community
RCS	Radar Cross Section
Recce	reconnaissance
RIO	Radar Intercept Officer, NFO position in fighter community
Rock	As in "The Rock", local term for Whidbey Island
ROE	Rules of Engagement
Rolling Thunder	Term used for missions into North Vietnam prior to 1969
Route Pack	Term used for six US military target areas in North Vietnam
RPM	Revolutions per Minute, usually used in terms of engine speed
RTF (1)	Return to Force, flight procedures used to proceed back to friendly ships
RTF (2)	Roll the Flick: Traditional time when squadron ready room starts the nightly movie
SAM	Surface to Air Missile
SAR	Search and Rescue
SARCAP	Search and Rescue Combat Air Patrol

SAREX	Search and Rescue Exercise
SATS	Short Airfield for Tactical Support; a Marine field base which uses catapults and arresting gear, as at Chu Lai Vietnam.
SDO	Squadron Duty Officer
SEAD	Suppression of Enemy Air Defenses
Secondaries	As in "secondary explosions", detonations subsequent to impact of ordnance
SINS	Ships Inertial Navigation System
SIOP	Single Integrated Operational Plan
SLEP	Service Life Extension Program
Slider	Navy slang for a hamburger
SLUF	Unofficial name for the Vought A-7 series, in polite terms "Short, Little Ugly Fellow"
SOJ	Sea of Japan
SOP	Standard Operating Procedure
Sour	Radio call to indicate tanker aircraft is unable to pass fuel
SPAD	Unofficial term for Douglas A-1 "Skyraider" series
Stash	Term for a very junior officer temporarily assigned to a unit while awaiting other duty
Stoof	Unofficial nickname for the Grumman S-2 (S2F) series
SuCap	Surface Combat Air Patrol: standing anti-ship duties
Sweet	Radio call that indicates a tanker is able to pass fuel
T&E	Test & Evaluation
TACAIR	Tactical Aircraft
TACAN	Tactical Aid to Navigation, standard US military method of electronic navigation during A-6's life
TARPS	Tactical Air Recon System, a pod carried on the F-14
TCS	Television Camera System, on later F-14 Tomcats
TF77	Task Force 77, the normal carrier organization off Vietnam
TFOA	Things Falling Off Aircraft
TFW	Tactical Fighter Wing; a USAF organization made up of squadrons
TINS	Traditional start of a Navy sea story or tall tail; 'This Is No S**t"
TOT	Time Over Target
TRACOM	Training Command, where aviators go through basic and advanced flight instruction
TRAM	Target Recognition Attack Multisensor
Trap	An arrested landing on a carrier
TRIM	Trails, Roads, Interdiction Multisensor
UDP	Unit Deployment Program; Marine rotation of squadrons to Japan
Unrep	Underway Replenishment
Urgent Fury	Operation Name for the invasion of Grenada
USAFE	United States Air Forces in Europe
VA	Attack Squadron (see also ATKRON). Flew A-4, A-6 and A-7 type aircraft.
VAH	Formal term for Heavy Attack Squadron, flew A-3 aircraft
VAL	Informal term for Navy Light Attack community
VAM	Informal term for Navy Medium Attack community
VC	Viet Cong, or a Navy Composite Squadron
VDI	Vertical Display Indicator
VF	Fighter Squadron (FITRON), flying F-8, F-4 or F-14s
VFR	Visual Flight Rules
VMA	Marine Attack Squadron, flying A-4 or AV-8 aircraft
VMA(AW)	Marine All Weather Attack Squadron, flying A-6 aircraft
VMAQ	Marine Tactical Electronic Warfare Squadron, flying EA-6B aircraft
VMC	Visual Meteorological Conditions
VMCJ	Marine Composite Squadron, flying EF-10B, EA-6A, RF-8 and/or RF-4 aircraft
VNAF	Vietnamese Air Force
West Pac	Western Pacific, as in "Going on a WestPac deployment"
WTFO	Unofficial, but widely recognized slang for "What the f**k, over?"
XO	Executive Officer, typically the second in command of a squadron
Yankee Station	Carrier operating area in the Gulf of Tonkin from which most strikes into North Vietnam were made from
Yo-yo	A carrier sortie which launches and recovers on the same cycle, typically as a tanker sortie.

Index

Selected Bases & Fields

Operations

Rick Morgan (L), Mark Morgan (R), Lincoln Nebraska, 1961

Intruder comes from Mark and Rick Morgan, a pair of Air Force brats that ended up in the Navy. They are award-winning authors, having both won the "Top Contributor" award from *The Hook*, the Journal of Carrier Aviation. This is their first book. Mark, 50, is a 1976 graduate of the University of New Mexico and that school's Naval ROTC program. Originally qualifying as an NFO, he later migrated into the intelligence field and served a total of 12 1/2 years in the Navy and Naval Reserve. He rates his first and last squadrons – VA-52 at NAS Whidbey Island and HC-9 at NAS North Island – as the best tours. His subsequent career moves included four years in aerospace as an operations analyst with the Lockheed Skunkworks in Burbank and General Dynamics-Fort Worth, nearly six years as a historian with the National Park Service at two parks; three years as a journalist and newspaper editor; and one year as a technical writer with Microsoft and Boeing in the Puget Sound region. Now with the Air National Guard, Mark serves as an Air Force historian under NORAD at McChord AFB, Washington. He and Carrie live in Issaquah, Washington. Rick, 48, served sixteen years in the Navy, all in the cockpit. He's a 1978 product of Naval ROTC at the University of Missouri and dabbled with A-3 *Skywarriors* at VAQ-33 before ending up in EA-6Bs at Whidbey Island. He subsequently deployed with VAQ-139 and flew 41 combat missions with VAQ-141 during *Operation Desert Storm*. He also had two tours as an instructor with VAQ-129, the *Prowler* RAG. Rick has over 2300 flight hours in the EA-6B and almost 500 arrested landings on eight different carriers as well as one ejection off *Constellation's* angle deck. Rick has authored over forty articles on naval and railroad history, and currently lives with his family, wife Julie, daughter Katherine and son Scott, in Woodbridge, Virginia.

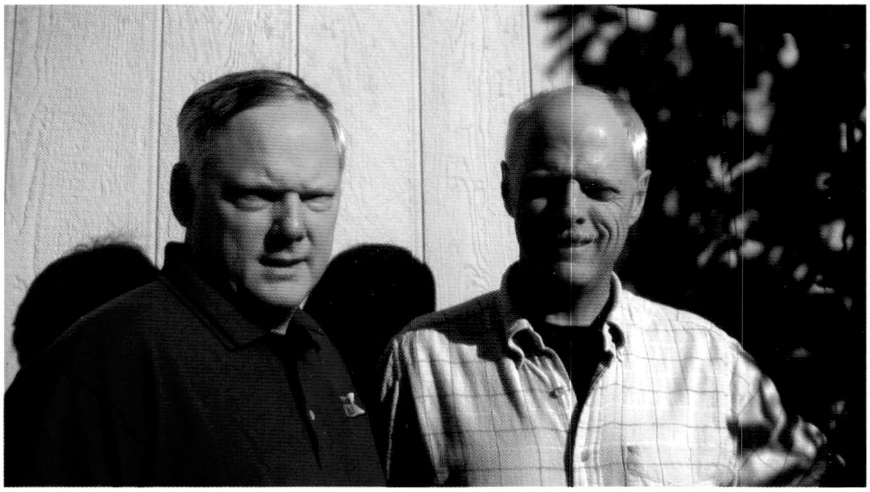

Rick Morgan (L), Mark Morgan (R) Oak Harbor, Washington 2003